SAMPLING METHODS FOR MULTIRESOURCE FOREST INVENTORY

SAMPLING METHODS FOR MULTIRESOURCE FOREST INVENTORY

Hans T. Schreuder
United States Department of Agriculture
Forest Service
Fort Collins, Colorado

Timothy G. Gregoire
Virginia Polytechnic Institute and State University
Blacksburg, Virginia

Geoffrey B. Wood
Australian National University
Canberra, Australia

JOHN WILEY & SONS, INC.

New York • Chichester • Brisbane • Toronto • Singapore

This text is printed on acid-free paper.

***Library of Congress Cataloging in Publication Data*:**

Schreuder, Hans T.
Sampling Methods for Multiresource Forest Inventory / Hans T. Schreuder, Timothy G. Gregoire, Geoffrey B. Wood.
p. cm.
Includes index.
ISBN 0-471-55245-3 (cloth)
1. Forest surveys--Statistical methods. 2. Forests and forestry--Statistical methods. 3. Sampling (Statistics) I. Gregoire, T. G. II. Wood, G. B. III. Title.
SD387.S86S37 1992
634.9′ 285--dc20 92-6752

Printed in United States of America

10 9 8 7 6 5 4 3 2

PREFACE

We have developed this book as a reference text for forest biometricians and for graduate students with a strong interest in becoming forest inventory specialists.

Many forest inventory methods appear to have developed almost independently of the corresponding theory on statistical survey sampling and associated methods. At the same time, there is a remarkable lack of awareness of the forestry literature by statisticians! Fortunately, some researchers in forest inventory are now publishing occasionally in the statistical literature, and researchers in survey sampling are gaining appreciation of the special opportunities forestry applications offer for developing new sampling strategies and analytical techniques. Students of the methodology of forest inventory are encouraged to scan both the natural resource and statistical literature on inventory methods. The cross-fertilization between survey sampling and forest inventory methodologies can pay dividends for both fields.

We hope to contribute to the cross-fertilization mentioned above by pointing out the interrelationships between sampling strategies, discussing forestry applications of efficient strategies, drawing attention to new developments in statistics that seem to have immediate applications in forestry, and discussing related developments that should have relevance to forest inventory specialists in the future.

To accomplish these goals, we introduce some general methods of collecting data and discuss the history of inventory methods in statistics and forestry in Chapter 1, give critical definitions and theory needed for statistical inference in Chapter 2, discuss design-based (or probabilistic) inference for single-level sampling strategies in Chapter 3, forestry applications in Chapter 4, multilevel sampling strategies in Chapter 5, model-based sampling strategies in Chapter 6, and measurement techniques in Chapter 7. Chapter 8 is devoted to related material of which forest inventory specialists should have some knowledge, and Chapter 9 concerns topics that could improve inference from survey data but considerable research and development are needed to make this possible. Finally, in Chapter 10, we identify future directions in multiresource sampling in forestry.

Fort Collins, Colorado — Hans T. Schreuder
Canberra, Australia — Timothy G. Gregoire
Blacksburg, Virginia — Geoffrey B. Wood

ACKNOWLEDGMENTS

It is a pleasure to acknowledge Dixie R. Smith and Clyde A. Fasick. Without their committment and support for quality research, this book would never have been written. We thank Michael Williams for reviewing the entire book and contributing several of the problems at the end of each chapter. We would also like to acknowledge the contributions of the following people for either reviewing parts of the manuscript or suggesting needed additions.

Barrett, Jim
Boboricken, Sherry
Bolstad, Paul
Bruce, Dave
Burnham, Ken
Chen, Mei-Ching
Czaplewski, Raymond L.
Dell, Tommy R.
Grosenbaugh, Lewis R.
Hazard, John
Iles, Kim
King, Rudy
LaBau, Jim (VJ)
Li, Jim (Hsin Gi)
Li, Junfang
Zhu, Zhiliang
Max, Tim
McRoberts, Ron
Nelson, Ross
Ouyang, Zhao
Pienaar, Leon
Reich, Robin M.
Rennolls, Keith
Roesch, Frank A.
Scott, Chip T.
Scott, Jeffrey L.
Smith, Jim
Snook, Paul W.
Thomas, Charles E.
Van Deusen, Paul C.
Wiant, Harry V.

We also appreciate the help of Penny Williams in typing and revising the several drafts of the manuscript; Margaret Spraker for assisting on several drafts; and Cathy Kochenderfer for carefully proofreading and correcting the manuscript when it was converted from one word processing system to another. We particularly appreciate the help of Tonya Fletcher for typing several early drafts, typing all of the final corrections, and then ensuring that the manuscript was in excellent condition.

CONTENTS

GLOSSARY

Accidental error. An unavoidable, compensating error that occurs due to inconstant environmental conditions, limitations, or deficiencies of instruments and methods, and rounding off.

Accuracy. Freedom from error or the closeness of a measurement or estimate to the true value. More broadly, it is the degree to which a statement or quantitative result approaches the truth.

Accurate estimate or measurement. An estimate or measurement for which the sum of the systematic and random errors is small. It is reliable and unbiased.

Angle gauge. An instrument used for deriving stand basal area rapidly and accurately.

Attribute. One type of discrete variable. Used when referring to individuals classified as having or not having some specific quality. Results are often expressed as a percent.

Bark gauge. An instrument for measuring the thickness of bark at any point along a tree stem, branch or log.

Basal area (tree). The cross-sectional area of a tree stem at breast height.

Basal area (stand). The sum of the cross-sectional areas at breast height of all trees on a defined area.

Basal area factor. The factor by which the count from an angle count sweep (variable radius plot) is multiplied to derive stand basal area.

Bias. A systematic error introduced into sampling, measurement, or estimation by selecting or encouraging, possibly unintentionally, one outcome or answer over others.

Biological volume. The volume of the tree stem with branches trimmed at their junctions with it.

Biomass (forest). The amount of living matter per unit area or volume of the forest habitat.

Bole area. The surface area of a tree stem under bark.

Borderline tree. A tree at the limiting distance for inclusion in the count of an angle count sweep (variable radius plot).

Breast height. The point on a tree stem 1.3 m (4′3″) or 1.4 m (4′6″) above ground.

Caliper. An instrument for measuring directly the diameter of a tree stem, branch or log.

Confidence interval. For practical purposes we consider it to mean the interval within which one may be confident that a parameter lies. This is technically not entirely correct.

Continuous variable. A variable characterized by being related to some numerical scale of measurement any interval of which may, if desired, be subdivided into an infinite number of values.

Count. One type of discrete variable. Used when describing a group of individuals by number only.

Covariate. Same as independent variable.

Dendrometer. An instrument for measuring the diameter and height above ground of points along the stem or branches of standing trees.

Dependent variable. A variable of interest in the inventory, which may be related to other variables of less interest (independent variables).

Discrete variable. A variable represented by integral values or ratios of integral values. It has two forms, attribute and count.

Efficiency. An estimator is said to be efficient if it predicts a parameter more reliably than competing estimators. Reliability is usually measured by the ratio of the mean square errors of the estimators.

Element. The basic sampling unit used. For example, that used in the last stage of multistage sampling.

Error. The difference between an observed or calculated value and the true value. There are four main types of error: mistake, accidental error, bias, and sampling error.

Estimate. The numerical value of an estimator computed from a sample.

Estimator. A real-valued statistic used for estimating a parameter based on a sample of units. Estimates are generated by an estimator.

Experiment. A planned inquiry to obtain new facts or to confirm or deny the results of previous experiments using a formal set of rules called the experimental design or experimental procedure.

Height sticks. Lightweight telescopic or extendable poles used to measure directly tree total height or height above ground of any point along the stem of a standing tree.

Hypsometer. An instrument based on geometric or trigonometric principles for measuring indirectly tree total height or height above ground of any point along the stem of a standing tree.

Imputation. A process of estimating missing data.

Independent variable. A variable that is of interest because it is relatively easy to measure and provides information on a more difficult to measure dependent variable that is of primary interest.

Instrumental error. The error arising in measurement due to limitations of the measuring instrument.

Line intercept sampling. Basically the same as line intersect sampling but the length of intercept of each object intersected by the transect is recorded.

Line intersect sampling. A method of sampling units (objects) in a population where a unit is sampled if a chosen line segment, called the transect, intersects the object.

Measurement. Determination of size in relation to some observed standard, for example, meter, kilogram, second, ampere, degree Kelvin, candela, and mole, or some unit derived from these seven basic units.

Merchantable height. The height from ground level to the point on the tree defining the utilisable length of stem.

Merchantable volume. See utilizable volume.

Mistake. An error caused by human carelessness, casualness or fallibility.

Multiphase sampling. A sampling design where the same size of sampling unit is used at each phase (level) of sampling but fewer units are selected at each succeeding phase.

Multistage sampling. A sampling design where the ultimate sampling units, the elements, are selected in stages and samples at each stage are taken from the sampling units or cluster of units selected in the previous stage.

Nested sampling. The same as multistage sampling.

Operator error. The error arising in measurement due to the operator.

Parameter. A characteristic that describes the whole or part of a population in some way, usually in numeric form.

Point estimation. A statistical term for making an educated guess about an unknown quantity for which some (sample) information is available, such as number of trees.

Population. An aggregate of items each with a common characteristic or common set of characteristics. In the statistical sense, a population is an assembly of individual units usually formed in order to describe the units quantitatively.

Pps sampling. Sampling with probability proportional to size. Sample units are selected proportional to a covariate x. Commonly implemented by a random systematic procedure, namely: order the units randomly, cumulate their x-values; and select the n-units corresponding to δ, $\delta + X/n$, $\delta + 2X/n, \ldots, S + (n-1)\, X/n$, where δ is a random number between 0 and X/n and X is the sum of all the x-values.

Ppswr sampling. Pps sampling with replacement.

Ppswor sampling. Pps sampling without replacement.

Precise measurement. A measurement for which the random errors are small relative to the measured value.

Precision. Relative freedom from random variation. In sampling, it is expressed as the standard error and relates to the degree of clustering of sample values about their own average or the reproducibility of an estimate in repeated sampling. It is also used to indicate the resolving power of a measuring device.

Predominant height. The average total height of a specified number of the tallest trees per unit area.

Pscx sampling: Sampling purposively from the cumulated x-values. Those sample units are selected that are within some prespecified percent p of (X_T/nN_i) $(i = 1, \ldots, n)$ where the units are sorted by a covariate x. N_i is number of units in class i where the boundaries of the i-classes are $(0, X_T/n)$, $(X_T/n, 2X_Tn), \ldots, (n-1\ X_T/n, X_T)$.

PsR sampling: Sampling purposively from the range of x. Those sample units are selected that are close to computed values $x_1^* = x_1 + (R/n) \cdot \delta \cdot i (i = 0, 1, \ldots, n-1)$ where δ is a random number between 0 and 1, x_1 is the smallest x-value in the population, and R is the range of the x-values.

RSRS sampling: Restricted simple random sampling. Samples are selected by SRS but samples are rejected if they do not meet certain specifications.

Sample. A part of a population used to obtain estimates of characteristics of that population.

Sample surveys. The design and execution of surveys to provide estimates of characteristics (parameters) of well-defined finite populations.

Sampling design. A formalized method of selecting a sample from a population.

Sampling error. The random error associated with an estimate purely due to sampling (synonymous with precision of the estimate).

Sampling plan. The same as sampling design.

Sampling strategy. Consists of a sampling design and an estimator.

Sampling unit. A determinate part of a population chosen as the unit for sampling (also see element).

Sectional area. The cross-sectional area of a tree stem at any position along it other than at breast height.

Significant digits. Those digits in a number when read from left to right, beginning with the first nonzero digit and ending with the last digit written which may be zero, but if zero, it must not result from rounding off.

Site index. The top or predominant height of an even-aged forest stand at a specified standard age or, in uneven aged forest, the mean total height attained by dominant trees at a standard DBHOB.

Site quality. In even-aged stands, volume production at a specified age.

SPPS sampling. Stratified pps sampling. Sample units are selected by pps sampling within specified strata. The strata are often set up in some manner based on a covariate.

SRS. Simple random sampling. All sample units and pairs of units in the population have equal probabilities of selection.

Stand mean height. The average total height of all trees in a forest stand.

Stand height curve. The relationship defined by plotting tree total height against diameter at breast height of all or a sample of trees from an even-aged forest stand.

Statistical decisions. Decisions based at least partially on inductive inference but which may also reflect personal information such as subjective priors specified so that Bayes theorem can be applied.

Statistical inference. The theory and methods concerned with the way background information and current data are used to draw conclusions concerning unknowns in a system under investigation. The context can be a scientific experiment, a survey, or a historical analysis.

Taper. The decrease in diameter of a tree stem from base to tip or the change in diameter per unit length between any two points along a tree stem.

Taper equation. An equation used to predict diameter at any point along a tree stem.

Taper line. The profile of a tree stem defined by plotting height at various points along the stem against the corresponding sectional area.

Tariff. A series of one-way tree volume functions or tables.

Top height. The average height of a specified number of the largest diameter trees per unit area.

Total height. The vertical distance from ground level to the uppermost point of a tree.

Unrepresentative point. A point on a tree stem where diameter is affected by insect attack, mechanical damage, nodal swelling, or other malformity whether natural or man-made.

Utilizable volume. The volume of the tree stem with branches trimmed at their junctions with it and excluding the normal irregularities of episodic growth, that is, the branch nodes.

Variable. A characteristic that may vary from one individual in a population to another. It relates to some feature or property of the individual.

Variable radius plot. The area sampled using an angle gauge.

Variate. Any one of the observed values of a particular variable.

Volume function (table). A mathematical function used to predict the volume of a tree or of a forest stand given certain easily measured tree or stand variables.

Volume line. The relationship between volume and basal area of trees in an even-aged forest stand.

Wedge prism. A particular type of angle gauge commonly used in variable radius plot (angle count) sampling.

1 Introduction

1.1 BACKGROUND

In this chapter, we discuss some general methods used to collect information so that resource managers can make decisions. We then follow with a brief history of how statistical methods of collecting information (called statistical designs) developed both outside and within the profession of forestry: many of the early applications of statistical designs and estimators were forestry-based and were directed at collecting more reliable information for managing forest resources.

1.2 COLLECTING INFORMATION

Resource managers need accurate and timely information to make decisions wisely. This information can be collected in many ways, but there is often a direct relationship between the quality of information and the cost of collecting it. The purpose of this section is to discuss the various approaches used to collect information and to point out the advantages and disadvantages of each.

Basic techniques for collecting data can be classified into four main groups, each of which is discussed below: (1) census, (2) mapping, (3) nonstatistical sampling, and (4) statistical sampling.

1.2.1 Census

By census, we mean complete enumeration of a population. If the operation can be done economically, then a census is highly desirable. When completed, everything of interest should be known about the population and decisions based on the information should be risk-free. Suppose, for example, we have a region to be classified into forested and nonforested land and that the information can be interpreted error-free by computer from available Landsat digital data. Thus the cost of obtaining the information will be relatively low per unit area and the census should be highly efficient. If, on the other hand, we wanted to establish by census the total volume of the timber resources in the same region, the cost would be astronomical and could not be justified.

A common disadvantage of a census is that implementation may be so difficult and take so long that serious errors are likely to occur and, when compiled, the information is already outdated. For example, in a large state such as Alaska, a well-executed statistical sample is likely to yield a more precise and reliable value of the total timber resource of the state than could be obtained by census.

1.2.2 Mapping

Mapping refers to the compilation of data on a map, which is a representation on a flat surface of the whole or part of an area. The big advantage of mapping is that both the values of individual elements in the population and their relationship to each other are displayed. Mapping is clearly one way of displaying census information, but good mapping, like conducting a census, is often too expensive to be practical. However, costs can be reduced by grouping variables of interest into coarser classes. For example, an area may be mapped into forest and nonforest by ignoring small forests on nonforested lands and small nonforested areas within forests. This allows quick classification of the population of interest into forested and nonforested subpopulations. The values of both the variables of interest and the land area covered by them are only approximate.

1.2.3 Nonstatistical Sampling

Nonstatistical sampling refers to the collection of information selected in a subjective manner. For example, two people may map an area into range and nonrange land. Differences may be either small or substantial, depending on the skill, experience, and conscientiousness of the mappers. The big appeal of the technique is its low cost, but the reliability of the assessment cannot be determined.

1.2.4 Statistical Sampling

Statistical sampling refers to the collection of data in a scientifically acceptable manner, which we take to mean the selection of a representative sample.

The concept of a representative sample is attractive even though no generally accepted definition is available (Kotz and Johnson 1988a, p. 77). We consider a representative sample generally to be one that is probabilistic, that is, every unit in the population has a positive probability of selection and these probabilities are known (usually before but sometimes only after sampling) and are independent of the person taking the sample. Such a representative sample provides the information of interest to the manager, and gives a measure of its reliability. However, the manager may find that the cost of attaining reliable estimates is greater than the funds available. For example, the cost of obtaining an estimate of the timber resources of Alaska at a precision of ±10 percent (95 times out of 100) may exceed the funds allocated for that purpose.

The ideal technique for collecting data in natural resource surveys is to map in a statistically valid manner. This is usually expensive, but it yields the most complete information about a population (both the "how much" and the "where" for each variable). Collecting data using aerial photography and digital remote sensing is attractive to many practitioners because much resource information can be mapped and measured accurately at low cost. For example, the land area of a region may be mapped for land cover classes from aerial photographs if the boundaries of the classes can be defined accurately on the photos.

In this book, we concentrate on statistically valid sampling methods for two reasons. First, assessment of the reliability of information is often as important as the information itself, and second, scientifically valid methods lend themselves to systematic exposition whereas nonstatistical techniques do not.

1.3 BRIEF HISTORY OF STATISTICAL SURVEY

The term statistical survey, or sample survey, refers to the design and execution of surveys that provide estimates of characteristics of well-defined finite populations (Hansen and Madow 1976). This definition may be changing because sample survey methodology is now often applied in the more general area of analysis and inference about causal systems, but in this chapter we follow the original definition.

The two earliest pioneers in sample surveys were A. N. Kiaer in Norway and Carroll D. Wright in the USA (Seng 1951). Kiaer was a theoretical statistician, while Wright was more interested in the practical, intuitive approach to sampling problems. According to Seng, Kiaer stated that:

> It is fundamental to observe that the accuracy of the results of a survey depends, not on the larger or smaller number of observations made, but on the method of obtaining correct representation.

Kiaer appreciated two important conditions of a successful sampling investigation, namely, proper representation and rational selection of the sample units. He observed in 1899 (Seng 1951) that:

- The representative method of investigation is applicable, not only in social and economic fields, but also in agriculture and forestry.
- To obtain a representative selection of sample units, it is necessary to group the different communities under investigation. Thus in social studies, the towns and country communes should be differentiated, and be further separated by size (large, medium, or small) and location

(coastal or inland). He pointed out that this principle of grouping (or stratifying) homogeneous parts of a country must be applied with care to obtain a representative sample.

- If possible, two or more different methods of selecting a representative sample should be used so that the results of the inquiry are credible, and proof of the usefulness of the methods can be obtained.
- It is important to study and develop the practical and theoretical aspects of the method so that proper limits can be set to representative statistics.

In an address to the British Association for the Advancement of Science, Bowley (1906) eloquently argued the value of sampling and the need to include an error estimate when sampling is used. He stated:

> The method of sampling is, of course, only one of many instances of the application of the theory of probability to statistics. I have taken it at length because the method is so persistently neglected, and even when it is used the test of precision is ignored. We are thus throwing aside a very powerful weapon of research.

He then demonstrated with some examples that if a normal distribution could be assumed, useful probability statements could be made about the estimates (also see Bowley (1926)).

"Student" (1908) developed the t-distribution both empirically and theoretically. Subsequent work by Pearson (1929) demonstrated the incredible robustness of tests based on the t-distribution. The t-statistic is also commonly used in constructing confidence intervals (which are closely related to t-tests).

Theoretical work on laws called the "weak and strong laws of large numbers" and the "central limit theorem" provided evidence that the assumption of normality applied to a large number of practical situations. Madow (1948) proved a very general theorem about this for sampling randomly without replacement from a finite universe:

> The limiting distribution of the mean is normal provided only that as the universe increases in size, the higher moments do not increase too rapidly as compared with the variance, and that for sufficiently large sizes of samples and populations, the ratio of size of sample to size of universe is bounded away from 1.

All these results made a strong case for random sampling.

The first known attempt at random sampling was made in 1921 in a survey of a rice crop (Hubback 1927). Other improved surveys were made in 1923 and 1925 and are described in some detail in Hubback's paper.

Hubback also realized the potential inefficiency and imprecision of sample units occurring in clusters (cluster sampling) because of the high correlation of individuals within a group. He may also have been the first to discuss different stages of sampling (multistage sampling). He stated that the only way in which a satisfactory estimate can be formed is to select samples by as close an approximation to random sampling as the circumstances permit. This, he pointed out, not only eliminates the personal element of choice by the sampler, but also makes it possible to state the probability that the result of a given number of samples will be within a given range from the true mean.

Tschuprow (1923) was the first to set stratified sampling on a sound theoretical basis. Neyman (1934), apparently unfamiliar with Tschuprow's work, obtained essentially the same results in a vastly simpler fashion. As noted by Smith (1976), Neyman's famous 1934 paper was stimulated by references such as Bowley (1926). Neyman's exposition of the logic of inference based on confidence intervals was of great importance to statistics in general and to sample survey theory in particular. To quote him:

> If we are interested in a collective character X of population π and use methods of sampling and of estimation allowing us to ascribe to every possible sample Σ, a confidence interval $X_1(\Sigma)$, $X_2(\Sigma)$ such that the frequency of errors in the statements
>
> $$X_1(\Sigma) \leq X \leq X_2(\Sigma)$$
>
> does not exceed the limit $1 - \epsilon$ prescribed in advance, *whatever the unknown properties of the population*, I should call the method of sampling representative and the method of estimation consistent.

The confidence statement was a frequency statement based on all possible random samples that could be drawn. As such, the probabilities were induced by the randomization and the confidence statement would be true regardless of the population structure.

The use of confidence intervals based on the probabilities from random sampling eliminated the need to use Bayes' theorem (Bayes 1763) or to make any assumptions about the distribution of a variable in the finite population except if the sample size is very small and the underlying population distribution is highly skewed. Random sampling allowed for the construction of confidence intervals that would give consistent inferences within the framework of repeated samples drawn according to the given method of random sampling. Inferences for sample surveys have been firmly tied to the distribution generated by randomization and the t-distribution, and are generally free of assumptions about the population (Smith 1976).

Neyman (1934) also introduced the idea of "most efficient estimation." Restricting himself to estimates linear in the sample values, he derived unbiased estimators with the smallest variances using the Gauss–Markov theorem. With large sample sizes, the distribution of estimates in repeated samples should be normal, and hence the minimum variance estimate should provide the shortest confidence interval. Neyman deduced that stratified sampling would usually be more efficient than simple random sampling, and that if costs were ignored, the optimal allocation of sampling effort was to choose sample size n_h in stratum h, proportional to the stratum size times standard deviation $N_h S_h$. This result was anticipated by Tschuprow (1923).

Neyman's (1934) paper compared stratified random sampling with purposive sampling using his new method of confidence intervals. To do this, he interpreted balanced sampling (purposive sampling) as a form of stratified sampling. In balanced sampling, the basic hypothesis is that the regression of $\bar{x}_i$ on y_i is linear where $\bar{x}_i$ and y_i are correlated. Neyman showed that a balanced sample could be efficient only under restrictive conditions concerning the linearity of the regression, which would rarely be met in practice. Neyman also showed that the balanced sample of Italian census records drawn by Gini and Galvani (1929) had failed to provide satisfactory estimates for many variables (Smith 1976).

Since Neyman's paper, random sampling has reigned supreme over balanced or purposive sampling. However, balanced sampling still survives in the form of quota sampling and is widely employed in market research (Stephan and McCarthy 1958). Also, some of the other papers in the 1926 bulletin of the International Statistical Institute made it clear that balanced sampling gives satisfactory results on many occasions (Smith 1976).

Implicit in Neyman's optimal allocation was the idea that units from different strata enter the sample with different probabilities. Equally important was the fact that confidence intervals for cluster samples could be computed using the selection probabilities for the groups. Since the groups were usually of unequal size, the estimator would be a ratio mean and this led Neyman to Geary's (1930) results on ratios.

Neyman (1934) developed a new theory of inference, introduced the ideas of efficiency and optimal allocation, provided a framework for inferences from cluster samples, and presented a powerful case for rejecting purposive sampling. In a later paper (Neyman 1938), he developed two-phase sampling for stratification. The only major features of current survey design that he failed to introduce were multistage sampling and variable probability (pps) sampling, but these followed logically from his work (Smith 1976).

In the (published) discussion that followed the presentation of Neyman's 1934 paper, discussants Bowley and Fisher were not convinced about the validity of the argument on confidence intervals, Bowley calling it a confidence trick, and Fisher arguing that randomization theory could help with questions of design but not with those of the analysis of experimental results after the sample is drawn (Smith 1976).

Fisher's (1925) technique of the analysis of variance made it possible to approach sampling problems from a new angle, namely estimating sampling errors from the results of the observations. This made it possible to determine the adequacy of the sampling, as well as increase the efficiency of future sampling of similar material. Clapham (1929), Wishart and Clapham (1929), and Yates and Zacopanay (1935) were the first to adapt the analysis of variance to sampling problems and to make a practice of estimating sampling errors from the results of their observations.

Cochran (1939) summarized what can be done with the analysis of variance in sampling. It is rarely possible to test the accuracy of two different sampling methods on the same population because it is so costly but, if suitably planned, sampling can usually provide information about the relative precision of alternative methods of sampling. This information can be extracted without difficulty from the analysis of variance of the results and is of considerable value in improving the accuracy of future samplings on the same type of material. Cochran also listed four of the most common devices for increasing the accuracy of a sample estimate: (1) subdivision, (2) subsampling, (3) choice of sample unit, and (4) double sampling:

1. *Subdivision (Now Commonly Known as Stratification, a Term Borrowed from Geology)*. The analysis of variance of the data provides estimates of the mean square error among groups and within groups. From these estimates it is possible to determine the sampling error to be expected from a sample of the same size if the population had not been subdivided. Thus the increase in precision due to subdivision may be estimated.
2. *Subsampling (Now Known as Cluster Sampling with Subsampling)*. This method might be regarded as a case of incomplete subdivision. The material is grouped by primary units, but not all primary units are sampled, so that the sampling error arises from variability among primary units and variation among secondary units within the selected primary units. The analysis of variance provides estimates of the sampling variance to be expected, in similar material, from any change in the amount of subsampling per primary unit or in the number of primary units.
3. *Choice of Sample Unit.* A thorough study of the relative effectiveness of different sizes of sample units generally requires a special investigation for that purpose. However, some useful information can be gained from the analysis of results of a particular sample plan. If, for instance, a four-section block is selected as a unit and data are recorded by single sections, it is possible with the analysis of variance to obtain an estimate of the sample error to be expected if a single section had been used as a sample unit.
4. *Double Sampling.* In double sampling, the analysis of variance approach can be used to obtain estimates of first-phase and second-phase variation. When combined with cost estimates for observing units at

the two phases, this approach makes it possible to choose between such alternative methods as double sampling with stratification or double sampling with regression estimators.

According to Yates (1946), sampling is largely a matter of common sense and the intuitive approach has often resulted in more rapid progress in technique than has the pure mathematical approach. Common sense alone, however, does not provide a method of estimating the efficiency of different sampling procedures. In Yates' opinion, the two main principles of sampling are: (1) to avoid bias, which requires that sample selection must not be influenced by the qualities of the objects sampled and must be free from any element of choice on the part of the observer, and (2) to provide a valid estimate of sampling error. This latter principle requires that two or more sample units be obtained from each population. These units must be selected randomly from the whole aggregate of units in the batch, and all units in the aggregate must be of approximately the same size and pattern and, together, must comprise the whole batch of material.

In some early applications, the sampled material was of such a nature that it could be divided into sample units of approximately the same size and shape, so that simple sampling schemes could be adopted (Yates 1946). New difficulties arose, however, when these methods were applied to estimating crop and timber resources and to surveying economic conditions and practices on farms. Here the natural units of the population under survey differed greatly in size. In particular, better methods were required for dealing with differences in variability of different parts of the material, and for handling sampling errors when the sample units differed in size. Although some of these methods (such as ratio and regression estimation) were already available, they were not always satisfactory as they generally provide biased estimates and require a knowledge of the population total for the covariate (which often is expensive to obtain). This led to two new developments in sampling: double sampling and unequal probability sampling.

Neyman was challenged to develop an efficient method that could be used for determining the average value of some variable that is costly to observe but is correlated with a second, more readily measurable variable. He responded by developing the theory of double sampling for stratification (Neyman 1938). At about the same time, Watson (1937) applied double sampling to regression estimation. He estimated the mean area per leaf and mean leaf area per plant using leaf weight as covariate. The mathematical basis of the method was worked out by Cochran (1939) and his formulation was used in Watson's paper. Double sampling is discussed more fully in Chapter 5 (Section 5.2.2).

The concept of sampling with unequal probabilities was introduced by Hansen and Hurwitz (1943), who recognized that unequal probabilities of selection of sample units could, under certain circumstances, materially improve the precision of results. This concept helped explain some of the older concepts and theory, such as optimum allocation in stratified sampling. Research on unequal probability sampling started during World War II and was pursued with increased vigor after the war. Horvitz and Thompson (1952) contributed a classic paper on the subject. In addition to providing an unbiased estimator, universally called the Horvitz–Thompson estimator, they also stimulated new basic research, for example, the paper by Godambe (1955), who showed that within the general class of linear estimators, no "best" (minimum variance unbiased) estimator could exist for all possible populations. Many other papers have been written about unequal probability sampling. Despite this, much remains to be done in this area, especially in adapting it to practical situations.

Much of the development in unequal probability sampling was based on one- or two-stage cluster sampling with stratification, particularly in the field of social science where lists of covariates are often available prior to sampling. Other investigations were made by mathematically oriented statisticians who were more concerned with mathematical niceties than the practical value of their results.

More recently, work has focused on model-based or model-dependent sampling. In this type of sampling, inference is dependent on an assumed underlying model and samples may be drawn in a purposive (nonprobabilistic) manner to improve the efficiency of estimates. A strong proponent and developer of such techniques has been Richard Royall (1970a,b). Model-based sampling has been controversial since it clashes with generally accepted statistical notions such as the need for representative sampling. Hansen *et al.* (1983), and discussion following presentation of their paper, give some feeling for the issues involved. Model-based sampling is discussed in Chapter 6.

Other relevant references on the history of survey sampling are Stephan (1948), Kruskal and Mosteller (1980), Murthy (1981), O'Muircheartaigh and Wong (1981), and Hansen (1987).

Survey sampling traditionally has focused on point estimation (population means or totals) with some statement about the precision of the estimates. More and more survey data are not being used for analytical purposes, i.e., to establish a basis for action on a cause-effect system or for the process, that is, to improve future products (Deming 1975). This topic is not well developed yet and is not covered in this book.

1.4 BRIEF HISTORY OF SAMPLING FOREST RESOURCES

Methods of inventorying forests and compiling resource data for whole management units were developed in Central Europe when modern statistical techniques did not exist. In the USA, formal forest inventories were not undertaken until statistical techniques were introduced into forestry science.

The extensive forest resources of the USA and the lack of maps gave rise to inventory techniques very different from those in Central Europe. Sampling was primarily concerned with large units and information on subunits of interest was collected as this was considered to be feasible. It is notable that notions such as representative sampling, stratified sampling, and unequal probability sampling were used in forestry well before they were accepted or published in the statistical literature. Statistics should be an aid to common sense and it is not surprising that practitioners sometimes develop sound and reasonable methods before theoreticians do.

Inventories are the basis for management planning. Hence, forest inventory techniques date back to the beginning of forest management, toward the end of the Middle Ages in Central Europe (Loetsch and Haller 1964). As early as the beginning of the 19th century, accurate maps and area registers existed for almost all the larger state and privately owned estates in Central Europe. The terrestrial survey of areas and the mapping of forests in this region are more detailed than anywhere in the world.

The early collection of area-related tree information aimed at assessing the total growing stock only. This information was satisfactory because during the 18th and the beginning of the 19th centuries, the demand for fuel was much greater than for sawtimber. The forest population of interest was divided into sample areas, which served as units of assessment. The total growing stock per hectare in each unit was visually estimated and checked by the yield from felled sample areas. Learning to visually estimate the growing stock per unit area was then part of good training in forestry. This cheap and simple method was used until the 1940's, for example, in Saxony in West Germany (Loetsch and Haller 1964). The method of visual estimation may have contributed to the early use of relatively small stands as ultimate units of assessment. The estimates of the single stands were combined into an estimate for the entire forest area. Cajanus (1913), Ilvessalo (1923), and Müller (1936) studied the bias arising from ocular estimation and showed it could be adjusted through regression analysis.

During the 19th century, techniques of collecting forest data were rapidly improved. Tree volume tables for individual species were compiled based on diameter at breast height, tree height, and form factor. A shift from fuel to timber production in forestry, due to the expanding system of transport and to the industrialization of Central Europe, made inadequate the simple area-related estimate of total volume per unit area. Estimates of higher quality timber products were required. The visual estimate was replaced by methods of measuring the growing stock. The possibility of reducing costs by adopting sampling methods was recognized early. The first sampling methods in forestry dated back to the end of the 19th century. Schmidt (1891) described a statistically sound sampling method developed by K. Zetzsche using circular plots.

During the 19th century, forest inventory techniques were developed in Scandinavia for large units of assessment, such as rural districts and provinces. Linear strip surveys were originated by Av Ström (Näslund 1930), who, according to Matérn (1982), described a random sample survey in a forestry handbook in 1830. Linear strip surveys were used as early as the 1840's in Sweden and at the end of the 19th century in Norway and Finland. Dalenius (1957) pointed out that during 1909–1911, a remarkably advanced (from a statistical point of view) sample survey was conducted in Värmland, Sweden. The method consisted of stratified sampling where trees were selected systematically from strips of land constituting the strata. The method supplied information on the area of forest types as well as area-related information on the trees. Forest maps of moderate accuracy were produced where the strips were evenly spaced and not too far apart, so that corresponding points on two lines, at which the forest type or stand changed, could be linked. This is a good method for assessing large forests in which detailed forest mapping has not been done. Basically, the method provides primary information on the whole forest from which secondary information for smaller subdivisions can be derived. Contrast this with the procedure in Central Europe, in which the primary information relating to individual stands was combined giving estimates for the total forest.

The 19th century also brought important developments in inventorying tropical forests. About 1850, D. Brandis introduced strip surveys in Burma at 5% intensity for inventorying and managing teak forests. The compartments were much larger than those used in Central Europe (Loetsch and Haller 1964). Strip sampling in modified form is still used to inventory tropical forests in many countries (Wood 1990a).

Aerial forest surveys were developed in Canada under the leadership of H. E. Seely in the 1920's and 1930's (Honer and Hegyi 1990). Inventories of forest conditions were conducted in Canada as early as 1911 (Honer and Hegyi 1990), but it is not clear when statistically valid surveys were first conducted. Wright's early monograph (Wright 1925) provides interesting historical insight into the growing awareness among forest researchers of the importance of statistical methods, but the emphasis was more on experimentation than on surveys.

Sampling of forests in the USA started in the 1930's. Survey sampling in a valid statistical manner was first reported by Hasel (1942a, b). Sophisticated sample schemes such as unequal probability sampling were used in forestry prior to their introduction in statistics. For example, the B-10 stick of Fred Miller for estimating tree volume from basal area at breast height and Walter Bitterlich's plotless cruising method were developed prior to 1940, but nothing was published until 1947 (Bitterlich 1947). Schumacher and Chapman (1942) wrote one of the first books discussing valid statistical survey methods.

Efficient, equal probability sampling techniques, such as strip sampling and double sampling, and unequal probability sampling schemes, such as point, line intercept, model-based, and point-Poisson sampling, have found early and sophisticated application in forestry. Corresponding development in instrumentation to implement efficient and nondestructive sampling methods has kept pace. The early use of double sampling, combining aerial photographic information with data derived from ground sampling, and the development of dendrometers, prisms, and range finders in ground sampling are examples. There is an awareness among forest samplers that efficient sample designs employing unequal probability sampling are useful for single purpose surveys for one or more correlated parameters but more robust, simple, and flexible sampling schemes need to be used for multipurpose surveys (Furnival 1979).

Low-altitude photography is used extensively in the USA as a basis for stratification and selecting samples in large scale surveys. This is coupled with a cluster of point samples on 0.4-ha (1-acre) ground plots by many of the US Forest Service Forest Inventory and Analysis (FIA) units on their state-wide surveys. Permanent plots are used so they can be resampled over time. Usually photo plots are located on a systematic grid covering a state (Forest Resources Economics Research Staff 1987).

Unequal probability sampling schemes, such as point-Poisson sampling, have been used for timber sales in the USA, and their use for inventorying the resources of both public and private plantations and native forests in Australia is expanding (Wood and Schreuder 1986). Model-based sampling procedures are being explored as alternatives in some parts of the USA. Trees are selected by point sampling and measured for diameter and height, and estimates of volume, the variable of primary interest, are derived using existing volume equations.

2 Focus, Fundamental Concepts, and Theory

2.1 FOCUS

This book deals almost exclusively with estimation and analyses within the framework of statistical survey sampling. The basic issues in this book relate to how best to draw samples from finite populations and then combine the information collected into meaningful statistical estimates of parameters of interest. Traditionally, the essence of survey sampling has been the selection of part of a finite collection of units (the sample) and the subsequent drawing of conclusions about the whole collection (the population) based on the sample.[1] Selection of the sample can be by random (also called probabilistic) or nonrandom (also called purposive or judgmental) sampling methods. Most of the discussion in this book centers on random sampling and estimation methods for which:

1. Each unit in the population has a known (positive) probability of being selected and the actual selection approximates these probabilities. This can be achieved by using some proper sampling scheme based on random numbers obtained from a table of random numbers or from a computing algorithm.
2. In the analysis of the sample, sample units are weighted inversely to their probabilities of selection.

A sampling design is a formalized way of selecting a sample from a population. Within the framework of random sampling, this means a clear formulation of all possible samples and their probabilities of selection.

A sampling strategy consists of both the sampling design and an estimator to be used in estimation. As indicated in Kotz and Johnson (1982b, p. 554), point estimation is the statistical term for an everyday activity, namely, the making of an educated guess about an unknown quantity for which some information is available, for example, the number of trees of a given species, age, or volume.

Typically, in simple situations of this type we have n measurements $y_1, \ldots, y_n$ that lead to the estimation of some quantity of interest θ, such as

[1]Analytical uses of survey data may change this emphasis of survey sampling. We limit our discussion of such uses because they are in many ways still poorly understood and controversial.

the average value in a population, and the objective is to obtain a good estimate of θ. Often the average value in a population is estimated by the average (or arithmetic mean) $\bar{y}$ of the n measurements. Estimators of a parameter, such as the sample mean, are called statistics. Alternative statistics that have been proposed as estimators of the population mean are the sample median (the value for which 50% of the observations lie on each side of it when arranged in order of magnitude) and the sample mode (the value of most frequent occurrence). All three can be considered as estimators of the population mean $\bar{Y}$.

If the Y_i ($i = 1, \ldots, n$) are independent and identically distributed (i.i.d) and have the same distribution, then the common distribution of the errors $Y_i - \theta$ does not depend on parameter θ. It has a probability density f and the joint density of the Y_i is

$$P_\theta(Y) = P_\theta(Y_1, \ldots, Y_n) = f(Y_1 - \theta) \cdots f(Y_n - \theta)\,. \tag{2.1}$$

The density function, f, is often assumed known except possibly for an unknown scale factor. Some of the more common distributions are described later in this chapter.

Statistical problems based either on eq. (2.1) or on a different model require that data $(y_1, \ldots, y_n)$ be taken that are the values of random variables $Y = (Y_1, \ldots, Y_n)$ with joint probability $P_\theta(y)$ depending on θ. The statistical estimator of θ or some function $g(\theta)$ is a function of the observations, say $\hat{\theta} = \delta(Y_1, \ldots, Y_n)$. For a given sample the estimate would be $\delta(y_1, \ldots, y_n)$.

In this book we are often concerned with questions such as which among alternative estimators is best for estimating the population mean or total θ under various circumstances. An early suggestion in the literature was to estimate θ by some value $\delta_1 = \delta_1(y_1, \ldots, y_n)$ that minimizes the sum of squared differences $\Sigma_{i=1}^n (y_i - \delta_1)^2$. The resulting estimate is called the least-squares estimate of θ and is the mean $\bar{y}$. If instead we decide to minimize the sum of absolute values $\Sigma_{i=1}^n |y_i - \delta_2|$, the solution is the median of the y.

These suggestions do not justify the use of the mean or median as reasonable estimates of θ, since no specific assumptions have been made relating the Y to θ. Such a connection can be made by considering the measurement and sampling process as an operation that is repeatable and can produce different values. This means considering the outcomes $(Y_1, \ldots, Y_n)$ of the process as random variables taking values $(y_1, \ldots, y_n)$ with frequencies defined by their joint probability distribution, which depends on θ. This is the cornerstone of the theory of estimation due to Simpson in 1755 (Kotz and Johnson 1982b, p. 555).

In survey sampling, the properties of (design) unbiasedness, (design) consistency, and (design) asymptotical unbiasedness have traditionally been emphasized for point estimation. Assume a sample of n units is selected with equal probability from the population of N units, with sample observations on the n units of $Y_1, Y_2, \ldots, Y_n$. If the statistic based on the n units (say the sample mean) is called $\hat{\theta}$, then the average value of $\hat{\theta}$ over all possible samples of size n from the population is called its expected value, denoted by $E(\hat{\theta})$. If $E(\hat{\theta}) = \theta$, then $\hat{\theta}$ is called a design-unbiased estimator of θ. If $E(\hat{\theta}) \neq \theta$, then $\hat{\theta}$ is called a design-biased estimator of θ and the bias is denoted by $B(\hat{\theta}) = E(\hat{\theta}) - \theta$.

Different samples generally give different estimates $\hat{\theta}_l$ $(l = 1, \ldots, S_n)$ where S_n indicates the total number of samples and $\mathbf{S}_n$ the set of all such samples. The difference between estimate $\hat{\theta}_l$ and θ is called the error of estimate $\hat{\theta}_l$.

An average measure of the dispersion of the different estimates around θ is the espected value of the squared error, called mean square error (MSE), and is $\text{MSE}(\hat{\theta}) = \Sigma_{l=1}^{S_n}(\hat{\theta}_l - \theta)^2/N$ for equal probability sampling. This is often used as a measure of the accuracy with which estimator $\hat{\theta}$ estimates θ.

We can write

$$\text{MSE}(\hat{\theta}) = [B(\hat{\theta})]^2 + \sum_{l=1}^{S_n} [\hat{\theta}_l - E(\hat{\theta})]^2/S_n$$

where $V(\hat{\theta}) = \Sigma_{l=1}^{S_n} [\hat{\theta}_l - E(\theta)]^2/S_n$ is called the variance of $\hat{\theta}$. $V(\hat{\theta})$ measures the precision of $\hat{\theta}$. Obviously precision and accuracy are the same when $\hat{\theta}$ is an unbiased estimator of θ.

We are often interested in comparing the performance of competing estimators of a parameter. If two estimators, $\hat{\theta}_1$ and $\hat{\theta}_2$, estimate θ, then $\hat{\theta}_1$ is more efficient than $\hat{\theta}_2$ if $\text{MSE}(\hat{\theta}_1) < \text{MSE}(\hat{\theta}_2)$.

The sampling distribution of estimator θ is the frequency distribution of estimates $\hat{\theta}_l$ $(l = 1, \ldots, S_n)$. If estimator $\hat{\theta}$ is unbiased and normally distributed, the interval $\{\hat{\theta} - k\sqrt{V(\hat{\theta})},\ \hat{\theta} + k\sqrt{V(\hat{\theta})}\}$ on average will include the parameter θ in $100P\%$ of the samples, where P is the proportion of the area between $-k$ and k of the distribution of the normal deviate $(\hat{\theta} - \theta)/\sqrt{V(\hat{\theta})}$ with mean zero and standard deviation one. The interval is said to be a confidence interval for θ with confidence coefficient P (in percent) and confidence limits $\hat{\theta} - k\sqrt{V(\hat{\theta})}$ and $\hat{\theta} + k\sqrt{V(\hat{\theta})}$. Note that for finite populations, the normal distribution is only an approximation to the distribution of $\hat{\theta}$, and then only when the size of both the sample and the population are large.

If $\hat{\theta}$ is biased, the confidence coefficient P associated with the specified confidence interval above decreases with an increase in the absolute value of the bias in $\hat{\theta}$ if $\hat{\theta}$ is normally distributed. Murthy (1967, p. 45) noted that the decrease in P is minimal for increases up to 20% in the ratio of the bias $B(\hat{\theta})$ to $\sqrt{V(\hat{\theta})}$, but becomes quite pronounced after that.

Usually $\sqrt{V(\hat{\theta})}$ is not known and has to be estimated from the sample. Then for unbiased, normally distributed $\hat{\theta}$, the statistic $t = (\hat{\theta} - \theta)/\sqrt{v(\hat{\theta})}$ follows the t-distribution where $v(\hat{\theta})$ is an estimator of $V(\hat{\theta})$.

Point estimation theory deals with the problem of finding an estimator $\hat{\theta}$ that not only tends to be close to θ (e.g., an estimator with no bias and very small variance) but that also has a measure of how close. Kotz and Johnson (1982b, p. 555) noted that many of the key ideas on the theory of estimation go back to Laplace and Gauss around 1820. The theory was developed extensively by Fisher (1922) and Wald (1950).

Intimately related to (point) estimation with confidence intervals is the idea of hypothesis testing (Kotz and Johnson 1983, p. 712). For example, we may be interested in an experiment with only two outcomes, namely, will chemical treatment cause death (yes) or not (no). Similarly, in coin tossing, we may be interested in determining whether the coin is fair (probability of a head $= p = .5$) or not ($p \neq .5$). The statement "this coin is fair ($p = .5$)" is a statistical hypothesis which can be tested by constructing a confidence interval. If the confidence interval contains the hypothesized parameter (for example $p = .5$ for a fair coin), then the hypothesis is not rejected. Otherwise it is. This requires the assumption of a specific statistical distribution for the estimator. Some of the more common distributions are discussed later in this chapter.

The field of hypothesis testing briefly discussed is closely linked to experimentation, that is, to formal statistical procedures to collect data and test one or more specific hypotheses. Many of these ideas were fully developed in Fisher (1925) within that context. Generally, sample surveys are not conducted specifically to collect data to test hypotheses, but may need to be used in this context in certain fields such as epidemiology and forestry. This creates special problems in testing hypotheses such as: does coffee drinking cause heart attacks or does acid rain cause a decline in tree growth? There is still considerable controversy in statistics and forestry concerning whether survey data can be used for such (analytical) purposes.

The remainder of this chapter is devoted to basic concepts and theory to allow fuller development of survey sampling theory and methodology. The notation used usually follows that of Cochran (1977), but most of the definitions below are taken from Cassel *et al.* (1977), and occasionally we use their notation to acquaint the reader with a second common notation. The definitions are presented in statistical terms so that the terminology is as unambiguous as possible. More practically useful definitions are given in the glossary. Some distributions important in both practical applications and development of theory are discussed, and theorems of fundamental importance to drawing conclusions (inference) based on sampling from populations are given.

2.2 DEFINITIONS: GENERAL

Definition 1. A finite population of size N is a collection of N units, where $N < \infty$ is the size of the population. For example, for a stand of ten trees, $N = 10$.

Definition 2. The units of a finite population of size N are said to be identifiable if they can be uniquely labeled from 1 to N and the label of each unit is known.

Identifiability would be easy for a stand of ten trees but would be difficult for large areas of forest.

The kth unit in the population is denoted by u_k and the finite population is denoted by

$$U = \{u_1, \ldots, u_k, \ldots, u_N\}.$$

When such practice is unambiguous, the label k is used to represent u_k. We define a parameter as follows:

Definition 3. A parameter is any real function of the vector $\mathbf{y}$.

The parameter space is denoted by Ω and contains all the possible values the function of $\mathbf{y}$ can take. Examples of parameters for the ten-tree stand in definition 1 are $y_1 + y_2 + \cdots + y_{10}$, $y_1 - y_2 + y_3 - y_4 + \cdots - y_{10}$, $y_1 \cdot y_2 \ldots y_{10}$, and so on. Clearly there is an infinite number of possible parameters.

Inference is usually made about parameters, especially the population mean and total given, respectively, by

$$\bar{Y} = \sum_{i=1}^{N} y_i/N, \tag{2.2}$$

and

$$Y = \sum_{i=1}^{N} y_i = N\bar{Y}. \tag{2.3}$$

Also important is the variance among units in the population

$$\sigma_y^2 = \sum_{i=1}^{N} (y_i - \bar{Y})^2/N. \tag{2.4}$$

In finite populations, $S_y^2 = N\sigma_y^2/(N-1)$ is used instead of σ_y^2. We often use σ^2 and S^2 instead of σ_y^2 and S_y^2 when no ambiguity is caused in doing so.

Definition 4. The rth population moment about zero, denoted by μ_r' is defined to be

$$\mu_r' = \sum_{i=1}^{N} y_i^r/N. \tag{2.5}$$

If $r = 1$, $\mu_1' = \overline{Y}$. The rth population moment about $\overline{Y}$, denoted by μ_r, is defined to be

$$\mu_r = \sum_{i=1}^{N} (y_i - \overline{Y})^r / N . \tag{2.6}$$

In particular, if $r = 2$,

$$\mu_2 = \sigma_y^2 .$$

Definition 5. A set $s = (k_1, \ldots, k_{n(s)})$ such that $k_i \in \{1, \ldots, N\}$ for $i = 1, \ldots, n(s)$ is called a sample. The label k_i is called the ith component of s. The number of components of s, $n(s)$, is called the sample size.

Definition 6. The rth sample moment about zero, denoted by M_r', is defined to be

$$M_r' = \sum_{i=1}^{n} y_i^r / n . \tag{2.7}$$

If $r = 1$, $M_1' = \overline{y}$. The rth sample moment about $\overline{y}$, denoted by M_r, is defined to be

$$M_r = \sum_{i=1}^{n} (y_i - \overline{y})^r / n . \tag{2.8}$$

In a sample, two or more components may be identical. A unit occurring more than once in s is called a repeat.

Sampling is called with replacement when units can occur more than once in a sample. In sampling without replacement, a unit can occur only once. For a sample of size n, sampling without replacement usually contains more information than a sample containing repeat information. If all units in the population of N units have positive probability of selection and the sample consists of n units, there are N^n possible samples that can be selected with replacement, and $\binom{N}{n} = [N!/\{n!(N-n)!\}]$ in without-replacement sampling.

Definition 7. The number of distinct components of a sample s is denoted by $\upsilon(s)$ and is called the effective sample size.

We do not consider ordered samples, so that in sampling we usually talk only about the set s with sample size $n(s)$ irrespective of the order of selection. When all the samples of interest in $\mathbf{S}_n$ have a fixed size n, we use n rather than $n(s)$.

For simplicity of notation

$$\sum_{s} y_i \qquad \sum_{i \in s} y_i , \qquad \text{and} \qquad \sum_{i=1}^{n(s)} y_i$$

are used in statistics to indicate summation over the $n(s)$ units. We generally use notation that includes explicit limits of summation. Further cases include

$$\sum_{i \neq j}^{n} y_i y_j\,, \qquad \sum_{i(\neq j)}^{n} y_i y_j\,, \qquad \text{and} \qquad \sum_{i<j}^{n} y_i y_j\,,$$

to denote, respectively, summation over i and j not equal to each other ($n(n-1)$ terms), summation over i not equal to j ($n-1$ terms), and summation over i and j with i less than j ($n(n-1)/2$ terms).

Definition 8. Representative sampling is a term with no single satisfactory definition. There are eight distinctly different interpretations of the concept: (1) rhetorical usage (claims that a representative sample backs up the statements made without an explanation of the sample obtained), (2) absence or presence of selective forces (does the inference apply to a population of interest or a badly distorted facsimile of it?), (3) miniaturization of the population (does the sample mirror the population in terms of all the variables of interest?), (4) typical or ideal cases (purposive sampling), (5) coverage of the population (can one or more sample units represent the entire class within the population?), (6) vague definition to be made more precise later (a specially designed sample to be explained in more detail later), (7) a specific sampling method (probability sampling such as simple random sampling), and (8) a sampling good enough for a particular purpose (despite obvious drawbacks it can still prove a point conclusively) (Kotz and Johnson 1988a, p. 77–78.)

As noted by Gilbert *et al.* (1977), we often attempt to make the sample a microcosm of the target population. Probabilistic techniques, the subject of much of the remainder of this book, often does this well on average, but not necessarily for the single sample a researcher takes. Much of the effort on probabilistic sampling design is devoted to making the set of allowable samples a microcosm of the population without making sample selection nonrandom.

In this book we tend to equate representative sampling with probabilistic sampling but, as we shall see, probability sampling is not always desirable; if considerable information is available about the population, one may be able and willing to do better if one is prepared to take certain risks. In other situations, the inference problem is so complex that the appropriate role of the sample selection method in the inferential procedure is obscure and may be ignored altogether.

2.3 DEFINITIONS: PROBABILISTIC SAMPLE DESIGN

Definition 9. A function $p(s)$ on $\mathbf{S}_n$ is called a sample design if it satisfies

$$p(s) \geq 0 \qquad \text{for all } s \in \mathbf{S}_n$$

and

$$\sum_{\mathbf{S}_n} p(s) = 1 .$$

Definition 10. The probability of selecting unit i (inclusion probability) for a design $p(s)$ is

$$\pi_i = \sum_{\mathbf{C}_i} p(s) \qquad (i = 1, \ldots, N)$$

where

$$\mathbf{C}_i = \{s : i \in s\} \qquad (i = 1, \ldots, N) ,$$

is the set of all samples containing i. The joint probability of selecting units i and j is

$$\pi_{ij} = \sum_{\mathbf{C}_{ij}} p(s) , \qquad (i \neq j = 1, \ldots, N)$$

where

$$\mathbf{C}_{ij} = \{s : i, j \in s\}, \qquad (i \neq j,\ i, j = 1, \ldots, N) ,$$

is the set of all samples containing both i and j.

2.4 DEFINITIONS: PROBABILISTIC ESTIMATION

Definition 11. The data d obtained in a set s of observed y, is defined as $d = (y_{i_1}, \ldots, y_{n(s)}) = \{(i, y_i) : i \in s\}$.

Definition 12. The sample space χ of the random variable D taking values d as above, is

$$\chi = \{d : s \in \mathbf{S}_n\} ,$$

Definition 13. A statistic $Z = f(D)$ is a function on χ such that for any given $s \in \mathbf{S}_n$, $f(\cdot)$ depends on $\mathbf{y}$ only through those y_i for which $i \in s$.

We focus on those statistics that are suitable in some sense as estimators of $\bar{Y} = \Sigma_{i=1}^{N} y_i/N$ or $Y = N\bar{Y}$. These estimators are denoted by $\bar{y}_{\text{sub}}$ or $\hat{Y}_{\text{sub}}$ where the "sub" often refers to some characteristic of the estimator used.

Definition 14. A linear estimator $\hat{Y}_{\text{sub}}$ of Y can be expressed as

$$\hat{Y}_{\text{sub}} = w_{os} + \Sigma_s w_{is} y_i ,$$

where w_{os} and w_{is} are constants that do not depend on the y values but that can depend on auxiliary information.

Definition 15. A linear homogeneous estimator is of the form

$$\hat{Y}_{\text{sub}} = \Sigma_s w_{is} y_i = \sum_{i=1}^{n} w_{is} y_i \, .$$

Definitions 16–19 refer to design-based properties of estimators for finite populations. With each population unit is associated a fixed but unknown real value, the value of the variable under study. Randomness derives from the design $p(s)$. The distribution of D or any function of D (say the estimator $\hat{Y}_{\text{sub}}$) is determined by $p(s)$ chosen by the sampler and the parameter being estimated. Definitions 16–19 relate to properties of interest for the distribution of an estimator $\hat{Y}_{\text{sub}}$ under a given design, $p(s)$.

Definition 16. An estimator $\hat{Y}_{\text{sub}}$ is called a design-unbiased estimator of Y if and only if

$$E(\hat{Y}_{\text{sub}}) = \Sigma_{S_n} \hat{Y}_{\text{sub}}(d)p(s) = Y \qquad \text{for all } \mathbf{y} \in \mathbf{\Omega}_y \, . \tag{2.9}$$

The design-bias of $\hat{Y}_{\text{sub}}$ is

$$B(\hat{Y}_{\text{sub}}) = E(\hat{Y}_{\text{sub}}) - Y \tag{2.10}$$

The design-variance of an estimator $\hat{Y}_{\text{sub}}$ is defined as

$$V(\hat{Y}_{\text{sub}}) = E\{\hat{Y}_{\text{sub}} - E(\hat{Y}_{\text{sub}})\}^2 = \Sigma_{S_n} [\hat{Y}_{\text{sub}}(d) - E(\hat{Y}_{\text{sub}})]^2 p(s) \, . \tag{2.11}$$

The design-MSE (where MSE stands for mean square error) of $\hat{Y}_{\text{sub}}$ is defined as

$$\text{MSE}(\hat{Y}_{\text{sub}}) = E(\hat{Y}_{\text{sub}} - Y)^2 = \Sigma_{S_n} [\hat{Y}_{\text{sub}}(d) - Y]^2 p(s) \, . \tag{2.12}$$

If $B(\hat{Y}_{\text{sub}}) = 0$, then $\text{MSE}(\hat{Y}_{\text{sub}}) = V(\hat{Y}_{\text{sub}})$.

Definition 17. An estimator $\hat{Y}$ is Cochran-consistent when its value is the same as the parameter of interest when the sample comprises the population.

This is called Cochran-consistent because at least one other definition of consistency is common, that of Definition 18.

Definition 18. $\hat{Y}$ is a consistent estimator of parameter Y if its value approaches Y, in the sense that the probability that the difference

$(\hat{Y} - Y)$ is less than any specified small quantity goes to certainty as n is increased indefinitely (Murthy 1967).

When N is finite, this definition must be modified to address the limiting behavior of a sequence of estimators as both n and N increase (Hansen *et al.* 1953; Hansen *et al.* 1983).

Generally (but not always) the property of Cochran-consistency ensures that the bias of the consistent estimator goes to 0 as $n \to N$. Consistency always ensures that, so it may be a more desirable property than Cochran-consistency. It is easier to check for Cochran-consistency however. Consistent estimators can be biased but the bias is often small, particularly in surveys involving large samples. Biased estimators are often used in surveys since they are commonly more reliable than unbiased estimators.

Definition 19. An estimator $\hat{Y}_l$ is an asymptotically design-unbiased (ADU) estimator of the parameter $Y_l = \Sigma_{k=1}^{N_l} Y_k$ if

$$\lim_{l \to \infty} E(\hat{Y}_l - Y_l) = 0 \tag{2.13}$$

where N_l and n_l are, respectively, the population size and sample size for a finite population l, and as $l \to \infty$, n_l and $N_l \to \infty$ such that n_l/N_l is bounded and the Y_k are fixed constants in each population. This is a specific case of the definition given in Robinson and Särndal (1983, p. 243) for Y_k as random variables. Basically, an estimator satisfying this definition approaches the population parameter of interest if n is large and N is quite large.

Definition 20. A sampling strategy consists of the pair $(p, \hat{Y}_{\text{sub}})$.

This means that to define a sampling strategy one must define both the sample selection procedure p and the estimator $\hat{Y}_{\text{sub}}$ to be used with the design. The basic purpose of (classical) survey sampling theory is to select a design and estimator that give a "best" estimate of a parameter of interest at least "cost."

By force of custom, a particular estimator may be associated with a particular design. But a design p does not dictate the estimator to be used nor does an estimator imply a design.

A sampling strategy is said to be design-unbiased if the estimator $\hat{Y}_{\text{sub}}$ is design-unbiased when used with design p. The design-variance and the design-MSE of a strategy $(p, \hat{Y}_{\text{sub}})$, denoted by $V(p, \hat{Y}_{\text{sub}})$ and MSE$(p, \hat{Y}_{\text{sub}})$, are defined, respectively, as the design-variance and the design-MSE of the estimator $\hat{Y}_{\text{sub}}$ under the design p.

Definition 21. A homogeneous linear estimator $\hat{Y}_{\text{sub}} = \Sigma w_{is} y_i$ is design-unbiased if, and only if,

$$\Sigma_{\mathbf{S}_n} \sum_s w_{is} y_i p(s) = Y \qquad \text{for all } \mathbf{y} \in \mathbf{\Omega}_y ,$$

or if, and only if, the set of N conditions $\Sigma_{\mathbf{C}_i} w_{is} p(s) = 1/N$ $(i = 1, \ldots, N)$ holds, where $\mathbf{C}_i = \{s : i \in s\}$.

Definition 22. A general linear estimator $\hat{Y}_{\text{sub}} = w_{os} + \Sigma_s w_{is} y_i$ is design-unbiased if, and only if, $\Sigma_{\mathbf{C}_i} w_{is} p(s) = 1/N$ $(i = 1, \ldots, N)$ and $\Sigma_{\mathbf{S}_n} w_{os} p(s) = 0$ are simultaneously satisfied.

Theorem 2.1. An estimator $\hat{Y} = N \, \Sigma_{j=1}^{n} \, y_j/n$ can be written as

$$\hat{Y} = N \sum_{i=1}^{N} \delta_i y_i / n \tag{2.14}$$

with variance

$$V(\hat{Y}) = E\hat{Y}^2 - [E\hat{Y}]^2 = \frac{N^2}{n^2} \left[E\Big(\sum_{i=1}^{N} \delta_i^2 y_i^2 \Big) + E\Big(\sum_{i \neq 1}^{N} \delta_i \delta_j y_i y_j \Big) \right] - \frac{(E\hat{Y})^2}{n^2} \tag{2.15}$$

where

$$\delta_i = \begin{cases} 1 & \text{if unit } i \text{ is in the sample ,} \\ 0 & \text{otherwise .} \end{cases}$$

The indicator δ_i is sometimes called the Bernoulli indicator of inclusion of u_i into the sample s. This result is very useful.

Theorem 2.2 (Hájek 1981, p. 22). If

$$\delta_i = \begin{cases} 1 & \text{if } i \in s \text{ for } 1 \leq i \leq N , \\ 0 & \text{otherwise ,} \end{cases}$$

then

$$E(\delta_i) = \pi_i \qquad 1 \leq i \leq N , \tag{2.16}$$

$$V(\delta_i) = \pi_i(1 - \pi_i) \qquad 1 \leq i \leq N , \tag{2.17}$$

and

$$\text{Cov}(\delta_i, \delta_j) = \pi_{ij} - \pi_i \pi_j \qquad 1 \leq i \neq j \leq N . \tag{2.18}$$

If sample size n is fixed,

$$n = \sum_{i=1}^{N} \pi_i , \tag{2.19}$$

and

$$\sum_{j(\neq i)}^{N} (\pi_i \pi_j - \pi_{ij}) = \pi_i(1 - \pi_i) \qquad 1 \leq i \leq N . \tag{2.20}$$

Theorem 2.3. For a without-replacement design p, with a sample of size $n(s)$,

$$E[n(s)] = \Sigma_{\mathbf{S}_n} n(s)p(s) \tag{2.21}$$

by definition. Furthermore:

1.
$$\sum_{i=1}^{N} \pi_i = E[n(s)] . \tag{2.22}$$

2.
$$\sum_{i \neq j}^{N} \pi_{ij} = E[n(s)]E[n(s) - 1] + V[n(s)]$$

$$\sum_{j(\neq i)}^{N} \pi_{ij} = E[n(s)] - \pi_i .$$

If $n(s) = n$ for all s such that $p(s) > 0$, then

$$\sum_{i \neq j}^{N} \pi_{ij} = n(n-1) \tag{2.23}$$

and

3.
$$\sum_{i(\neq j)}^{N} \pi_{ij} = v_j - \pi_j , \tag{2.24}$$

where

$$v_j = \sum_{n(s)=1}^{N} n(s)[P\{(\delta_j = 1), n(s)\}] .$$

If $n(s) = n$, for all $s \in \mathbf{S}_n$, then

$$\sum_{i(\neq j)}^{N} \pi_{ij} = (n-1)\pi_j .$$

***Proof*:**

1. Since

$$n(s) = \sum_{i=1}^{N} \delta_i ,$$

then

$$E[n(s)] = \sum_{i=1}^{N} E[\delta_i] = \sum_{i=1}^{N} \pi_i .$$

2. Since

$$\begin{aligned}\{n(s)\}^2 &= \left(\sum_{i=1}^{N} \delta_i\right)^2 \\ &= \sum_{i=1}^{N} \delta_i^2 + \sum_{i \neq j}^{N} \delta_i \delta_j \\ &= \sum_{i=1}^{N} \delta_i + \sum_{i \neq j}^{N} \delta_i \delta_j \\ &= n(s) + \sum_{i \neq j}^{N} \delta_i \delta_j ,\end{aligned}$$

then

$$E[\{n(s)\}^2] = E[n(s)] + \sum_{i \neq j}^{N} \pi_{ij} .$$

Substituting $V\{n(s)\} + \{E[n(s)]\}^2$ for the left side and rearranging gives the required result. For the case when $n(s) = n$, $V\{n(s)\}$ is zero.

3. Let $\pi_i[n(s)]$ denote the first-order inclusion probability of u_i in a sample of $n(s)$ units. Note that $\pi_i = \Sigma_{n(s)=1}^{N} \pi_i[n(s)]$. Let $\pi_{ij}\{n(s)\}$ denote the second-order inclusion probability for units u_i, u_j $(i \neq j)$. Then,

$$\begin{aligned}\sum_{i(\neq j)}^{n} \pi_{ij} &= \sum_{i(\neq j)}^{n} [\pi_{ij}(2) + \pi_{ij}(3) + \cdots + \pi_{ij}(N)] \\ &= \pi_j(2) + 2\pi_j(3) + 3\pi_j(4) + \cdots + (N-1)\pi_j(N) \\ &= \pi_j(1) + 2\pi_j(2) + \cdots + N\pi_j(N) \\ &\quad - \{\pi_j(1) + \pi_j(2) + \cdots + \pi_j(N)\} \\ &= \nu_j - \pi_j .\end{aligned}$$

(K. R. W. Brewer, personal correspondence, 1986). When $n(s) = E\{n(s)\}$ for all $s_n \in \mathbf{S}_n$, then $\pi_j(k) = 0$, unless $k = E\{n(s)\}$; then $\pi_j(n) = \pi_j$. Therefore, $\nu_j = n\pi_j$, which upon substitution leads to the required result.

An inequality that is often useful in survey sampling is the Cauchy–Schwarz inequality:

$$\left(\sum_{i=1}^{n} a_i b_i\right)^2 \le \sum_{i=1}^{n} a_i^2 \sum_{i=1}^{n} b_i^2 \tag{2.25}$$

where a_i and b_i $(i = 1, \ldots, n)$ are real numbers. Equality holds only if $a_i = cb_i$ for all $i = 1, \ldots, n$.

Also, for real-valued square integrable functions $g(\mathbf{x})$ and $h(\mathbf{x})$ with respect to a given measure P over a set $\mathbf{S}_n$ in R_n,

$$\left\{\int_{\mathbf{S}_n} gh \, dP\right\}^2 \le \int_{\mathbf{S}_n} g^2 \, dP \cdot \int_{\mathbf{S}_n} h^2 \, dP \tag{2.26}$$

so that for bivariate random variables (X, Y),

$$E[|XY|^2] \le E[|X|^2]E[|Y|^2] , \tag{2.27}$$

which is called Hölder's inequality (Kotz and Johnson 1982a, p. 387). (A function $g(x)$ is real when for every real value x, $g(x)$ is real. Such a function is square integrable if

$$\int_{-L}^{L} g^2(x) \, dx < \infty \qquad \text{on } [-L, L] .$$

An important consideration in selecting an estimator for any design is that all the pertinent information in the sample be utilized. This is embodied in the statistical criterion of sufficiency, which is discussed fully on pages 35–44 of Cassel *et al.* (1977). For our purposes, this means limiting ourselves to classes of estimators that are functions of a sufficient statistic. Our main interest in this is that, as Cassel *et al.* (1977) show, we can safely ignore information about the order and multiplicity of sample units in estimation.

Another statistical consideration with considerable intuitive appeal is that of the likelihood function. Basically, the likelihood function measures the probability (likelihood) of obtaining the data d if $\mathbf{y}$ were the true parameter value. This offers an opportunity to estimate the likelihood function for various possible values of $\mathbf{y}$ and pick that value of $\mathbf{y}$ that gives the largest likelihood.

In survey sampling, the likelihood function is given by the following definition (Godambe 1966):

Definition 23. For any design p, the likelihood function is given by

$$L_d(\mathbf{y}) = \begin{cases} p(s) & \text{for all } y \in \mathbf{\Omega}_y \\ 0 & \text{otherwise} . \end{cases} \tag{2.28}$$

This is a flat function in that there is no unique maximum. It only tells us that all possible values of the unobserved components of $\mathbf{y}$ in $\boldsymbol{\Omega}_y$ have the same likelihood, which does not give us much information. The likelihood function is more informative if superpopulations are assumed.

Estimators can be label-dependent or -independent. Data $d = \{(i, y_i) : i \in s\}$ usually consist of values of the variable of interest and population number (Definition 11) denoted, respectively, by y and i ($i = 1, \ldots, N$). Covariate information x may also be part of the data. The data d can be reduced in varying degrees by ignoring part of the information in d. For example, the unlabeled data, say $\{y_s\}$, are a reduction of d because knowledge of $\{y_s\}$ does not permit reconstruction of d.

Definition 24. An estimator is called label-independent if, and only if, it can be constructed without knowledge of the set U of labels that correspond to the data d. An estimator that can be constructed from d or from a reduction of d is said to be label-dependent as long as U can be reconstructed from the reduction of d. For example, the sample mean of distinct units, $\bar{y}$, is label-independent. It can be constructed from knowledge of the unlabeled data y_s alone and does not contain information about U. The issue of whether labels may be informative in estimation and should or should not be included in estimators has raised some controversy (Hartley and Rao 1975 and references therein).

2.5 DEFINITIONS: SUPERPOPULATIONS

In superpopulation inference, as opposed to the earlier probabilistic inference, the vector of population values $\mathbf{y} = (y_1, \ldots, y_N)$ is considered to be a realized outcome of a vector random variable $\mathbf{Y} = (Y_1, \ldots, Y_N)$ and the joint distribution of $(Y_1, \ldots, Y_N)$ is denoted by ξ. A superpopulation model refers to a specified set of conditions defining a class of distributions to which ξ belongs. Briefly then, statistical analysis in the superpopulation approach treats $\mathbf{y}$ as an outcome of $\mathbf{Y}$ with distribution ξ for which some characteristics are known (Cassel *et al.* 1977). The class of distributions ξ may range from simple specification of the features of the means, variances, and covariances of ξ to highly detailed specifications of ξ.

For ξ, an arbitrary superpopulation distribution, we define the model-based expected value and model-based variance as follows.

Definition 25. If $Q = Q(Y_1, \ldots, Y_N)$ is a function of $Y_1, \ldots, Y_N$, the model-based expected value of Q is denoted by

$$E_m(Q) = \int Q \, d\xi \,, \tag{2.29}$$

the model-based variance of Q is denoted by

$$V_m(Q) = \int \{Q - E_m(Q)\}^2 \, d\xi \,, \tag{2.30}$$

and the model-based covariance of two functions Q_1 and Q_2 of $\mathbf{y}$ is denoted by

$$\mathrm{Cov}_m(Q) = \int \{Q_1 - E_m Q_1\}\{Q_2 - E_m(Q_2)\} \, d\xi \,. \tag{2.31}$$

Cassel *et al.* (1977, p. 81) list six superpopulation concepts in survey sampling. We primarily use the concept that treats the finite population actually sampled as representing a larger universe of finite populations. A principal purpose of this is to allow for analytical comparisons of probabilistic sampling strategies when the comparative results for the latter are still too complex to yield meaningful results. Analytical inferences with survey data basically dictate the same concept.

2.6 DISTRIBUTIONS

Populations so far have been defined to be discrete or finite ($N < \infty$), but often they may be assumed to be infinitely large ($N = \infty$) so that continuous distributions are more likely to approximate the real population. It is sometimes easier to work under an assumption of an infinitely large population. Theorems exist for such populations that are crucial to the field of statistical inference. From the many discrete and continuous statistical distributions (Johnson and Kotz 1969, 1970a,b, 1972) that are in common use a few are discussed that are used or mentioned in later chapters and we cite some key results from classical statistical theory based on continuous distributions.

Definition 26. If Y is a random variable with probability density $f(y)$, then the expected value of e^{ty} is called the moment generating function (mgf) of Y if the expected value exists for every value of t in some interval $-h^2 < t < h^2$. This is denoted by

$$m(t) = E(e^{ty}) = \sum_y e^{ty} f(y) \, dy \tag{2.32}$$

where for discrete distributions, $f(y)$ is the probability mass function, and

$$m(t) = E(e^{ty}) = \int_{-\infty}^{\infty} e^{ty} f(y) \, dy \tag{2.33}$$

where for continuous distributions, $f(y)$ is the probability density function (Mood *et al.* 1974). The logarithm of the mgf, called the cumulant generating function, is often used. The moments of this function are called the cumulants.

$m(t)$ generates the moments of a distribution that are very important in both applied and theoretical statistics. For example, in survey sampling we are often interested in estimating the first two moments (of the normal distribution). The nth moment may be obtained from the mgf by differentiating with respect to t n times and setting $t = 0$.

2.6.1 Continuous Distributions

Three examples are given: normal, gamma, and multivariate distributions.

Normal Distribution. De Moivre in 1733 appears to have been the first person to derive the normal distribution (as an approximation to a binomial distribution) (Johnson and Kotz 1970a). The distribution is defined as

$$F(y) = P[Y \leq y] = \int_{-\infty}^{y} \frac{1}{\sqrt{2\pi}\sigma} \exp\left[-\frac{1}{2}\left[\frac{y-\mu}{\sigma}\right]^2\right] dy = \int_{-\infty}^{y} f(y; \mu, \sigma)\, dy \tag{2.34}$$

with parameters μ and σ^2 ($\sigma > 0$), and $-\infty < Y < \infty$. The parameters μ and σ^2 are called the mean and variance of the distribution, respectively.

The cumulative distribution is denoted by $F(y) = P(Y \leq y)$ and $Y \sim F$ denotes that Y follows the cumulative distribution function F. The word "cumulative" is usually dropped.

Distributions are often characterized by their moment generating function. The mgf of the normal distribution is

$$m(t) = \exp(t\mu + t^2\sigma^2/2) \tag{2.35}$$

with mean

$$m'(0) = \mu_1' = \mu \tag{2.36}$$

and variance

$$m''(0) - [m'(0)]^2 = \mu_2 = \sigma^2 \tag{2.37}$$

so that the mean and variance are in fact the two parameters of the distribution.

The standard normal distribution has mean $\mu = 0$ and variance $\sigma^2 = 1$. The standard normal distribution is denoted by $\Phi(Z) = P[Z \leq z]$, and $\phi(Z)$ denotes the standard normal density. Although there are numerous situations where the normal distribution approximates the distribution of individuals in a population (such as the heights of all trees in a large plantation), it is most commonly used as a convenient approximation to other distributions (Johnson and Kotz 1970a). The normal distribution is important in probability theory because it is the limiting distribution of almost any standardized sums (or means) of random variables as the number of variables in the sum increases (see Theorem 2.6).

If a statistic $\hat{\theta}$ is an unbiased estimator of a parameter θ, with estimated variance $v(\hat{\theta})$, and is approximately normally distributed, then the statistic $t = (\hat{\theta} - \theta)/\sqrt{v(\hat{\theta})}$ follows Student's t-distribution with density

$$f(t) = \Gamma\left(\frac{v+1}{2}\right)(1/\sqrt{v})(1 + t^2/v)^{-(v+1)/2}/\Gamma(1/2)\Gamma(v/2) \qquad (2.38)$$

where v is the number of degrees of freedom on which the estimate of the standard error is based and $\Gamma(v/2) = \int_0^\infty t^{v/2-1}e^{-t}\,dt$. The t-distribution is fundamental to constructing confidence intervals, and tables for this distribution are widely available (Mendenhall *et al.* 1986) as are the computing algorithms, for example, subroutine TDF (IMSL Version 1.0 1987. International Mathematical Statistical Library, 2500 Park West Tower 1, 2500 City West Blvd, Houston, TX 77042-3020, USA.)

Gamma Distribution. Laplace first derived the gamma distribution in 1836 (Johnson and Kotz 1970a).

This distribution appears naturally as the distribution of the sum of squares of independent, standard, normally distributed random variables, $Z_1, Z_2, \ldots, Z_n$. Then $\Sigma_{i=1}^n Z_i^2$ has a χ^2-distribution with parameter n where χ^2 is a special case of the gamma distribution and n is the number of degrees of freedom. The gamma distribution is often used in survey sampling to enable comparisons of sampling strategies (Cassel *et al.* 1977). The distribution is (Johnson and Kotz 1970a):

$$F(y) = P[Y \le y] = \int_0^y y^{\alpha-1}\exp(-y/\beta)/\{\beta^\alpha(\Gamma(\alpha-1)\}\,dy = \int_0^y f(y)\,dy \qquad (2.39)$$

with parameters α and $\beta > 0$ and $y > 0$, and $\Gamma(\alpha - 1) = \int_0^\infty t^{\alpha-2}e^{-t}\,dt$.

The gamma distribution has the mgf

$$m(t) = (1 - \beta t)^{-\alpha} \qquad (2.40)$$

with mean

$$\mu_1' = \alpha\beta \qquad (2.41)$$

and variance

$$\mu_2 = \alpha\beta^2\,. \qquad (2.42)$$

Multivariate Distributions. The multivariate normal distribution has been studied much more extensively than other multivariate distributions and is used more frequently for inference among multivariate continuous distributions than is the normal distribution among univariate continuous distributions (Johnson and Kotz 1972, Ch. 34). Of the continuous distributions used in forestry, the normal, Weibull, gamma, S_B, beta, and lognormal distributions (Hafley and Schreuder 1977), apparently only the normal and the S_B generalize to a multivariate situation in a reasonable manner (Hafley and Schreuder 1976; Schreuder *et al.* 1982, Knoebel and Burkhart 1991). We briefly discuss only the bivariate normal distribution, which is defined as (Johnson and Kotz 1972, Ch. 36)

$$F(x, y) = P(X \le x, Y \le y) = (2\pi\sigma_x\sigma_y\sqrt{1-\rho^2})^{-1}$$

$$\times \exp\left\{\left[\frac{-1}{\{2(1-\rho^2)\}}\right]\left\{\frac{(x-\mu_x)^2}{\sigma_x^2} - 2\rho\left(\frac{x-\mu_x}{\sigma_x}\right)\left(\frac{y-\mu_y}{\sigma_y}\right) + \frac{(y-\mu_y)^2}{\sigma_y^2}\right\}\right\} \tag{2.43}$$

with $-\infty < x < \infty$, $-\infty < y < \infty$, where

$$\mu_x = EX\,, \qquad \mu_y = EY\,, \qquad \sigma_x^2 = V(X)\,, \qquad \sigma_y^2 = V(Y)\,,$$

and

$$\rho = \sqrt{\frac{E(X-\mu_x)(Y-\mu_y)}{\sigma_x^2\sigma_y^2}} = \frac{\sigma_{xy}}{\sigma_x\sigma_y} \qquad (-1 < \rho < 1)$$

is called the correlation between X and Y, and σ_{xy} is the covariance between X and Y. Often one deals with the standardized bivariate normal distribution obtained by $\mu_x = \mu_y = 0$, and $\sigma_x = \sigma_y = 1$.

The bivariate normal distribution was used as early as the beginning of the 19th century (Johnson and Kotz 1972). For the standardized bivariate normal distribution, the conditional distribution of Y given X, is normal with expected value ρX and variance $(1-\rho^2)$. The cumulants of order >2 of bivariate (and multivariate) normal distributions are zero.

2.6.2 Discrete Distributions

Well known examples of discrete distributions are the binomial, hypergeometric, Poisson, and multinomial distributions.

Binomial Distribution. The binomial distribution was first derived by James Bernoulli in 1713 and was one of the first distributions studied (Johnson and Kotz 1969).

If n independent trials are made (such as whether a load of logs should be sampled or not) and each trial has the probability p of outcome I occurring, then the number of times in which I occurs may be represented as a random variable Y following the binomial distribution with parameters n and p. This distribution is defined as the distribution of a random variable Y (=number of occurrences of I) for which

$$P[Y=y]=\binom{n}{y}p^y(1-p)^{n-y} \qquad (y=0,1,2,\ldots,n). \tag{2.44}$$

The mgf of the binomial distribution is

$$m(t)=(1-p+pe^t)^n. \tag{2.45}$$

From the mgf, the mean and variance are derived as

$$\mu_1'=np \tag{2.46}$$

and

$$\mu_2=\mu_2'-(\mu_1')^2=p(1-p) \tag{2.47}$$

where $\mu_2'=np+n(n-1)p^2$.

There are many approximations to the binomial distribution. Such approximations often involve limiting distributions that arise when one or both parameters converge to a specific value. Limiting distributions are the (discrete) Poisson distribution discussed below ($n\to\infty$, $p\to 0$) with np = a constant θ, and the (continuous) normal distribution (which is a special case of the standardized binomial variable $(Y-np)/\sqrt{np(1-p)}$ as $n\to\infty$).

Hypergeometric Distribution. This distribution was used as early as 1896 by G. G. J. Petersen for estimating animal populations from capture-recapture sampling (see Johnson and Kotz 1969, Ch. 6). The classic situation in which it arises in forestry is outlined as follows.

Suppose we have a population with N trees, M of which are dead, and $N-M$ of which are alive. If n trees are drawn at random without-replacement, then the probability of selecting y dead trees is

$$P[Y=y]=\binom{M}{y}\binom{N-M}{n-y}\Big/\binom{N}{n} \tag{2.48}$$

for $\max(0, n-N+M)\le y\le \min(n, M)$ with parameters M, N, and n.

The mgf for the hypergeometric distribution is

$$m(t)=\frac{(N-n)!(N-M)!}{N!}\,H(-n,-M;N-M-n+1;e^t) \tag{2.49}$$

where

$$H(\alpha,\beta;\gamma;z)=1+\frac{\alpha\beta}{\gamma}\frac{z}{1!}+\frac{\alpha(\alpha+1)\beta(\beta+1)}{\gamma(\gamma+1)}\frac{z^2}{2!}+\cdots$$

is a hypergeometric function which converges for $|z|<1$.

The mean is

$$\mu_1' = nM/N \tag{2.50}$$

and variance

$$\mu_2 = [(N-n)/(N-1)]n(M/N)(1 - M/N)\,. \tag{2.51}$$

Several approximations to the hypergeometric distribution exist, a simple one being the binomial distribution

$$P[Y=y] = \binom{n}{y}(M/N)^y(1 - M/N)^{n-y}\,, \tag{2.52}$$

which is usually adequate when $n < 0.1N$ (Johnson and Kotz 1969). Confidence intervals can be constructed as described under the binomial distribution assuming either the binomial or normal approximation to the hypergeometric distribution.

Poisson Distribution. If the future lifetime of an item of equipment (say a hatchet) is independent of its present age, then the lifetime can be represented by a random variable Y with distribution

$$P[Y=y] = e^{-\theta}\theta^y/y! \qquad y = 0, 1, 2, \ldots; \theta > 0\,, \tag{2.53}$$

where θ is the only parameter ($=np$ in the binomial distribution with $n \to \infty$ as $p \to 0$), $y! = y(y-1)\cdots 1$, and $0! = 1$.

The Poisson distribution was first derived by Poisson in 1837 (Johnson and Kotz 1969). A widely quoted application of this distribution concerns the number of soldiers annually kicked to death by mules in the Prussian army. (An analogous situation might be the number of loggers killed by falling trees in a forest.) The probability of death was small and the number of soldiers exposed was large. It is doubtful that the conditions of independence and constant probability (p) were satisfied but the data available were satisfactorily fitted by this distribution.

The moment generating distribution is

$$m(t) = \exp\{\theta(e^t - 1)\} \tag{2.54}$$

with mean

$$\mu_1' = \theta \tag{2.55}$$

and variance

$$\mu_2 = \theta\,. \tag{2.56}$$

Multinomial Distribution. As indicated in Johnson and Kotz (1969, Ch. 11), multivariate discrete distributions are often closely related to univariate ones, for example, the marginal distribution of individual variables is generally a simple binomial, Poisson, or hypergeometric distribution or a distribution obtained by modifying or compounding one of the univariate distributions. For example, the joint distribution of the random variables $n_1, \ldots, n_k$ representing the number of occurrences of events $O_1, O_2, \ldots, O_k$ in n trials is the multinomial distribution

$$P(n_1, n_2, \ldots, n_k) = n! \prod_{j=1}^{k} (p_j^{nj}/n_j!) \tag{2.57}$$

with $0 \leq n_j$ for all $j = 1, \ldots, k$ and $\Sigma_{j=1}^{k} n_j = n$. This distribution is a natural extension of the binomial distribution, which is a special case for $k = 2$. J. Bernoulli worked with the multinomial as well as the binomial distribution around 1713.

The joint distribution of any subset $s < k$ of eq. (2.57) above is also a multinomial, hence the important subset where we have two classes, that is, "class 1" and "all others," is also a binomial.

The mgf of the multinomial is

$$m(t_1, \ldots, t_k) = (p_1 e^{t_1} + \cdots + p_k e^{t_k})^n \tag{2.58}$$

where the moments of the n_i, $(i = 1, \ldots, k)$ are simply

$$\mu_1'(t_i) = p_i\,, \qquad i = 1, \ldots, k \tag{2.59}$$

and

$$\mu_2(t_i) = p_i(1 - p_i)\,. \tag{2.60}$$

Note that the covariance of n_i and n_j is

$$\text{Cov}(n_i, n_j) = -np_i p_j \tag{2.61}$$

and that the correlation between n_i and n_j is

$$\text{Corr}(n_i, n_j) = -\sqrt{p_i p_j / \{(1 - p_i)(1 - p_j)\}}\,. \tag{2.62}$$

Confidence intervals for any p_i or $n_i p_i$ can be constructed by treating the multinomial as a binomial with probability of selecting class i, p_i, and of all other classes, $1 - p_i$. Then the discussion for constructing confidence intervals under the binomial is appropriate.

Multivariate Hypergeometric Distribution. A generalization of the hypergeometric distribution is the multivariate hypergeometric distribution defined as follows: If there is a population of N units, N_i of which are of type i ($i = 1, \ldots, k$) so that $\Sigma_{i=1}^{k} N_i = N$ and a sample of size n is taken without replacement from the N units, then

$$P(n_1, n_2, \ldots, n_k) = \prod_{i=1}^{k} \binom{N_i}{n_i} \Big/ \binom{N}{n} \tag{2.63}$$

is the multivariate hypergeometric distribution with

$$\sum_{i=1}^{k} n_i = n; \qquad 0 \le n_i \le N_i \qquad i = 1, \ldots, k .$$

The moments of the multivariate hypergeometric distribution are analogous to those for the hypergeometric distribution and the correlation between n_i and n_j is

$$\mathrm{Corr}(n_i, n_j) = -\sqrt{N_i N_j / \{(N - N_i)(N - N_j)\}} , \tag{2.64}$$

which clearly is similar to eq. (2.62).

2.7 LAWS OF LARGE NUMBERS

Inductive inference seeks to determine something about a population of interest, say its mean, by examining a sample from the population. Random samples of n sample units possess important inferential properties as shown below. A finite number of values of Y can be used to make reliable inferences about $E(Y)$, the average of an infinite (or very large finite) number of values of Y (i.e., $\Sigma_{i=1}^{N} y_i/N$, the average for the whole population). For simple random samples, the following three theorems apply (Mood *et al.* 1974).

Theorem 2.4 (Tchebysheff's Inequality). Let there be a distribution $F(y)$ with mean μ and finite variance σ^2. Let $\bar{y}$ be the mean of a random sample of size n from this distribution and let α be any positive number. Then

$$P[-\alpha\sigma/\sqrt{n} \le \bar{y} - \mu \le \alpha\sigma/\sqrt{n}] \ge 1 - 1/\alpha^2 . \tag{2.65}$$

Note that this is a special case of Markov's inequality below:

$$P[-k \le \bar{y} - \mu \le k] \ge 1 - \frac{\mathrm{MSE}(\bar{y})}{k^2} \qquad (k > 0) . \tag{2.66}$$

When $k = \alpha\sigma$, the inequality can be written as

$$P[-\alpha\sigma \le \bar{y} - \mu \le \alpha\sigma] \ge 1 - \frac{1}{\alpha^2} - \frac{1}{\alpha^2}\left(\frac{\mathrm{B}(\bar{y})}{\sigma}\right)^2 . \tag{2.67}$$

When Bias = 0, this inequality yields Tchebysheff's inequality given in the preceding theorem.

Theorem 2.5 (Weak Law of Large Numbers). Let $F(y)$ be a distribution with mean μ and finite variance σ^2. Let ϵ and δ be two specified small numbers where $\epsilon > 0$ and $0 < \delta < 1$. If n is any integer greater than $(\sigma^2/\epsilon^2\delta)$ and $\bar{y}_n$ is the mean of a random sample of size n from $F(y)$, then

$$P[-\epsilon < \bar{y}_n - \mu < \epsilon] \geq 1 - \delta . \tag{2.68}$$

Thus this theorem states that for any two small numbers ϵ and δ, where $\epsilon > 0$ and $0 < \delta < 1$, there is an integer n such that for a random sample of size n or larger from the distribution of the population of y-values ($F(y)$), the probability that the mean of the sample of y-values ($\bar{y}_n$) is arbitrarily close to the population mean μ can be made as close to one as desired. The weak law of large numbers is an example of convergence in probability. The following theorem exemplifies the notion of convergence in distribution.

Theorem 2.6 (Central Limit Theorem). Let $F(y)$ be a distribution with mean μ and finite variance σ^2 and let $\bar{y}_n$ be the mean of a random sample of size n from $F(y)$. If $z_n = (\bar{y}_n - \mu)\sqrt{n}/\sigma$, then the distribution of z_n approaches the standard normal distribution as n increases without bound.

This theorem states that the mean $\bar{y}_n$ of a random sample from any distribution with finite variance σ^2 and mean μ is approximately distributed as a normal variate with mean μ and variance σ^2/n. This result underlies the wide application of the t-distribution in constructing confidence intervals around estimates of parameters of interest and highlights the critical importance of the normal distribution.

Note that since we usually deal with populations of size N with N considerably less than ∞, n cannot increase without bound, so the applicability of this theorem to finite populations is arguable. In many cases, when n is not too small and the distribution of y is not too far from symmetry (little skew), the distribution of z_n will be approximately normal for inference.

2.8 MODELS

A very general model used in inventories and in many places throughout this monograph is the following linear model between a variable of interest y and an independent (or covariate) variable x:

$$y_i = \alpha + \beta x_i + e_i, \qquad i = 1, \ldots, N \tag{2.69}$$

where

$$E_m e_i = 0,$$

$$E_m e_i e_j = \begin{cases} \sigma^2 v_i & \text{if } i = j \\ 0 & \text{otherwise ,} \end{cases}$$

and α and β are the intercept and slope of the equation between y and x. The usual assumption that x_i is a fixed constant may not often be true, so inference is often conditional on the observed x_i. To make this notation explicit, one should write $E_m(e_i|x_i) = 0$ and $E_m(e_i^2|x_i) = \sigma^2 v_i$, but we dispense with this when no confusion should result.

The suggestion has been made that A and B be used for finite populations and α and β for superpopulation models of type (2.69). We do not follow that convention. The e_i indicates errors in the model and these are assumed to have a zero mean (i.e., $E_m e_i = 0$), to be independent (i.e., $E_m e_i e_j = 0$ for $i \neq j$), and to have a variance structure that may or may not be a function of values associated with i (i.e., $E_m e_i e_j = \sigma^2 v_i$ for $i = j$). The error in the model is not a misspecification of the model, but instead indicates the deviation of the y_i from their conditional expected value, $e_i = y_i - E_m\,(y_i|x_i)$.

If all the v_i have the same value, then the model has homogeneous or constant variance. If the v_i are not constant, then the model has heterogeneous variance. In many cases it is reasonable to model v_i as a function of x_i, namely:

$$y_i = \alpha + \beta x_i + e_i, \qquad i = 1, \ldots, N \tag{2.70}$$

where

$$E_m e_i = 0$$

$$E_m e_i e_j = \begin{cases} \sigma^2 x_i^k & \text{if } i = j \\ 0 & \text{otherwise,} \end{cases}$$

and k is usually a prespecified value, often considered to be $1 < k < 2$. Evidently, (2.70) is a special case of (2.69) with $v_i = x_i^k$.

2.9 PROBLEMS

1. Prove that

$$\text{MSE}(\hat{\theta}) = \text{Bias}^2(\hat{\theta}) + \sum_{l=1}^{S_n} [(\hat{\theta}_l - E(\hat{\theta})]^2/S_n\,.$$

2. Prove that the mode and median are usually not unbiased estimators of the population mean under simple random sampling.

3. Construct a population for which the median is unbiased for the population mean.

4. For each of the eight representative sampling techniques described, give a persuasive reason why the technique is undesirable.

5. Construct a small population and write down an estimator $\bar{y}_1$ that is (design) unbiased, another one, $\bar{y}_2$, that is Cochran-consistent but not consistent, another one, $\bar{y}_3$, that is consistent, and a fourth one, $\bar{y}_4$, that is asymptotically unbiased.

6. Describe some realistic forestry populations where:

(a) It may be reasonable to assume a normal, gamma, or multivariate normal distribution.

(b) The binomial, hypergeometric, Poisson, or multinomial distributions could be assumed.

(c) It would not be reasonable to assume any of the distributions described in (a) or (b).

3 Probabilistic Sampling Strategies

3.1 INTRODUCTION

Probabilistic sampling strategies rely on some type of random selection of units into the sample. The design-based statistical properties of an estimator, described in definitions 16–19 of Chapter 2, pertain to the distribution of all possible estimates under the design. In this chapter we present various sampling strategies and discuss design-based characteristics of corresponding estimators of population parameters. An advantage of basing inference about population parameters upon the design-based characteristics of estimators of these parameters is that very little needs to be assumed about the population. In Chapter 6 we discuss alternative bases for inference that rely on stronger assumptions about the structure of the population.

Because some sampling strategies possess desirable, perhaps even optimal, design-based characteristics when a stochastic model describes the underlying population, the two approaches to inference are sometimes closely linked. Indeed a class of asymptotically design-unbiased estimators has been devised that is optimal, in the sense discussed in Chapter 6, when the presumed model is valid, yet that has desirable design-based characteristics even when the model is not valid. This chapter is restricted to considerations that arise with design-based inference, and when comparisons among estimators are made with regard to design-based properties. To this end, the implication of population models to design-based inference is discussed where relevant, but model-based inference per se is not considered.

We begin by assuming that a basic list of all units in the population is available. Interest is assumed to be in the population total Y, although occasionally we assume interest in the population mean $\bar{Y}$, too, when it is not straightforward to go from Y to $\bar{Y}$. Using the simplest probabilistic sampling technique, simple random sampling (SRS), we discuss the notion of sample selection, estimator and variance, expectation, bias, mean square error (MSE), and asymptotic unbiasedness. Some useful theorems are then derived and illustrated. Next we discuss related sampling designs for which a list of all population units is available prior to sampling (called list sampling). Samples of fixed size n can still be implemented without a list prior to sampling (Pinkham 1987), but this presents some different problems. Therefore we treat separately populations for which no a priori list is available. Next, we discuss in considerable detail point estimation and variance estimation with covariates.

3.2 DESIGNS INCORPORATING A SAMPLING FRAME

Suppose in a population of N units certain characteristics (parameters) are of interest, for example, the mean value per unit or the aggregate value over all units. For simplicity, we usually confine our attention to a single characteristic, denoted by the variable y. Because we assume a sampling frame is available, samples of a given size n can be selected easily so n is considered fixed in this section. Fixed sample sizes are more desirable than random sample sizes for both planning purposes and efficiency.

3.2.1 Simple Random Sampling—Continuous Variables

This is the method whereby n units are selected out of the population of N units in such a way that each of the possible samples (or combinations) has an equal probability of selection.

A simple random sample can be selected by numbering the units in the population from 1 to N and then generating n numbers between 1 and N (inclusive) using random number tables or an appropriate computer program. A random sample can also be selected by considering each element in the population in sequence with a probability conditional on the number of units already selected, which ensures an equal probability of selection of each unit. Sampling can be with or without replacement. Sampling with replacement allows a unit to occur more than once in the sample, that is, a random integer is allowed to occur more than once in the n random numbers. In sampling without replacement, the n units are all distinct and estimation is more efficient. Intuitively, this is reasonable because nothing is gained by repeating information already in the sample. Because SRS with replacement hardly ever makes sense, from now on SRS denotes SRS without replacement.

Bissell (1986) gave an efficient method for selecting an SRS of size n when N is known. Sometimes, in more complex designs, only sampling with replacement can be implemented. For some of these designs, variances are simpler to compute when sampling is with replacement. When without-replacement sampling can be implemented, the with-replacement variance is often easier to compute and can serve as a useful approximation.

Because SRS is simple to understand, it is ideal for illustrating statistical concepts and properties of estimators. We use it below to illustrate estimation of Y.

Theorems 3.1–3.3 give the basic estimation theory for SRS. (Apply Theorem 2.5 in the proofs for these three theorems; for SRS $\pi_i = n/N$ and $\pi_{ii'} = n(n-1)/\{N(N-1)\}$.)

Theorem 3.1. In SRS, the sample total given by

$$\hat{Y} = N \sum_{s} y_i/n = N \sum_{i=1}^{n} y_i/n \tag{3.1}$$

is a design-unbiased and consistent estimator of Y. Clearly if this is true, then the sample mean $\bar{y}$ is an unbiased estimator of the population mean $\bar{Y}$.

In practical terms, the design-unbiasedness of $\hat{Y}$ implies that if l independent samples of size n are taken from the N units and their means $\hat{Y}_j$ $(j = 1, \ldots, l)$ computed, then $\Sigma_{j=1}^{l} \hat{Y}_j/l$ tends to Y as $l \to \binom{N}{n}$. Although it is unlikely that more than one sample is taken from a population in practice, unbiasedness and consistency are desirable properties in many instances.

Proof Design-Unbiasedness: Write $\hat{Y} = N \, \Sigma_{i=1}^{N} \delta_i y_i/n$ (Theorem 2.1) where

$$\delta_i = \begin{cases} 1 & \text{if } i \text{ is in the sample ,} \\ 0 & \text{otherwise .} \end{cases}$$

Clearly,

$$P(\delta_i = 1) = \frac{n}{N} \qquad \text{and} \qquad P(\delta_i = 0) = \frac{N-n}{N}$$

(see Theorem 2.4); hence

$$\begin{aligned} E(\hat{Y}) &= N \sum_{i=1}^{N} E\, \delta_i y_i/n = N \sum_{i=1}^{N} y_i/n[1 \cdot n/N] \\ &\quad + N \sum_{i=1}^{N} y_i/n[0 \cdot (N-n)/N] = \sum_{i=1}^{N} y_i = Y . \end{aligned}$$

The estimator is consistent, too.

Theorem 3.2. The variance of $\hat{Y}$ from SRS is

$$\begin{aligned} V(\hat{Y}) &= (1-f)N^2 \sum_{i=1}^{N} (y_i - \bar{Y})^2/[n(N-1)] \\ &= (1-f)N^2 S^2/n \end{aligned} \tag{3.2}$$

where $f = n/N$ is called the sampling fraction and $(1-f)$ is generally called the finite population correction.

Proof: Since $V(\hat{Y}) = N^2 V(\bar{y})$, we derive the result for $V(\bar{y})$ for simplicity. Using $\delta_i = 0$, 1 as under the proof for Theorem 3.1,

$$V(\bar{y}) = E(\bar{y} - \bar{Y})^2 = E(\bar{y}^2) - 2E\bar{y}\bar{Y} + E(\bar{Y})^2$$

$$= E\left(\sum_{j=1}^{n} y_j/n\right)^2 - 2\bar{Y}^2 + \bar{Y}^2$$

$$= E\sum_{j=1}^{n} y_j^2/n^2 + E\sum_{j\neq j'}^{n} y_j y_{j'}/n^2 - \bar{Y}^2$$

$$= E\sum_{i=1}^{N} \delta_i y_i^2/n^2 + E\sum_{i\neq i'}^{N} \delta_{ii'} y_i y_{i'}/n^2 - \bar{Y}^2$$

$$= \sum_{i=1}^{N} y_i^2/(nN) + \sum_{i\neq i'}^{N} n(n-1)y_i y_{i'}/\{n^2N(N-1)\} - \bar{Y}^2$$

$$= \sum_{i=1}^{N} y_i^2/(nN) + n(n-1)/\{n^2N(N-1)\}\left[\left(\sum_{i=1}^{N} y_i\right)^2 - \sum_{i=1}^{N} y_i^2\right] - \bar{Y}^2$$

$$= \sum_{i=1}^{N} y_i^2/\{nN\}\{1-(n-1)/(N-1)\}$$

$$+ (n-1)N^2/\{nN(N-1)\}\bar{Y}^2 - \bar{Y}^2$$

$$= \sum_{i=1}^{N} \{y_i^2(N-1-n+1)\}/\{nN(N-1)\}$$

$$- \bar{Y}^2(nN - n - nN + N)/\{n(N-1)\}$$

$$= \sum_{i=1}^{N} y_i^2(N-n)/\{nN(N-1)\} - \bar{Y}^2(N-n)/\{n(N-1)\}$$

$$= \sum_{i=1}^{N} [(y_i - \bar{Y})^2/\{n(N-1)\}](N-n)/N$$

$$= (S^2/n)(1-f)\,.$$

Since $\hat{Y} = N\bar{y}$, then $V(\hat{Y}) = N^2(S^2/n)(1-f)$. Here $S^2 = \Sigma_{i=1}^{N}(y_i - \bar{Y})^2/(N-1)$ is used instead of $\sigma^2 = \Sigma_{i=1}^{N}(y_i - \bar{Y})^2/N$ to denote the population variance. S^2 is a convenient approximation to σ^2. To distinguish among variances, if necessary, subscripts will be appended, that is, S_x^2 and S_y^2. As S^2 is used for population variance, s^2 is used for sample variances with subscripts such as s_y^2 or s_x^2 appended as needed.

The standard error of an estimator is the square root of the variance.

Theorem 3.3. An unbiased estimator of $V(\hat{Y})$ is

$$v(\hat{Y}) = N^2(1-f)\sum_{i=1}^{n}(y_i - \bar{y})^2/n(n-1)$$

$$= N^2(1-f)s^2/n\,. \qquad (3.3)$$

Proof: As before, we derive $Ev(\bar{y})$ for simplicity's sake.

$$Ev(\bar{y}) = (1-f)E\sum_{j=1}^{n}\frac{[(y_j-\bar{Y})-(\bar{y}-\bar{Y})]^2}{n(n-1)}$$

$$= (1-f)E\sum_{j=1}^{n}\frac{(y_j-\bar{Y})^2}{n(n-1)} - (1-f)E\sum_{j=1}^{n}\frac{(\bar{y}-\bar{Y})^2}{n(n-1)}$$

$$= \frac{(1-f)}{n(n-1)}\left[E\sum_{i=1}^{N}\delta_i(y_i-\bar{Y})^2 - E(\bar{y}-\bar{Y})^2\right]$$

$$= \frac{(1-f)}{n(n-1)}\left[\sum_{i=1}^{N}\frac{n}{N}(y_i-\bar{Y})^2 - nV(\bar{y})\right]$$

$$= \frac{1}{n(n-1)}\left[\frac{n^2(N-1)}{N}\,\frac{\sum_{i=1}^{N}(y_i-\bar{Y})^2(1-f)}{n(N-1)} - n(1-f)V(\bar{y})\right]$$

(from Theorem 3.2)

$$= \frac{1}{n(n-1)}\left[\frac{n^2(N-1)}{N}V(\bar{y}) - n(1-f)V(\bar{y})\right]$$

(from Theorem 3.2)

$$= V(\bar{y})\,.$$

3.2.2 Simple Random Sampling—Proportions

Sometimes interest is in the proportion or percentage of units in a population possessing a specific characteristic, for example the percentage of trees in a forest that can serve as den trees for song birds or that are dead. Although this can be considered a special example of the continuous variable case considered in Section 3.2.1 (by setting $y_i = 0$ if unit i does not have the characteristic, and $y_i = 1$ if it does), it is simpler in certain circumstances, important in its own right, and can serve well to illustrate the utility of some of the distributions described in Chapter 2. For simplicity we say that unit i is in class C if it has the characteristic of interest, otherwise it is in the complement of C, called C'.

The following theorem gives the relevant theoretical results for proportions derivable directly from Theorems 3.1–3.3.

Theorem 3.4. Given SRS, the sample proportion

$$\hat{p} = m/n \tag{3.4}$$

is an unbiased estimator of the population proportion

$$p = M/N$$

with variance

$$V(\hat{p}) = \frac{p(1-p)}{n}(N-n)/(N-1)\,, \tag{3.5}$$

and an unbiased variance estimator is

$$v(\hat{p}) = \hat{p}(1-\hat{p})(N-n)/\{(n-1)N\}\,, \tag{3.6}$$

where M is the number of units in C, and m is the number of units in C that are included in the sample.

The sampling distribution of m is hypergeometric (eq. 2.48) with mean $E[m] = np$ and variance $V(m) = [(N-n)/(N-1)]np(1-p)$. Consequently, $E[\hat{p}] = p$ and $V(\hat{p}) = [(N-n)/(N-1)]\,p(1-p)/n$. Whereas the inclusion probability of each population unit is $\pi_i = n/N$, the probability of including a unit from class C in the sample depends on the number of units from C also chosen for the sample. However, if n is sufficiently small relative to M and N, then this latter probability is approximately constant and equal to p. In such cases, the distribution of m is approximately binomial (eq. 2.44), as it is, too, under SRS with replacement. Cochran (1977) summarized conditions under which the normal approximation to the binomial distribution can be used to provide a reasonably accurate estimate of the $(1-\alpha)$ 100% confidence intervals on p. Alternatively, one can use the binomial distribution directly to find lower and upper confidence bounds p_u and p_l such that the following equalities are satisfied, at least approximately.

$$\sum_{s=0}^{m} \binom{n}{s} p_u^s (1-p_u)^{n-s} = \alpha/2$$

$$\sum_{s=m}^{n} \binom{n}{s} p_l^s (1-p_l)^{n-s} = \alpha/2\,.$$

Walters (1985) explained the conservative nature of intervals formed in this fashion, and he presented a Bayesian procedure that may be useful whenever reasonable prior knowledge about the likely range of p exists.

3.2.3 Sampling with Arbitrary Probabilities

Arbitrary probability sampling covers cases both where individual units may have different probabilities of selection and where joint probabilities of selection may differ between pairs of units. It includes such well-known sampling designs as stratified random sampling, cluster sampling, systematic sampling and sampling with probability proportional to (some measure of) size (pps). Obviously SRS is a special case too.

If π_i is the probability that unit i enters the sample and π_{ij} the probability that both units i and j do, then in sampling without replacement, the following well-known theorem applies (Stuart 1964):

Theorem 3.5. The Horvitz–Thompson estimator

$$\hat{Y}_{HT} = \sum_{s} y_i/\pi_i \tag{3.7}$$

is an unbiased estimator of Y. The variance of $\hat{Y}_{HT}$ is

$$V(\hat{Y}_{HT}) = \frac{1}{2} \sum_{i \neq j}^{N} (\pi_i \pi_j - \pi_{ij}) \left(\frac{y_i}{\pi_i} - \frac{y_j}{\pi_j} \right)^2 . \tag{3.8}$$

An unbiased estimator of $V(\hat{Y}_{HT})$ is given by

$$v_1(\hat{Y}_{HT}) = \frac{1}{2} \sum_{i \neq j}^{n} \frac{\pi_i \pi_j - \pi_{ij}}{\pi_{ij}} \left(\frac{y_i}{\pi_i} - \frac{y_j}{\pi_j} \right)^2 \tag{3.9}$$

(Yates and Grundy 1953) or by:

$$v_2(\hat{Y}_{HT}) = \left[\sum_{i=j}^{n} \frac{1 - \pi_i}{\pi_i^2} y_i^2 + \sum_{i \neq j}^{n} \frac{\pi_{ij} - \pi_i \pi_j}{\pi_{ij}} \left(\frac{y_i y_j}{\pi_i \pi_j} \right) \right] \tag{3.10}$$

(Horvitz and Thompson 1952) if all π_i and $\pi_{ij} > 0$.

Proof:

Let:

$$\delta_i = \begin{cases} 1 & \text{if } i \text{ is in the sample,} \\ 0 & \text{otherwise .} \end{cases}$$

Then

$$\begin{aligned} E\hat{Y}_{HT} &= E \sum_{i=1}^{N} \delta_i y_i/\pi_i \\ &= \sum_{i=1}^{N} y_i \\ &= Y \end{aligned}$$

since $E\delta_i = \pi_i$ by Theorem 2.4. Similarly, the other formulas can be derived directly by using Theorems 2.3 and 2.4.

If sampling is with replacement, and the probability of selecting unit i at a single draw is p_i, then the estimated total is

$$\hat{Y}_{\text{pps}}(wr) = \frac{1}{n} \sum_{i=1}^{n} y_i/p_i \tag{3.11}$$

with variance

$$V\{\hat{Y}_{\text{pps}}(wr)\} = \frac{1}{n} \sum_{i=1}^{N} p_i(y_i/p_i - Y)^2 \tag{3.12}$$

and unbiased estimator of variance

$$v\{\hat{Y}_{\text{pps}}(wr)\} = \sum_{i=1}^{n} \{y_i/p_i - \hat{Y}_{\text{pps}}(wr)\}^2/\{n(n-1)\}\ . \tag{3.13}$$

The formula for $V(\hat{Y}_{HT})$ is useful in explaining when unequal probability sampling might be better than SRS (Stuart 1964). SRS is a special case with $\pi_i = n/N$ and $\pi_{ij} = n(n-1)/\{N(N-1)\}$ (for all units i and j) so that each unit has an equal probability of selection and all pairs of units have the same joint probability of selection.

Improving on the efficiency of SRS sampling requires sample designs that make $V(\hat{Y}_{HT})$ smaller (Stuart 1964). If prior knowledge of the y-values exists so that one can accept that $\pi_i = ky_i$ where k is a constant, then $V(\hat{Y}_{HT})$ in eq. (3.8) is 0. Although we usually do not know the y-values, we may have information on a variable x closely correlated with y so that we can use $\pi_i = kx_i$ and make $V(\hat{Y}_{HT})$ small. This is the principle behind pps sampling discussed later. If the π_i are fixed and only π_{ij} can be varied (all $\pi_{ij} > 0$), then the term $(y_i/\pi_i - y_j/\pi_j)^2$ is fixed. Let $w_{ij} = \pi_i\pi_j - \pi_{ij}$. Then if π_{ij} can be made equal to $\pi_i\pi_j$, $w_{ij} = 0$ and the term involving i and j disappears from the variance. In SRS, $\pi_i\pi_j - \pi_{ij} = n(N-n)/\{N^2(N-1)\}$ (for all units i and j), which is positive so that all of the $\frac{1}{2}N(N-1)$ terms in the summation contribute to $V(\hat{Y}_{HT})$.

If the π_i are all equal to n/N and if some π_{ij} are made equal to n^2/N^2 so that the corresponding $w_{ij} = 0$, then those π_{ij} have increased from $n(n-1)/(N(N-1))$. This means that some of the other π_{ij} have to be reduced correspondingly because $\Sigma_{i \neq j}^{N} \pi_{ij} = n(n-1)$. This will increase some of the w_{ij}. To reduce the variance V, it would clearly be advantageous if $w_{ij} = 0$ for large values of $(y_i/\pi_i - y_j/\pi_j)^2$ or equivalently for $|y_i - y_j|$, even if w_{ij} is increased for small values. Similarly, if the π_i are unequal, the zero weights w_{ij} should be attached to larger values of $|y_i/\pi_i - y_j/\pi_j|$ for similar reductions in $V(\hat{Y})$ (Stuart 1964).

Stratified Random Sampling. From the preceding, it can be seen that manipulation of the probabilities and joint probabilities of selection can result in a smaller variance V and correspondingly improved precision of the estimator. A zero w_{ij} means that $\pi_{ij} = \pi_i \pi_j$, and hence y_i and y_j must be selected independently. This leads to splitting the selection process into two or more independent operations or grouping the N individuals into L subpopulations, with n_h $(h = 1, \ldots, L)$ selected from the hth subpopulation independent of what occurs in the other subpopulations. Ideally, $w_{ij} = 0$ for widely different y_i and y_j, so that members of different groups should be as different (heterogeneous) as possible and members of the same group should be as similar (homogeneous) as possible. This is the principle of stratification (Stuart 1964).

In stratified sampling, the population of N units is divided into subpopulations of $N_1, N_2, \ldots, N_L$ units, respectively. These subpopulations are nonoverlapping and together they comprise the whole population. Subpopulations are often called strata. The strata sizes $N_1, N_2, \ldots, N_L$ are assumed known here. If SRS is used for selecting samples in each stratum, the procedure is called stratified random sampling. Stratified random sampling is a very important procedure used in many applications. The beauty of stratified random sampling is that it is simple both in implementation and estimation (SRS is used within each subpopulation) and is usually very efficient (the sampling effort is guaranteed to be spread out over the whole population and the subpopulations are often much more homogeneous than the whole population).

Strata may be set up for the following reasons:

- To derive information for certain parts of the population, for example, we may want estimates not only for a given region but also for the plains (subpopulation 1) and the mountains (subpopulation 2) comprising the region.
- For convenience, for example, the agency conducting the survey may have field offices in the two subregions mentioned above.
- To apply different sampling procedures in different parts of the population, for example, sampling for timber volume could be quite different in the mountains and plains. (Then, of course, it would not be stratified random sampling.)
- To improve sampling efficiency by making the strata more homogeneous (less variable) than the whole population. Sampling intensity may also need to be varied, for example, in sampling for timber volume, the variability in volume per hectare on the plains may be less than that for the whole region. Sampling intensity could also be expected to be higher in the mountains because of the greater volume and value of timber, but this may be affected by an expected higher cost of sampling due to more difficult access.

Stratification can be expected to yield considerable improvement in efficiency over SRS if the values of the variable(s) of interest (y) vary tremendously, if a good measure of size (x) is available to set up the strata, and if this measure of size is correlated well with the variable(s) of interest (but not necessarily known). If the relationship between y and x is fairly well known, it may be feasible to improve upon stratification as discussed later under ratio and regression estimation.

Theorem 3.6 summarizes the estimation theory for stratified random sampling.

Theorem 3.6. Given stratified random sampling,

$$\hat{Y}_{st} = \sum_{h=1}^{L} N_h \bar{y}_h \tag{3.14}$$

is an unbiased estimator of $\hat{Y}$ with variance

$$V(\hat{Y}_{st}) = N^2 \sum_{h=1}^{L} W_h^2 \frac{\sum_{i=1}^{N_h} (y_{hi} - \bar{Y}_h)^2}{(N_h - 1)n_h} (1 - f_h) . \tag{3.15}$$

An unbiased estimator of $V(\hat{Y}_{st})$ is given by

$$v(\hat{Y}_{st}) = N^2 \left[\sum_{h=1}^{L} W_h^2 (1 - f_h) \frac{\sum_{i=1}^{n_h} (y_{hi} - \bar{y}_h)^2}{(n_h - 1)n_h} \right] \tag{3.16}$$

where

$$\bar{y}_h = \sum_{i=1}^{n_h} y_{hi}/n_h , \qquad W_h = N_h/N , \qquad f_h = n_h/N_h ,$$

y_{hi} = value of unit i in stratum h, and n_h = size of sample in stratum h.

Proof: The result for $\hat{Y}$ ($\hat{Y}_{HT}$) in Theorem 3.5 can be rewritten as:

$$\hat{Y} = \sum_{h=1}^{L} \sum_{i=1}^{n_h} y_{hi}/\pi_{hi}$$

where y_{hi} and π_{hi} are the value and probability of selection of unit i in stratum h. Note that $\Sigma_{h=1}^{L} n_h = n$ and $\pi_{hi} = n_h/N_h$. Hence

$$\hat{Y} = \sum_{h=1}^{L} \sum_{i=1}^{n_h} y_{hi} N_h/n_h = \sum_{h=1}^{L} N_h \bar{y}_h = \hat{Y}_{st} .$$

Similarly $V(\hat{Y}_{HT})$ and $v(\hat{Y}_{HT})$ in eq. (3.8) and (3.9) can be shown to equal $V(\hat{Y}_{st})$ and $v(\hat{Y}_{st})$ since $\pi_{hi} = n_h/N_h$ for all units i in stratum h,

$\pi_{ij} = n_h(n_h - 1)/N_h(N_h - 1)$ for all units i and j in stratum h, and $\pi_{ij'} = n_h n_{h'}/N_h N_{h'}$ for units in different strata h and h'.

In stratified sampling, a different sampling intensity can be used in each stratum. Commonly, samples are allocated proportional to stratum size (say area) so that the same sampling intensity is used in each stratum, that is, $n_h/N_h =$ constant for all strata $h = 1, \ldots, k$. Swindel and Yandle (1972) showed that when information on within-stratum variability is unavailable, proportional allocation is a sound strategy. This may also be true in multiresource inventories where there may be many variables of interest that are not well correlated, so that allocation among strata designed to reduce the variance of an estimator of one parameter may inflate that of another. Proportional allocation in this case may be a useful compromise to ensure that all characteristics are estimated well.

In situations where some strata are of more interest than others, warranting different intensities of sampling (e.g., when estimating timber volume in a stratum containing commercial forest versus one containing primarily scrubland and pasture), an approximation to optimum allocation may be desirable. In optimum allocation, either the overall variance $V(\hat{Y}_{st})$ in eq. (3.15) or the overall cost of sampling C is minimized. A simple linear cost function is often assumed, namely,

$$C = C_o + \sum_{h=1}^{L} C_h n_h \tag{3.17}$$

where C_o is an overhead cost and C_h is the cost of sampling a unit in stratum h. Then $V(\bar{y}_{st})$ is a minimum (Cochran 1977, p. 98) for optimal sample sizes,

$$n_h = n(N_h S_h/\sqrt{C_h} \Big/ \sum_{l=1}^{L} N_l S_l/\sqrt{C_l}) \,, \qquad h = 1, \ldots, L \,.$$

If C is the limiting constraint, then by substitution of n_h into eq. (3.17) one gets

$$n = \frac{(C - C_o) \sum_{h=1}^{L} N_h S_h/\sqrt{C_h}}{\sum_{h=1}^{L} N_h S_h \sqrt{C_h}} \,. \tag{3.18}$$

where S_h^2 is the variance within stratum h. Conversely, if $V(\bar{Y}_{st})$ is fixed at some desired level, V say, to be achieved, then by substitution into eq. (3.15), one gets

$$n = \frac{\left(\sum_{h=1}^{L} W_h S_h \sqrt{C_h}\right) \sum_{h=1}^{L} W_h S_h/\sqrt{C_h}}{V + \sum_{h=1}^{L} W_h S_h^2/N} \,. \tag{3.19}$$

If the cost per sampling unit is the same in each stratum, then $V(\bar{y}_{st})$ is minimized for $n_h = nN_hS_h/\Sigma_{l=1}^{L} N_lS_l$. This can often be a useful approximation in the frequent situation where little information is available on costs. Although eq. (3.17) is usually not realistic, the above results may still be quite useful for several reasons, namely: (1) it may be difficult to develop a better cost function; (2) available estimates of the variance and cost of S_h^2 and C_h are usually approximate only; and (3) often the variance is relatively insensitive to the allocation of samples (Cochran 1977).

Often an important consideration when designing a survey is the number of strata to use. This is a function of sample size (usually one should select at least two sample units per stratum), the quality of the information available to determine the strata, the number of subpopulations for which information is desired, and the number expected to give the best sampling efficiency in estimating the population total or mean. An empirical rule is that it is often not profitable to have more than six strata (see, e.g., Cochran 1977, section 5A.8).

Sometimes it is fairly clear what the number of strata should be. An excellent example of this is provided by Frayer (1978), who showed that when stratifying landscape into forest and nonforest strata based on air photo interpretation, a tremendous increase in efficiency can be achieved by stratifying the resource into forest, nonforest, and "not obvious which" strata. In Frayer's example, the homogeneity within the initial two strata was dramatically increased by assigning all the areas that were difficult to interpret confidently into a third stratum. Going from two to three strata increased the heterogeneity between strata and therefore gave a lower overall variance. The "not obvious which" stratum is a category that can often be used very effectively in forestry. Trying to classify sample units falling in this category into other strata is often time-consuming and unproductive.

Cluster Sampling. Note that we can modify $V(\hat{Y}_{HT})$ in eq. (3.8) even more drastically and make some of the $w_{ij} = \pi_i\pi_j - \pi_{ij} < 0$. This means making some of the π_{ij} as large as possible instead of equal to $\pi_i\pi_j$. No π_{ij} can exceed π_i or π_j so that, for example, $\pi_{ij} \leq n/N$ if $\pi_i = n/N$ for all units. If some of the $\pi_{ij} = n/N$, then y_i is selected when y_j is. Thus all units y_i and y_j for which $\pi_{ij} = n/N$ are members of a group selected as a whole, that is, part of a cluster of units. To make some of the $w_{ij} < 0$, we want the π_{ij} that are equal to n/N attached to the largest values of $|y_i - y_j|$, which means that the members of the cluster should differ among themselves as much as possible. This clustering principle tells us then to make the clusters as variable (heterogeneous) as possible, which is clearly antithetical to the stratification principle (Stuart 1964).

The clustering idea is more risky than stratification. There will be sharp gains if clusters are chosen well but sharp losses if negative w_{ij} values are attached to small values of $|y_i - y_j|$. Stratified sampling is less risky because the w_{ij} values are changed much less. In cluster sampling, the population of M_o units is divided into N groups with sizes M_i ($i = 1, \ldots, N$) of at least one unit per group. For clarity, this type of cluster sampling is discussed first.

Cluster sampling is used when no reliable list of the units in the population is available, and it would be very expensive to construct such a list. However, it is also used even when a list of population units is available, because it is usually much cheaper to visit and sample n clusters with $\Sigma_{i=1}^{n} M_i = m$ units than to sample m randomly selected units. A list of the number of clusters is assumed known.

The potential gains to be achieved with cluster sampling by having heterogeneous clusters is rarely a reason for using cluster sampling in natural resource surveys. Clusters are usually homogeneous because travel expenses dictate that units in close proximity comprise the clusters. If units within a cluster are heterogeneous, this is a bonus.

Considerable work has been done on the problem of whether to select a few large clusters or more smaller clusters, concentrating on selecting the combination of number and size of clusters that yields the lowest variance for a given cost or the lowest cost for a specified variance. Hansen *et al.* (1953) described the work in detail for specific cost functions.

Theorem 3.7, which can be derived from Theorem 3.5, summarizes some estimation theory for cluster sampling. Both estimation of the mean/unit and the overall total are considered.

Theorem 3.7. When n out of N clusters are selected randomly and each selected cluster is measured completely for the variable of interest, a biased estimator of $\bar{Y}$ (mean/unit) and an unbiased estimator of the overall total Y are

$$\bar{y}_{cl1} = \frac{1}{n} \sum_{i=1}^{n} Y_{i\cdot}/M_i \tag{3.20}$$

and

$$\hat{Y}_{cl1} = N \sum_{i=1}^{n} Y_{i\cdot}/n \tag{3.21}$$

where $Y_{i\cdot}$ is the total for cluster i with variances

$$V(\bar{y}_{cl1}) = \frac{(1-f)}{n(N-1)} \sum_{i=1}^{N} \left[\frac{Y_{i\cdot}}{M_i} - \bar{Y} \right]^2 \tag{3.22}$$

$$V(\hat{Y}_{cl1}) = \frac{N^2(1-f)}{n(N-1)} \sum_{i=1}^{N} (Y_{i\cdot} - \bar{Y})^2 , \tag{3.23}$$

and unbiased variance estimators

$$v(\bar{y}_{cl1}) = \frac{(1-f)}{n(n-1)} \sum_{i=1}^{n} \left[\frac{Y_{i\cdot}}{M_i} - \bar{y}_{cl1} \right]^2 \tag{3.24}$$

and

$$v(\hat{Y}_{cl1}) = \frac{N^2(1-f)}{n(n-1)} \sum_{i=1}^{n} (Y_{i.} - \bar{y}_{cl1})^2 . \tag{3.25}$$

The problem with estimator $\bar{y}_{cl1}$ is that it is not only biased but also inconsistent if the M_i vary. To help understand why, note that $\hat{Y}_{cl1}$ as given in eq. (3.21), is an unbiased estimator of $Y = \Sigma_{i=1}^{N} \Sigma_{j=1}^{M_i} y_{ij}$, since we are taking a SRS of n out of N clusters, but it is not an unbiased estimator of $\bar{Y} = Y/M_o$ where $M_o = \Sigma_{i=1}^{N} M_i$. An unbiased estimator of $\bar{Y}$ is

$$\bar{y}_{cl2} = \frac{1}{n} \sum_{i=1}^{n} Y_{i.}/\bar{M} , \tag{3.26}$$

where $\bar{M} = M_o/N$ is the average number of units per cluster in the population. The difference between $\bar{y}_{cl2}$ and $\bar{y}_{cl1}$ is that we divide by M_i for each cluster sampled in $\bar{y}_{cl1}$ but by the average of the M_i ($\bar{M}$) in $\bar{y}_{cl2}$.

Note that $\bar{y}_{cl2}$ requires knowledge of $\bar{M}$. Often $\bar{M}$ is not known and is estimated by $\bar{M}_n = \Sigma_{i=1}^{n} M_i/n$, in which case the resulting estimator is not unbiased. This problem does not occur if we are estimating Y rather than $\bar{Y}$. Clearly $\bar{y}_{cl1} = \bar{y}_{cl2}$ only if all the M_i are the same.

The variance of $\bar{y}_{cl2}$ is

$$V(\bar{y}_{cl2}) = \frac{1-f}{n} \sum_{i=1}^{N} \left(\frac{M_i \bar{y}_{i.}}{\bar{M}} - \bar{Y} \right)^2 \Big/ (N-1) \tag{3.27}$$

where $\bar{M} = M_o/N$ and $\bar{y}_{i.} = Y_{i.}/M_i$.

An unbiased estimator of the variance is (Sukhatme *et al.* 1984)

$$v(\bar{y}_{cl2}) = \frac{N-n}{N} \frac{1}{n(n-1)} \sum_{i=1}^{n} \left(\frac{M_i \bar{y}_i}{\bar{M}} - \bar{y}_{cl2} \right)^2 . \tag{3.28}$$

Again, $\bar{M}_n$ may need to be substituted for $\bar{M}$ in which case $v(\bar{y}_{cl2})$ is not unbiased.

A biased, but consistent, estimator of $\bar{Y}$ is a weighted mean of the cluster means or the ratio of two random variables (ratio estimators in general are discussed later in this chapter). This estimator is

$$\bar{y}_{cl3} = \sum_{i=1}^{n} M_i \bar{y}_{i.} \Big/ \sum_{i=1}^{n} M_i \tag{3.29}$$

with approximate variance

$$V(\bar{y}_{cl3}) = \frac{1-f}{n} \sum_{i=1}^{N} \frac{M_i^2}{\bar{M}^2} (\bar{y}_{i.} - \bar{Y})^2/(N-1) \tag{3.30}$$

and approximate variance estimator

$$v(\bar{y}_{cl3}) = \frac{1-f}{n} \sum_{i=1}^{n} \frac{M_i^2}{M_n^2} (\bar{y}_{i\cdot} - \bar{y}_{cl3})^2/(n-1) . \tag{3.31}$$

Systematic Sampling. In systematic sampling, the population of N units is sampled by selecting every kth unit so that $N/n = k + r$ where $0 \leq r < n$ since k is an integer and N/n usually is not. If the first unit is selected arbitrarily between 1 and k, we have systematic sampling; if it is selected randomly, we have systematic sampling with a random start. We consider only the latter case, because it is most common, and we call it systematic sampling.

To ensure a constant sample size in systematic sampling, the circular systematic procedure of Murthy (1967) can be followed. That is, list the N units and let k be the nearest integer to N/n. Then select a random number between 1 and N and take every kth unit until the desired n units have been chosen. If r is the random number between 1 and N, the sample consists of units corresponding to

$$\begin{cases} \{r + jk\} & \text{if } r + jk \leq N \\ \{r + jk - N\} & \text{if } r + jk > N \end{cases} \quad \text{for } j = 0, 1, 2, \ldots, (n-1) .$$

Systematic sampling presumes that the population can be arrayed in some order, which may be natural (e.g., a time sequence implicit in the population such as days in the use of a campground during the summer) or artificial (such as numbered quadrants on a map). In the first case the value y_i of the ith population is likely to be correlated with its position in the ordered frame (weekend versus weekday use) but not in the latter. The ordering may be considered haphazard in the latter case but needs to be carefully considered in the former.

Systematic sampling is not generally endorsed by theoretical statisticians, but practitioners and applied statisticians have continued using it because it is a convenient way of collecting information and avoids the problem of poorly distributed sample units, which can occur in random sampling. Assurance from theoreticians that a random sample is still a representative sample does not allay the fears of practitioners, who intuitively feel that a well-spaced sample, guaranteed with a systematic sample, ought to be more informative than a less-well-distributed random sample.

There is no assurance that a systematically chosen sample will be well distributed over the population of interest, but practical experience suggests that in many cases a good distribution will be obtained. Systematic sampling does well relative to SRS if there is a linear trend in the ordered population. When the ordering is purely haphazard, the systematic sampling estimator is as precise as SRS. Generally, no penalty is incurred by choosing systematic sampling for practical convenience.

Systematic sampling can be considered as a special case of either stratified sampling or cluster sampling. For example, each group of k units is equivalent to a stratum with one unit selected per stratum and each potential sample of n units is equivalent to a cluster with one cluster selected randomly from the k clusters. Since only one unit per stratum or one out of k clusters is selected, it is clear that although unbiased estimation of the parameter of interest (Y) is possible, design-unbiased estimation of the variance is not. This is confirmed by Theorem 3.5 (some of the π_{ij} are zero). Systematic sampling then results in a well-dispersed sample because the probabilities of selecting adjacent sampling units at the same time are set to zero, while the probabilities of selecting certain combinations of distant units is equal to their individual probabilities of selection. When the random starting value is selected from the first k population units, where k is the stipulated sampling interval, k possible samples exist. In contrast, regardless of the sampling interval, N possible samples exist with the circular systematic sampling design described above.

Let Y_i and $\bar{y}_i$ denote the total and the mean, respectively, of the ith possible sample, and let Y^* denote the total of the sample that is selected. For the case where the random start is selected among the first k units only, the following results hold.

Theorem 3.8. An unbiased estimator of Y is

$$\hat{Y}_{sy} = kY^* \tag{3.32}$$

and its variance is

$$V(\hat{Y}_{sy}) = \frac{1}{k} \sum_{i=1}^{k} (kY_i - Y)^2 \,. \tag{3.33}$$

Proof: Write Y^* as $Y^* = \Sigma_{i=1}^{k} Y_i \delta_i$, where δ_i is an indicator variable whose value is $\delta_i = 1$ if the ith sample is chosen and $\delta_i = 0$ otherwise. Since each possible sample is equally likely, $E[\delta_i] = 1/k$. Thus

$$\begin{aligned} E[\hat{Y}_{sy}] &= E\left[k \sum_{i=1}^{k} Y_i \delta_i\right] \\ &= k \sum_{i=1}^{k} Y_i E[\delta_i] \\ &= \sum_{i=1}^{k} Y_i \\ &= Y \end{aligned}$$

and

$$\begin{aligned} V(\hat{Y}_{sy}) &= E[(kY^* - Y)^2] \\ &= E\left[\left(k \sum_{i=1}^{k} Y_i \delta_i - Y\right)^2\right] \\ &= k^2 \sum_{i=1}^{k} Y_i^2 E[\delta_i^2] - 2kY \sum_{i=1}^{k} Y_i E[\delta_i] + Y^2 \\ &= k \sum_{i=1}^{k} Y_i^2 - Y^2 \\ &= \frac{1}{k} \sum_{i=1}^{k} (kY_i - Y)^2 . \end{aligned}$$

An unbiased estimator of $\bar{Y}$ is $\bar{y}_{sy} = \hat{Y}_{sy}/N$. Unless $N = nk$, $\bar{y}_{sy}$ generally will not coincide with the sample mean $\bar{y}$.

Let S_i^2 denote the sum of squared deviations around $\bar{y}_i$ for the ith sample. In the case where $N = nk$, the variance of $\hat{Y}_{sy}$ can be expressed as

$$\begin{aligned} V(\hat{Y}_{sy}) &= \frac{N^2}{k} \sum_{i=1}^{k} (\bar{y}_i - \bar{Y})^2 \\ &= N(N-1)S^2 - N^2\left(\frac{n-1}{n}\right)S_w^2 \end{aligned}$$

where $S_w^2 = (1/k(n-1)) \Sigma_{i=1}^{k} S_i^2$ denotes the pooled within-group variance. Clearly as S_w^2 increases, $V(\hat{Y}_{sy})$ decreases. In other words the precision of $\hat{Y}_{sy}$ increases directly with the increasing heterogeneity within the possible systematic samples. In the event that $S_w^2 = S^2$, $V(\hat{Y}_{sy}) = N^2(1-f)S^2/n$, which one recognizes as the variance of $\hat{Y}$ under SRS. This result is comforting in situations where systematic sampling is undertaken solely for administrative convenience, because unless there is a fortuitous correlation between the characteristic of interest and the characteristic by which the population is ordered, S_w^2 will be close to S^2. Consequently, no loss of precision is suffered by choosing to sample systematicallly rather that randomly in this situation.

When circular systematic sampling is employed and the random starting value is selected from among any of the N population elements, the following results hold.

Theorem 3.9. An unbiased estimator of Y is $\hat{Y}'_{sy} = (N/n)Y^*$ and its variance is

$$V(\hat{Y}'_{sy}) = \frac{1}{N} \sum_{i=1}^{N} (Y_i - Y)^2 . \tag{3.34}$$

Proof: Write Y^* as $Y^* = \Sigma_{i=1}^{N} y_i \delta_i$ where δ_i is an indicator variable whose value is $\delta_i = 1$ if the ith element of the population, y_i is in the sample, and $\delta_i = 0$ otherwise. Since sample size is constant and set at n, y_i will appear in n of the N equally likely samples. Thus $E[\delta_i] = n/N$ and

$$\begin{aligned} E[\hat{Y}'_{sy}] &= E\left[\frac{N}{n}\sum_{i=1}^{N} y_i \delta_i\right] \\ &= \frac{N}{n}\sum_{i=1}^{N} y_i E[\delta_i] \\ &= \sum_{i=1}^{N} y_i \\ &= Y\,. \end{aligned}$$

The variance in eq. (3.34) follows directly from the definition.

Since no design-unbiased variance estimator for systematic sampling exists without taking replicated samples, considerable work has been done to determine estimators that have small bias and perform well across different populations (Buckland 1951; Iachan 1982; Wolter 1985; Bellhouse 1988). If the population is haphazardly ordered then the variance estimator used for the simple expansion estimator under SRS performs well; however, for orderings that exhibit greater within-sample variance than is exhibited in the population as a whole, this estimator tends to overestimate. Wolter (1985) recommended an estimator based on successive differences when little is known about the population of interest. This estimator is given by

$$v(\hat{Y}_{sy}) = N^2\left(1 - \frac{n}{N}\right)\left(\frac{1}{n(n-1)}\right)\sum_{j=2}^{n}\frac{\Delta^2 y_j}{2} \tag{3.35}$$

where $\Delta y_j = y_j - y_{j-1}$ for those values of y observed in the sample in the sequence in which the procedure is implemented.

Sampling with Probability Proportional to Size (pps). In the previous sections, sample selection procedures were discussed for which little quantitative prior information was available about the population of interest. In pps sampling,[1] it is assumed that there is information on a covariate x that is positively correlated with the variable of interest.

To take full advantage of the covariate when designing a sampling strategy, one can select each unit proportional to the covariate. This clearly can be very efficient, for example if $y_i = cx_i$ for every unit in the population

[1]Many authors have distinguished between inclusion probabilities proportional to x (πps sampling) and probability of selection at each draw proportional to x (pps sampling). We do not follow this convention but group both under pps sampling.

where c is a constant, then $V(\hat{Y}_{HT}) = 0$ in eq. (3.8). But one can argue that pps sampling is even more risky than cluster sampling. If the w_{ij} are held constant, it is clear that $(y_i/\pi_i - y_j/\pi_j)^2$ can be very large if the probabilities π_i are negatively correlated with the y_i. It seems unlikely that anyone would purposely implement pps sampling to estimate a parameter with so little knowledge of the population that x_i and y_i would be negatively correlated, but this is more likely to happen when estimating characteristics of secondary interest in multiresource inventories.

Although pps sampling without replacement is usually more efficient than pps sampling with-replacement, the latter is preferred in some instances. This is because selection of the sample is simple; it has an unbiased estimator of variance that is simple to compute; and it yields easily obtained, unbiased estimators of total variance and components of variance for more complex designs such as multistage designs (Brewer and Hanif 1983). A simple way of implementing pps with-replacement sampling is to take n independent draws from the population, selecting each unit proportional to its covariate value x_i $(i = 1, \ldots, N)$ (Hansen and Hurwitz 1943). The appropriate estimation theory for pps with-replacement sampling is given in eq. (3.11)–(3.13).

If every unit in the population is selected with probability $\pi_i = nx_i/X$ without replacement, then Theorem 3.5 gives an unbiased estimate of Y and its variance. One of two problems can occur when implementing pps without replacement sampling: (1) the sampling schemes may be difficult to design for $n > 2$ with preassigned inclusion probabilities $\pi_i = nx_i/X$, or (2) for schemes that are easy to implement, the inclusion and/or joint inclusion probabilities of selection may be difficult to determine.

Many pps sampling schemes without replacement have been introduced. Hanif and Brewer (1980) and Brewer and Hanif (1983) discussed 50 of them, and Sunter (1986) and Chaudhuri and Vos (1988) added more. Many of these are of academic interest only. The reason for the large number is that there are many desirable criteria, such as ease of implementation, fixed sample size, selection probabilities exactly proportional to size, and joint probabilities of selection that are easy to compute. An example of the difficulties encountered in satisfying all of these criteria is evident in Hájek (1981, p. 24), who showed that it is difficult to find a design of fixed sample size for which the joint probabilities can be computed easily without approximations. He noted that (see Theorem 2.4)

$$\begin{aligned} V(\delta_i) &= \pi_i(1 - \pi_i) \\ &= \mathrm{Cov}(\delta_i, \delta_i) = -\mathrm{Cov}(\delta_i, n - \delta_i) = -\sum_{j(\neq i)}^{N} \mathrm{Cov}(\delta_i, \delta_j) \\ &= \sum_{j(\neq i)}^{N} (\pi_i \pi_j - \pi_{ij}) , \end{aligned}$$

which proved eq. (2.20) and he then proves the following theorem, which illustrates the difficulty of computing joint probabilities.

Theorem 3.10. For designs with fixed sample size and

$$d = \sum_{i=1}^{N} \pi_i(1 - \pi_i) ,$$

let there exist numbers $c_1, c_2, \ldots, c_N$ such that

$$\pi_i \pi_j - \pi_{ij} = c_i c_j , \qquad 1 \le i \ne j \le N . \tag{3.36}$$

Then if $d \ge \frac{3}{4}$, eq. (3.36) is satisfied for

$$c_i = C/2 - \sqrt{C^2/4 - \pi_i(1 - \pi_i)} , \qquad 1 \le i \le N$$

where C is a solution of

$$(N - 2)C = \sum_{i=1}^{N} \sqrt{C^2 - 4\pi_i(1 - \pi_i)} ,$$

and for each $d \ge \frac{3}{4}$,

$$\left(1 - \frac{1}{4d - 1}\right) I_i \le c_i c_j \le I_i \left(1 + \frac{1}{4d - 2}\right)^2 , \qquad 1 \le i \ne j \le N$$

where $I_i = \{\pi_i(1 - \pi_i)\pi_j(1 - \pi_j)/d\}$. Hájek (1981) did not address this issue if $d < \frac{3}{4}$.

Finding simple solutions for $c_i c_j$, and hence being able to compute π_{ij}, is difficult. Hájek gave some illustrations of this issue.

As an example, Hartley and Rao (1962) gave approximations for joint probabilities of selection for Goodman and Kish's (1950) random systematic sampling procedure. In this design, units are arranged in random order and a uniform variate δ with $0 \le \delta < 1$ is selected at random. Those n units satisfying $\Pi_{j-1} \le \delta + k < \Pi_j$ are selected where $\Pi_{j-1} = \Sigma_{i=1}^{j} np_i$, and $\Pi_0 = 0$, Π_j being the progressive totals of the np_i ($p_i = x_i/X$) in that order and k being some integer between 0 and $n - 1$. To order $O(N^{-4})$,[2] Hartley and Rao (1962) showed that the probability of both units i and j being in the sample is

$$\pi_{ij} \doteq \frac{n - 1}{n} \pi_i \pi_j \left[1 + \frac{1}{n} (\pi_i + \pi_j) - \frac{1}{n^2} \sum_{l=1}^{N} \pi_l^2 + \frac{2}{n^2} (\pi_i^2 + \pi_j^2 + \pi_i \pi_j) \right.$$
$$\left. - \frac{3}{n^3} (\pi_i + \pi_j) \sum_{l=1}^{N} \pi_l^2 + \frac{3}{n^4} \left(\sum_{l=1}^{N} \pi_l^2 \right)^2 - \frac{2}{n^3} \sum_{l=1}^{N} \pi_l^3 \right] , \tag{3.37}$$

[2]A sequence of numbers R_n is said to be $o(1/k_n)$ as $n \to \infty$ if $k_n R_n \to 0$, and to be $0(1/k_n)$ if there exist M and n_o such that $|k_n R_n| < M$ for all $n > n_o$ or, equivalently, if $k_n R_n$ is bounded (Lehman 1983, p. 161).

assuming that: (1) $\pi_i \leq CN^{-1}$ for all i and N, and (2) $C_1 N^{-2} \leq S_{ij}^2 \leq C_2 N^{-2}$ for all pairs (i, j) and N, where C, C_1, and C_2 are positive constants not dependent on N such that $N^2 S_{ij}^2$ is bounded below and above for all N and S_{ij}^2 is the mean square of the $\pi_{i'}$ $(i' \neq i, j)$. The upper limit of the second assumption is implied by the first. The lower limit of the second can be circumvented but that of the first is vital.

Brewer and Hanif (1983, p. 60) noted that the efficiency of the Horvitz–Thompson estimator (eq. 3.7) varies little in practice from one selection procedure to another and the difference can basically be ignored. But classifying the procedures by method of selection and in other ways leads to a better understanding of the methods and their differences. Various types of sample selection procedures and examples are discussed in the following sections.

Draw by Draw Procedure. At each of n successive draws, one unit is selected from those not previously selected. These procedures are often hard to implement because probabilities of selection are defined for each draw and depend on the units already selected.

Rao *et al.* (1962) described one of these procedures, called RHC after the initials of the authors, as follows: Randomly assign the N population units to n groups making group sizes as equal as possible. Select one unit from each of the n groups with probability proportional to its size in the group. With this method, each unit t is selected with probability p_t/π_i where π_i = the sum of the probabilities of the group in which unit t occurs, that is, $\pi_i = \sum_{\text{group } i} p_t$.

Hence, an unbiased estimator of Y is

$$\hat{Y}_{RHC} = \sum_{i=1}^{n} y_i/(p_i/\pi_i) . \tag{3.38}$$

Notice that we do not have to separate subscripts t and i any more since only one unit is selected per group. This estimator is simply a sum of Horvitz–Thompson estimators.

The variance is

$$V(\hat{Y}_{RHC}) = \frac{n\left(\sum_{i=1}^{N} N_i^2 - N\right)}{N(N-1)} \left[\sum_{i=1}^{N} \frac{y_i^2}{np_i} - \frac{Y^2}{n}\right] \tag{3.39}$$

and an unbiased variance estimator, which is always nonnegative, is

$$v(\hat{Y}_{RHC}) = \frac{\left(\sum_{i=1}^{n} N_i^2 - N\right)}{N^2 - \sum_{i=1}^{n} N_i^2} \sum_{i=1}^{n} \pi_i \left(\frac{y_i}{p_i} - \hat{Y}_{RHC}\right)^2 . \tag{3.40}$$

Note that this variance estimate can be derived easily because it does not contain the joint probabilities of selection π_{ij}. This was one of the primary reasons for developing the RHC procedure. The probability of selecting y_i is p_i/π_i, which is unlikely to be proportional to y_i. The procedure is likely to be less efficient than others that select units exactly proportional to size. On the other hand, $V(\hat{Y}_{RHC})$ is always smaller than the variance of $\hat{Y}_{pps}(wr)$ in eq. (3.11) where *ppswr* indicates pps sampling with replacement using per-draw selection probability p_i. When all sizes of the n groups in the RHC strategy are the same, eq. (3.39) is minimized and can be reexpressed as

$$V(\hat{Y}_{RHC}) = \left(1 - \frac{n-1}{N-1}\right) V(\hat{Y}_{ppswr}) .$$

The preliminary randomization into groups has the effect of reducing the variance by an amount equal to the finite population correction factor $(1 - (n-1)/(N-1))$ (Berg 1974).

Brewer (1975) proposed an alternative method of selection, namely: draw unit i at the lth draw (given that it has not been selected yet) with probability

$$p_i(l) = \frac{p_i(1 - p_i)/(1 - lp_i)}{\sum_{j=1}^{N} p_j(1 - p_j)/(1 - lp_j)}$$

for $l = 2, \ldots, n$ and $p_i(1) = p_i$ = probability of selecting unit i at the first draw.

This procedure selects exactly n units proportional to size. The joint probabilities of selection can be computed, but it is cumbersome.

Ordered Systematic Procedure. The population is ordered in some convenient order (usually by size) and sizes are accumulated to total size X. An example of this procedure, given in Madow (1949), is as follows: Pick a random number r $(0 < r \leq X/n)$. The n units chosen are those for which the cumulated probabilities are equal to or just greater than $r, r+1, \ldots, r+n-1$. (If a unit has a x-value greater than X/n, that unit is selected with certainty and deleted from this process.) This ordered systematic procedure may be very efficient if there is a linear trend among the y_i/x_i values in the ordered population (Wolter 1985). The Horvitz–Thompson estimator is unbiased but the usual variance estimator may seriously underestimate the variance since many joint probabilities of selection are zero (Brewer and Hanif 1983). Goodman and Kish (1950) proposed a random systematic procedure that is the same as the ordered systematic approach except that the population units are listed in random order prior to sample selection. This eliminates the problem of underestimating the variance to a large degree but some of the joint probabilities of selection can still be 0 (Brewer and Hanif 1983).

Using the Hartley–Rao approximation for π_{ij} in eq. (3.37), we get

$$V(\hat{Y}_{HT}) \doteq \sum_{i=1}^{N} \pi_i\left(1 - \frac{n-1}{n}\pi_i\right)\left(\frac{y_i}{\pi_i} - \frac{Y}{n}\right)^2 - (n-1)/n^2$$
$$\times \sum_{i=1}^{N}\left(2\pi_i^3 - \frac{\pi_i^2}{2}\sum_{j=1}^{N}\pi_j^2\right)\left(\frac{y_i}{\pi_i} - \frac{Y}{n}\right)^2$$
$$+ 2(n-1)/n^3\left(\sum_{i=1}^{N}\pi_i y_i - \frac{Y}{n}\sum_{j=1}^{N}\pi_j^2\right)^2 \quad (3.41)$$

with variance estimator

$$v(\hat{Y}_{HT}) = \frac{1}{2(n-1)}\sum_{i \neq i'}^{n}\left(1 - \pi_i - \pi_{i'} + \sum_{j=1}^{N}\frac{\pi_j^2}{n}\right)\left(\frac{y_i}{\pi_i} - \frac{y_{i'}}{\pi_{i'}}\right)^2 . \quad (3.42)$$

This solution is correct to order $O(N^1)$. Connor (1966) gave an exact solution that requires tremendous computation for large populations.

Wolter (1985, Ch. 7) gave additional reliable variance estimators. One, called the successive difference estimator, treats the sample as if $n_h = 2$ units (in order of selection) were selected from within each of $n/2$ equal-sized strata. This variance estimator is

$$v_{sd}(\hat{Y}_{HT}) = \sum_{i=1}^{n/2}(y_{2i}/p_{2i} - y_{2i-1}/p_{2i-1})^2/n^2 . \quad (3.43)$$

A similar estimator based on overlapping differences is

$$v_{od}(\hat{Y}_{HT}) = \sum_{i=2}^{n}(y_i/p_i - y_{i-1}/p_{i-1})^2/\{2n(n-1)\} . \quad (3.44)$$

Both of these variance estimators performed well in simulations discussed in Wolter (1985, Ch. 7). The big attraction of these procedures is that they are probably the easiest to implement.

Rejective Procedure. This procedure, discussed in considerable detail by Hájek (1964), resembles draw by draw procedures in that a single unit is selected at each of n draws. But if, at a given draw, a unit is selected that has been drawn previously, the partial sample is abandoned and a new sample is drawn. This is continued until a sample of n distinct units is selected. If the probabilities of selection are constant, the rejective procedure equates to SRS. Rejective sampling could be implemented fairly easily in some practical situations such as computer selection of samples, but the idea of having to start again if a unit appears more than once has little appeal.

Whole Sample Procedure. A probability is specified for each possible sample of n distinct units and one of these samples is selected using these probabilities (Lahiri 1951). This may require enumeration of all possible samples, and hence the procedure is often difficult to implement.

Other Procedures. Two of these, Poisson and collocated sampling, are discussed later in this chapter.

Brewer and Hanif (1983) also classified procedures by whether they have the ratio estimator property with the Horvitz–Thompson estimator ($\hat{Y}_{HT}$). This means that if the y_i are all exactly proportional to the corresponding π_i with fixed sample size n, the variance of $\hat{Y}_{HT}$ is zero. All selection procedures can be classified accordingly into three groups:

1. Those for which the $\hat{Y}_{HT}$ estimator possesses the ratio property exactly. An example is Madow's (1949) ordered systematic procedure.
2. Those for which there is no estimator that has this property exactly but for which $\hat{Y}_{HT}$ possesses it approximately. Brewer and Hanif (1983) gave several examples of this.
3. Those for which one or more other estimators possess this property. An example is the RHC estimator.

Brewer and Hanif (1983) showed for the linear stochastic model (eq. 2.70) with $1 \le k \le 2$ that

$$V(\hat{Y}_{HT}) \doteq \sum_{i=1}^{N} \pi_i\left(1 - \frac{n-1}{n}\,\pi_i\right)\left(\frac{y_i}{\pi_i} - \frac{Y}{n}\right)^2 \tag{3.45}$$

for all pps procedures asymptotically to order $O(N^1)$.

Wolter (1985, pp. 286–290) gave a series of alternative variance estimators for unequal probability systematic sampling.

3.3 DETERMINATION OF SAMPLE SIZE

What sample size does one need? Although this appears to be a difficult question to answer, and certainly is one of the most freqently asked questions of statisticians, the standard answer: take the largest sample you can take, given the cost constraints you have, answers the question most of the time. Since this is often not perceived to be a very scientific answer, we cover existing theory in some detail.

3.3.1 Introduction

The task of specifying the size of sample required for a particular survey is often difficult, due to the many aspects that need to be considered, including the resources available to conduct the survey, the expected benefits, and possibly the expected liabilities of inaccurate estimates. The first of these factors is usually the easiest to quantify.

Given the multifaceted nature of the planning task, it is not surprising that various approaches to determining sample size are possible. Regardless of which approach is adopted, one needs, at the very least, to consider the purpose of the survey. If it is to estimate the aggregate or mean value of some quantitative characteristic or the proportion of some qualitative attribute, then the importance or need for that estimate may dictate the precision that is desired. In large scale surveys, it is unlikely that a single purpose will be served, or more germane, that a broadly stated purpose will be achieved, by the precise estimation of a single characteristic. Consequently, the precision of estimation that is desirable for some characteristic may be insufficient for others; conversely, letting sample size be governed by the most stringent requirement of precision may result in a survey effort that cannot be afforded. Even when a single-purpose sample survey is planned by an agency whose members agree upon the importance of the outcome of the survey, there could be a difference in informed judgment as to the precision that is needed and the compromise that will be necessary.

For the single-level sample designs considered in this chapter, the minimum sample size needed to assure a stipulated degree of precision will minimize the expected cost of sampling. If one instead aims to ensure that the cost of sampling will not exceed some maximum amount, then the maximum sample size that can be afforded will provide the greatest precision for that cost. Whether one proceeds under a cost constraint or a precision requirement makes little difference, and no generality is lost by focusing on one in preference to the other. In most of the following discussion, we examine various ways to determine sample size in order to achieve a desired level of precision.

3.3.2 The Approach with Simple Random Sampling (SRS)

Cochran (1977) identifies five key steps in choosing a sample size n for SRS:

Step 1. Specify the objectives of sampling, for example, estimate the total volume of a tract of timber within a specified interval or with a precision of $\pm 5\%$ with a specified degree of confidence. The timber will be sold if the estimate exceeds a specified amount.

Step 2. Develop an equation that expresses n in terms of the desired precision of estimate. This is feasible in probability sampling, for example, in SRS.

If the estimated population total for $\hat{Y}$ is assumed to be distributed approximately normally, then using confidence intervals based on the normal distribution, we have

$$P[\hat{Y} - zS_y/\sqrt{n} \leq Y \leq \hat{Y} + zS_y/\sqrt{n}] = 1 - \alpha \ , \tag{3.46}$$

where z is a standard normal percentile selected to ensure a high probability $(1 - \alpha)$ and $S_y/\sqrt{n}$ is the standard error of $\hat{Y}$. This statement says that the parameter Y is expected to be within the constructed range on average $(1 - \alpha)$ 100% of the time if the stated assumptions are correct.

Especially in sampling, where estimation of parameters is usually the objective, narrow intervals are more desirable than wide intervals for a given confidence level. If one wishes to achieve a width of length $2\lambda_t$, then the sample size n should be large enough to ensure that $\lambda_t \geq zS_y/\sqrt{n}$. In other words, the sample size n should be the smallest integer value that satisfies

$$n \geq z^2 S_y^2/\lambda_t^2 \tag{3.47}$$

for the stipulated value λ_t. This yields the minimum sample size that will give the desired interval at the stated confidence level if our assumptions are reasonable.

Sometimes it may be easier to specify λ_t in units expressed as a percent of the target parameter Y. For example, when the size of Y is quite uncertain, it may be easier to reason that, whatever its value, one wishes to have a confidence interval whose width is no larger than a stipulated percentage of Y. To this end, define the allowable error as $\lambda_t^* = (\lambda_t/Y)$ 100%. With this specification, an alternative computation of the needed sample size is

$$n \geq z^2 (CV_y/\lambda_t^*)^2 \tag{3.48}$$

where $CV_y = (S_y/Y)$ 100% is the population coefficient of variation.

Step 3. Estimate the unknown population parameters in the equations used to estimate the desired sample size. Here, prior estimates of N, S_y, CV_y, and an approximation to z (since S_y is unknown) are needed.

Regardless of whether S_y or CV_y is correctly specified in the computation of n with eq. (3.47) or eq. (3.48), the interval (3.46) has fixed width. However, if an incorrect value of S_y is used to construct eq. (3.46), it will either exceed or fail to achieve its nominal $1 - \alpha$ coverage, depending on whether the specified value exceeds or understates S_y.

Substituting an estimate s_y^2 for S_y^2 and an approximation (t) for z, we obtain

$$n \geq t^2 s_y^2/\lambda_t^2 \tag{3.49}$$

where t is the $1 - \alpha/2$ quantile of the central t-distribution with $n - 1$ degrees of freedom. If instead we prefer specifying λ_t^* and an estimate of CV_y, $\widehat{CV}_y$, we calculate

$$n \geq t^2(C\hat{V}_y/\lambda_t^*)^2 . \tag{3.50}$$

After sampling has concluded, an alternative interval is possible that uses the estimate of s_y provided by the sample. This interval is

$$\hat{Y} \pm ts_y/\sqrt{n} . \tag{3.51}$$

In contrast to the interval (3.46), the half-width of eq. (3.51) does not have a fixed value λ_t, but instead is a random variable, say λ, where $\lambda = ts_y/\sqrt{n}$, whose value varies among samples. For large samples, $E[\lambda] \approx \lambda_t$, where λ_t continues to denote the desired half-width of a $(1-\alpha)$ 100% confidence interval. This means that "t-intervals," as shown by eq. (3.51), achieve the desired width on average, but that no assurance is given about the relative frequency with which λ will exceed the desired limit. Kupper and Hafner (1989) demonstrated that λ may exceed the desired λ_t with a probability greater than 0.5.

The intervals (3.46) and (3.51) ignore the $(1-f)$ term that appears in $V(\hat{Y})$ shown in eq. (3.2). Its exclusion has little effect on the computation of n when f is small. If, after computing n by either eq. (3.47) or eq. (3.48), it is found that n exceeds 0.1, say, then the alternate formula

$$n = \left(\frac{zS_y}{\lambda_t}\right)^2\left(1 + \frac{1}{N}\left(\frac{zS_y}{\lambda_t}\right)^2\right)^{-1} \tag{3.52}$$

will provide a correspondingly smaller estimate of the required sample size because:

$$\lambda_t = z\sqrt{1 - \frac{n}{N}}\,\frac{S_y}{\sqrt{n}}$$

$$\lambda_t^2 = \frac{z^2S_y^2}{n} - \frac{z^2S_y^2}{N}$$

$$\frac{\lambda_t^2}{z^2S_y^2} = \frac{1}{n} - \frac{1}{N}$$

$$\frac{\lambda_t^2}{z^2S_y^2} + \frac{1}{N}\,\frac{(\lambda_t^2/z^2S_y^2)}{(\lambda_t^2/z^2S_y^2)} = \frac{1}{n}$$

$$\left(\frac{\lambda_t}{zS_y}\right)^2\left(1 + \frac{1}{N}\left(\frac{zS_y}{\lambda_i}\right)^2\right) = \frac{1}{n}$$

$$n = \left(\frac{zS_y}{\lambda}\right)^2\left(1 + \frac{1}{N}\left(\frac{zS_y}{\lambda}\right)^2\right)^{-1} .$$

We may want to account for the fact that λ is a random variable, where λ

denotes the half-width of eq. (3.51). That is, for the desired or expected half-width λ_t, we may wish to assure that $\lambda \leq \lambda_t$ with high probability:

$$P(\lambda \leq \lambda_t) = 1 - \gamma \tag{3.53}$$

for some suitably small value of γ. The bracketed term in eq. (3.53) can be reexpressed as $ts_y/\sqrt{n^*} \leq \lambda_t$. Squaring both sides, multiplying by $n^*(n^* - 1)/S_y^2$, and rearranging gives

$$P\left(\frac{s_y^2(n^* - 1)}{S_y^2} \leqslant \frac{n^*(n^* - 1)\lambda_t^2}{S_y^2 F}\right) \geq 1 - \gamma \tag{3.54}$$

where F denotes the percentile from an F-distribution with 1 and $n^* - 1$ degrees of freedom (d.f.) at the α confidence level. Under a normally distributed population, the left side of the bracketed quantity in eq. (3.54) is a central χ^2 random variable with $n^* - 1$ d.f. Consequently for eq. (3.54) to be true, the right side of the bracketed quantity must equate to the $1 - \gamma$ quantile of the χ^2-distribution with $n^* - 1$ degrees of freedom:

$$\frac{n^*(n^* - 1)}{F}\left(\frac{\lambda_t}{S_y}\right)^2 = \chi^2(n^* - 1, 1 - \gamma)\,. \tag{3.55}$$

From eq. (3.47) we have

$$\left(\frac{\lambda_t}{S_y}\right)^2 = \frac{(z_{1-\alpha/2})^2}{n} = \frac{\chi^2(1, \alpha)}{n}\,.$$

Substituting this result into eq. (3.55) and rearranging gives the relation

$$n^*(n^* - 1) = nF\chi^2(n^* - 1, \gamma)/\chi^2(1, \alpha)\,, \tag{3.56}$$

which may be solved for the nearest integer value of n^* (Guenther 1965).

Suppose, for example, that λ_t is stipulated such that upon applying eq. (3.48) with what is believed to be a plausible value of the population CV_y, a value of $n = 20$ is obtained for a nominal 90% confidence interval on Y. It is desired to draw a sample of size n^* such that the achieved interval half-width λ will be less than or equal to λ_t with a probability of 0.80 (i.e., $\gamma = 0.20$). Iterative solution of eq. (3.56) yields a value $n^* = 27$ as the requisite sample size. Had the value $n = 20$, obtained from the traditional formula, been used, the relative frequency with which $\lambda \leq \lambda_t$ would have been 0.42! Kupper and Hafner (1989) supplied extensive tabulated results that provide convincing evidence of the tendency of the traditional sample size formula to provide confidence intervals as short as the desired width with discouragingly low probability.

If a prior estimate of S_y, say $\tilde{s}_y$, is available independently from a prior sample of size m, a slightly different approach may be taken to assure eq. (3.53). As derived by Harris *et al.* (1948), the sample size n^* is computed by the nearest integer that solves

$$n^* = F_1 F_2 / \left(\frac{\lambda_t}{\tilde{s}_y} \right)^2 \tag{3.57}$$

where F_1 and F_2 are, respectively, the F percentiles for 1 and $n^* - 1$ d.f. at the $1 - \alpha$ level and for $n^* - 1$ and $m - 1$ d.f. at the γ level.

One disadvantage and possible limitation of the two methods considered in this section on SRS is their reliance on normality of the sampled population, not just of the distribution of the estimator $\hat{Y}$. Their sensitivity to departures from normality is unknown. The pertinent issue in this regard is the difference between the targeted level $1 - \gamma$ and the actual level that is achieved when sampling from nonnormal populations. Realistic simulations could perhaps shed some light on whether lack of normality substantially decreases the relative frequency with which $(1 - \alpha)$ 100% confidence intervals of desired width are achieved.

The second disadvantage is the reliance upon a value of population S_y that is considered to be plausible, or equivalently, upon a plausible value of population CV_Y. In Guenther's method this value is stipulated by the planner; in the method of Harris *et al.* it is estimated by a prior sample. There is evidently no way to determine sample size in order to achieve a specified precision without some information about the heterogeneity of the underlying population.

Step 4. Set priorities on the objectives of sampling. Suppose, for example, that more than one characteristic of the population is of interest. Obviously, calculating the desirable sample size for each characteristic is likely to lead to conflict, and compromise will be needed. If, for example, $n = 30$ for total volume of the tract but $n = 60$ for total volume in sawlogs, a decision will have to be made about which estimate is the more important.

Step 5. Ensure that the value of n chosen is consistent with the resources available to take the sample. For example, if only a sample of size $n = 20$, rather than $n = 30$ or 40, is affordable, one has three choices: (1) accept $n = 20$; (2) seek out a more efficient sampling scheme; or (3) decide not to sample at all because the precision of estimate will be unacceptable.

In our experience, the affordable size of sample (step 5) should be considered at the outset. Often n is determined solely on this basis. The other steps should still be taken but they are primarily useful in deciding whether or not sampling should be done at all. If, for example, we require a reliable estimate of the percentage of trees infected with fusiform rust, and we can afford to sample only 10 out of 10000 trees, then if the incidence of rust is expected to be 1% or less, it is pointless proceeding with the project until more resources are available. This is a relatively simple example but it is a common occurrence.

SRS for proportions provides an interesting illustration of the tremendous variability in sample size requirements that can occur, depending on what the value of the parameter of interest is expected to be and on the type of error specification desired.

Suppose the percentage or proportion p is to be estimated by $\hat{p}$, which is assumed to be normally distributed about p, and the precision of the estimate required has probability $1 - \alpha = .95$, then

$$\sigma_{\hat{p}} \doteq \sqrt{p(1-p)/n} \,,$$

and accepting $t \doteq 2$, we obtain

$$n \doteq 4p(1-p)/\lambda_t^2 \,, \tag{3.58}$$

which ignores the finite population correction $(N-n)/N$. Note that n increases as $p \to .50$. Thus, for $p = .01$ or $.99$, $n = 15.8 \doteq 16$; for $p = .10$ or $.90$, $n \doteq 144$; and for $p = .50$, $n \doteq 400$. If interest is in relative error, that is, the standard error of $\hat{p}$ expressed as a percent of the estimate p, much larger sample sizes are required for rare proportions (say $p \leq .10$) than for common proportions ($p \doteq .50$). For example, ignoring the finite population correction gives

$$\frac{\sqrt{p(1-p)/n}}{p} = RE \qquad \text{or} \qquad n = \frac{1-p}{RE^2 p} \,,$$

where RE is the desired relative error. Thus, if $RE = .10$, for example, then for $p \doteq .5$, $n = 100$, and for $p = .1$, $n = 900$. Clearly, the expected p and level of precision desired markedly affect the size of sample needed. Often, a preliminary sample is justified to assess the sample size needed and determine whether sampling is feasible given the resources available.

We have seen then when p approximates 0 or 1, that is, when the condition is rare, the sample size required to obtain a small relative error is large. In this situation, it may be worthwhile using inverse sampling (Cochran 1977, p. 76), in which one continues sampling until m of the rare elements are found. If n is the sample size attained when the mth rare unit is found ($m > 1$), then for very large N, small p, and $m \geq 10$, an unbiased estimate of p is given by

$$\hat{p} = (m-1)/(n-1) \tag{3.59}$$

with

$$V(\hat{p}) \doteq mp^2(1-p)/(m-1)^2 \,. \tag{3.60}$$

Inverse sampling can be a desirable approach under the conditions specified above because the coefficient of variation of $\hat{p}$, $cv(\hat{p}) \doteq \sqrt{m(1-p)}/(m-1) < \sqrt{m}/(m-1)$, which is a close upper limit if p is small. Thus $cv(\hat{p})$ can be controlled without prior knowledge of p.

3.3.3 Conditional Control of Interval Width

It is well known that there is a correspondence between the nonrejection region for an α-level test of significance and a $(1-\alpha)$ 100% confidence interval. When testing is a relevant concern, as it is, for example, when fitting a model for analytic purposes of inference, the desired power of the test can be incorporated into the determination of sample size. Bristol (1989), for example, derived exact results for the two-sided test case, and many textbooks of mathematical statistics have given approximate solutions for the same.

When estimation is the relevant concern, setting sample size so as to control the width of a confidence interval is analogous to setting sample size so as to control the power of a test. As the power to detect a specific alternative hypothesis increases, the length of the corresponding interval decreases. However, as Bristol (1989) showed, the close relation between Type I error rate and confidence level does not carry over to a similarly close correspondence linking Type II error rate with interval width. Nonetheless, the methods discussed in the previous section, which determine sample size by probabilistically restricting the width of the interval, are attempts to accomplish in an estimation setting what is accomplished in a test setting by considering power when determining sample size. A further restriction is outlined below.

Let I_α denote the random interval given by eq. (3.51) , and let $Y \in I_\alpha$ denote the event that the target parameter Y is within I_α. In general, the probability that the length λ of the interval is less than the desired length λ_t will vary depending on whether or not $Y \in I_\alpha$. The relation (3.53) can be reexpressed in terms of the probability of $\lambda \le \lambda_t$ conditional upon the events $Y \in I_\alpha$ or $Y \notin I_\alpha$ as follows:

$$\begin{aligned} 1-\gamma \le \mathrm{P}(Y \in I_\alpha)\,\mathrm{P}(\lambda \le \lambda_t \mid Y \in I_\alpha) \\ + \mathrm{P}(Y \notin I_\alpha)\,\mathrm{P}(\lambda \le \lambda_t \mid Y \notin I_\alpha)\,. \end{aligned} \tag{3.61}$$

Beal (1989) has suggested that one's interest should be focused upon the length of I_α given the event $Y \in I_\alpha$ only, because if Y lies outside the interval then its length becomes an irrelevant concern. Following this suggestion, we wish to determine the sample size n' such that

$$\mathrm{P}(\lambda \le \lambda_t \mid Y \in I_\alpha) \ge 1-\gamma' \tag{3.62}$$

for some suitably small value γ'. Because $\mathrm{P}(Y \in I_\alpha) = 1-\alpha$, the conditional probability in eq. (3.62) can be evaluated by dividing $1-\alpha$ into the joint probability $\pi(n') = P(\lambda \leq \lambda_t,\ Y \in I_\alpha)$, which depends on n'.

Following Beal (1989), one can show that, when sampling from a normally distributed population, the joint probability of $\lambda \leq \lambda_t$ and $Y \in I_\alpha$, conditional upon the estimate S_y, is

$$P(\lambda \leq \lambda_t,\ Y \in I_\alpha | s_y) = 2\Phi\left(\frac{ts_y}{S_y}\right) - 1 \tag{3.63}$$

where $\Phi(\bullet)$ is the standard normal distribution function. The unconditional joint probability $\pi(n')$ is the integral of the product of eq. (3.63) and the density of the χ^2-distribution, where the range of integration is the region of the parameter space of s_y that maintains $\lambda \leq \lambda_t$. Beal (1989) provided algorithmic detail as well as extensive tables for different combinations of values of α, γ', λ_t, and S_Y.

The example of the previous section used $\alpha = 0.10$ and $\gamma = 0.20$. The traditional sample size formula gives a value $n = 20$ for $S_y = 1$, $\lambda_t = 0.38$, whereas $n^* = 27$ when the requirement that $P(\lambda \leq \lambda_t) \geq 1 - \gamma$ is imposed. Under the more restrictive approach outlined above, using $\gamma' = 0.20$, the required sample size is $n = 30$.

3.3.4 Sample Size Determination for Stratified Sampling

It was shown in Section 3.2.3 how the overall sample can be allocated among strata in order to minimize the overall cost for a specified precision or to maximize precision for a specified cost. Whichever constraint is used, overall sample size is determined by it, as evidenced by formulas (3.18) and (3.19).

Bankier (1988) pointed out that Neyman allocation minimizes the relative variance $V(\hat{Y}_{st})/Y^2$, but as a consequence, may cause the relative variances in an individual stratum, $V(\hat{Y}_h)/Y_h^2$, to be unacceptably large. Using W_h to denote some measure of importance attached to stratum h, he suggested an alternative allocation scheme, namely

$$n_h = \frac{nS_h W_h^q / Y_h}{\sum_{h=1}^{H} S_h W_h^q / Y_h}$$

for a selected value of q $(0 \leq q \leq 1)$. This allocation scheme permits one to make estimates for individual strata of about equal precision. Or if there are some strata which have large S_h but which are relatively unimportant, one can downweight the sampling effort expended in these strata by a judicious choice of W_h and q. For a fixed value of q, it is easy to show that this allocation leads to

$$\frac{V(\hat{Y}_h)/Y_h^2}{V(\hat{Y}_{h'})/Y_{h'}^2} = \frac{CV_{Y_h}}{CV_{Y_{h'}}}\left(\frac{W_{h'}}{W_h}\right)^2$$

where CV_{Y_h} is the coefficient of variation in stratum h. Thus the relative variance in stratum h compared to that of stratum h' is directly affected by the comparative heterogeneity within each stratum, but can be offset by the comparative importance attached to the strata. For a particular measure of importance W_h, one can graph the relative variance at different values of q $(0 \leq q \leq 1)$ for each stratum. The visual evidence of the effect that alternative values of q have upon the precision of estimation among strata can help determine the appropriate value of q to use in determining sample allocation among strata.

3.3.5 Minimizing Cost Plus Expected Loss

It may at times be possible to quantify not only the cost of conducting a survey but also the cost associated with the error of an estimate. For the former, linear cost functions of the type $C = C_0 + nC_1$ are commonly used because of their simplicity. More complicated functions may more realistically portray the actual costs, especially when the differential costs of field travel are considered. Functions that describe the costs associated with estimation error are usually referred to as loss functions. Hamilton (1979) discussed situations wherein linear, quadratic, and absolute-value loss functions may be plausible. For example, an absolute-value function may be appropriate when the monetary loss due to the error in estimation depends linearly on the magnitude of the error but not its sign. The expected loss is the product of the loss function and the probability function of the estimator, integrated over the real line. The sample size, n, can be computed as that value that minimizes the combined cost of conducting the survey plus the expected loss.

As an example, consider the absolute-value loss function given by $L = D|\hat{Y} - Y|$ where D is a known constant. Assuming that $\hat{Y}$ is approximately normally distributed with variance S_Y^2/n, the expected loss is

$$E[L] = \int_{-\infty}^{+\infty} \frac{D|\hat{Y} - Y| \exp(-(\hat{Y} - Y)^2/(2S_y^2/n))\, d\hat{Y}}{\sqrt{2\pi S_y^2/n}}$$

$$= D\left(\int_{-\infty}^{Y} \frac{-(\hat{Y} - Y) \exp(-(\hat{Y} - Y)^2/(2S_Y^2/n))\, d\hat{Y}}{\sqrt{2\pi S_y^2/n}}\right.$$

$$\left. + \int_{Y}^{+\infty} \frac{(\hat{Y} - Y) \exp(-(\hat{Y} - Y)^2/(2S_y^2/n))\, d\hat{Y}}{\sqrt{2\pi S_y^2/n}}\right)$$

$$= \frac{2DS_y^2}{n}\left(\int_{-\infty}^{0} \frac{-z \exp(-z^2)\, dz}{\sqrt{2\pi S_y^2/n}} + \int_{0}^{+\infty} \frac{z \exp(-z^2)\, dz}{\sqrt{2\pi S_y^2/n}}\right)$$

$$= 2DS_y(2\pi n)^{-1/2}.$$

Under a linear cost function the combined cost plus expected loss is

$$\Lambda = C_0 + C_1 n + 2DS_Y(2\pi n)^{-1/2},$$

which is minimized at the value $n = (D^2 S_Y^2 / 2\pi C_1^2)^{1/3}$. Hamilton (1970, 1979) discussed applications of this method in the context of specifying sample sizes in natural resource inventories.

For example, suppose that it costs $C_1 = \$100$ to measure a 0.2-ha forest plot. From previous surveys it is known that the standard deviation of volume among plots is 0.25 m^3, so that over a 400-ha tract, $S_Y = 500$ m^3. Suppose further that the loss function is assumed to be $L = \$50|\hat{Y} - Y|$, that is, \$50 is lost for each cubic meter underestimated or overestimated. The expected cost plus loss function thus is minimized when $n = 22$ plots are sampled.

3.4 DESIGNS WITHOUT A SAMPLING FRAME

In these designs, the number of units in the population, N, is assumed known only after sampling is completed (if a list of N is available before sampling, list sampling should be used). This situation may occur, for example, when a stand of trees is to be thinned and interest is both in marking the trees to be thinned and in estimating the volume of the marked timber. The number of trees to be removed will be known only after marking is completed. In this situation, methods similar to those discussed in Section 3.2 can be implemented, the only real difference being that in this case there is often no guarantee that the desired sample size will be achieved, that is, sample size is a random variable. This is not always true however. For example, Pinkham (1987) discussed an algorithm for drawing a simple random sample of fixed size n (without replacement) when N is unknown. Only $2n$ random uniform deviates have to be selected. The algorithm works as follows:

1. Label the units in the population $-(n-1), \ldots, -1, 0, 1, \ldots$.
2. Initially consider only units with labels $-(n-1), \ldots, 0$. This is called the potential sample.
3. (a) If $n = N$, quit.
   (b) If number of units selected at random $\geq n$, quit.
   (c) Select random numbers δ from a uniform distribution $U[0, 1]$ and determine j such that

$$\frac{(i+1)\cdots(j-1)}{(n+i+1)\cdots(n+j-1)} > \delta > \frac{(i+1)\cdots(j-1)j}{(n+i+1)\cdots(n+j-1)(n+j)}$$

   where i is the label of the last unit selected from the population for the potential sample. (If $\delta > (i+1/n+i+1)$, set $\delta = 1$).

Randomly delete one unit from the potential sample and replace with unit j if $j \le N - n$. If $j > N - n$, stop.

4. Repeat step 3.

Pinkham's method can be extended to unequal probability sampling too (Chao 1982), but it is unlikely that such methods can be implemented readily in practice. Hence we assume in the following discussion that probabilistic sampling schemes will result in a random sample size when no list of the population is available. Preliminary samples are often taken in such surveys (as well as in other types of surveys) to determine sample size needed in the actual survey. Such preliminary samples can sometimes be utilized to even greater advantage to estimate the number of population units or some other useful parameter, when implementing the sample procedures discussed below.

3.4.1 Equal Probability Sampling

In equal probability sampling, all units are selected with equal probability and all pairs of units are selected with equal joint probability. One method that is simple to implement meets these criteria, namely, binomial sampling discussed by Goodman (1949) and Yandle (1982). In binomial sampling, each unit in the population ($i = 1, \ldots, N$) enters the sample with probability $\pi = n_p / \hat{N}$ where n_p = planned sample size and $\hat{N}$ is a prior estimate of the population size. This method is called binomial sampling because the sample size n, actually achieved, is binomially distributed. The estimation theory is summarized in Theorem 3.11.

Theorem 3.11. For binomial sampling, an unbiased estimate of Y is given by

$$\hat{Y}_b = \sum_{i=1}^{n} y_i / \pi = \sum_{i=1}^{n} (y_i / n_p) \hat{N} \tag{3.64}$$

with variance

$$V(\hat{Y}_b) = \sum_{i=1}^{N} y_i^2 (1 - \pi) / \pi \tag{3.65}$$

and unbiased variance estimator

$$v(\hat{Y}_b) = \sum_{i=1}^{N} y_i^2 (1 - \pi) / \pi^2 \,. \tag{3.66}$$

Since n is a random variable, we can also estimate the expected sample size n_e and the variance of n. However, since we know what N is after sampling, we can then compute n_e and the variance of n exactly, as can be seen from Theorem 3.12.

Theorem 3.12. The expected value n_e of the random sample size n in binomial sampling is

$$n_e = N(n_p/\hat{N}) \tag{3.67}$$

and the variance of n is

$$V(n) = N\pi(1-\pi) . \tag{3.68}$$

3.4.2 Unequal Probability Sampling

Each of the sampling methods discussed in Section 3.2 (sampling with a frame) has an analog under unequal probability sampling without a frame and is a special case of the more general procedures discussed in this section. Note also that although the results of Theorem 3.5 are applicable here too, v_1 (eq. 3.9) and v_2 (eq. 3.10) are only approximately valid because of the random sample size. Unequal probability sampling procedures include post-stratified, post-cluster, post-systematic and post-pps sampling:

Post-Stratification Following Simple Random Sampling. In some situations it may be unreasonable or impossible to stratify a population prior to sampling, but nonetheless worthwhile to stratify afterwards in order to increase the precision of the estimator of a population parameter. This situation may arise, for example, in retrospective analyses of survey data. In what follows we assume that the survey data were collected under SRS, and we assume that the strata sizes N_h, or equivalently the strata weights W_h, are known without error, but not necessarily known prior to sampling.

The SRS design ensures that the number of elements n_h post-stratified into the hth stratum will be selected proportional to N_h except for sampling error. Since n_h is hypergeometrically distributed with mean $E[n_h] = nW_h$ and variance

$$V(n_h) = nW_h(1-W_h)\left(\frac{N-n}{N-1}\right) \doteq nW_h(1-W_h) \qquad \text{when } n \ll N$$

then

$$E\left(\frac{1}{n_h}\right) \doteq \frac{1}{nW_h} + \frac{1-W_h}{n^2W_h^2} .$$

The post-stratified estimator is

$$\hat{Y}_{pst} = N \sum_{h=1}^{L} w_h \bar{y}_h \tag{3.69}$$

where $\bar{y}_h = (1/n_h) \sum_{i=1}^{n_k} y_{hi}$.

Theorem 3.13. $\hat{Y}_{pst}$ is an unbiased estimator of Y with variance

$$V(\hat{Y}_{pst}) \doteq N^2\left[\left(\frac{N-n}{nN}\right)\sum_{h=1}^{L} W_h S_h^2 + \frac{1}{n^2}\sum_{h=1}^{L}(1-W_h)S_h^2\right]. \tag{3.70}$$

Proof: Since

$$\bar{y}_h = \frac{1}{n_h}\sum_{i=1}^{N_h} y_{hi}\delta_{hi}$$

where δ_{hi} is the usual indicator of inclusion, then $E[\delta_{hi}|n_h] = n_h/N_h$. Therefore the conditional expectation of $\bar{y}_h$, namely $E[\bar{y}_h|n_h]$, is unbiased for $\bar{Y}_h$, and is also unconditionally unbiased. Now

$$V(\bar{y}_h) = E[V(\bar{y}_h|n_h)] + V\{E[\bar{y}_h|n_h]\}$$

(Murthy 1967). The latter term on the right is obviously zero since $E[\bar{y}_h|n_h]$ is constant. Considering the inner part of the first term on the right, we have

$$\begin{aligned}V(\bar{y}_h|n_h) &= \frac{(1-f_h)W_h^2 S_h^2}{n_h}\\ &= \left(\frac{1}{n_h} - \frac{1}{N_h}\right)W_h^2 S_h^2 .\end{aligned}$$

Therefore upon averaging over n_h we have

$$\begin{aligned}E[V(\bar{y}_h|n_h)] &= E\left[\frac{1}{n_h} - \frac{1}{N_h}\right]W_h^2 S_h^2\\ &\doteq \left[\frac{1}{nW_h} + \frac{1-W_h}{n^2 W_h^2} - \frac{1}{N_h}\right]W_h^2 S_h^2\\ &\doteq \left(\frac{1}{n} - \frac{1}{N}\right)W_h S_h^2 + \frac{(1-W_h)}{n^2} S_h^2 .\end{aligned}$$

Thus

$$\begin{aligned}V(\bar{y}_{pst}) &= \sum_{h=1}^{L} V(\bar{y}_h)\\ &= \left(\frac{N-n}{nN}\right)\sum_{h=1}^{L} W_h S_h^2 + \frac{1}{n^2}\sum_{h=1}^{L}(1-W_h)S_h^2 .\end{aligned}$$

Hence, since $V(\hat{Y}_{pst}) = N^2 V(\bar{y}_{pst})$, eq. (3.70) follows.

There are two aspects of $V(\hat{Y}_{pst})$ that deserve emphasis. The second term is obviously the inflation in variance, or loss of precision, over stratified sampling with proportional allocation, which is incurred by the randomness of the strata sample sizes. Furthermore, this term diminishes at n times the rate that the first term diminishes, so that for moderately large SRS samples a relatively small price is paid by post-stratification.

Since the number of units in the population is not known, it is reasonable to assume that the strata sizes are also unknown prior to sampling. However, it may be feasible to obtain a rough estimate of both by taking a preliminary sample during a reconnaissance survey.

Schreuder (1975) described two post-stratification selection procedures for a population of trees, one of which is a modification of the SRS selection procedure proposed by Schreuder *et al.* (1971). This procedure is as follows:

1. Decide on the number of strata L desired and on the size limits to the strata such that the covariate values in stratum 1 are less than or equal to $X_{(1)}$, those in stratum 2 are less than or equal to $X_{(2)}$, and so on until those in stratum L are less than or equal to $X_{(L)}$. $X_{(h)}$ ($h = 1, \ldots, L$) indicates the largest covariate value in stratum h (hence, $X_{(L)}$ is the largest covariate value in the population).
2. Estimate the number of trees in each stratum by $\hat{N}_h$ ($h = 1, \ldots, L$), preferably $\hat{N}_h > N_h$. Call $\hat{N} = \Sigma_{h=1}^{L} \hat{N}_h$.
3. Select n_{sh} ($h = 1, \ldots, L$) random numbers from 1 to $\hat{N}_h$ ($h = 1, \ldots, L$), where n_{sh} is the planned sample size for stratum h.
4. Visit the population in the field and record the covariate values for all trees in the population. Record the number of trees in each stratum and select for sampling those trees that correspond to the random numbers allocated to the stratum. For these trees, record both the values of the variable of interest and the covariate.

If $\hat{N}_h \le N_h$ for all strata, then this procedure yields a sample of size $n_h \ge n_{sh}$, such that $\Sigma_{h=1}^{L} n_h = n \ge n_s$, where n_h ($h = 1, \ldots, L$) is the actual sample size attained in stratum h and $n_s = \Sigma_{h=1}^{L} n_{sh}$.

If $\hat{N}_h < N_h$ for one or more strata, but there are only a few units left to visit in the population, then it is desirable to measure all remaining trees for the variable of interest and, in estimation, treat these trees as an additional, fully measured stratum. Alternatively, the covariate values of the remaining trees can be incorporated into the original L strata, ignoring the slight bias caused by the fact that $N_h - \hat{N}_h$ trees could not enter into the sample.

If $N_h - \hat{N}_h$ is positive and large for one or more strata, then further sampling is indicated. It is best to treat the units that had no chance of being selected as a separate stratum and proceed as indicated in Schreuder *et al.* (1971). Generally, it is more efficient to use weights N_h/N rather than $\hat{N}_h/\hat{N}$ in the stratified sampling estimators.

Post-Cluster Sampling. A common problem in cluster sampling is the lack of information about the composition of the clusters. The clusters can be formed from an initial random sample of n out of N units and classifying the selected units into M clusters (several of which are likely to be empty since they are not sampled), and then selecting m out of M clusters at random.

Ghosh (1963) proved that for this situation

$$\hat{y} = \frac{M}{mn} \sum_{i=1}^{m} Y_{i\cdot} \tag{3.71}$$

where $Y_{i\cdot}$, the total for cluster i, is an unbiased estimator of the population mean per unit,

$$\bar{Y} = \sum_{i=1}^{N} y_i/N = \sum_{i=1}^{M} \sum_{j=1}^{N_i} y_{ij}/N$$

where N_i = number of units in cluster i with variance

$$V(\bar{y}) = \frac{M(N - mn/M)}{mn(N-1)} \sigma_u^2 + \frac{M^2(M-m)(n-1)}{Nmn(N-1)(M-1)} \sigma_c^2 + \frac{M-m}{m(N-1)N} \left(Y^2/n - \sum_{i=1}^{M} \sum_{j=1}^{N_i} y_{ij}^2 \right). \tag{3.72}$$

Here

$$\sigma_u^2 = \sum_{i=1}^{M} \sum_{j=1}^{N_i} y_{ij}^2/N - \bar{Y}^2$$

is the variance between the units, and

$$\sigma_c^2 = \sum_{i=1}^{M} (Y_{i\cdot} - Y/M)^2/M$$

is the variance between the cluster totals (Ghosh 1963). The corresponding result for $\hat{Y} = N\bar{y}$ is straightforward. Ghosh also presented results for an alternative ratio estimator, and the appropriate theory if the post clusters are selected proportional to size with replacement.

Post-Systematic Sampling. This is nearly identical to ordinary systematic sampling. Estimate N by $\hat{N}$, and if a sample of size n_p is desired, select a systematic sample of $\hat{N}/n_p$ units starting with a unit randomly selected between 1 and integer value $\hat{N}/n_p$. The actual sample size attained, $n_a = (N/n_p)$ or $n_a = (N/n_p) + 1$, is a fixed value, but it is likely to differ from n_p depending on how good our estimate $\hat{N}$ of N is. The estimation theory for systematic sampling with a random start (Theorem 3.8) is applicable, but the attained sample size n_a rather than n is used. We are not aware of any references on this topic, probably because it is essentially the same as systematic sampling.

Post-pps Sampling. The methods discussed under pps sampling when there is a frame can all be extended in an obvious way to sampling without a frame. This requires substituting an estimate for the known parameter needed in sampling with a frame. Usually this parameter is X, the sum of the population values of the covariate.

Hájek (1949), Lahiri (1951), and Midzuno (1951) introduced a simple version of pps sampling (with the intent of generating an unbiased ratio estimator) that can be adapted to pps sampling without a frame. The procedure is as follows:

1. Draw one unit with probability proportional to x_i, $\pi_i = x_i/X$, $1 \leq i \leq N$.
2. Draw an SRS of size $n - 1$ from the remaining $N - 1$ units.

Grosenbaugh (1964) developed a procedure of particular interest in forestry, which is related to that described above. The procedure, Poisson sampling (or 3P sampling as Grosenbaugh called it), is the direct, unequal probability extension of binomial sampling. It was introduced first by Hájek (1957, 1964), who called it Poisson sampling because of the connection to the Poisson process. In this procedure, unit i is selected with probability $\pi_i = n_t x_i / X^*$.

One way to implement Poisson sampling is to visit every unit in the population. The value of the covariate x is obtained for unit i (x_i, $i = 1, \ldots, N$) and compared to a random number generated uniformly between 0 and (X/n_t) where n_t is the desired (target) sample size. If the random number for unit i is less than or equal to x_i, the unit becomes a sample unit, otherwise it does not. Clearly, if $x_i > X/n_t$, the unit is selected with certainty. In actual practice X is not known and has to be estimated by X^* and random numbers are generated between 0 and $L = X^*/n_t$. Theorem 3.14 summarizes the estimation theory.

Theorem 3.14. For Poisson sampling, an unbiased estimator of Y is

$$\hat{Y}_u = \sum_{i=1}^{n_a} y_i/\pi_i = \sum_{i=1}^{n_a} (y_i/\{n_t x_i\})X^* \tag{3.73}$$

with variance

$$V(\hat{Y}_u) = \sum_{i=1}^{N} y_i^2(1 - \pi_i)/\pi_i \tag{3.74}$$

and unbiased estimator of variance

$$v(\hat{Y}_u) = \left[\sum_{i=1}^{n_a} y_i^2(1 - \pi_i)/\pi_i^2\right] \tag{3.75}$$

where n_a = achieved sample size (Schreuder *et al.* 1971a).

Hájek (1981, p. 132) noted that by considering only samples of a specific size, SRS, stratified sampling, and rejective sampling are special cases of Poisson sampling. A problem with Poisson sampling is that the size of sample achieved (n_a) is a random variable about the expected size (n_e). This helps explain why $\hat{Y}_u$ is an excellent example of why the property of unbiasedness is not always desirable. Although unbiased, $\hat{Y}_u$ can be a spectacularly inefficient estimator!

Grosenbaugh (1967a) suggested an estimator much more efficient generally than $\hat{Y}_u$, but slightly biased:

$$\hat{Y}_a = \hat{Y}_u n_e / n_a \,. \tag{3.76}$$

An approximate variance of this is

$$V(\hat{Y}_a) \doteq \sum_{i=1}^{N} [p_i(y_i/p_i - Y)^2/n_e][1 + V(n_a)/n_e^2] \tag{3.77}$$

where $p_i = x_i/X$, n_e = expected sample size = X/L and $V(n_a) = \Sigma_{i=1}^{N} \pi_i - \Sigma_{i=1}^{N} \pi_i^2$.

An approximate variance estimator is

$$v(\hat{Y}_a) = (1/n_e) \sum_{i=1}^{n_a} (y_i/p_i - \hat{Y}_a)^2/(n_a - 1) \tag{3.78}$$

(Grosenbaugh 1967a). Van Deusen (1987) recommended a more stable variance estimator suggested by Schreuder *et al.* (1971)

$$v_1(\hat{Y}_a) = \frac{X^2}{n_e} \sum_{i=1}^{n_a} \left(y_i/x_i - \sum_{i=1}^{n_a} y_i/n_a x_i \right)^2 /(n_a - 1). \tag{3.79}$$

Sunter (1986) proposed a generalization of Poisson sampling in a list-sampling context. We discuss it here because of its close relationship to Poisson sampling. Sunter makes the following premises:

1. Units are selected with probabilities π_i proportional to x_i.
2. The joint inclusion probabilities π_{ij} of units indexed i and j are positive for all i, j.
3. $\pi_i \pi_j - \pi_{ij} > 0$ for all i, j $(i \neq j)$.
4. π_{ij} is known or can be calculated for all i, j.

Poisson sampling satisfies all these conditions except that $\pi_{ij} = \pi_j \pi_j$ so that $\pi_i \pi_j - \pi_{ij} = 0$, for all $i \neq j$.

As Sunter (1986) indicated, under any set of positive selection probabilities, a Horvitz–Thompson estimator of the population total Y is given by

$$\hat{Y}_{HT} = \sum_{i=1}^{n_a} y_i/\pi_i \tag{3.80}$$

where n_a is the achieved sample size, with variance

$$V(\hat{Y}_{HT}) = \sum_{i=1}^{N} y_i^2(1-\pi_i)/\pi_i - \sum_{i \neq j}^{N} y_i y_j(\pi_i\pi_j - \pi_{ij})/(\pi_i\pi_j)$$

and unbiased variance estimator (if $\pi_{ij} > 0$ for all i, j)

$$v(\hat{Y}_{HT}) = \sum_{i=1}^{n_a} y_i^2(1-\pi_i)/\pi_i^2 - \sum_{i \neq j}^{n_a} y_i y_j(\pi_i\pi_j - \pi_{ij})/(\pi_{ij}\pi_i\pi_j) . \tag{3.81}$$

For fixed n_a

$$\sum_{i=1}^{N} \pi_i = n_a, \qquad \sum_{j(\neq i)}^{N} \pi_{ij} = (n_a - 1)\pi_i, \qquad \sum_{i \neq j}^{N} \pi_{ij} = n_a(n_a - 1) ,$$

so that

$$V(\hat{Y}_{HT}) = \frac{1}{2}\sum_{i \neq j}^{N} (\pi_i\pi_j - \pi_{ij})(y_i/\pi_i - y_j/\pi_j)^2 \tag{3.82}$$

and

$$v(\hat{Y}_{HT}) = \frac{1}{2}\sum_{i \neq j}^{n_a} (\pi_i\pi_j - \pi_{ij})(y_i/\pi_i - y_j/\pi_j)^2/\pi_{ij} , \tag{3.83}$$

which is the same as eq. (3.9).

Sunter's first premise arose from the desire to make the squared term in eq. (3.82) small. This will be true if there is proportionality between the true value of the variate and the size (x) measured. Satisfying the second and fourth premises ensures that either eq. (3.81) or eq. (3.83) can be computed as appropriate. The third premise implies that the variance estimate in eq. (3.83) cannot be negative and ensures some stability in the estimate. But since it can be advantageous to have $\pi_i\pi_j - \pi_{ij} < 0$ for the largest $(y_i/\pi_i - y_j/\pi_j)^2$ (Stuart 1964), there are pps schemes that can be more efficient than either Poisson or Sunter sampling. However, it is also advantageous to be able to compute all joint probabilities of selection and to have estimates of the variance that are always positive.

Assuming (without loss of generality) that the size measures (x) are scaled so that $\Sigma_{i=1}^{N} p_i = 1$ where $p_i = x_i/X$ and that there are no units in the population for which $np_i \geq 1$, then Sunter's procedure is as follows:

1. Select a number p^* such that

$$n_t p_i \le P_i^* = \sum_{j=i}^{N} p_j + p^* \qquad (i = 1, 2, \ldots, N) \tag{3.84}$$

where n_t is the target sample size. In effect, sampling is proportional to x.

2. Select the ith unit with conditional probability

$$P(U_i | n_i) = n_i p_i / P_i^* , \tag{3.85}$$

where U_i is the event that unit i is selected and n_i denotes the number of units remaining to be selected when the ith unit is reached.

Sampling stops when n units are selected or when the population of units has been exhausted. Thus the achieved sample size n_a is no greater than the target sample size n_t and units are selected with probability

$$\pi_i = np_i / P^* \qquad (i = 1, 2, \ldots, N)$$

and joint probabilities

$$\pi_{ij} = n_t(n_t - 1) p_i p_j \gamma_i^* / P^* \qquad (j > i = 1, 2, \ldots, N-1)$$

where

$$P^* = P_1^* = 1 + p^* , \qquad P_i^* = \sum_{j=i}^{N} p_j + p^* , \qquad \gamma_1^* = 1/P_2^*,$$

and

$$\gamma_i^* = (1/P_{i+1}^*)(1 - p_1/P_2^*) \cdots (1 - p_{i-1}/P_i^*) \qquad (i = 2, 3, \ldots, N-1) .$$

Note that the expected value of the sample size, n_a, is

$$E(n_a) = n_t / P^* = n_e ,$$

with variance

$$V(n_a) = \sum_{i=1}^{N} \pi_i (1 - \pi_i) - \sum_{i \ne j}^{N} (\pi_i \pi_j - \pi_{ij})$$

and

$$P(n_a < n_t) \le n_t p^* / P^* .$$

This sampling scheme reduces the variance in the sample size n_a relative to Poisson sampling by

$$\sum_{i \neq j}^{N} (\pi_i \pi_j - \pi_{ij}) .$$

The selection algorithm will work for any ordering of the population, but will be easier for certain ones. An ideal ordering would be close to the requirements of Corollary 1.1 in Sunter (1986) which states: If

$$p_{N-n_t+1} = p_{N-n_t+2} = \cdots = p_N \quad \text{and} \quad n_t p_i \leq P_i = \sum_{j=1}^{N} p_j$$

with $i = 1, 2, \ldots, N - n_t$ for some ordering of the population, then $p^* = 0$ will work and a sample of exactly n_t units will be obtained. Satisfying such an ordering may be hard to implement in practice. The next best ordering is to arrange the population units in descending order of x. If a sampling list is available, p^* can be computed easily. If no sampling list is available, there does not seem to be an obvious way to implement Sunter's sampling scheme.

The Horvitz–Thompson estimator for Sunter sampling given in eq. (3.80) can be rewritten as (using $\hat{Y}_{us}$ instead of $\hat{Y}_{HT}$ to avoid confusion)

$$\hat{Y}_{us} = (X/n_e) \sum_{i=1}^{n_a} y_i/x_i , \tag{3.86}$$

where X is the population total of the auxiliary variable x. Analogous to the biased adjusted estimator $\hat{Y}_a$, suggested by Grosenbaugh (1967a) for Poisson sampling, a more efficient but biased estimator for Sunter sampling is

$$\hat{Y}_{as} = \hat{Y}_{us} n_e / n_a . \tag{3.87}$$

(Schreuder *et al.* 1990a).

The variance of $\hat{Y}_{as}$ is very complicated. An approximate variance similar to one suggested by Grosenbaugh (1976) for Poisson sampling is given by

$$V_1(\hat{Y}_{as}) \doteq \left[\sum_{i=1}^{N} p_i \left(\frac{y_i}{p_i} - Y \right)^2 / n_e \right] \left[1 + \frac{V(n_a)}{n_e^2} \right] \tag{3.88}$$

or, similar to one suggested by Brewer and Hanif (1983, p. 83), by

$$V_2(\hat{Y}_{as}) = \sum_{i=1}^{N} \pi_i (1 - \pi_i) \left(\frac{y_i}{\pi_i} - \frac{Y}{n_e} \right)^2 . \tag{3.89}$$

Corresponding variance estimators are

$$v_1(\hat{Y}_{as}) \doteq \frac{n_a}{n_a - 1} \left(\sum_{i=1}^{n_a} y_i^2 / \pi_i - \hat{Y}_{as}^2 \right) \tag{3.90}$$

and

$$v_2(\hat{Y}_{as}) \doteq \sum_{i=1}^{n_a} (1-\pi_i)\left(\frac{y_i}{\pi_i} - \frac{\hat{Y}_{as}}{n_e}\right)^2 \left(\frac{n_e}{n_a}\right). \tag{3.91}$$

Brewer and Hanif (1983, p. 84) discussed collocated sampling, a technique similar to Poisson sampling but with reduced variability in the sample size. This is accomplished by generating a random ordering of the population units, say $M_i = 1, 2, \ldots, N$, selecting a random variable θ from the uniform distribution over the interval $[0, 1]$, and selecting unit i for the sample if $x_i \leq r_i = [(M_i + \theta - 1)/N]\ X/n_e$. Brewer and Hanif's (1983, p. 94) results show that for their simulations, collocated sampling is slightly more efficient than Poisson sampling.

3.5 ESTIMATION WITH COVARIATES

As we saw in the simple situation of a variable of interest with no covariates, estimation comprises the procedures used to estimate a specific value of interest (parameter) in the population and its reliability. How well one estimates a parameter depends on the population of interest, the amount of information available about the population prior to and after sampling, and (sometimes) the willingness on the part of the investigator to take risks.

The primary emphasis in survey sampling has been on point estimation and a great deal of research has concentrated on finding "good" estimators for the target parameter. Thus in this section, we discuss point estimation and estimation of reliability within finite, and to a lesser extent infinite populations. The assumption of finite populations is clearly more realistic in traditional estimation, but certain concepts and properties make more sense and can be investigated more readily under the superpopulation concept. For ease of understanding, we use one variable of interest y and one covariate x, often assuming eq. (2.70) or special cases of it. If such a model can be assumed, we can often do better in sampling efficiency (although not estimation bias) than the simple estimators presented earlier.
In the following discussion a sampling frame is assumed so that the sample size n is a constant.

3.5.1 Classes of Estimators

Covariate information can be combined with information on the variable of interest in many ways. We discuss here the most common methods (estimators) used in survey sampling.

Simple Estimators. To recapitulate briefly, we encountered in Theorem 3.5 the unbiased Horvitz–Thompson estimator of the population total

$$\hat{Y}_{HT} = \sum_{i=1}^{n} y_i/\pi_i \,. \tag{3.92}$$

Selecting units in the population with different probabilities implies that something is known about these units to make this desirable but, until now, this has not been stated explicitly. If there is a covariate x that is correlated with y, we can sample units proportional to their x-values. Frequently, given a definitive relationship between y and x, more conclusive results can be obtained.

For example, Godambe and Joshi (1965) show for eq. (2.70) with $\alpha = 0$ and $1 \leq k \leq 2$ that the expected variance of $\hat{Y}_{HT}$ achieves the lower bound of the expected variance of any design-unbiased estimator. This bound is

$$V_{GJ} = N^2\left[\sum_{i=1}^{N} \sigma^2 x_i^k/\pi_i - \sum_{i=1}^{N} \sigma^2 x_i^k\right]. \tag{3.93}$$

This is the design-based expectation of the model-based variance (Brewer and Hanif 1983, p. 16).

Another example: Two unbiased variance estimators v_1 $(\hat{Y}_{HT})$ and v_2 $(\hat{Y}_{HT})$ are given in eq. (3.9) and eq. (3.10). Estimator v_2 is generally considered to be inferior to v_1 since it is negative more frequently, but even v_1 is often unstable and can be negative (Brewer and Hanif 1983). Thus Brewer and Hanif (p. 68) proposed an approximate variance estimator based on the underlying stochastic eq. (2.70) with $\alpha = 0$ and $1 \leq k \leq 2$, namely,

$$v_3(\hat{Y}_{HT}) = \frac{n}{(n-1)}\left\{1 - \left(\sum_{i=1}^{N} \pi_i^k\right)\Big/\left(\sum_{i=1}^{N} \pi_i^{k-1}\right)\right\}\sum_{i=1}^{n}\left\{\frac{y_i}{\pi_i} - \frac{\hat{Y}_{HT}}{n}\right\}^2 \tag{3.94}$$

where k is the best "guesstimate" of the parameter k in eq. (2.70) but v_3 is not very sensitive to the value of k chosen. If an intelligent estimate cannot be made, Brewer and Hanif recommended the compromise value $k = 1.50$. In this example, assuming a model, such as eq. (2.70), may not be necessary since other approximations are available. For example, a simpler approximation is to use the variance estimator for pps with-replacement sampling. This should be a good approximation when N and n are large and small, respectively. This would give

$$v_4(\hat{Y}_{HT}) = \frac{n}{n-1}\left(\sum_{i=1}^{n} \frac{y_i^2}{\pi_i^2} - \frac{\hat{Y}_{HT}^2}{n}\right). \tag{3.95}$$

An estimator based on jackknifing (discussed shortly) is

$$v_J(\hat{Y}_{HT}) = [(N-n)/N]v_4(\hat{Y}_{HT}) \,. \tag{3.96}$$

Assumptions required for v_4 and v_5 to be satisfactory approximations are less clearly stated than eq. (2.70) for v_3. Estimator v_J is intuitively more satisfying than v_4 since it takes into account the sampling fraction $(N-n)/N$.

More Complex Estimators. Other estimators are also commonly used because they possess good properties for certain populations and seem to perform well for many other types of populations too, that is, they are robust. These include the ratio-of-means estimator (also called the quotient estimator), mean-of-ratios estimator, product estimator, difference estimators, mixed estimators, and regression estimators (linear and weighted).

One feature shared in common by these estimators is the adjustment of the expansion estimator of the population total, $\hat{Y}$, based on the sample total and population total of a covariate x. The direction and magnitude of the adjustment depends on information contained in the covariate(s), the extent to which that information correlates with the characteristic of interest, and the departure of the observed sample total from the population total X of the auxiliary variate. The feature that distinguishes these estimators is the manner in which the auxiliary information is used to make the adjustment.

Ratio-of-Means (rm) Estimator. This estimator adjusts the estimator $\hat{Y}_{HT}$ by the ratio of the population total X and the estimator $\hat{X}_{HT}$. In the general case of arbitrary probability sampling, this estimator is

$$\hat{Y}_{rm} = \left[\sum_{i=1}^{n} y_i/\pi_i \Big/ \sum_{i=1}^{n} x_i/\pi_i\right] X = \hat{R}X \tag{3.97}$$

or

$$\hat{Y}_{rm} = (\hat{Y}_{HT}/\hat{X}_{HT})X = \hat{R}X$$

where $\hat{R} = \hat{Y}_{HT}/\hat{X}_{HT}$, with approximate variance

$$V(\hat{Y}_{rm}) = V(\hat{Y}_{HT}) - 2R\,\text{Cov}(\hat{Y}_{HT}, \hat{X}_{HT}) + R^2 V(\hat{X}_{HT}) \tag{3.98}$$

where $R = Y/X$.

For SRS, the finite population correction $(1-f) = (1-n/N)$ is usually added to both $V(\hat{Y}_{rm})$ and $v(\hat{Y}_{rm})$. A similar correction is desirable for unequal probability sampling, but it is not obvious what it should be. The bias of $\hat{Y}_{rm}$ is approximately

$$B(\hat{Y}_{rm}) = [RV(\hat{X}_{HT}) - \text{Cov}(\hat{X}_{HT}, \hat{Y}_{HT})]/X\,. \tag{3.99}$$

The bias $B(\hat{Y}_{rm})$ decreases with an increase in sample size so that $\hat{Y}_{rm}$ is consistent. If $R = \text{Cov}(\hat{X}_{HT}, \hat{Y}_{HT})/V(\hat{X}_{HT})$, the bias in $\hat{Y}_{rm}$ is zero, which is true if the regression line of $\hat{Y}_{HT}$ on $\hat{X}_{HT}$ is a straight line through the origin.

If a sample of size n is selected so that the first unit i is selected proportionally to x_i and the remaining $(n-1)$ units are selected by SRS from the $(N-1)$ other units in the population, then the whole sample is selected proportional to $\bar{x}$ and an unbiased ratio-of-means estimator $\hat{Y}_{rm1}$ is

$$\hat{Y}_{rm1} = (\bar{y}/\bar{x})X \tag{3.100}$$

(Midzuno 1951) with variance

$$V(\hat{Y}_{rm1}) = N^2\left[\bar{X}^2 \sum_c \left(\frac{\bar{y}^2_{(c)}}{\bar{x}^2_{(c)}}\right)\binom{N}{n} - \bar{Y}^2\right] \tag{3.101}$$

where Σ_c denotes summation over all possible samples and $\bar{y}_{(c)}$ and $\bar{x}_{(c)}$ are the simple sample means for sample c. An unbiased estimator of $V(\hat{Y}_{rm1})$ is

$$v(\hat{Y}_{rm1}) = N^2\left\{\hat{Y}^2_{rm1} - \left(\frac{\bar{X}}{\bar{x}}\right)\left[\bar{y}^2 - s_y^2\,\frac{N-n}{Nn}\right]\right\} \tag{3.102}$$

(Sukhatme *et al.* 1984).

So far we have primarily addressed the general form of arbitrary probabilities of selection with $\hat{Y}_{rm}$. $\hat{Y}_{rm}$ and $\hat{R}$ are simple to compute for a sample with arbitrary probabilities of selection for the units, but deciding what variance estimator to use for $V(\hat{Y}_{rm})$ is not straightforward. Several estimators are available for $V(\hat{Y}_{HT})$ and $V(\hat{X}_{HT})$ and it is not always obvious how to estimate $\text{Cov}(\hat{X}_{HT}, \hat{Y}_{HT})$. Generally $\hat{Y}_{rm}$ is used with SRS and most of the literature deals with that situation. We follow this convention in the remainder of this section on $\hat{Y}_{rm}$ since the interesting results have been derived for SRS.

It is possible to show that the bias in $\hat{R}$ as an estimator of $R = Y/X$, and hence the bias in $\hat{Y}_{rm}$ as an estimator of Y, is bounded by its standard error whenever the standard error of $\hat{X}$ is less than X (Cochran 1977). The bias of $\hat{Y}_{rm}$ has been negligibly small in numerous empirical and simulation studies using SRS, except when extremely small samples are selected. In these cases, both the standard errors of $\hat{Y}$ and $\hat{X}$ are large, as are the biases of $\hat{R}$ and $\hat{Y}_{rm}$. If $\hat{Y}$ and $\hat{X}$ are restated in delta notation for SRS, that is, if

$$\hat{Y} = Y(1+\delta_y)\,, \qquad \text{where } \delta_y = \frac{\hat{Y}-Y}{Y}$$

$$\hat{X} = X(1+\delta_x)\,, \qquad \text{where } \delta_x = \frac{\hat{X}-X}{X}$$

and $(1+\delta_x)^{-1}$ is expanded as a Taylor series about 1, then $\hat{Y}_{rm}$ can be rewritten as

$$\hat{Y}_{rm} = Y(1 + \delta_y)(1 - \delta_x + \delta_x^2 - \delta_x^3 + \delta_x^4 \ldots) \qquad (3.103)$$

if $|\delta_x| < 1$. Subtracting Y from both sides, squaring the result, and then taking expected values on both sides yields an expression for MSE $(\hat{Y}_{rm})$ as detailed in Sukhatme *et al.* (1984). Upon dropping higher order terms, an approximation to MSE $(\hat{Y}_{rm})$ is obtained, which in the absence of appreciable bias, also serves as an approximate variance expression for the variance of $\hat{Y}_{rm}$. In particular, by truncating the series to include only terms in δ_x^2, δ_y^2, $\delta_x\delta_y$, one gets

$$V(\hat{Y}_{rm}) \doteq N^2\left(\frac{1-f}{n}\right)(S_y^2 + R^2S_x^2 - 2RS_{xy}) \qquad (3.104)$$

$$\doteq N^2\left(\frac{1-f}{n}\right)\sum_{i=1}^{N}(y_i - Rx_i)^2/(N-1)\,. \qquad (3.105)$$

We emphasize the approximate nature of the variance expressions given above. In situations where bias cannot be expected to be small, these expressions will be poor approximations to the variance. What may be less obvious, however, is that they will also be poor approximations to MSE $(\hat{Y}_{rm})$.

The conventional estimator of $\mathrm{Var}(\hat{Y}_{rm})$ is its sample analog:

$$v_1(\hat{Y}_{rm}) = N^2\left(\frac{1-f}{n}\right)(s_y^2 + \hat{R}^2s_x^2 - 2\hat{R}s_{xy}) \qquad (3.106)$$

$$= N^2\left(\frac{1-f}{n}\right)\sum_{i=1}^{n}(y_i - \hat{R}x_i)^2/(n-1) \qquad (3.107)$$

where

$$s_{xy} = \sum_{i=1}^{n}(x_i - \bar{x})(y_i - \bar{y})/(n-1)\,.$$

An alternative estimator is

$$v_2(\hat{Y}_{rm}) = (\bar{X}^2/\bar{x}^2)v_1(\hat{Y}_{rm})\,, \qquad (3.108)$$

which obviously involves a similar adjustment to $v_1(\hat{Y}_{rm})$ that is made to $\hat{Y}$ in $\hat{Y}_{rm}$. In a number of empirical and simulation studies (Rao and Rao 1971, Rao 1981, Wu 1985), $v_2(\hat{Y}_{rm})$ is a less biased estimator of $V(\hat{Y}_{rm})$ than $v_1(\hat{Y}_{rm})$, and both tend to underestimate $V(\hat{Y}_{rm})$, especially in small samples.

The jackknifing technique, which is discussed at more length later in this chapter, has been studied extensively in situations calling for ratio estimation. In variance estimation, it often provides a less biased estimator than either $v_1(\hat{Y}_{rm})$ or $v_2(\hat{Y}_{rm})$. The jackknife estimator can be written in various ways, such as

$$v_J(\hat{Y}_{rm}) = N^2(1-f)\bar{X}^2(n-1)\sum_{i=1}^{n} D_{(j)}^2/n \tag{3.109}$$

where for every j in the sample, $D_{(j)}$ is the difference between the ratio $(n\bar{y} - y_j)/(n\bar{x} - x_j)$ and the average of these n ratios. The estimator often provides an overestimate of the variance of $\hat{Y}_{rm}$; hence it has been called a conservative estimator of variance (Gregoire 1984). It will, however, often have larger sampling variation than $v_1(\hat{Y}_{rm})$ or $v_2(\hat{Y}_{rm})$.

Royall and Cumberland (1981a) offered alternative robust variance estimators:

$$v_3(\hat{Y}_{rm}) = v_1(\hat{Y}_{rm})(\bar{X}\bar{x}_2/\bar{x}^2)/(1 - c^2/n) \tag{3.110}$$

where $c^2 = \Sigma_{i=1}^{n} (x_i - \bar{x})^2/\{\bar{x}^2(n-1)\}$, and $\bar{x}_2$ is the mean of the units not sampled; and

$$v_4(\hat{Y}_{rm}) = N^2[(1-f)/n^2](\bar{X}\bar{x}_2/\bar{x}^2)\sum_{i=1}^{n} r_i^2/[1 - (x_i/n\bar{x})] \tag{3.111}$$

with $r_i = y_i - (\bar{y}/\bar{x})x_i$.

Wu and Deng (1983) compared the performances of v_1, v_2, v_3, v_4, v_J, and others in terms of estimation bias, estimating conditional and unconditional MSE of $\bar{y}_{rm}$, and success rates of interval estimates that cover the population mean $\bar{Y}$. They recommended that v_1 not be used much anymore since it was consistently the least satisfactory estimator. The performance of v_1 and v_2 in estimating MSE depended on the underlying population and had no direct bearing on the performance of interval estimates. v_J gave the most reliable confidence intervals and v_3 and v_4 were almost as good in that respect. v_J estimates the unconditional MSE poorly because it estimates the conditional MSE well, which usually varies greatly with sample means $\bar{x}$. We recommend that v_J, or perhaps v_3 or v_4, should be used for estimating $V(\hat{Y}_{rm})$.

At the conclusion of sampling, one has a choice between $\hat{Y}$ or $\hat{Y}_{rm}$ as estimators of Y. Comparing the relative efficiency of these estimators is difficult, but for SRS if we ignore bias, the relative efficiency of $\hat{Y}_{rm}$ to $\hat{Y}$ is

$$\begin{aligned} RE &= \frac{V(\hat{Y})}{V(\hat{Y}_{rm})} \\ &\doteq \left(1 + \frac{R^2S_x^2}{S_y^2} - \frac{2\rho RS_x}{S_y}\right)^{-1} \end{aligned} \tag{3.112}$$

where ρ indicates the correlation coefficient, $S_{xy}/(S_xS_y)$. A sufficient condition that eq. (3.112) be greater than unity is that $\rho > \frac{1}{2}\,(CV_x/CV_y)$ where

CV indicates the coefficient of variation, that is, $CV_x = S_x/\bar{X}$ and $CV_y = S_y/\bar{Y}$. This result is intuitively pleasing, for it implies that, if the auxiliary information contained in the covariate is to be helpful in estimating Y, then the x_i, y_i values must be positively correlated. If inherent variation among the x_i, y_i, as measured by their respective coefficients of variation, is roughly equal, then $\rho > \frac{1}{2}$ suffices to ensure that $\hat{Y}_{rm}$ is more precise than $\hat{Y}$.

The Mean-of-Ratios (*mr*) *Estimator*. This estimator adjusts each sample value y_i by the covariate value x_i as follows:

$$\hat{Y}_{mr} = X\left[\sum_{i=1}^{n} (y_i/\pi_i)/(x_i/\pi_i)\right]\Big/ n = X \sum_{i=1}^{n} y_i/(nx_i)\,. \tag{3.113}$$

Murthy (1967, Sect. 10.9) showed that the variances for $\hat{Y}_{rm}$ and $\hat{Y}_{mr}$ are the same to the first order of approximation. For large sample sizes Murthy found that the bias of $\hat{Y}_{mr}$ is approximately

$$B(\hat{Y}_{mr}) \doteq nB(\hat{Y}_{rm})\,. \tag{3.114}$$

As $\hat{Y}_{mr}$ is not consistent, the bias does not decrease with n. This bias is at least a partial explanation of why, in SRS, $\hat{Y}_{mr}$ is often highly erratic, as shown in simulation experiments, and why $\hat{Y}_{rm}$ is used in preference to $\hat{Y}_{mr}$. However, if sampling is proportional to x, $\hat{Y}_{mr}$ becomes the Horvitz–Thompson estimator $\hat{Y}_{HT}$, which is unbiased and often highly efficient.

Various unbiased and nearly unbiased ratio estimators are feasible, such as the Tin estimator (Murthy 1967, Sect. 10.9–10.10), (Kotz and Johnson 1981, p. 270), but they are generally less efficient than $\hat{Y}_{mr}$. Of course, in situations where unbiasedness is critical, efficiency considerations among ratio-type estimators become secondary. To illustrate how these unbiased or nearly unbiased ratio estimators are obtained, we briefly discuss one proposed by Hartley and Ross (1954) for SRS. The Hartley-Ross estimator adjusts the estimator in eq. (3.113) by its estimated bias to obtain

$$\begin{aligned}\hat{Y}_{HR} &= \hat{Y}_{mr} + (N-1)s_{rx} \\ &= \hat{Y}_{mr} + \frac{n(N-1)}{(n-1)}\,(\bar{y} - \tilde{R}\bar{X})\end{aligned} \tag{3.115}$$

where $\tilde{R} = \Sigma_{i=1}^{n} y_i/(nx_i)$ and $s_{rx} = \Sigma_{i=1}^{n} (y_i/x_i)(x_i - \bar{x})/(n-1)$.

Generally speaking, $\text{MSE}(\hat{Y}_{HR}) > \text{MSE}(\hat{Y}_{rm})$. The small-sample performance of $\hat{Y}_{HR}$ and other unbiased or nearly unbiased ratio-type estimators was evaluated by Rao (1969).

As mentioned above, competitors to $\hat{Y}_{HR}$ are numerous. Detailed discussions about them can be found in Cochran (1977), Sukhatme *et al.* (1984), and Chaudhuri and Vos (1988). The mean-of-ratios estimator does not appear to be used much in practice with SRS because of its inconsistency. It is important to remember, however, that under other sampling designs, for example, pps with $\pi_i \propto x_i$, $\hat{Y}_{mr}$ is unbiased.

Ratio estimators are widely used due to their simplicity. They have been widely exploited as estimators to motivate model-based inference, which is addressed at length in Chapter 6.

Product (p) *Estimators*. Product estimators complement the ratio estimators in many respects. They are biased but the bias is small if y is strongly positively correlated with $1/x$ and the sample size is not small. One example of a product estimator is

$$\hat{Y}_{p1} = \sum_{i=1}^{n} (x_i/\pi_i) \sum_{i=1}^{n} (y_i/\pi_i)/X . \tag{3.116}$$

Note that $\hat{Y}_{p1}$, like $\hat{Y}_{rm}$, is a composite of Horvitz–Thompson estimators, that is, $\hat{X}_{HT} = \Sigma_{i=1}^{n} x_i/\pi_i$ and $\hat{Y}_{HT} = \Sigma_{i=1}^{n} y_i/\pi_i$, so that

$$\hat{Y}_{p1} = \hat{X}_{HT}\hat{Y}_{HT}/X . \tag{3.117}$$

Murthy (1967) gave an approximate formula for the variance of $\hat{Y}_{p1}$ as

$$V(\hat{Y}_{p1}) = V(\hat{Y}_{HT}) + 2(Y/X)\,\text{Cov}(\hat{X}_{HT}, \hat{Y}_{HT}) + (Y^2/X^2)V(\hat{X}_{HT}) , \tag{3.118}$$

which can be estimated by

$$v(\hat{Y}_{p1}) = v(\hat{Y}_{HT}) + 2(\hat{Y}_{HT}/\hat{X}_{HT})\,\text{cov}(\hat{X}_{HT}, \hat{Y}_{HT}) + (\hat{Y}^2_{HT}/\hat{X}^2_{HT})v(\hat{X}_{HT}) . \tag{3.119}$$

The bias of $\hat{Y}_{p1}$ is

$$B(\hat{Y}_{p1}) = \text{Cov}(\hat{Y}_{HT}, \hat{X}_{HT})/X , \tag{3.120}$$

which can be estimated by

$$b(\hat{Y}_{p1}) = \text{cov}(\hat{Y}_{HT}, \hat{X}_{HT})/\hat{X}_{HT} . \tag{3.121}$$

Murthy (1967, p. 380) mentioned that since the covariance of $\hat{Y}_{HT}$ and $\hat{X}_{HT}$ is usually of the form θ/n, where θ is independent of the sample size n, $\hat{Y}_{p1}$ is usually consistent. $\hat{Y}_{p1}$ is always Cochran-consistent.

Another product estimator, analogous to the mean-of-ratios estimator, is

$$\hat{Y}_{p2} = \sum_{i=1}^{n} (y_i x_i/\pi_i)/\bar{X} . \tag{3.122}$$

Little is known about the utility of $\hat{Y}_{p2}$ but clearly it, like $\hat{Y}_{mr}$, is not consistent. Product estimators may be unbiased under certain conditions that are not known at this time. These estimators have not been used as frequently as ratio estimators probably because in many situations the user is interested in the ratio of two parameters. Thus it is an easy step to ratio estimation. But note that product estimators such as $\hat{Y}_{p1}$ naturally complement ratio-of-means estimators when y is inversely proportional to x.

Goodman (1960) derived exact results for the variance of the product of any number of random variables, and Bohrnstedt and Goldberger (1969) derived them for the covariance. The results presented in these papers involve considerable computation.

Mixed Estimators. For SRS, Vos (1980) suggested generalizations of the above estimators and discussed earlier generalizations by Srivenkataramana and Tracy (1979). The latter authors suggested that, in SRS, only the nonsampled part of the population total $Y - n\bar{y}$ needs to be estimated and added to $n\bar{y}$, the known value for the sampled part of the population. Using ratio or product estimators to estimate the unsampled part of the population, Srivenkataramana and Tracy suggested the following four estimators:

$$\hat{Y}_{m_{1p}} = (1+f)\hat{Y} - f\hat{Y}_{p1} , \tag{3.123}$$

$$\hat{Y}_{m_{1q}} = f\hat{Y} + g\hat{Y}_{rm} , \tag{3.124}$$

$$\hat{Y}_{m_{2q}} = f\hat{Y} + g\hat{Y}^{*}_{rm} , \tag{3.125}$$

and

$$\hat{Y}_{m_{2p}} = f\hat{Y} + g\hat{Y}_{p1} , \tag{3.126}$$

where $\hat{Y}^{*}_{rm} = \hat{Y}_{rm}\bar{x}/(\bar{X} - \bar{x})$. To this Vos (1980) added

$$\hat{Y}_{m_{3q}} = (1+f)\hat{Y} - f\hat{Y}_{rm} \tag{3.127}$$

where, for all given estimators, $g = 1 - f$ and $f = n/N$.

These five estimators are all linear combinations of the simple mean $\hat{Y}$ and either product or ratio-of-means estimators. Each has a fixed weighting coefficient, either f or $1-f$, one of which is generally very small. All the estimators are approximately equal to either $\hat{Y}$, $\hat{Y}_{rm}$, $\hat{Y}_{p1}$, or $\hat{Y}^{*}_{rm}$. Thus a substantial improvement over the standard estimators is unlikely if f is small.

Vos (1980) proposed estimators he called "mixing estimators," a generalization of the estimators above, namely

$$\hat{Y}_{m_1} = p_1\hat{Y} + q_1\hat{Y}_p , \qquad \text{with } p_1 + q_1 = 1 , \tag{3.128}$$

$$\hat{Y}_{m_2} = p_2\hat{Y} + q_2\hat{Y}_{rm} , \qquad \text{with } p_2 + q_2 = 1 . \tag{3.129}$$

For $q_2 = -q_1 = \sigma_{xy}/(R\sigma_x^2)$, the variances of $\hat{Y}_{m1}$ and $\hat{Y}_{m2}$ are minimized and both are approximately equal to

$$V(\hat{Y}_m) = N(g/f)S_y^2(1-\rho^2)\,, \tag{3.130}$$

which is recognizable as the approximate variance of the regression estimator $\hat{Y}_{lr}$ discussed later. Thus since $V(\hat{Y}_{lr}) \le V(\hat{Y}_{m1})$ or $\le V(\hat{Y}_{m2})$, the usefulness of either $\hat{Y}_{m1}$ or $\hat{Y}_{m2}$ is limited (Chaudhuri and Vos 1988).

When $q_2 = \sigma_{xy}/(R\sigma_x^2) < 1/2$, $\hat{Y}_{m1}$ is preferred to $\hat{Y}_{m2}$ since it is less biased under these conditions; otherwise $\hat{Y}_{m2}$ is preferred.

Asymptotically equivalent to $\hat{Y}_{m1}$ and $\hat{Y}_{m2}$ are estimators discussed by Reddy (1974):

$$\hat{Y}_{RE} = \bar{X}\,\frac{\hat{Y}}{(1-\theta)\bar{X} + \theta\bar{x}}\,; \tag{3.131}$$

Sahai (1979):

$$\hat{Y}_{SA} = \hat{Y}\,\frac{(1-\theta)\bar{x} + \theta\bar{X}}{(1-\theta)\bar{X} + \theta\bar{x}}\,; \tag{3.132}$$

and Srivastava (1967):

$$\hat{Y}_{SR} = \hat{Y}\left(\frac{\bar{X}}{\bar{x}}\right)^{\theta}. \tag{3.133}$$

Vos (1980) presented a tabular summary of the approximate bias and variance of these estimators. Each is optimal for a certain choice of θ, the choice requiring good prior knowledge of population characteristics. Although they are asymptotically equivalent in efficiency, their relative performance in small sample situations may differ, but too little is known at present about their small sample behavior to be able to distinguish between them. They are generally less precise than $\hat{Y}_{lr}$ so they would only be preferred if smaller bias is important.

Regression Estimators. Regression estimators use the auxiliary information x in a manner similar to the ratio estimators but the adjustment is more complex. The increased complexity is due to the fact that we are estimating both an intercept (α) and a slope (β) for the relationship between y and x. This seems reasonable if one assumes model (2.70) with $\alpha \neq 0$.

Särndal (1980a, 1982) proposed a general regression estimator $\hat{Y}_{gr}$:

$$\hat{Y}_{gr} = \sum_{i=1}^{n} \frac{y_i}{\pi_i} + a_{gr}\left(N - \sum_{s} \frac{1}{\pi_i}\right) + b_{gr}\left(X - \sum_{i=1}^{n} \frac{x_i}{\pi_i}\right) \tag{3.134}$$

$$= \sum_{i=1}^{N} \hat{y}_i + \sum_{i\in s} \frac{e_i}{\pi_i} \tag{3.135}$$

where $\hat{y}_i = a_{gr} + b_{gr}x_i$, $e_i = y_i - \hat{y}_i$

$$a_{gr} = \frac{\sum_{i=1}^{n} y_i/(\pi_i v_i) - b_{gr} \sum_{i=1}^{n} x_i/(\pi_i v_i)}{\sum_{i=1}^{n} 1/(\pi_i v_i)}, \tag{3.136}$$

and

$$b_{gr} = \frac{\sum_{i=1}^{n} 1/(v_i \pi_i) \sum_{i=1}^{n} x_i y_i/(v_i \pi_i) - \sum_{i=1}^{n} y_i/(v_i \pi_i) \sum_{i=1}^{n} x_i/(v_i \pi_i)}{\sum_{i=1}^{n} 1/(v_i \pi_i) \sum_{i=1}^{n} x_i^2/(v_i \pi_i) - \left(\sum_{i=1}^{n} x_i/\{v_i \pi_i\}\right)^2} \tag{3.137}$$

and two other regression estimators $\hat{Y}_{pi}$ (using probabilities π_i) and $\hat{Y}_{wr}$ (using variance weights). These estimators are

$$\hat{Y}_{pi} = N[a_{pi} + b_{pi}\bar{X}], \tag{3.138}$$

$$\hat{Y}_{wr} = N[a_{wr} + b_{wr}\bar{X}] \tag{3.139}$$

where the b coefficients are

$$b_{wr} = \frac{\sum_{i=1}^{n} 1/v_i \sum_{i=1}^{n} x_i y_i/v_i - \sum_{i=1}^{n} y_i/v_i \sum_{i=1}^{n} x_i/v_i}{\sum_{i=1}^{n} 1/v_i \sum_{i=1}^{n} x_i^2/v_i - \left(\sum_{i=1}^{n} x_i/v_i\right)^2}, \tag{3.140}$$

if the probability weights are uniform, and

$$b_{pi} = \frac{\sum_{i=1}^{n} 1/\pi_i \sum_{i=1}^{n} x_i y_i/\pi_i - \sum_{i=1}^{n} y_i/\pi_i \sum_{i=1}^{n} x_i/\pi_i}{\sum_{i=1}^{n} 1/\pi_i \sum_{i=1}^{n} x_i^2/\pi_i - \left(\sum_{i=1}^{n} x_i/\pi_i\right)^2}, \tag{3.141}$$

if the variance weights are constant. Another estimator, similar to $\hat{Y}_{gr}$, is

$$\hat{Y}_{piwr} = N[a_{piwr} + b_{piwr}\bar{X}], \tag{3.142}$$

with

$$b_{piwr} = b_{gr} \tag{3.143}$$

(Royall and Cumberland 1978). The corresponding a coefficients are

$$a_{piwr} = a_{gr}\,, \tag{3.144}$$

$$a_{wr} = \frac{\sum_{i=1}^{n} y_i/v_i - b_{wr}\sum_{i=1}^{n} x_i/v_i}{\sum_{i=1}^{n} 1/v_i}\,, \tag{3.145}$$

and

$$a_{pi} = \frac{\sum_{i=1}^{n} y_i/\pi_i - b_{pi}\sum_{i=1}^{n} x_i/\pi_i}{\sum_{i=1}^{n} 1/\pi_i}\,. \tag{3.146}$$

Note that $\hat{Y}_{pi}$ and $\hat{Y}_{wr}$ are special cases of $\hat{Y}_{piwr}$.

An approximation to the variance of $\hat{Y}_{gr}$ is

$$V(\hat{Y}_{gr}) \doteq \frac{1}{2}\sum_{i\neq j}^{N}(\pi_i\pi_j - \pi_{ij})\left(\frac{e_i}{\pi_i} - \frac{e_j}{\pi_j}\right)^2 \tag{3.147}$$

where $e_i = y_i - \alpha - \beta x_i$ and e_j is defined similarly (Ouyang *et al.* 1991).

Särndal (1982) and Särndal *et al.* (1989) proposed and discussed two approximate variance estimators for $\hat{Y}_{gr}$:

$$v_1(\hat{Y}_{gr}) = \frac{1}{2}\sum_{i\neq j}^{n}\left(\frac{\pi_i\pi_j - \pi_{ij}}{\pi_{ij}}\right)\left(\frac{e_i}{\pi_i} - \frac{e_j}{\pi_j}\right)^2 \tag{3.148}$$

and

$$v_2(\hat{Y}_{gr}) = \frac{1}{2}\sum_{i\neq j}^{n}\left(\frac{\pi_i\pi_j - \pi_{ij}}{\pi_{ij}}\right)\left(\frac{e_i'}{\pi_i} - \frac{e_j'}{\pi_j}\right)^2 \tag{3.149}$$

where

$$e_i = y_i - \tilde{y}_s - b_{gr}(x_i - \tilde{x}_s)\,,$$

and

$$\begin{aligned} e_i' = e_i - e_i\Bigg\{ & \frac{(\hat{N}-N)\sum_{l=1}^{n} x_l^2/(v_l\pi_l) - (\hat{X}-X)\sum_{l=1}^{n} x_l/\pi_l v_l}{v_i} \\ & + \frac{\left[-(\hat{N}-N)\sum_{l=1}^{n} x_l^2/(v_l\pi_l) + (\hat{X}-X)\sum_{l=1}^{n} 1/(\pi_l v_l)\right]x_i}{v_i}\Bigg\} \\ & \times\left\{\frac{1}{\left\{\sum_{l=1}^{n} x_l^2/(\pi_l v_l)\sum_{l=1}^{n} 1/(\pi_l v_l) - \left[\sum_{l=1}^{n} x_l/(\pi_l v_l)\right]^2\right\}}\right\} \end{aligned}$$

with

$$\hat{N} = \sum_{l=1}^{n} 1/\pi_l , \qquad \tilde{N}_s = \sum_{l=1}^{n} 1/(\pi_l v_l) , \qquad \hat{X} = \sum_{l=1}^{n} x_l/\pi_l , \qquad \tilde{x}_s = \frac{\sum_{l=1}^{n} x_l/(\pi_l v_l)}{\tilde{N}_s} ,$$

and

$$\tilde{y}_s = \frac{\sum_{l=1}^{n} y_l/(\pi_l v_l)}{\tilde{N}_s} .$$

Note the similarity between $v_1(\hat{Y}_{gr})$ and $v_2(\hat{Y}_{gr})$ with $v_1(\hat{Y}_{HT})$ in eq. (3.9). $v_1(\hat{Y}_{gr})$ and $v_2(\hat{Y}_{gr})$ vanish when $\pi_{ij} = \pi_i \pi_j$ for units i and j in the sample. Hence, alternative forms of $v_1(\hat{Y}_{gr})$ and $v_2(\hat{Y}_{gr})$ are necessary for methods such as stratified sampling with one unit per stratum where $\pi_{ij} = \pi_i \pi_j$ for all i and j in the sample. Alternative approximations are (Ouyang *et al.* 1991)

$$v_3(\hat{Y}_{gr}) = \sum_{i=1}^{n} \frac{(1-\pi_i)}{\pi_i^2} e_i^2 + \sum_{i \neq j}^{n} \left(\frac{\pi_{ij} - \pi_i \pi_j}{\pi_i \pi_j \pi_{ij}} \right) e_i e_j , \tag{3.150}$$

and

$$v_4(\hat{Y}_{gr}) = \sum_{i=1}^{n} \frac{(1-\pi_i)}{\pi_i^2} \acute{e}_i^2 + \sum_{i \neq j}^{n} \left(\frac{\pi_{ij} - \pi_i \pi_j}{\pi_i \pi_j \pi_{ij}} \right) \acute{e}_i \acute{e}_j , \tag{3.151}$$

which for $\pi_i \pi_j = \pi_{ij}$ reduce to

$$v_3(\hat{Y}_{gr}) = \sum_{i=1}^{n} \frac{(1-\pi_i) e_i^2}{\pi_i^2} \tag{3.152}$$

and

$$v_4(\hat{Y}_{gr}) = \sum_{i=1}^{n} \frac{(1-\pi_i) \acute{e}_i^2}{\pi_i^2} . \tag{3.153}$$

An alternative approximate estimator of the MSE for $\hat{Y}_{gr}$ is

$$v_5(\hat{Y}_{gr}) = \left\{ \sum_{i=1}^{n} \frac{e_i^2}{\pi_i^2} + \frac{1}{2} \sum_{i \neq j}^{n} \frac{e_i e_j}{\pi_{ij}} \right\} \frac{(n-1)}{n} \tag{3.154}$$

where $e_i = y_i - a_{gr} - x_i b_{gr}$ and e_j is defined equivalently (Ouyang *et al.* 1991).

These same variance estimators v_1–v_5 (using the appropriate coefficients a and b) may also be useful for $\hat{Y}_{wr}$, $\hat{Y}_{pi}$, and $\hat{Y}_{piwr}$. As we shall see in Chapter 4, bootstrap variance estimators may be more appropriate for the regression estimators (based on a simulation study). Robinson and Särndal (1983) proved that $\hat{Y}_{gr}$ is asymptotically unbiased. $\hat{Y}_{wr}$, $\hat{Y}_{pi}$, and $\hat{Y}_{piwr}$ are also asymptotically design unbiased. An approximate expression for the bias of the weighted regression estimators is

$$B(\hat{Y}_{gr}) = -\text{Cov}(\hat{X}_{HT}, \hat{\beta}_{gr}) \tag{3.155}$$

(Särndal 1980a).

Survey samplers have traditionally used SRS with the simple linear regression estimator $\hat{Y}_{lr}$ one obtains when the variance weights v_i used for the coefficients in eq. (3.139) are uniform. Considerable information is available about this relatively simple regression estimator, and we discuss this special situation in detail in the remainder of this section. The estimator is

$$\hat{Y}_{lr} = N(a + b\bar{X}) = \hat{Y} + b(X - \hat{X}) \tag{3.156}$$

with

$$b = \frac{\sum_{i=1}^{n} (y_i - \bar{y})(x_i - \bar{x})}{\sum_{i=1}^{n} (x_i - \bar{x})^2} \tag{3.157}$$

and

$$a = \bar{y} - b\bar{x} . \tag{3.158}$$

Let $\delta_y = (\hat{Y} - Y)/Y$, $\delta_x = (\hat{X} - X)/X$, $\delta_{sxy} = (s_{sxy} - S_{xy})/S_{xy}$, $\delta_{sxx} = (s_x^{\,2} - S_x^{\,2})/S_x^{\,2}$, and $\beta = S_{xy}/S_x^{\,2}$. Then $\hat{Y}_{1r}$ can be rewritten as

$$\hat{Y} = Y(1 + \delta_y) + \frac{S_{xy}(1 + \delta_{sxy})}{S_x^2(1 + \delta_{sxy})} \{X - X(1 + \delta_x)\} .$$

Consequently

$$\begin{aligned}(\hat{Y} - Y)^2 \doteq{}& Y^2\delta_y^2 - 2XY\beta\delta_y\delta_x(1 + \delta_{sxy} - \delta_{sxx} - \delta_{sxy}\delta_{yxx} \cdots) \\ &+ \beta^2X^2\delta_x^2(1 + \delta_{sxy})^2(1 - \delta_{sxx} \cdots)^2 .\end{aligned}$$

Disregarding higher order terms in δ_x, δ_y, δ_{sxy}, and δ_{sxx}, and using Sukhatme's (1944) results yields an approximate variance of $\hat{Y}_{lr}$:

$$V(\hat{Y}_{lr}) = \frac{N^2(1-f)}{n} [S_y^2 - 2\beta S_{xy} + \beta^2 S_x^2] \tag{3.159}$$

$$= \frac{N^2(1-f)}{n} S_y^2(1 - \rho^2) \tag{3.160}$$

where ρ^2 is the correlation coefficient squared between y and x, and

$$\begin{aligned}\beta &= \sum_{i=1}^{N} (y_i - \bar{Y})(x_i - \bar{X}) \Big/ \sum_{i=1}^{N} (x_i - \bar{X})^2 \\ &= S_{xy}/S_x^2 . \end{aligned} \tag{3.161}$$

Deng and Wu (1987) gave several approximate variance estimators for $\hat{Y}_{lr}$:

$$v_1(\hat{Y}_{lr}) = \left[\frac{N^2(1-f)}{n}\right] \sum_{i=1}^{n} \frac{d_i^2}{(n-2)} \tag{3.162}$$

where $d_i = y_i - \bar{y} - b(x_i - \bar{x})$;

$$v_2(\hat{Y}_{lr}) = \frac{N^2 f(1-f)}{n} \left\{\sum_{i=1}^{n} \frac{d_i^2}{(n-2)}\right\} + N^2 \frac{(1-f)^2}{n^2} \sum_{i=1}^{n} t_i d_i^2 \tag{3.163}$$

where

$$t_i = r_i^2 \Big/ \left(1 - \sum_{i=1}^{n} w_i k_i / n\right),$$

$$r_i = 1 + (x_i - \bar{x})(\bar{x}_2 - \bar{x}) / s_x^2 ,$$

$$w_i = r_i^2 \Big/ \sum_{i=1}^{n} r_i^2 ,$$

$$k_i = [1 + (x_i - \bar{x})^2 / s_x^2] / n ,$$

and $\bar{x}_2 = (X - n\bar{x})/(N - n)$ is the mean of the units not sampled;

$$v_3(\hat{Y}_{lr}) = \frac{N^2(1-f)^2}{n(n-1)} \sum_{i=1}^{n} c_i d_i^2 \tag{3.164}$$

where

$$c_i = \left(r_i^2 + \frac{f}{1-f}\right) \Big/ [1 - (x_i - \bar{x})^2 / \{(n-1)s_x^2\}] ;$$

and a jackknife variance estimator

$$v_J(\hat{Y}_{lr}) = N^2[(1-f)/n] \sum_{i=1}^{n} (\hat{Y}_{lr}^{(i)} - \hat{Y}_{lr}^{(J)})^2 / (n-1) \tag{3.165}$$

where $\hat{Y}_{lr}^{(i)}$ is the simple linear regression estimator of $\hat{Y}$ (known as pseudovalue) obtained by deleting the ith observation from the sample, and $\hat{Y}_{lr}(J) = \Sigma_{i=1}^{n} \hat{Y}_{lr}^{(i)}/n$. Royall and Cumberland (1981a) derived v_2 and v_3 as estimators that are robust against misspecification of model (2.70) with $k = 0$.

Deng and Wu (1987) gave additional variance estimators for $\hat{Y}_{lr}$ but the above appear to be most useful. Based on asymptotic expansions they found that $v_J \geq v_3 \geq v_2$ where v_J and v_2 generally over- and underestimate the MSE of $\hat{Y}_{lr}$, respectively. Based on simulations, v_J gave the most reliable coverage probabilities when confidence intervals were constructed, followed by v_3 and v_2. The traditional estimator v_1 should not be used any longer in our opinion. We recommend v_J for general use.

In situations where simple linear regression estimation may be appropriate and where the regression coefficient β is known reasonably well, the difference estimator is a simple version of the regression estimator, $\hat{Y}_{lr}$,

$$\hat{Y}_d = N(\bar{y} + c\bar{X}) \tag{3.166}$$

where c is a known constant. In fact $\hat{Y}_d$ is a special case of $\hat{Y}_{lr}$ since the optimum value of c is equivalent to β in eq. (3.161) (Sukhatme *et al.* 1984).

This estimator can have utility, for example, in updating a previous census where X represents the census value, $\bar{y}$ and $\bar{x}$ are the current and previous estimates of the sample means, and $c = 1$ can be assumed.

For c (or β) known, regression estimation is much simpler, $V(\hat{Y}_d) = V(\hat{Y}_{lr})$ in eq. (3.159), and an unbiased variance estimator is

$$V(\hat{Y}_d) = N^2\{(1-f)/n\}[s_y^2 - 2cs_{yx} + c^2s_x^2]\,. \tag{3.167}$$

It is possible to define general estimators of which the above are special cases. For example, Basu (1971) gave the generalized difference estimator $\hat{Y}_{gd}$,

$$\hat{Y}_{gd} = \sum_{i=1}^{n} \frac{y_i - e_i}{N\pi_i} + \bar{e}\,, \tag{3.168}$$

where e_i is an arbitrary but known number attached to unit i ($i = 1, \ldots, N$), and $\bar{e} = \Sigma_{i=1}^{N} e_i/N$. For $e_i = 0$, $\hat{Y}_{gd}$ is the Horvitz–Thompson estimator $\hat{Y}_{HT}$ and for SRS, $\hat{Y}_{gd}$ is the difference estimator $\hat{Y}_d$ (eq. 3.166).

Estimators $\hat{Y}_{rm}$, $\hat{Y}_{mr}$, $\hat{Y}_p$, $\hat{Y}_{lr}$, $\hat{Y}_d$ and $\hat{Y}_{gd}$ are all improvements over $\hat{Y}$ (eq. 3.1) if variables x and y are correlated in some manner and if proper advantage is taken of the relationship between the variables. The extent of knowledge about the actual relationship and the type of relationship determine the choice of estimator. All of these estimators perform best if there is some linear relationship between y and x (or some transformation of x such as $1/x$ in the case of $\hat{Y}_{p1}$). Specifically, for SRS, $\hat{Y}_{rm}$ is the best linear unbiased estimator (BLUE) if the underlying population model (2.70) holds with $\alpha = 0$ and $k = 1$. Similarly, with SRS, the mean-of-ratios estimator $\hat{Y}_{mr}$ is BLUE if model (2.70) holds with $\alpha = 0$, $\beta > 0$, and $k = 2$. The regression estimator $\hat{Y}_{lr}$ is BLUE if $k = 0$ in model (2.70).

Srivenkataramana (1980) indicated that for SRS, $\hat{Y}_{p_1}$ is nearly optimal if

$$y_i = \beta/x_i + e_i\,, \qquad Ee_i = 0\,, \qquad Ee_ie_j = \begin{cases} 0 & \text{for } i \neq j \\ \sigma^2/x_i & \text{for } i = j\,, \end{cases} \tag{3.169}$$

and if $\bar{x}/\bar{X} = \tilde{x}/\tilde{X}$, for variable x where

$$1/\tilde{x} = \sum_{i=1}^{n} 1/(nx_i) \tag{3.170}$$

and $\tilde{X}$ is similarly defined. $\tilde{x}$ and $\tilde{X}$ are the sample and population harmonic means for χ.

Little work has been done on optimality properties of the product estimator $\hat{Y}_{p_2}$. However, Singh *et al.* (1985) discussed it and other ratio and product (R & P) estimators as special cases of a more general R & P estimator. They also discussed unbiased and optimum estimators of the R & P type.

3.5.2 Srivastava Estimation

Srivastava (1985) introduced a new concept in estimation that seems promising for many survey sampling applications. The idea is to introduce a weight function $r(s)$ defined over samples s such that application of r leads to more reliable estimates. The function r can capture information collected at any level of sampling as long as the process of selecting it is independent of the sample drawn.

The idea may be introduced as follows: Let $p(\cdot)$ be a sampling design and $r(\cdot)$ a real function. If

$$\pi_r(i) = \sum_{s \ni i} p(s)r(s)\,, \tag{3.171}$$

where $s \ni i$ indicates summation over all samples containing i, then the Srivastava estimator is defined as

$$\hat{Y}_{sr} = r(s) \sum_{i \in s} y_i/\pi_r(i)\,, \tag{3.172}$$

with variance

$$\begin{aligned} V(\hat{Y}_{sr}) = &\sum_{i=1}^{N} y_i^2 \left(\frac{\pi_{r^2}^{(i)} - [\pi_r(i)]^2}{[\pi_r(i)]^2} \right) \\ &+ \sum_{i \neq j}^{N} y_i y_j \left(\frac{\pi_{r^2}^{(i,j)} - \pi_r(i)\pi_r(j)}{\pi_r(i)\pi_r(j)} \right) \end{aligned} \tag{3.173}$$

where

$$\pi_{r^2}^{(i)} = \sum_{s \ni i} p(s)[r(s)]^2\,,$$

and

$$\pi_{r^2}^{(i,j)} = \sum_{s \ni i,j} p(s)[r(s)]^2\,, \qquad i \neq j\,.$$

The merit of the Srivastava estimator is that the "sample weight function" $r(s)$ can be determined after the sample is drawn; hence $r(s)$ can be chosen to improve estimation using information gathered during the survey. Note that any weighting of a random sample will be unbiased if the weighting is done independently of the sample itself. A primary example of this is the Horvitz–Thompson estimator in sampling with pps in eq. (3.7) where the weight, the probability of sampling unit i, π_i, is set prior to sampling. Note that, if all the π_i are equal as in SRS, then $\hat{Y}_{HT}$ becomes the simple estimator of the total, $\hat{Y} = N\bar{y}$, which is an equal-weighting scheme, often used by default when little prior information is available on the units in the population.

A set of weights has been suggested by Srivastava and Ouyang (1992a) based on relating the sample observations to corresponding theoretical observations if we know the approximate underlying distribution. Often the solutions are not easily calculated and iterative approximations to the true solution have to be used.

When the y_i are all ≥ 0, a necessary and sufficient condition for $\hat{Y}_{sr}$ to have zero variance at some values $(y_1^*, \ldots, y_N^*)$ is that the sample weight have the following structure for a set of possible r:

$$\sum_{i \in s} 1/r_i = 1/r(s) , \tag{3.174}$$

and the relation between the sample weight function and y_i^* be

$$y_i^*/Y^* = \pi_r(i)/r_i , \tag{3.175}$$

where Y^* is the population mean of y_i^*.

If $\hat{Y}_{sr}$ has zero variance at $(y_1^*, \ldots, y_N^*)$, which is close to the true value of $(y_1, \ldots, y_N)$, then $\hat{Y}_{sr}$ is considered to provide a "good" estimate of Y (Ouyang 1990). As noted by Srivastava and Ouyang (1992a), to use the r_i we only need estimates for all the y_i. This can be done after sampling and may be less demanding than pps sampling, which requires some such estimates before sampling. Note that $\hat{Y}_{sr}$ is quite general; for example, if all $r_i = 1$, then $\hat{Y}_{sr} = \hat{Y}_{HT}$. But for the theory to be strictly valid, the information on the r_i should be independent of the sample y-values.

Since it is often not easy to solve eq. (3.174) to obtain r_i $(i = 1, \ldots, N)$, some approximation is needed. If pps sampling is used, the Srivastava estimator can be approximated by the "general ratio estimator":

$$\hat{Y}_{gre} = \left\{ \frac{\sum_{i \in s} y_i/\pi_i}{\sum_{i \in s} y_i^*/\pi_i} \right\} Y^* \tag{3.176}$$

where the y_i^* and Y^* can, for example, be obtained from a regression model prediction, that is, $y_i^* = a + bx_i$ $(i = 1, \ldots, N)$.

Ouyang and Schreuder (1992) give a simple illustration of an approximate Srivastava estimate. To use $\hat{Y}_{sr}$ in practice, the nonlinear equations (3.174) and (3.175) need to be solved for some estimated value y^*. Srivastava and Ouyang (1992a) gave an approximate iterative solution. Let

$$\pi_r^o(i) = \sum_{s \ni i} p(s)/n \qquad (i = 1, \ldots, N), \tag{a}$$

$$r_i^m = \pi_r^m(i)/(y_i^*/Y^*) \quad (i = 1, \ldots, N), \tag{b}$$

$$r^m(s) = \left[\sum_{i \ni s} (r_i^m)^{-1}\right]^{-1}, \tag{c}$$

and

$$\pi_r^{m+1}(i) = \sum_{s \ni i} r^m(s)p(s) \qquad (i = 1, \ldots, N). \tag{d}$$

The iterative procedure begins with $m = 0$. From $\pi_r^o(i)$ given in (a), we obtain r_i^o from (b), and $r^o(s)$ from (c). Then we get $\pi_r^1(i)$ from (d) and use it in (b) to start the next iteration, and so on.

Srivastava and Ouyang (1992b) discussed some other applications of Srivastava estimation. Some potential applications in forestry are discussed in Chapters 4 and 5.

3.6 JACKKNIFE AND BOOTSTRAP VARIANCE ESTIMATION

Variance estimates are used to indicate the reliability of point estimates and to allow the confidence interval, indicating the range within which the parameter of interest is expected to lie, to be specified. To obtain design-unbiased point and variance estimates, respectively, each unit in the population needs to have a positive probability of selection and all joint probabilities of selection need to be known. Hence unbiased variance estimation is not feasible, for example, for stratified sampling when only one unit per stratum is selected. Even when all probabilities and joint probabilities are larger than zero, the sampling strategy may be too complex to derive unbiased or reliable variance estimators. For example, with the Horvitz–Thompson estimator in Theorem 3.5, both the Yates and Grundy and Horvitz–Thompson variance estimates v_1 and v_2 can be negative. Thus situations where unbiased variance estimators are either unavailable or undesirable are quite common. In such cases, overestimates of the variance are preferred to underestimates, since this yields results that are likely to be at least as reliable as predicted.

Jackknifing and bootstrapping techniques are based, respectively, on the principle of splitting the sample into subsamples and sampling with replacement from the sample treating the subsamples as independent samples.

Since the subsamples are not independent, variance estimates based on them are approximations. However, they can be computed for even the most complex designs and estimators, and hence are likely to be suitable for complex linear and nonlinear estimators. In addition, they guarantee positive variance estimates. Both of these methods are resampling methods that have become feasible with the advent of cheap and efficient computing machinery. The original purpose of jackknifing was to reduce estimation bias (Quenouille 1949, 1956) whereas Efron (1979) introduced bootstrapping as a general nonparametric method to solve various statistical problems. The central idea of both the jackknife and the bootstrap is the construction of an empirical distribution of estimates from which moments or other parameters of distributions can be calculated directly. To the extent that this empirical distribution successfully mimics the sampling distribution $F(\hat{\theta})$ of an estimator $\hat{\theta}$, then the empirical moments are nonparametric estimators of the moments of $F(\hat{\theta})$. Jackknife or bootstrap estimators of variance can often be used in situations where design-based estimators fail to exist or where the conditions are lacking that ensure that design-based variance estimators perform well. A case in point is the estimator of $V(\hat{Y}_{rm})$ (eq. 3.98), based on data from small samples. Examples of how jackknifing and bootstrapping work (Efron and Gong 1983) are given below.

Consider the simple situation where the data set consists of a random sample of size n from an unknown probability distribution F with

$$Y_1, Y_2, \ldots, Y_n \sim F$$

and

$$-\infty < Y_i < \infty \qquad \text{for all } Y\,. \tag{3.177}$$

After observing $Y_1 = y_1, Y_2 = y_2, \ldots, Y_n = y_n$, compute the sample average $\bar{y} = \Sigma_{i=1}^{n} y_i/n$ where $\bar{y}$ is an estimate of the expectation $E(y)$ of F. The sample also gives an estimate of the accuracy of $\bar{y}$ (i.e., the estimated standard error of the mean), which is

$$\hat{\sigma}_{\bar{y}} = \left[\sum_{i=1}^{n} (y_i - \bar{y})^2/\{n(n-1)\}\right]^{1/2}. \tag{3.178}$$

This "classical" type estimator is based on simple derivations using moment estimators and substituting appropriate unbiased or consistent estimates. This is appropriate in many cases but not, for example, in estimating the median of a distribution even with SRS. For such more complicated situations, jackknifing and bootstrapping may be appropriate. Both jackknifing and bootstrapping generate an estimator and a variance estimator, but in most cases we are exclusively interested in the variance estimators.

To illustrate jackknifing, let

$$\bar{y}_{(i)} = \frac{n\bar{y} - y_i}{n-1} = \sum_{j(\neq i)}^{n} \frac{y_i}{(n-1)} \tag{3.179}$$

be the sample mean of the data set with the ith observation deleted and let $\bar{y}_{(\cdot)} = \Sigma_{i=1}^{n} \bar{y}_{(i)}/n$ be the average of the deleted means. Then the standard jackknife standard error is

$$\hat{\sigma}_J = \left[[(n-1)/n] \sum_{i=1}^{n} (\bar{y}_{(i)} - \bar{y}_{(\cdot)})^2 \right]^{1/2} . \tag{3.180}$$

$\hat{\sigma}_J$ is the same as $\hat{\sigma}$ in eq. (3.178) in this case, but $\hat{\sigma}_J$ can be generalized to any estimator $\hat{\theta} = \hat{\theta}(y_1, y_2, \ldots, y_n)$ by substituting $\hat{\theta}_{(i)} = \hat{\theta}(y_1, \ldots, y_{i-1}, y_{i+1}, \ldots, y_n)$ for $\bar{y}_{(i)}$ and $\theta_{(\cdot)} = \Sigma_{i=1}^{n} \hat{\theta}_{(i)}/n$ for $\bar{y}_{(\cdot)}$. Efron and Stein (1981) proved that the standard jackknife estimator of variance has a positive bias.

Deleting one unit at a time is the standard procedure generally, although more than one unit can be deleted. Rao and Webster (1966) showed that for the ratio-of-means estimator, the sample should be split into the largest number of distinct groups for both bias reduction and variance estimation. This means that deleting one unit at a time is best for this estimator. But uniform guidelines on the optimal number of units to delete are not available.

The bootstrap works differently. Let $\hat{F}$ be the empirical probability distribution of the data with probability mass $1/n$ at each y_i, and if y_1^*, $y_2^*, \ldots, y_n^*$ is a random sample from $\hat{F}$, then

$$y_1^*, y_2^*, \ldots, y_n^* \sim \hat{F} . \tag{3.181}$$

Each y_i^* is drawn independently with replacement and with equal probability from $\{y_1, \ldots, y_n\}$. Then $\bar{y}^* = \Sigma_{i=1}^{n} y_i^*/n$ has variance

$$v(\bar{y}^*) = \sum_{i=1}^{n} (y_i - \bar{y})^2/n^2 . \tag{3.182}$$

The bootstrap standard error estimator for $\hat{\theta}(y_1, y_2, \ldots, y_n)$ is

$$\hat{\sigma}_B = [V\{\hat{\theta}(y_1^*, y_2^*, \ldots, y_n^*)\}]^{1/2} . \tag{3.183}$$

For this simple case, $\hat{\sigma}_B$ could be made exactly equal to $\hat{\sigma}$. Efron and Gong (1983) showed that $\hat{\sigma}_J$ is almost equivalent to $\hat{\theta}_B$ for a linear approximation of $\hat{\theta}$. Hence $\hat{\sigma}_J$ is a reasonable approximation for the actual $\hat{\sigma}_B$ only if the linear approximation is good.

Jackknifing and bootstrapping can be fruitfully applied in regression estimation. The following discussion focuses on bootstrapping but much of it would be applicable to jackknifing too.

In regression estimation, n independent real-valued quantities $Y_i = y_i$,

$$y_i = g_i(\beta) + e_i\,, \qquad i = 1, 2, \ldots, n \tag{3.184}$$

are observed where the $g_i(\cdot)$ are of known form, usually $g_i(\beta) = g(\beta; x_i)$ where x_i is an observed covariate and β is a vector of unknown parameters to be estimated. The e_i are independent and identically distributed (i.i.d) random variables (r.v.) with, for example, $E(e) = 0$ or $P(e < 0) = 0.5$ and distribution F. Model (2.70) is an example of model (3.184).

Two bootstrap methods are feasible for regression model (3.184) (Wu 1986). For simplicity we assume one parameter β.

Method 1. Draw an i.i.d. sample $\{e_i^*\}$, $i = 1, \ldots, n$ from the normalized residuals $\{r_i/(1 - p/n)^{1/2}\}$ $(i = 1, \ldots, n)$ where $r_i = y_i - \hat{\beta}x_i$ is the ith residual. Define bootstrap observations $y_i^* = \hat{\beta}x_i + e_i^*$ using $\hat{\beta}$ as the true parameter and $\{r_i/(1 - p/n)^{1/2}\}$ as all possible errors $(i = 1, \ldots, n)$. The bootstrap least square estimator is

$$\beta^* = \sum_{i=1}^{n} x_i y_i^* \Big/ \sum_{i=1}^{n} x_i^2\,. \tag{3.185}$$

For a nonlinear estimator $\hat{\theta} = g(\hat{\beta})$, the bootstrap variance estimator is

$$v_b = E_*(\theta^* - \hat{\theta})(\theta^* - \hat{\theta})' \tag{3.186}$$

where $*$ indicates bootstrap sampling from the normalized residuals. For

$$\theta = \beta, \qquad E_*\beta^* = \hat{\beta}, \qquad v_b = \hat{v} = \hat{\sigma}^2 \Big/ \sum x_i^2\,, \qquad \text{and} \qquad \hat{\sigma}^2 = \sum_{i=1}^{n} r_i^2/(n - p)\,,$$

v_b is unbiased if $\sigma_i^2 = \sigma^2$ for all i. But generally $\sigma_i^2 \neq \sigma^2$; thus v_b is generally biased and inconsistent since

$$V(\hat{\beta}) = \sum_{i=1}^{n} \sigma_i^2 x_i^2 \Big/ \left(\sum x_i^2\right)^2\,. \tag{3.187}$$

This bootstrapping approach is dangerous in our view since it assumes that the residuals r_i are exchangeable. Their heterogeneity is lost in the process of i.i.d. sampling and is not reflected in v_b.

Method 2. Draw i.i.d. samples $\{y_i^*, x_i^*\}$ $(i = 1, \ldots, n)$ from $\{(y_i, x_i)\}$ $(i = 1, \ldots, n)$ and compute the bootstrap least squares estimator

$$\beta^* = \sum_{i=1}^{n} x_i^* y_i^* \Big/ \sum_{i=1}^{n} x_i^{*2} \tag{3.188}$$

and the unweighted bootstrap variance estimator

$$v^* = E_*(\theta^* - \hat{\theta})(\theta^* - \hat{\theta})' , \qquad (3.189)$$

where $\theta^* = g(\beta^*)$ and $*$ denotes i.i.d. sampling from $\{(y_i, x_i)\}$. v^* neglects the unbalanced nature of the regression data $\{y_i, x_i\}$ and is inconsistent and biased even for $\sigma_i^2 = \sigma^2$. This method does not take advantage of the special structure of the regression model and does not yield the classical variance estimate for ordinary least squares. But it does yield reliable estimates for the variability of $\hat{\beta}$ even if the regression model (3.184) is incorrect. This is a desirable property.

In complex sample surveys, ordinary with-replacement bootstrapping may produce inadmissible simulated samples (Hinkley 1988). Special bootstrap techniques have been proposed for these situations and McCarthy and Snowden (1985) appraised some of them. They concluded that simply increasing the size of the bootstrap sample from n to $n/(1-f)$, where f is the sampling fraction, may be satisfactory. They noted also that selections are not independent in sampling finite populations without replacement since the correlation between the variants of the elements selected at the ith and jth draws is $-1/(N-1)$. Such sampling designs are frequently characterized by the sampling fraction, $f = n/N$.

There are difficulties experienced when applying the bootstrap to finite populations: if the sample values $y_1, \ldots, y_n$ refer to n elements drawn from a finite population of N elements, then it is not possible to select n elements from the sample that will duplicate a sample drawn from the original population. If with-replacement sampling is used in the bootstrap, then repetitions of elements can occur in the sample that, by definition, are not possible in an original sample from the population. On the other hand, without-replacement sampling would simply yield the original sample again.

To adapt the bootstrap idea to sampling finite populations, McCarthy and Snowden (1985) proposed two options:

1. *The Without-Replacement Bootstrap*. For simplicity assume that $N = kn$. (McCarthy and Snowden gave the simple extension when this is not true.) First, replicate each of the sample elements k times to create a population of $N = kn$ elements. Then draw samples of n elements without replacement from this artificial population. To obtain the desired variance $v(\bar{y})$, multiply the variance of the sample means $\bar{y}^*$ of the repeated samples drawn from the artificial population by $(n - 1/k)/(n-1)$.
2. *The With-Replacement Bootstrap*. This emphasizes variance estimation. Choose $n^* = (n-1)/(1-f) = \{k/(k-1)\}(n-1)$ elements with replacement. This yields a variance equal to $v(\bar{y})$.

Of these two options, the second is more closely related to the classical bootstrap approach and the first to traditional, finite population theory. The with-replacement bootstrap generalizes directly to stratified and stratified cluster sampling situations, whereas the without-replacement procedure does so only in very special circumstances.

Rao and Wu (1988) and references therein discussed resampling inference with complex survey data. They concluded that the i.i.d. bootstrap can be extended to such complex designs. The variance estimators for nonlinear statistics, $\hat{\theta}$, reduce to the classical variance estimates for linear statistics. These methods also allow for improved confidence intervals accounting for skewness in the distribution of $\hat{\theta}$ (see also Kovar *et al.* 1988). The authors have had considerable experience with jackknifing and bootstrapping in forestry and this is discussed in Chapters 4 and 5.

3.7 PROBLEMS

1. Using the population given, determine which units would be selected using the SRS method with $n = 3$ if the following random numbers were generated: 0.0110, 0.5623, 0.9281.

Unit	y
1	433.8
2	431.9
3	1153.8
4	1860.9
5	2016.2
6	1754.8
7	1889.6
8	2553.3
9	2568.2
10	1796.4

2. Determine which $y -$ elements would be selected using pps sampling from the cumulated, sorted $x -$ values of the population given. Use $n = 3$ and the random number $r = 0.96$. (Hint: Check for selection with certainty)

Unit	x	y
1	786.3	854.7
2	811.7	864.3
3	822.7	1314.5
4	1052.4	1147.3
5	1382.8	799.8
6	1513.6	1276.2
7	1660.5	1657.4
8	1684.9	1799.5
9	2424.2	2046.6
10	7168.7	6949.0

3. Using a sample size $n_t = 3$, determine which units of the population would be selected using Poisson sampling.

Unit	x	y	Random Number
1	237.3	259.3	0.0132
2	1350.6	1122.5	0.8296
3	1513.6	1276.2	0.5912
4	1524.6	1371.3	0.1325
5	2199.4	1156.9	0.7374
6	2306.8	1825.9	0.7272
7	2361.1	2715.9	0.9999
8	2667.2	3346.3	0.4574
9	3152.7	3043.9	0.6434
10	4180.0	3262.3	0.4700

4. The following data comprise a sample drawn from a population. Use this information to determine the Horvitz–Thompson estimate $\hat{Y}_{HT}$.

Unit	x	y	π
1	985.4	1573.2	.0225
2	1537.5	2426.3	.0351
3	2036.1	1511.9	.0468
4	2106.2	1754.8	.0477
5	2199.4	1156.9	.0504
6	2420.2	1489.2	.0549
7	3152.7	3043.9	.0720
8	3327.1	2397.9	.0756
9	4039.9	3406.4	.0918

5. Assume the sample in the previous problem was drawn using SRS with $N = 275$ and $n = 9$ and the population total is 379 424.7. Use the SRS probabilities of selection to compute $\hat{Y}_{HT}$. Compare the estimates from this problem and the previous problem with the actual solution. Explain why the two estimates differ.

6. For the following sample, compute $\bar{y}_{HT}$, $\bar{y}_{rm}$, $\bar{y}_{mr}$, $\bar{y}_{p1}$, and $\bar{y}_{p2}$ where the population size is $N = 275$. Poisson sample selection was used with $\bar{X} = 1434.46$ and $\bar{Y} = 1379.8$. Compute the estimates and explain the results as best you can.

Unit	x	y	π_i
1	1035.5	1716.6	0.0208
2	1084.5	1300.6	0.0216
3	1557.5	1697.1	0.0312
4	1884.2	1623.4	0.0384
5	1911.3	1345.6	0.0385
6	2883.7	3431.8	0.0584
7	3121.5	2568.2	0.0632
8	7168.7	6949.0	0.1456

7. Calculate the regression coefficients a, b, a_{pi}, b_{pi}, a_{piwr}, b_{piwr}, a_{wr}, and b_{wr} and the corresponding estimates $\bar{y}_{lr}$, $\bar{y}_{pi}$, $\bar{y}_{piwr}$, and $\bar{y}_{wr}$ for the sample given in the previous problem. Use $v_i = x_i^{0.75}$ and the π_i shown above.
8. In SRS, prove that $\pi_i = n/N$ and $\pi_{ij} = n(n-1)/\{N(n-1)\}$ for all units i, j $(i, j = 1, \ldots, N)$.
9. Verify in SRS that when $y_i = 0$ or 1, that $S^2 = Np(1-p)/(N-1)$ so that $V(\hat{p}) = \{(N-n)/N\}S^2/n$.
10. It is possible in binomial sampling to obtain a sample of size 0. Compute the probability of $n = 0$ and give the unbiased estimator and variance of the estimator when a sample of $n = 0$ is not permitted, in which case another binomial sample is drawn.
11. Prove the results in eq. (3.8) in Theorem 3.5 using Theorems 2.4 and 2.5.
12. Prove the results in Theorem 3.7 from the results in Theorem 3.5.
13. What is the difference between the following two simulation procedures in selecting at random ten samples of size $n = 20$ from a population?
    (a) For each of the ten samples, sort and cumulate the $x-$ values. Generate 20 random numbers between 1 and X and select the corresponding $x-$ values. Draw a new random number between 1 and X if a unit is selected more than once until all 20 units in the sample are distinct.
    (b) As above but, for each sample, randomly arrange the population before cumulation.

    Run a simulation using an artificial population to support your answer.
14. Use the results of Theorem 3.5 to derive the results for binomial sampling in Theorem 3.11. Why is one of the variance estimators in Theorem 3.5 not appropriate here?
15. What are the differences between stratified random sampling and SRS in terms of probabilities of selection if all units in stratified random sampling have the same probabilities of selection (i.e., proportional allocation)?
16. Derive the proper estimation theory when units within strata are selected by pps sampling.
17. Why in general is it difficult to select samples with preassigned inclusion probabilities $\pi_i = nx_i/X$ for $n > 2$?
18. A cruiser takes a sample of 7 plots from a population of 28 plots, obtaining values of $y = 0, 0, 0, 1, 2, 3, 4$. What is the estimated total for the population? He then decides he is interested in one part of the population, called A. As it turns out, 16 of the plots fall within A and 12 do not. Also, 4 sample plots fall in A, those with $y = 1, 2, 3, 4$. What is the estimate for subpopulation A? What is the variance estimate using only the 7 of the 28 units? What is the variance estimate if we treat it as stratified sampling with 4 units in subpopulation A? Why is there such a

large difference in variance estimates? Assume that 7 units were randomly selected in the population as above but now $y = 0, 0, 0, 0, 1, 2, 3$ with (1, 2, 3) falling in A. What are the $\hat{Y}$ estimates now? Why are they different?

19. Use the results of Theorem 3.5 to derive the results for Poisson sampling in Theorem 3.14.

20. Prove that the unadjusted Poisson estimator is unbiased. Also show by example that this can be a very unreliable estimator.

21. Prove Theorems 3.1, 3.2, and 3.3 from the results in Theorem 3.5.

22. Show why stratified sampling with proportional allocation should still be considered unequal probability rather than equal probability sampling.

23. If n_h/N_h is constant across all strata, why is proportional stratified sampling still an unequal probability sampling scheme?

24. Show that systematic sampling is generally more efficient than simple random sampling. (Hint: Set π_{ij} close to 0 for units not in the same cluster sample.)

25. Prove for pps sampling that if all the π_i are small ($i = 1, \ldots, N$), then $\Sigma_{i=1}^{N} \pi_i(1 - \pi_i) \doteq n$.

26. Prove for pps with replacement that eq. (3.11), (3.12), and (3.13) become (3.7), (3.8) and (3.9), respectively, for suitable definitions of the π_i. What is π_i for pps with replacement sampling if $p_i = x_i/X$?

27. Prove for post-cluster sampling that $\hat{Y} = (M/nm)\,\Sigma_{i=1}^{m} Y_{i.}$ is an unbiased estimator of $\bar{Y}$ and derive its true variance and an estimator of its variance (Ghosh 1963).

28. Write a computer program using SRS and pps sampling to select a sample from a population and compute $\hat{Y}_{HT}$ and $v(\hat{Y}_{HT})$. Use your program to select 1000 samples from each of the two given populations. Use a sample size of $n = 3$. For both sampling methods, compare the variance and bias of the estimators and discuss the results.

Population 1		Population 2	
x	y	x	y
1	2.3	1	51.3
2	4.3	2	43.1
3	7.2	3	45.6
4	9.9	4	43.0
5	9.1	5	43.5
6	11.2	6	41.2
7	12.7	7	38.6
8	12.3	8	34.2
9	16.8	9	36.9
10	22.7	10	31.3
11	20.3	11	30.8

Population 1		Population 2	
x	y	x	y
12	23.4	12	30.8
13	25.0	13	37.6
14	31.0	14	29.4
15	32.3	15	24.9
16	34.1	16	25.5
17	33.4	17	22.2
18	39.2	18	14.6
19	37.6	19	9.8
20	46.0	20	10.2

29. Discuss when cluster sampling may be superior to stratified sampling. Include in your discussion the availability of a population list, the heterogeneity and homogeneity of data within strata and clusters, and cost considerations.

4 Forest Sampling—Single Level

4.1 INTRODUCTION

We are now in a position to discuss some single-level sampling strategies. More sophisticated designs involving several levels of sampling are discussed in Chapter 5. Though the strategies we discuss have applications in many fields, we restrict our discussion to estimating forest parameters. We focus particularly on total standing volume but clearly this can be a surrogate for any other tree-related forest parameter. Set out below is the notation we use:

A	Size of the forest stand to be inventoried, usually expressed in hectares.
a	Area of the sample plot expressed in the same units as A in fixed-area plot designs.
N	Number of trees in the forest population.
u_i	ith tree in the population.
l_{1i}, l_{2i}	Coordinates of u_i on the plane of the forest.
l_{1s}, l_{2s}	Coordinates of a sampling location (e.g., a plot center) on the plane of the forest.
m	Number of distinct sampling locations established in an inventory.
$n(s)$	Number of (not necessarily distinct) trees in the sample.
N_k	Number of distinct trees in the sample at the kth sampling location.
a_i	Inclusion area associated with u_i expressed in the same units as A.
a_{ij}	Joint inclusion area associated with u_i, u_j expressed in the same units as A.
π_i	Inclusion probability associated with u_i at a single sampling location.
π_{ij}	Joint inclusion probability associated with u_i, u_j at a single sampling location.
v_{oi}	Bole volume of u_i.
V_o	Total bole volume of forest, $\Sigma_{i=1}^{N} v_{oi}$.
g_i	Basal area of u_i, usually expressed in square meters.
G	Total basal area, $\Sigma_{i=1}^{N} g_i$.
dbh_i	Breast height diameter outside-bark of u_i, usually expressed in centimeters.

$vbar_i$	Volume : basal area ratio, v_{oi}/g_i, of u_i.
h_i	Total height of u_i.
h'_i	Merchantable height of u_i.
h_{ci}	Critical height of u_i at l_{1s}, l_{2s}.
F	Basal area factor (BAF) in variable radius plot (VRP) sampling. It is the number of m^2/ha (or ft^2/acre) of basal area represented by a sample tree each time it is tallied.[1]
$c_i(h)$	Cross-sectional area of the stem of u_i at a height h above ground (outside-bark).
$h_i(d)$	Height above ground of the stem of u_i where the diameter is d (outside-bark).
m_t	Number of sample locations on a tree.

The meaning of many of the terms in the preceding list is elaborated in succeeding sections. We are ambiguous about the definition of v_{oi}. Depending on context, it may be the total stem-wood volume of the tree or its volume up to h'_i, and it may be expressed in cubic meters or in some unit of measure of anticipated product yield such as board feet.

4.2 EARLY SAMPLING METHODS

As indicated in Chapter 1, use of equal probability sampling procedures started early in the 20th century. For sampling small stands, systematic sampling selecting (say) every 20th tree or randomly selecting one out of every 20 trees by some method (e.g. from a set of numbered beads in one's pocket) was commonly used. Sampling large forested areas involved systematic patterns of plots, line plots or strips (Schumacher and Bull 1932, Hasel 1942b). For example, a 10% sample comprising two strips per 40 acres (16.2 ha), each strip one chain wide (20.1 m), was used to sample virgin stands in the national forests of California. (Clearly this is an example of systematic sampling as described in Section 3.2.3, where the strip is the sampling unit.) On cut-over land, a single 1- by 2-chain plot (0.08 ha) was cruised in the center of each 20 acres (8.1 ha) (1% sample). Other intensities of sampling were obtained by varying plot size.

The purpose of Hasel's (1942b) article was to popularize random selection of units and provide guidance on sampling intensity. His conclusions are reasonable. He noted that estimates of total forest volume tended to be more precise than separate estimates by species; that the sampling intensities required did not increase directly with size of the area but depended on other factors; that less intensive but carefully executed cruises tended to be less biased than more intensive, poorly executed cruises; and that

[1] There is some advantage in using $1/K^2$ instead of F where $K^2 g_i$ represents the inclusion area of tree i. Simply substituting $1/K^2$ for F should make the equations more readily understood by statisticians.

sampling using elongated plots cutting across the contours (which was already standard practice) was efficient because it maximized variability within the plots. Hasel's observation that an estimate of total volume is more reliable than separate estimates by species is reasonable because more variability exists between sample units for estimating volume by species than for total volume. The notion that lower sampling intensities can be used for large areas is apparent from the formula in Theorem 3.2. The sampling fraction plays less of a role for larger populations because forest type and condition largely determine the variability of the variable of interest. This is the idea behind stratification. Variation within strata, hopefully, is less variable than in the unstratified population. Elongated plots across contours tend to maximize variability within plots. The plots can be considered clusters of trees and, as indicated in Section 3.2.3, cluster sampling is statistically quite efficient if within cluster variability is maximized.

Canfield (1941) proposed a randomly selected line intercept method as an improved version of the line transect method. He used it to measure density and composition of herbaceous vegetation and shrubs but suggested that it could serve equally well to measure minor vegetation and regeneration in forests. The lines are visualized as having length and vertical dimensions only. Intercepted plants are measured. It is not clear whether Canfield appreciated that this is equivalent to selecting elements proportional to some function of their dimensions, but his reference to Klapp (1934) suggests that he might have been. In another early application of line intersect sampling, Schumacher and Bull (1932) discussed the estimation of forest area using the proportion of the length of line in a line-plot survey that fell within the type.

4.3 FIXED-AREA PLOT SAMPLING

Sampling using fixed-area plots is the oldest and still the most widespread method of sampling, certainly in tropical mixed-moist forests (Wood 1990a). It relies on establishing plots or quadrats of fixed dimensions, shape, and perhaps even orientation. The plots may be temporary (used once only) or permanent (revisited and measured in subsequent inventories).

Trees, or whatever other objects comprise the forest population to be inventoried, are included in the sample tally if they are within the boundaries of the sample plot. Under a random or random systematic arrangement of plots, trees of a certain type will be tallied in the sample with the same relative frequency as those which occur in the forested population. In other words, with a fixed-area plot sampling design, inclusion probability is proportional to frequency. It is worthwhile emphasizing that tree locations are fixed, as are the values of the characteristics associated with them. In a probabilistic framework, the only uncertainty associated with a tree prior to sampling is its inclusion in the sample.

To evaluate the inclusion probability of a tree in a randomly located plot, it is convenient to designate the plot's location by a single point; for example, when working with circular plots, the center of the plot serves as a convenient reference point for location. With rectangular plots, a corner point may be used as the locational reference. At each plot, the inclusion probability of a tree in these cases is proportional to the area that is defined by the locus of points where a plot can be located such that the tree will be tallied. We refer to the latter area as the tree's inclusion area a_i. For circular plots, it is straightforward to verify that this area is a tree-concentric circle of the same radius as the circular plot. The inclusion probability of a tree is its inclusion area divided by the area defined by the universe of points where plots can be located. Usually the latter area is just the forested area A itself. Thus, if a_i denotes the area of the plot for unit i in the same units of areal measure as A, then $\pi_i = a_i/A$, for all $i = 1, \ldots, N$. Exceptions to this scenario occur when tree inclusion areas are truncated by the stand boundary or when the plot can be located in an area larger than A. Consideration of these cases is postponed to the section of Chapter 7 dealing with boundary overlap and edge effect. For the present we assume that plots can be established only within the forested area to be inventoried, and we postpone the simplification that results when $a_i = a$ $(i = 1, \ldots, N)$.

The statistical characteristics of an estimator based on fixed-area plots are affected by both plot size and shape. For plots of the same size but different shape, the joint inclusion probabilities π_{ij} differ, but the first order inclusion probabilities do not. In practice the statistical characteristics may be affected, too, in that the shape of the plot affects accurate field work and the magnitude of measurement error. The choice of plot shape (circular, rectangular, square, etc.) usually depends on the inventory objective, terrain, forest composition, and tradition. In its favor, a circular plot minimizes the length of plot edge compared to any other shaped plot of equal area, thereby in principle minimizing the error associated with incorrectly tallying trees. However, on sloping terrain, the boundaries of rectangular plots may be easier to establish than those of circular plots. Rectangular plots were, and still are, often used in systematic strips surveys (Dalenius 1957). Line-plot surveys predate Schumacher and Bull (1932) at least. Other shapes of plots are not in widespread use, presumably because of the difficulty of the field procedure that would be required.

For a fixed number of plots (m), it stands to reason that the larger the size of the plot, the better is the precision of estimated characteristics. The sampling intensity (%) is reckoned in terms of the forest area sampled, namely $(ma/A)100\%$. Obviously as a increases, so does the sampling intensity and hence the precision. For a fixed intensity of sampling, one must choose m and a to minimize variance subject to cost constraints. For multiresource surveys, the optimal plot size for one or more characteristics may differ greatly from that for other characteristics, and a compromise solution must be made. Of necessity, this is often based on subjective criteria.

With fixed area plots, and the implied equal probability sampling design, we use the homogeneous linear estimator (Chapter 2). Specifically, if $Y = \Sigma_{i=1}^{N} y_i$ is the target parameter, then

$$\hat{Y} = \frac{1}{m} \sum_{k=1}^{m} \sum_{i=1}^{N_k} w_{ki} y_{ki} \tag{4.1}$$

where w_{ki} is a weight attached to the ith tree in the kth sample plot. For the present discussion we assume that w_i is a fixed constant, so that

$$E[\hat{Y}] = \frac{1}{m} \sum_{k=1}^{m} \sum_{i=1}^{N_k} w_i y_{ki} \pi_{ki} .$$

For unbiasedness, it is sufficient to make $w_i = \pi_{ki}^{-1}$, so that

$$\hat{Y} = \frac{1}{m} \sum_{k=1}^{m} \sum_{i=1}^{N_K} \frac{y_{ki}}{\pi_{ki}} .$$

Whereas the number of plots, m, in an inventory is usually fixed in advance, the size of the sample, $n(s)$, in terms of the number of population elements included, is random. Whether a tree is precluded from being tallied from more than one plot in a systematic design depends on the distance between plots. In random designs, plot boundaries may overlap, thus giving rise to the probability that a tree may be tallied from more than one plot.

Alternatively, the estimator $\hat{Y}$ may be expressed as the arithmetic mean of the individual plot estimates, that is,

$$\hat{Y} = \frac{1}{m} \sum_{k=1}^{m} \hat{Y}_k \tag{4.2}$$

where

$$\hat{Y}_k = A \sum_{i=1}^{N_k} \frac{y_{ki}}{a_{ki}} .$$

Since $\hat{Y}_k$ is a special case of $\hat{Y}_{HT}$,

$$\begin{aligned} V(\hat{Y}_k) &= \sum_{i=1}^{N} y_i^2 \left(\frac{1 - \pi_i}{\pi_i} \right) + \sum_{i \neq j}^{N} y_i y_j \left(\frac{\pi_{ij} - \pi_i \pi_j}{\pi_i \pi_j} \right) \\ &= \sum_{i=1}^{N} y_i^2 \frac{(A - a_i)}{a_i} + \sum_{i \neq j}^{N} y_i y_j \frac{(A a_{ij} - a_i a_j)}{a_i a_j} . \end{aligned} \tag{4.3}$$

Note that $N_k = \Sigma_{i=1}^{N} \delta_{ki}$ is random where δ_{ki} is the indicator of inclusion corresponding to the ith tree on the kth plot in the design, such that

$E[\delta_{ki}] = \mathrm{P}[\delta_{ki} = 1] = \pi_{ki}$. Then $V(\hat{Y}_k)$ can be unbiasedly estimated by

$$v(\hat{Y}_k) = \sum_{i=1}^{N_k} y_{ki}^2 \left(\frac{1 - \pi_{ki}}{\pi_{ki}}\right) \frac{1}{\pi_{ki}} + \sum_{i \neq j}^{N_k} y_{ki} y_{kj} \left(\frac{\pi_{kij} - \pi_{ki}\pi_{kj}}{\pi_{ki}\pi_{kj}}\right) \frac{1}{\pi_{kij}} \tag{4.4}$$

$$= A \sum_{i=1}^{N_k} y_i^2 \frac{(A - a_i)}{a_i} + A \sum_{i \neq j}^{N_k} y_{ki} y_{kj} \frac{(Aa_{kij} - a_{ki}a_{kj})}{a_{ki}a_{kj}a_{kij}}$$

if all $\pi_{kij} > 0$.

This variance estimator can be evaluated only by calculating the overlap areas among pairs of trees included in the tally of the kth plot.

If plots are located independently, the variance of $\hat{Y}$ is

$$V(\hat{Y}) = \frac{1}{m} \sum_{k=1}^{m} V(\hat{Y}_k).$$

In this situation, the observed variation among the $\hat{Y}_k$ $(k = 1, \ldots, m)$ estimates serves as an alternative unbiased estimator of $V(\hat{Y})$, namely

$$v(\hat{Y}) = \frac{1}{m(m-1)} \sum_{k=1}^{m} (\hat{Y}_k - \hat{Y})^2. \tag{4.5}$$

If plots are located systematically, then $V(\hat{Y})$ differs from the expression given above, and given its dependence on the spatial arrangement of trees in the forest, no general statement about the relative magnitudes of the variances of $\hat{Y}$ under random and systematic placement can be made (Palley and Horvitz 1961). Regardless of plot layout, $V(\hat{Y})$ is affected by:

1. Variation among the y_i values.
2. Spatial arrangement of the trees, u_i, and its consequent effect upon the π_{ij}.
3. Variation in N_k, the number of trees tallied in the kth sample plot.

These three sources of variation are, of course, confounded.

The traditional estimator of Y is

$$\tilde{Y} = \frac{A}{m} \sum_{k=1}^{m} \sum_{i=1}^{N_k} \frac{y_{ki}}{a}, \tag{4.6}$$

which differs from $\hat{Y}$ by the substitution of a for a_{ki}. $\tilde{Y}$ is biased to the extent that $\pi_{ki} \neq a/A$. The magnitude of the bias, which is never positive, depends on the spatial distribution of trees near the forest boundary.

It is informative to examine $\tilde{Y}$ when $y_{ki} = 1$, that is, when $Y = N$ and $\tilde{Y}$ is an estimator of tree frequency. In this situation, the variation among y_{ki} values is obviously removed from $V(\tilde{Y})$. Furthermore, since

$$\tilde{Y} = \frac{A}{a} \sum_{k=1}^{m} \sum_{i=1}^{N_k} \delta_{ki}$$
$$= An(s)/(ma) ,$$

then

$$V(\tilde{Y}) = \frac{A^2}{a^2 m^2} V\{n(s)\} .$$

This last expression makes it clear that the variation in $\tilde{Y}$ among plots is due solely to variable sample size.

4.4 VARIABLE RADIUS PLOT SAMPLING

Horizontal variable radius plot sampling (Bitterlich 1947) was introduced in forestry as a quick, low cost method of obtaining estimates of total basal area G (m^2/ha) of a forest by a simple counting technique. The procedure is also called angle count sampling, point sampling, plotless cruising, and Bitterlich sampling. We call it variable radius plot (VRP) sampling because this seems to be most descriptive of the method and does not conflict with terminologies used elsewhere. The widely used expression, angle count sampling, is not comprehensive enough since VRP is only a count sample if we are sampling with probability proportional to the variable of interest. The theory of VRP sampling was conceived by Bitterlich in the early 1930's but was not published until after the Second World War (Bitterlich 1947). The history of VRP sampling, its theory, significance, applications, developments, and future prospects are all covered in a fascinating book, *The Relaskop Idea: Relative Measurements in Forestry* (Bitterlich 1984). The method works as follows: An observer visits each of a number of sample points in the forest and counts the number of trees that, when observed with some instrument from the point, subtend an angle greater than some fixed critical angle α generated by the instrument. The instrument, called an angle gauge, can be one's thumb at a given distance from the eye, a simple rod with crosspiece or, most commonly now, a wedge prism or an instrument called a Relaskop (see Ch. 7). Trees are selected proportional to their cross-sectional area at the sighted point on the tree (usually breast height). The principle behind horizontal VRP sampling is illustrated in Figure 4.1. The small circles represent the cross-sectional areas of trees 1 and 2; the larger circles (a_1 and a_2) represent the areas within which the trees would be selected if a sample point fell within them barring truncation by the forest boundary (see Ch. 7). The area of overlap (a_{12}) of the two large circles represents the area within which, from any sample point, both trees would be selected.

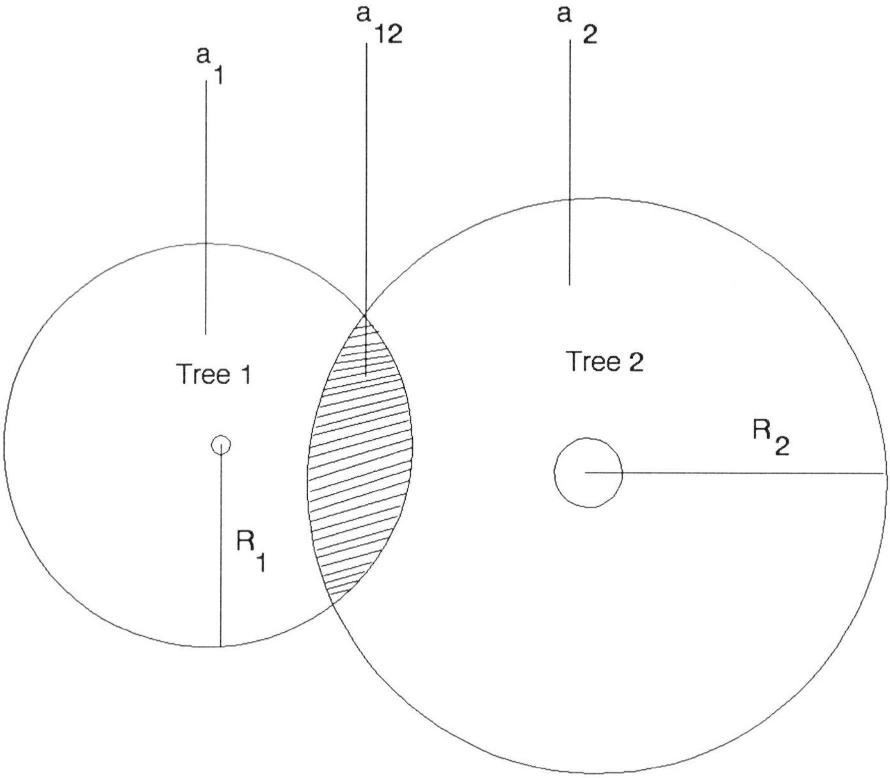

Figure 4.1 An illustration of inclusion areas with horizontal variable radius plot sampling. a_1, a_2 = areas where tree 1 and tree 2 will be selected in the sample; a_{12} = area where both trees 1 and 2 will be selected in the sample; R_1, R_2 = radii of VRP plots for trees 1 and 2 which are proportional in size to the tree diameters of the two trees.

Grosenbaugh (1952) was the first to formally realize that horizontal VRP sampling is an example of pps sampling. The estimator for any parameter of interest Y is

$$\hat{Y}_{HT} = \frac{1}{m} \sum_{k=1}^{m} \sum_{i=1}^{N_k} y_{ki}/\pi_{ki} \tag{4.7}$$

where $\pi_{ki} = g_{ki}/(FA)$. Note that the estimator can also be written as

$$\hat{Y} = \sum_{k=1}^{m} \hat{Y}_k/m \tag{4.8}$$

where

$$\hat{Y}_k = FA \sum_{i=1}^{N_k} y_{ki}/g_{ki}$$

is the Horvitz-Thompson estimator of Y at the kth point. Clearly, if $y_i = g_i$, then $\hat{Y} = FAn(s)/m$.

The variance of $\hat{Y}$ is

$$V(\hat{Y}) = \left[\sum_{i=1}^{N} y_i^2/\pi_i + \sum_{i \neq i'}^{N} y_i y_{i'} \pi_{ii'}/(\pi_i \pi_{i'}) - Y^2 \right] \Big/ m \tag{4.9}$$

with unbiased variance estimator

$$v(\hat{Y}) = \sum_{k=1}^{m} (\hat{Y}_k - \hat{Y})^2/\{m(m-1)\} . \tag{4.10}$$

Note that the inclusion probability of a tree being selected in both fixed-area (circular) plot sampling and horizontal VRP sampling depends only on the distance between the tree and the sample point. This distance is constant for all trees in the fixed-area sample but depends on g (tree basal area) in horizontal VRP sampling. Clearly a tape measure could be used for either but since the distance is tree-dependent for horizontal VRP sampling, it is more effective to use an angle gauge or other ocular device.

The variance and variance estimators of this estimator can be derived readily by treating horizontal VRP sampling as a special case of pps sampling where $n(s)$ is a random variable (Palley and Horwitz 1961) or by treating it as SRS of m out of M clusters (Schreuder 1970). In the first case, the variances can be obtained directly from Theorem 3.5 and in the latter from Theorems 3.2 and 3.3. For trees to be selected proportional to their cross-sectional area, a number of assumptions are required (Grosenbaugh 1958), namely:

- Trees are vertical and their cross sections are truly circular.
- Terrain is level. Note: compensation for slope is automatic with the Relaskop. With other angle gauges, compensation is easily made by dividing the estimate of G by the cosine of the maximum angle of slope at the sample point. Other methods of compensating for slope are discussed by Barrett and Nevers (1967) but all give the same result. The method outlined is considered the most practical by most users. Correction for slope is often ignored with slopes less than 5° (error $< 0.5\%$).
- Sample trees are visible from the sample point.
- Calibration constant of the angle gauge is known and the instrument is used correctly.
- The area within which each potential sample tree can be selected with the gauge lies entirely within the forested area or, if not, this is compensated for properly.

Grosenbaugh discussed how the failure of these assumptions affects the probability of selecting a tree and suggested how to avoid problems when an assumption is known to be false (see Ch. 7 for more detail).

It was quickly realized in forestry that the principles of horizontal VRP sampling could be applied to sample proportional to other variables of interest. For example, Hirata (1955) proposed vertical VRP sampling, in which trees are selected proportionally to height squared, to estimate mean squared tree height (clinometers with percent scales can be used for this—Abney, Haga, Suunto, etc.); and Strand (1957) introduced line sampling where random line segments are used instead of points so that a tree is selected by angle gauge proportional to diameter (horizontal gauge) or height (vertical gauge) on both sides of sample line segments located at random. Often continuous and parallel lines, analogous to strips, or short discontinuous segments, similar to rectangular plots, are used.

Zöhrer (1978, 1979), as noted in Bitterlich (1984), extended VRP sampling to a method called sampling with programmed probability (SPP) where sample trees are selected proportional to g^c for $-\infty < c < \infty$. His basic concept involves the computation of the area (a_i) of the limiting circle using the equation

$$a_i = g^c g_{\text{norm}}^{1-c}\, 10^4/F \tag{4.11}$$

where c = basal area exponent, and

g_{norm} = variable radius plot related basal area (m^2) for which F is valid.

This method includes horizontal VRP sampling ($c = 1$), fixed-area sampling ($c = 0$), all transitional forms between these two ($0 < c < 1$), "g-overproportional" ($c > 1$), and "g-inverse proportional" ($c < 0$) sampling. The Telerelaskop, in combination with pocket computers, can readily be used to implement sample-tree selection (Bitterlich 1984).

Schreuder (1978), using the sampling technique of Fender and Brock (1963), suggested a count sampling technique to estimate number of trees (N), basal area (G), and a linear combination of N, G, and sum of tree diameters (D). The method uses small (circular) fixed-area plots to estimate N, points with VRP sampling to estimate G, and VRP sampling on the fixed-area plots to estimate a linear combination of N, G and D. Steiner's relationship between two parallel convex areas (Masuyama 1953) underlies this idea. Basically, Steiner's relationship implies that the average count $\bar{Z}$, the average number of trees counted using a critical angle α over fixed-area plots with radius KC centered at the sampling point, is an unbiased estimate of $G/(AF) + \pi CD/(AF) + \pi C^2 N/(AF)$. Here C = radius of the fixed-area circular plot, and A = total area of stand.

An interesting simplification of Steiner's formula that deserves further investigation is to take the plot size as a multiple of the tree circumference c (Satyamurthi 1979). To implement the method in the field, a rope is used. Bitterlich (1984) thought that this may be a practical suggestion under some conditions, for example, in tropical forests where only a small number of large, definitely noncircular trees are to be measured, and use of optical instruments is impractical.

Van Deusen and Grender (1989) suggested an unbiased estimator for clusters of horizontal VRP samples. All sample trees occur only once in the estimator. This estimator's efficiency increases, relative to the traditional estimator, as the percentage of multiple-occurring trees in the sample increases. They gave the necessary formulas to compute the areas of overlap for clusters of three equidistant points with a fixed orientation.

Yang and Wang (1987) noted that horizontal line sampling using variable-width rectangular plots was first used on a large scale in Taiwan. They derived the expectations and variances of G and growth in G.

Schreuder *et al.* (1975) introduced instant-count sampling, an exact analog of VRP sampling, into the forest recreation literature. Visitors to a recreation site are sampled from a single vantage point and are selected proportional to their length of stay. This simple method is particularly efficient for estimating total use of scenic overlooks, wayside parks, historical sites, visitor information centers, and small picnic grounds and campgrounds.

A modification of this technique, called interval-count sampling, can be used for large recreation areas where one-way vehicular traffic is mandatory and travel time through the area cannot be ignored (Schreuder et al. 1975). Assume that l_1 is the time required to travel through the sample site against the traffic (to avoid double counting); then the count estimate in interval-count sampling,

$$\hat{Y}_{ic} = \sum_{j=1}^{n} y_j/n , \tag{4.12}$$

is an unbiased estimate of $(U + l_1N)/L_0$, where

y_j = count at sample interval j,
U = total use in visitor days,
N = total number of users, and
L_0 = length of recreation season.

Hence, an estimate of either U or N is required to allow estimation of U and N separately.

In VRP sampling, some points may yield none or only one or two sample trees. Some practitioners consider this undesirable since it is costly to travel to and establish a sample location. Beers and Miller (1964) recommended selecting that F for a population that will yield *on average* a certain number of trees (say 5–8) per point. A modification of their technique is to change F at a given point until the number of trees included is within the desired range, but this can cause appreciable bias (Wensel *et al.* 1980).

Schreuder *et al.* (1981) derive the expectation of the estimator of G under the assumption that the change in tree count resulting from a change in F is independent of the number of trees counted with the original F. One switch only is allowed to a smaller or larger F. This expectation for a sample of one point is:

$$E(\hat{G}) = G + G_1 + G_2 \tag{4.13}$$

where

$$G_1 = (F_2 - F_1)P(n_o \leq n_1)[G/F_2 - E(n_o|n_o \leq n_1)] \tag{4.14}$$

$$G_2 = (F_3 - F_2)P(n_o \geq n_2)[E(n_o|n_o \geq n_2) - G/F_2]\,, \tag{4.15}$$

where n_o = observed number of trees with standard $F = F_2$; F_1 and F_3 are, respectively, the F used when switching to a smaller and larger F; n_1 and n_2 are the number of trees at which we switch from F_2 to F_1 or F_3; $P(n_o \leq n_1)$ and $P(n_o \geq n_2)$ are, respectively, the probabilities that the number of trees observed (n_o) with F_2 is less than or equal to n_1 and greater than or equal to n_2; and $E(n_o|n_o \leq n_1)$ and $E(n_o|n_o \geq n_2)$ are the average number of trees, n_o observed if only points with less than $n_1 + 1$ and more than $n_2 - 1$ trees, respectively, are considered. Generally B_1 and B_2 are positive so that switching to both smaller and larger BAF's introduces positive bias. Thus estimates of G tend to be high with this procedure. Iles and Wilson (1988) provide further insights into the effects that changing F has upon tree tallies.

4.5 POISSON SAMPLING AND ALTERNATIVES

Poisson sampling (Hajek 1957) was introduced into the forestry literature as 3-P sampling by Grosenbaugh (1964). This method, discussed earlier in Section 3.4.2 under post-pps sampling, was proposed by Grosenbaugh for timber sales where trees to be cut must be selected and marked and therefore can be sampled for volume at the same time. In the original application, sampling was done proportional to a covariate, usually an ocular estimate of tree volume made by highly skilled cruisers. Properly done, Poisson sampling is highly efficient for estimating total volume since the covariate is highly correlated with tree volume. As indicated by Schreuder *et al.* (1968), Poisson sampling has the following disadvantages of which the forest manager should be aware:

1. The sample size desired is seldom achieved exactly, so the cost of the survey may be larger than expected (if the actual size attained is greater than planned) or the actual precision may be poorer than that desired (if actual size is smaller than planned). Of course, one can also view these disadvantages as advantages (respectively, better precision and lower cost than planned).

2. The adjusted estimator proposed by Grosenbaugh (1967a) is biased, but with proper precautions the bias is small.
3. Variance approximations proposed by Grosenbaugh (1967a) and Schreuder *et al.* (1971) based on pps-with-replacement sampling are good approximations of the variance (Grosenbaugh 1976) despite findings by Schreuder *et al.* (1971) to the contrary.

Furnival *et al.* (1987) (see also Ogus and Clark 1971 and Brewer *et al.* 1972) rewrote $\hat{Y}_a$ as

$$\hat{Y}_a = (\hat{Y}_u/\hat{X}_u)X \tag{4.16}$$

where $\hat{X}_u = \Sigma_{i=1}^{n(s)} x_i/\pi_i$ (analogous to $\hat{Y}_u$ in eq. 3.73).

They noted that this shows that $\hat{Y}_a$ is basically a ratio-of-means estimator $\hat{Y}_{rm}$ (eqn. 3.97). The bound on the bias of $\hat{Y}_{rm}$ (eq. 3.99) should be applicable to $\hat{Y}_a$, as should the approximate variance estimator for $\hat{Y}_{rm}$ such as $v_J(\hat{Y}_{rm})$ in eq. (3.109).

If an estimator based on the conditional probabilities of selection is desired, these probabilities need to be estimated. Hajek (1981, p. 132–133) suggested that simple or quadratic regression be used. Furnival *et al.* (1987) did this, defining a regression estimator of the conditional probability of selecting unit i in a sample of size $n(s)$ with expected sample size n_e as

$$\hat{\pi}_i^c = \pi_i + \hat{\beta}\{n(s) - n_e\} , \tag{4.17}$$

where $\hat{\beta} = \pi_i(1 - \pi_i)/V\{n(s)\}$ is the least squares estimator of $\mathrm{Cov}\{\delta_i, n(s)\}/V\{n(s)\}$.

The estimator of Y is

$$\hat{Y}_H = \sum_{i=1}^{n(s)} y_i/\hat{\pi}_i^c . \tag{4.18}$$

To investigate the effectiveness of the adjustment $\pi_i' = \pi_i\, n(s)/n_e$ in $\hat{Y}_a$ relative to π_i^c, Furnival *et al.* (1987) expressed $\hat{\pi}_i^c$ and π_i' as relatively adjusted probabilities, namely,

$$\Delta\pi_i^c = \frac{\hat{\pi}_i^c - \pi_i}{\pi_i} = (1 - \pi_i)\{n(s) - n_e\}/V\{n(s)\} \tag{4.19}$$

and

$$\Delta\pi_i' = \{n(s) - n_e\}/n_e . \tag{4.20}$$

They noted that the sign of the relative adjustment in these equations depends only on the term $n(s) - n_e$ and hence is the same for $\Delta\hat{\pi}_i^c$ and $\Delta\pi_i'$.

Sample inclusion probabilities are increased when $n(s)$ exceeds n_e and are decreased when n_e exceeds $n(s)$. Both adjustments are zero when $n(s) = n_e$. When $n(s) \neq n_e$, the relative adjustment of π_i to $\hat{\pi}_i^c$ depends on the values of the π_i. For a given $n(s)$, the relative adjustment is less for larger values of π_i. For $\Delta\pi_i'$, the relative adjustment for a given sample size $n(s)$ is independent of π_i. Hence, the differences in adjustments are most evident for larger values of π_i. For example, π_i approaches unity for the largest units in the population when X/n_e approximates the maximum x-value in the population. The regression adjusted probability is sensitive to this and does not correct much for such units, whereas $\Delta\pi_i'$ (and hence π_i') does if $n(s)$ is very different from n_e.

Furnival *et al.* (1987) compared the adjusted estimator $\hat{Y}_a$ (eq. 4.16) with $\hat{Y}_H$ (eq. 4.18) and with a ratio-of-means type estimator using regression-based corrected probabilities, namely,

$$\hat{Y}_{Hr} = (\hat{Y}_H/\hat{X}_H)X \tag{4.21}$$

where $\hat{Y}_H$ is defined in eq. (4.18) and $\hat{X}_H$ can be defined similarly. Estimator (4.21) combines the benefits of regression adjustments of the inclusion probabilities with the stability of the ratio estimator ($\hat{Y}_{rm}$). As a benchmark, Furnival *et al.* also used the impractical but unbiased estimator $\hat{Y}_c$, which uses the true conditional probabilities π_i^c ($i = 1, \ldots, N$) where

$$\hat{Y}_c = \sum_{i=1}^{n(s)} y_i/\pi_i^c\ . \tag{4.22}$$

In a simulation study of populations ranging in size from 9 to 144 units and exhibiting a large range of proportionality between the variable of interest and the variable used for probability selection, their results showed that the adjusted estimator $\hat{Y}_a$ is as efficient or more efficient than its competitors except when the unconditional probabilities of inclusion approached unity for some population units or when y_i and x_i were far from proportional. In the latter case $\hat{Y}_H$ and $\hat{Y}_c$ offer substantial improvement over $\hat{Y}_a$ but their performance is erratic relative to $\hat{Y}_a$ and the degree of improvement over $\hat{Y}_a$ decreases with increasing population size.

Another unbiased estimator of Y with Poisson sampling is $\hat{Y}_{s1}$, a special case of $\hat{Y}_{sr}$ in eq. (3.172):

$$\hat{Y}_{s1} = \frac{1}{n(s)} \sum_s y_i/\pi_i^* \tag{4.23}$$

where

$$\pi_i^* = \sum_{s \ni i} p(s)/\{n(s) + 1\}$$

(Ouyang and Schreuder 1992) where $s \ni i$ indicates all samples containing unit i.

When sampling with a fixed sample size $\hat{Y}_a = \hat{Y}_{s1} = \hat{Y}_{HT}$ but, for Poisson sampling, $\hat{Y}_a \neq \hat{Y}_{s1}$. $\hat{Y}_{s1}$ is unbiased but the π_i^* are complex and need to be approximated, for example,

$$\pi_i^* = \pi_i \sum_{s \in \Omega_{-1}} p(s)/(n+1)$$

$$= \pi_i \sum_{k=0}^{N-1} \frac{1}{|k+1|} P(n(s) = k | s \in \Omega_{-i}) \tag{4.24}$$

where Ω_{-i} is the population of units with i deleted, and $P\{n(s) = k | s \in \Omega_{-i}\}$ is the probability of selecting a Poisson sample of size k from Ω_{-i},

$$P\{n(s) = k | s \in \Omega_{-i}\} \doteq (d_i^{-1/2}) \phi(r) \left[1 + \frac{1 - 2\bar{\bar{\pi}}_i}{6\sqrt{d_i}} (r^3 - 3r) + \frac{1}{4!} \right.$$

$$\left. \times \frac{1 - 6d_i^*}{d_i} (r^4 - 6r^2 + 3) + \frac{10}{6!} \frac{(1 - 2\bar{\bar{\pi}}_i)^2}{d_i} (r^6 - 15r^4 + 45r^2 - 15) \right]$$

with

$$d_i = \sum_{j(\neq i)}^{N} \pi_j (1 - \pi_j)$$

$$n_{e-i} = \sum_{j(\neq i)}^{N} \pi_j$$

$$r = \frac{(k - n_{e-i})}{\sqrt{d_i}}$$

$$\phi(r) = \exp\left(\frac{-r^2}{2}\right) \Big/ \sqrt{2\pi}$$

$$\bar{\bar{\pi}}_i = \frac{\sum_{j(\neq 1)}^{N} \pi_j^2 (1 - \pi_j)}{\sum_{j(\neq i)}^{N} \pi_j (1 - \pi_j)} \quad \text{and} \quad d_i^* = \frac{\sum_{j(\neq i)}^{N} \pi_j^2 (1 - \pi_j)^2}{\sum_{j(\neq i)}^{N} \pi_j (1 - \pi_j)}$$

(Ouyang and Schreuder 1992). In addition to $\hat{Y}_{s1}$, special cases of the general ratio estimator $\hat{Y}_{gre}$ in eq. (3.176) can also be used as alternatives to $\hat{Y}_a$, that is,

$$\hat{Y}_{gre1} = \left[\hat{Y}_a \Big/ \left\{ \sum_{i=1}^{n(s)} y_i^* / \pi_i (n_e / n_a) \right\} \right] Y^* \tag{4.25}$$

where $y_i^* = \hat{\alpha} + \hat{\beta} x_i$, and $\hat{\alpha}$ and $\hat{\beta}$ are weighted regression coefficients estimated from the sample, for example eq. (3.137) and (3.140) can be used.

Ouyang and Schreuder (1992) found in a simulation study across 21 populations that $\hat{Y}_{s1}$ is about equally as efficient as $\hat{Y}_a$. $\hat{Y}_{grel}$ tended to be considerably more efficient and only slightly more efficient, respectively, than $\hat{Y}_a$ when sample-based information and independent information were used to compute y_i^* $(i = 1, \ldots, N)$.

Schreuder *et al.* (1971) proposed pps sampling alternatives to Poisson sampling based on sampling from the cumulated x procedures discussed in Section 3.4.2. Their first alternative, based on systematic sampling from the cumulated measures of size, shares with Poisson sampling the properties that sampling is without replacement and that the probabilities of selection are proportional to x_i. The procedure is as follows:

1. Decide on a sampling interval L.
2. Choose a random integer δ from the interval $0 < \delta \leq L$, and list the n_t numbers $\{\delta + hL\}$ corresponding to $h = 0, 1, 2, \ldots, (n_t - 1)$.
3. Visit each unit in the population and observe x_i on it. The units may be visited in any order that is entirely independent of the list of numbers $\{\delta + hL\}$. As enumeration proceeds, cumulate the x_i so that the cumulated value at the jth unit visited is $T_j = \Sigma_{i=1}^{j} x_i$ and T_0 is defined to be 0.
4. Unit j is included in the sample (and y_j is measured on it) if, and only if, one of the set of numbers $\{\delta + hL\}$ falls in the interval between the integers $T_{j-1} + 1$ and T_j (inclusive).
5. When all units have been visited, the actual sample size will be $n(s) = X/L$. To carry out this systematic selection, however, X need not be known in advance; it is only necessary to choose L.

To apply the procedure in the field, visit the first tree, observe x_1, and calculate $T_1 = x_1$. If one previously listed element of the set of numbers $\{\delta + hL\}$ is between 1 and x_1 (inclusive), then measure y_1. Otherwise, move to tree 2, observe x_2, and calculate $T_2 = x_1 + x_2$. If one of the set of numbers $\{\delta + hL\}$ is between $x_1 + 1$ and $x_1 + x_2$ (i.e., between $T_1 + 1$ and T_2 inclusive), then measure y_2. Proceed in this way until all trees are visited.

If $x_i \leq L$ for $i = 1, 1, \ldots, N$, the probability that the ith unit is included in the sample is x_i/L and sampling is without replacement. If any unit has $x_i > L$, it should be selected in the sample with certainty, and its values of x are omitted in the cumulation. Then, for the remaining units, sampling is with probabilities proportional to the x_i and without replacement.

One would ordinarily choose L to attain a prespecified value of the sample size, n_t (say), by guessing the value of X (say, $\tilde{X}$) and taking $L = \tilde{X}/n_t$. The actual value of n will be $n_t X/\tilde{X}$. If we guess X correctly, then

$n(s) = n_t$. With Poisson sampling, one might select L to try to attain a prespecified sample size, n_t^*, by guessing the value $\tilde{X}$ and choosing $L = \tilde{X}/n_t^*$. The expected value of $n(s)$ will be $n_t X/\tilde{X}$, which is equal to the actual value of $n(s)$ for the systematic method if $n_t = n_t^*$. However, with Poisson sampling, deviations between the actual sample size, $n(s)$, and the expected sample size, n_e, cause a possible additional discrepancy between planned and actual sample sizes. In practice, this is bothersome but acceptable (Seppälä 1971). Wood and Wiant (1992a) pointed out that in most inventories, particularly large ones, the discrepancy is of minor importance. However, it can be of concern in small inventories because if the size of the second-stage sample deviates by, say, 20% from the planned size (which is not improbable), then either the precision of the final estimate will be less than planned or the total cost of the inventory will be greater than anticipated.

With the systematic procedure, $\hat{Y}_{HT}$ (eq. 3.7) is an unbiased estimator of Y if $x_i \le L$ for all units. (Otherwise, units with $x_i > L$ are considered to form a separate stratum, and estimation proceeds in the usual way.) Approximations for $V(\hat{Y}_{HT})$ and $v(\hat{Y}_{HT})$ are given in eq. (3.12) and eq. (3.13), respectively. These approximations are based on the assumption that the distribution of the estimator under systematic pps sampling approximates the distribution of (3.11) under ordinary pps sampling with replacement.

A second alternative, based on sampling from the cumulated measures of size, selects units with replacement, but with single-draw probabilities proportional to the x_i. The procedure is as follows:

1. Make an estimate $\tilde{X}$ of X. It is desirable to choose $\tilde{X}$ so that $\tilde{X} \ge X$ but as close to X as possible.
2. Select n_0 integers $(d_1, d_2, \ldots, d_{n0})$ at random and with replacement, such that $1 \le d_i \le \tilde{X}$, and arrange them in ascending order.
3. Cumulate the x_i as the units are observed in arbitrary order, defining the cumulated value at the jth unit as

$$T_j = \sum_{i=1}^{j} x_i \quad \text{with} \quad T_0 = 0 .$$

4. Include unit j in the sample (and measure y_j on it), if, and only if, at least one of the random integers $(d_1, d_2, \ldots, d_{n0})$ falls in the interval between the integers $T_{j-1} + 1$ and T_j (inclusive).
5. If $\tilde{X} < X$, then adopt modifications analogous to those described in Schreuder *et al.* (1971) for equal probability sampling when $\tilde{N} < N$.

If $\tilde{X} \ge X$, this alternative will yield a with-replacement sample of size $n \le n_0$, and the ith unit will be selected with single-draw probability proportional to x_i.

Assuming that $\tilde{X} \geq X$, the estimator $\hat{Y}_{HT}$ given in eq. (3.7) is an unbiased estimator of Y, eq. (3.12) is the exact expression for the variance of $\hat{Y}_{HT}$, and eq. (3.13) is an unbiased estimator of the variance. Sampling with this design is with replacement, which will be a disadvantage if the sampling fraction is moderately large. For many forestry applications, however, the sampling fraction is likely to be very small. Thus this random method may prove useful. Stratified sampling alternatives to Poisson sampling are given in Schreuder (1975).

4.6 CRITICAL HEIGHT SAMPLING

The critical height of a tree included in a horizontal VRP sample may be defined as the height above ground of the cross section of the bole whose diameter is borderline when viewed by the angle gauge (e.g., Relaskop) from the sample point on the ground.

Critical height sampling (CHS) extends VRP sampling to the third dimension and enables an unbiased estimate of stand volume to be determined directly using a simple formula derived by Kitamura (1962), namely,

$$\hat{V}_o = \frac{FA}{m} \sum_{k=1}^{m} \sum_{i=1}^{N_k} h_{cki} \tag{4.26}$$

where $\hat{V}_o$ = stand volume (m^3/ha) and h_{cki} = critical height of ith tree on the kth point included in the VRP sample.

Although it eliminates the need for volume tables or equations, Kitamura's method received little support from practicing foresters, who found it difficult to locate and measure the critical height of trees included in a VRP sample because the sighting angles are steep on trees near the sample point and, for all trees, vision is often obscured by branches and foliage. Another disadvantage is that CHS makes no allowance for bark thickness. Thus from 1962 until the mid-1980's, the technique was considered to have little practical value. In recent years, however, there has been a resurgence of interest in the method because of the increasing availability of small, lightweight, hand-held computers and the realization that if a reliable underbark taper function is available for the species and locality, the critical height of a sample tree can be derived indirectly from the function, thus avoiding the problems associated with measurement. Alternatively, if an established taper function is not available, one can be derived on site by Telerelaskop using the "three point method" (Bitterlich 1984), but this is more time consuming and makes no allowance for bark thickness.

CHS was popularized in English by Iles (1979, 1981, 1988, 1990). Iles views each tree included in a VRP sample as surrounded by a three-dimensional space, called the expanded tree stem, whose profile is determined by the diameter and height of the tree, taper of the bole, and

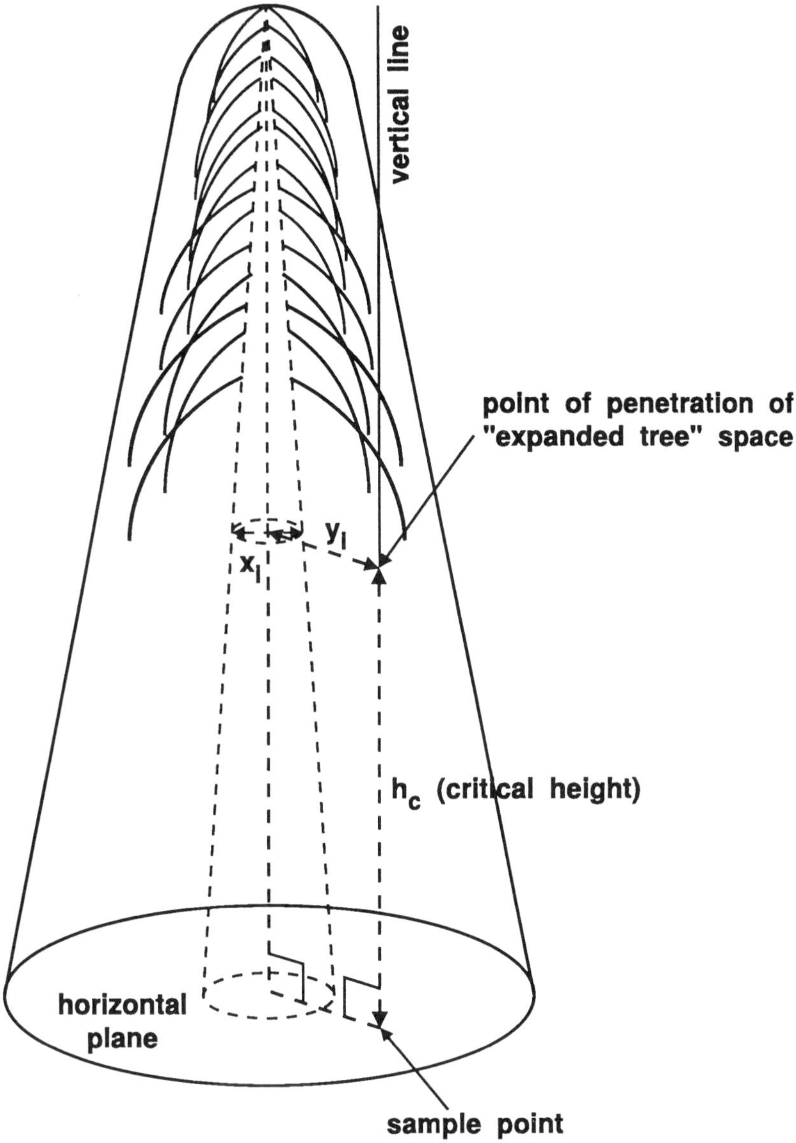

Figure 4.2 Formula for generating surface of "expanded tree" space $y_i = (50x_i/\sqrt{F})$ (y_i and x_i in meters, F in m^2/ha). Adapted from Iles 1988, 1990.

strength (F) of the angle gauge (Figure 4.2). The expanded tree stem is such that, at any height h, the horizontal cross-sectional area of the expanded stem is equal to $c(h)/F$, where $c(h)$ is the cross-sectional area of the actual stem. The observed critical height is an estimate of the volume-basal area ratio (*vbar*) of the tree (Iles 1979). In effect, the expanded tree stem is the space occupied by the solid of revolution that would be generated about the tree axis if each point along the bole from base to tip were viewed with the angle gauge to establish the borderline distance (i.e., the horizontal radius of revolution at each point). Because the tree stem consistently decreases in size with increasing height, the critical height of a sample tree is the vertical distance from the sample point on the ground (assumed level with the base of the tree) to the point of penetration of the surface of the 'expanded tree stem' or enveloping solid of revolution.

There are several ways to show that eq. (4.26) is unbiased without assuming a particular shape or geometric form for the tree profile. One can envisage the distribution of wood over the landscape as a two-dimensional surface. Measuring the depth of wood at a randomly located point provides an estimate of the average depth of the surface over the landscape. Such an estimate may be poor because of the spiked nature of the surface and the expanded boles are a means to smooth the surface somewhat.

Statistical proof of unbiasedness can be obtained by using indicator functions $\delta_{ki}(h)$ for a point in the plane, where

$$\delta_{ki}(h) = 1 \quad \text{if the } i\text{th tree is ``in'' at a height } h \text{ above ground}$$
$$= 0 \quad \text{otherwise.}$$

If the point on the plane (l_{1s}, l_{2s}) is randomly selected, then $\delta_{ki}(h)$ is a random function. The critical height for the ith tree is

$$h_{cki} = \int_0^{h_v} \delta_{ki}(h)\, dh$$

where h_v is any value at least equal to the tree height. Since $E[\delta_{ki}(h)] = \mathrm{P}(\delta_{ki}(h) = 1) = c_{ki}(h)/(AF)$, then

$$\begin{aligned} E(h_{cki}) &= \int_0^{h_v} E[\delta_{ki}(h)]\, dh \\ &= \int_0^{h_v} c_{ki}(h)\, dh/(AF) \\ &= \frac{v_{oki}}{AF}\,. \end{aligned}$$

Consequently, $E[FA\, \Sigma_{i=1}^{N_k} h_{cki}] = \Sigma_{i=1}^{N} v_{oi} = V_o$. Since $E[h_{cki}|\phi_{ki}] = E[h_{cki}]/\mathrm{P}(\phi_{ki})$, where ϕ_{ki} = {event that the ith tree is tallied at the kth point}, then it is straightforward to derive

$$E(h_{cki}|\phi_{ki}) = v_{oki}/g_i\,,$$

which is the *vbar* of the ith tree.

Usually, the decrease in taper of a tree bole is fairly uniform from base to tip and only one vertical distance has to be measured; but if a tree should have abnormal taper causing penetration of the enveloping "expanded tree" space at two points, it is necessary to sum two vertical lengths to derive the critical height.

Estimation of volume by the critical height method is appealing because a tree's probability of selection in the sample is proportional to its volume. This property is sacrificed when a taper equation is used instead of direct measurement of critical heights. If the taper function is correct on average, trees of the same size are sampled proportional to volume, on average, but this introduces more variability into the estimator of stand volume. Any bias in the taper function used to describe the average taper in the stand will be reflected in the stand volume estimator, too.

Bitterlich (1984) describes another unbiased method for estimating stand volume that simplifies Kitamura's basic concept of critical height sampling. The method, developed by Ueno (1979, 1980), is a two-stage sampling process based on the premise that tree volume density (m^3/ha) reflects the probability (density ratio) that a point in a specified three-dimensional space will fall within the space occupied by tree boles. The specified total possible space is generated by imagining a "box" enveloping the forest stand of interest and just higher than the tallest tree within it. Put simply, the volume of the stand is given by multiplying the volume of the "box" by the density ratio. Ueno's formula is

$$\hat{V}_o = F(h_{\max}/m) \sum_{k=1}^{m} \sum_{i=1}^{N_k} \delta_{ki} \tag{4.27}$$

where V_o = stand volume (m^3/ha),

F = BAF of angle gauge (m^2/ha),

$h_{\max}$ = chosen height (m), a little greater than the maximum tree height in the stand,

m = number of sample points in the stand,

N_k = number of sample trees associated with sample point k ($k = 1, \ldots, m$), and

δ_{ki} = an indicator variable that takes the value 1 or 0 depending on whether the tree is in the sample or not.

The field procedure involves applying horizontal VRP sampling at each sample point and subsampling the included trees using a random point that can move vertically along their central axes, that is, a random height ($0 < h < h_{\max}$) is generated for each tree. Then, instead of measuring the critical height of a given sample tree, the observer estimates by eye whether it is greater (count 1) or smaller (ignore) than the generated random height. If the decision cannot be made ocularly, the critical height must be measured, say by Relaskop, but only accurately enough to enable a decision to be made. This method resembles point-Poisson sampling (Bitterlich 1984), discussed in Chapter 5.

Bitterlich (1984) points out that the subsample can be taken a number of times (r) on each sample tree, in which case m in Ueno's formula is replaced by mr. The sampling error of the volume estimate derived by Ueno's method is calculated from the variability of the estimates determined at the m first-stage sample points.

Little has been done with Ueno's technique. It appears to be counter-productive from a practical point of view to select a tree in the first stage sample only, even though ocular heights are also available for trees excluded from the second-stage sample. More than one subsample may need to be taken for each sample tree to offset a likely decrease in precision relative to critical height sampling.

4.7 IMPORTANCE SAMPLING

Valentine *et al.* (1984) introduced importance sampling (IS) into the forestry literature in connection with the estimation of tree biomass and mineral content. It was further developed by Gregoire *et al.* (1986) to estimate bole volume of standing trees and by Furnival *et al.* (1986) to estimate volumes of bucked logs. The common theme in all these applications is that the characteristic to be estimated is the integral of some other characteristic. For example, bole volume between two points, say h_l and h_u, is the integral of bole cross-sectional area over that range. Importance sampling can be regarded as a continuous analog of sampling discrete units by pps with replacement sampling. Its genesis is in the Monte Carlo and simulation literature (see Hammersley and Handscomb 1964 or Rubinstein 1981).

To illustrate its use we consider its application to estimate tree bole volume v_o.

Letting $c(h)$ denote bole area at height h ($h_l \le h \le h_u$), then

$$v_o = \int_{h_l}^{h_u} c(h)\, dh \,. \tag{4.28}$$

Given a taper function $t(h)$ that approximates $c(h)$, then

$$v_{o1} = \int_{h_l}^{h_u} t(h)\, dh \tag{4.29}$$

is an approximation, generally biased, to v_o. In the IS literature, $t(h)$ is known as a proxy function, and it is used to control the point(s) at which the integrand $c(h)$ is to be measured. If a tree is to be sampled independently at m_t points, where $h = \theta_i$ ($i = 1, \ldots, m_t$) denotes the heights of these points on the stem, then the θ_i are solutions to

$$\begin{aligned} \mu_i &= \int_{h_l}^{\theta_i} \frac{t(h)}{v_{oi}}\, dh \\ &= \int_{h_l}^{\theta_i} f(h)\, dh \,, \qquad \text{where} \qquad \mu_i \sim U[0, 1] \,. \end{aligned} \tag{4.30}$$

As explained by Rubinstein (1981), the θ_i are selected according to density $f(h)$. The IS estimator of v_0 is thus

$$\hat{v}_{oIS} = \frac{1}{m_t} \sum_{i=1}^{m_t} \frac{c(\theta_i)}{f(\theta_i)} = \frac{v_{o1}}{m_t} \sum_{i=1}^{m_t} \frac{c(\theta_i)}{t(\theta_i)} . \tag{4.31}$$

For illustration, consider the case when $m_t = 1$. Then $\hat{v}_{oIS}$ is an adjustment of the proxy volume v_{o1} up or down, depending on whether and by how much the measured cross-sectional area is greater or lesser than that predicted by the proxy function. Its variance when $m_t = 1$ is

$$\begin{aligned} V(\hat{v}_{oIS}) &= E[\hat{v}_{oIS}^2] - (E[\hat{v}_{oIS}])^2 \\ &= \int_{h_l}^{h_u} \frac{c^2(h)}{f^2(h)}\, f(h) dh - \left(\int_{h_l}^{h_u} \frac{c(h)}{f(h)}\, f(h)\, dh \right)^2 \\ &= \int_{h_l}^{h_u} \frac{c^2(h)}{f(h)}\, dh - v_o^2 . \end{aligned} \tag{4.32}$$

When $m_t > 1$ independent sample points are taken, then

$$V(\hat{v}_{0IS}) = \frac{1}{m} \left[\int_{h_l}^{h_u} \frac{c^2(h)}{f(h)}\, dh - v_o^2 \right] , \tag{4.33}$$

which is estimated unbiasedly by

$$v(\hat{v}_{oIS}) = \frac{1}{m_t(m_t - 1)} \sum_{i=1}^{m_t} \left(\frac{c(\theta_i)}{f(\theta_i)} - \hat{v}_{oIS} \right)^2 . \tag{4.34}$$

Barring measurement error, $\hat{v}_{oIS}$ is unbiased regardless of how well or poorly $t(h)$ approximates $c(h)$; however, the better the approximation the better the precision of $\hat{v}_{oIS}$. For estimating volume of standing trees, IS has an important practical advantage. Points of measurement tend to be selected in the lower bole where volume per unit length is accumulated more quickly than it is higher up. However, points occasionally still fall within the crown.

For estimating stand volume, IS can constitute the last stage of a multistage design. Wiant *et al.* (1989) and Wood and Wiant (1992b) successfully applied IS to estimate the volume of sample trees of radiata pine and several species of *Eucalyptus*, respectively. When the procedure was combined with Poisson sampling, it gave reliable estimates of stand volume with a reduction in dendrometry of 96% (radiata pine) and 84% (*Eucalyptus* species) compared to the conventional procedure. Importance sampling has considerable potential for inventorying forests for which tree and stand volume functions are unavailable. A practical guide for applying the procedure has been prepared by Wiant *et al.* (1992b).

Gregoire *et al.* (1987) and Van Deusen (1990) considered control-variate sampling as an alternative to IS for estimation of bole volume, and Van

Deusen and Lynch (1987) considered the use of antithetic-variate sampling as yet another alternative. In control-variate sampling, an additive adjustment is made to the integrated proxy function, whereas IS involves a multiplicative adjustment. For estimation of volume increment, Gregoire *et al.* (1987) explained how to apply both IS and a modification of IS known as correlated sampling. Wood *et al.* (1990) used IS sampling ideas as the basis for centroid sampling, which is a variant of IS where cross-sectional area is measured at the centroid of the bole, the height of which is estimated from $t(h)$ (see Sect. 4.11.2).

4.8 LINE INTERSECT SAMPLING

Basically, line intersect sampling is a sampling method where units are sampled if they are intersected by a randomly chosen line segment called a transect. It has an advantage over plot sampling in that no plot needs to be established nor does one need to determine what units fall in each plot. The method has particular use for sampling the frequency of plants, shrubs, tree crowns, nearly stationary animals, animal dens or signs, roads, logs, forest debris, particles on a microscope slide, particles in a plane section of a rock or metal sample, and so on (Kaiser 1983). A detailed review of use of the method in forestry was given by De Vries (1986, Ch. 13).

Warren (1990) noted that line intersect sampling provides a natural measure of the total length of objects of interest in a stand (such as logs) just as VRP sampling does for stand basal area. However, VRP sampling requires measurement of the basal area of all selected trees to estimate parameters other than basal area, whereas line intersect sampling requires measurement of the length of objects to estimate parameters other than their total length. Warren cited Kotz and Johnson (1985, p. 70–74) as giving a clear description of the distinction between line intersect, line intercept, and line transect sampling, but then noted that De Vries (1979) showed that the line transect estimators used can be derived from line intersect theory. Line intersect sampling is covered in more detail in Chapter 8, Section 8.2.1.

4.9 RANKED SET SAMPLING

Ranked set sampling (RSS) was introduced by McIntyre (1952) to estimate pasture yields of forage. The procedure is as follows:

1. Select n units at random from the population and rank these n units from 1 to n by inspection or judgment regarding the variable of interest. Measure the variable of interest y on the lowest ranked unit in the group.

2. Repeat the above with an additional $(n-1)$ randomly selected sets of size n but measure the second, third, . . . , nth ranked units in these groups.
3. Repeat steps 1 and 2 m times so that a total of mn units is measured. A logical modification of this is to use differential sampling rates so that m_i units of rank i are measured, so

$$\sum_{i=1}^{n} m_i = mn \, .$$

Thus RSS employs n strata ($=n$ rank classes) where m_i ($i=1, \ldots, n$) units are selected from stratum i. It is usually difficult to rank many observations, so n is usually kept small and m (or m_i) increased if a large sample of measured units is wanted. Assuming perfect ranking, McIntyre examined the relative precision of the RSS and SRS estimators of the mean using the same sample size n. He defined the relative precision as RP = variance of SRS estimator/variance of RSS estimator. He found that RP was generally slightly less than $(n+1)/2$ for several distributions. Takahasi and Wakimoto (1968) showed that for any distribution with finite variance, RP is bounded between 1 and $(n+1)/2$ and only the uniform distribution achieves the upper bound.

Halls and Dell (1966) visually ranked three closely grouped quadrats of pasture into high, intermediate, and low forage weight classes and concluded that RSS was more efficient than SRS for estimating pasture yield. Martin *et al.* (1980) arrived at a similar conclusion when estimating shrub phytomass in forests, their ranking being based on ocular estimates. They found little gain in using variable m_i over constant m for each group. Cobby *et al.* (1985) conducted four experiments to assess the performance of RSS relative to SRS to estimate herbage mass and clover content in grazed swards. They did not observe the expected theoretical gains in efficiency, due to the method of selection of quadrats and the nature of the distribution of herbage biomass.

RSS is actually a post-stratification sampling scheme where sample size in each stratum can be controlled (Dell and Clutter 1972). For example, Schreuder (1975) suggested RSS as a stratified sampling alternative to Poisson sampling for estimating volume in timber sales. Dell and Clutter (1972) extended ranked set theory to the situation in which the ordered observations are judgments, which means that the ordering can be in error. When there are errors it is essential that the erroneous rankings are not related to the process of selecting sample units. Dell and Clutter showed that, regardless of the presence of errors in ranking, the arithmetic mean $\hat{\mu}$ of the mn ranked set observations is an unbiased estimator of $\bar{Y}$, and that $V(\bar{y})/V(\hat{\mu}) \geq 1$ where $\bar{y}$ is the sample mean of a SRS of mn observations. In contrast to the conclusion of Martin *et al.* (1980) above, Dell and Clutter's simulations showed that variable m_i increased precision considerably over constant m. Both Dell (1969) and Dell and Clutter (1972) note that ranking without error leads to the well-known topic of order statistics (David 1981). Stokes (1980) proposed a consistent but biased variance estimator $v(\hat{\mu})$ of $V(\hat{\mu})$, namely,

$$v(\hat{\mu}) = \sum_{i=1}^{m} \sum_{r=1}^{n} [y_{[r]i} - \hat{\mu}]^2/[mn(mn-1)] \tag{4.35}$$

where $\hat{\mu} = \Sigma_i \Sigma_r y_{[r]i}/(mn)$ and $y_{[r]i}$ = rth measured value of the ith cycle (out of the m possible cycles). The expected value of this estimator is

$$Ev(\hat{\mu}) = V(\hat{\mu}) + \sum_{r=1}^{n} \gamma_{[r]}^2/[n^2 m(nm-1)]$$

where $\gamma_{[r]} = \mu_{[r]} - \bar{Y}$ and $\mu_{[r]} = E(X_{[r]})$.

Stokes gives a simple proof that $\hat{\mu}$ is generally more efficient than $\bar{y}$. If $m = 1$ and if $x_{(1)}, \ldots, x_{(n)}$, $y_1, \ldots, y_n$, and $y_{(1)}, \ldots, y_{(n)}$ are the ranked set, random sample, and ordered random samples, respectively, then

$$\begin{aligned} n^2 V(\bar{y}) &= V\left(\sum_{r=1}^{n} y_{(r)}\right) \\ &= \sum_{r} \sigma_{[r]}^2 + \sum_{r<s} \operatorname{cov}(y_{(r)}, y_{(s)}) \\ &= n^2 V(\hat{\mu}) + 2 \sum_{r<s} \operatorname{cov}(y_{(r)}, y_{(s)}) \\ &\geq n^2 V(\hat{\mu}) \end{aligned}$$

because order statistics from a single sample are positively correlated. Since measures of dispersion such as the variance are generally functions of differences rather than sums of random variables, the reliability of estimating these measures may suffer (Stokes 1980). This is because the variance of a difference requires subtraction rather than addition of the covariance terms. Stokes proves the following theorem

Theorem 4.1. Let $y_{[r]i}$ $(r = 1, \ldots, n,\ i = 1, \ldots, m)$ denote a ranked set sample and let y_j $(j = 1, \ldots, m_n)$ be a random sample each with mn measured observations from a population whose fourth moment $E(y - \bar{Y})^4$ exists. Then there exists a sample size N^* such that for $mn > N^*$, $V[v(\bar{y})]/$ MSE $[v(\hat{\mu})] \geq 1$ with equality occurring if and only if $E(y_{[r]} - \bar{Y})^2 = E(y - \bar{Y})^2$ for all r.

For ranking with perfect judgment using $m = 1$, the relative precision of the small and large samples is shown to be affected by the underlying distribution of the population. For each distribution studied, the inequality in the above theorem was achieved for $n = 8$.

RSS is often more efficient than SRS, as stated earlier. In some cases the information needed to apply RSS, or a slight generalization of it, can be used more efficiently. For example, in Schreuder (1975) the variable used for ranking was also used in regression estimation. This is unlikely to improve efficiency in the Halls and Dell (1966) case with only three ranks but might improve efficiency considerably in the Martin *et al.* (1980) situation where the ranking variable was continuous.

McConnell and David (1976) assumed an infinite population and a relationship between y_i and x_i $(i = 1, \ldots, n)$ such that $y_i = \mu_y + \beta(x_i - \mu_x) + e_i$ where e_i is an error variate independent of x_i with expected value $E(e_i) = 0$, variance $\sigma_y^2(1 - \rho^2)$, $\beta = \rho\sigma_y/\sigma_x$, and ρ the correlation coefficient between variables x and y with standard deviations σ_x and σ_y. They determined the k (x, y) values selected based on the ordered x-values $x_{[1]}, \ldots, x_{[n]}$, which would minimize the variance of the double sampling regression estimator $\tilde{y}_{[k]} = \bar{y}_{[k]} - b\,(\bar{x}_{[k]} - \bar{x}_n)$, where n and k were the large and small sample sizes. Thus

$$\bar{y}_{[k]} = \frac{1}{k}\sum_{i=1}^{k} y_{[i]},\ \bar{x}_{[k]} = \frac{1}{k}\sum_{i=1}^{k} x_{[i]},\ \bar{x}_n = \frac{1}{n}\sum_{i=1}^{n} x_i\,,$$

and

$$b = \frac{\sum_{i=1}^{k} (x_{[i]} - \bar{x}_{[k]})(y_{[i]} - \bar{y}_{[k]})}{\sum (x_{[i]} - \bar{x}_{[k]})^2}\,.$$

They compared $\tilde{y}_{[k]}$ and $\bar{y}_{[k]}$ and the ranked set estimator $\hat{\mu}$ by simulating the procedure using artificial populations. The comparisons showed that the simple average of k suitably selected values, $\bar{y}_{[k]}$, was approximately as efficient as $\tilde{y}_{[k]}$ unless x and y were highly correlated. The comparison also showed that $\bar{y}_{[k]}$ was considerably more efficient than the RSS estimator $\hat{\mu}$ unless the correlation between x and y was quite low.

Ridout and Cobby (1987) noted that the need to rank the elements in a set may preclude the use of random sampling in forming the sets. For example, sets of quadrats have to be clustered together when estimating herbage biomass to make visual comparisons possible. Ridout and Cobby showed that the introduction of both ranking errors and systematic differences between sets reduces the precision of RSS relative to SRS, the precision being dependent on the ratios of the "between set" variance σ_u^2 and the error variance σ_z^2 to the "within set" variance σ_v^2.

Stokes and Sager (1988) noted that RSS works by creating an artificially stratified sample and they illustrated this with an example. Improvement of RSS over SRS in estimating the mean is explained in the case of perfect ranking by noting that an RSS sample is made up of order statistics that are independent but those for SRS are always positively correlated. Distributions, such as the uniform, which have more highly correlated order statistics will benefit most from RSS estimation of the mean. Stokes and Sager characterized an RSS sample to clarify why RSS sampling provides better estimators of both the mean and variance than SRS sampling, even when the ranking is imperfect. They applied this result to show that the empirical distribution function of a ranked set sample is an unbiased estimator of the population distribution and it is a more precise estimator than that of a random sample. Ranked set sampling is generalized by introducing a selection matrix of probabilities using optimal estimation (Yanagawa and Shirahata 1976; Yanagawa and Chen 1980; Shirahata 1982). Muttlak and McDonald (1990) proposed a two-phase sampling scheme where size-biased (pps) sampling is used at the first phase sample. Based on simulation results, they concluded that efficiency is gained whether there are errors in ranking or not. It is doubtful whether this gain in efficiency compensates for the increased complexity of estimation.

It is likely then that RSS is a useful technique for selecting samples in those cases where the covariate is only somewhat positively correlated with the variable of interest or where the covariate is basically an order-statistic variable only.

4.10 ESTIMATING GROWTH

A forest stand is a dynamic system; thus its size and the aggregate value of any characteristic, say its volume, change over time. In forestry, the term growth is often used synonymously with change despite the implication that growth involves positive change. Estimation of growth is an important part of many forest surveys associated with forest management. The following is extracted from Gregoire (1992).

To be precise about the meaning of growth, we restrict our consideration to the change that occurs between two occasions, t_1 and t_2. Let N_1 and N_2 denote the number of trees that exist in the population at t_1 and t_2, respectively. Let v_{o1i} denote the value at t_1 of the characteristic of interest on the ith element, u_{1i}, and let $V_{o1} = \Sigma_{i=1}^{N_1} v_{o1i}$ denote the parameter of interest at t_1. Let v_{o2i} and V_{o2}, and so on, be similarly defined. Sample selection may be restricted to only those trees that exceed some threshold size. It is understood that smaller trees have a value of zero even though they have finite size. For example, if the characteristic of interest is merchantable volume, where merchantability requires that tree diameter at breast height (DBH) must be greater than some lower limit, then a tree with a smaller DBH has zero merchantable volume. Finally, by growth we mean the net change $\Delta V_o = V_{o2} - V_{o1}$. This definition of growth ignores the growth on trees that die or are harvested after t_1 and prior to t_2, and it ignores the growth on trees that remain smaller than the threshold size at t_2. With these definitions there is no necessary equivalence among the labels between occasions, that is, for any stipulated index value i, u_{1i} need not be the same element as u_{2i}.

Overall growth of the forest can be decomposed into components that are meaningful to managers and components that are important for the monitoring of growth trends. We define a component of growth as that portion of ΔV_o that is attributable to a subset of $\{u_{1i},\ i=1,\ldots,N_1\} \cup \{u_{2i},\ i=1,\ldots,N_2\}$. Estimation of growth components is problematic because of the changing composition of the component sets. An added complication is injected when sampling is conducted with pps, due to the altered inclusion probability of an individual from t_1 to t_2, which results from the individual's growth over the sampling interval. For example, a tree that is merchantable but out at time t_1 may become a sample tree at t_2 because of its increase in size.

The problems related to the estimation of change and its components from repeated forestry surveys are not unlike those encountered in demographic or epidemiologic surveys (see, e.g., Holt and Skinner 1989), although there has been practically no crossover among disciplines. There does appear to be cross-disciplinary interest in preserving additivity of the estimators of the components of growth without unduly compromising precision of either the overall estimator or that of the individual components. This subject has been the topic of ongoing debate in the forestry literature especially with VRP sample designs.

In accordance with the standard decomposition in a forestry context, we let

$$V_{01} = S_1 + M + C \tag{4.36}$$

and

$$V_{02} = S_2 + I\,, \tag{4.37}$$

where:

S_1 and S_2 are the survivor-tree components of V_{o1} and V_{o2}, respectively, that are attributable to the set **S** of trees existing at t_1, that are larger than some threshold size, and that survive to t_2.

C is the cut-tree component of V_{o1} attributable to the set **C** of trees that exist at t_1 but are cut or harvested prior to t_2;

M is the mortality-tree component of V_{o1} attributable to the set **M** of trees that exist at t_1 but are removed for reasons other than harvesting operations prior to t_2.

I is the ingrowth-tree component of V_{o2} that is attributable to the set of trees **I** that cross the size threshold after t_1 and prior to t_2.

We emphasize that these population sets **S**, **M**, **C**, and **I** are defined irrespective of any sampling design applied at t_1 and t_2, that is, a tree's inclusion or exclusion from the sample at either occasion has no bearing on the set it is in. At t_1, it is impossible to know with certainty in which set of trees u_{1i} is a member. Likewise, at t_2, it is impossible to assign u_{2i} to a set unless the size of u_{2i} at t_1 is discernible. If the same sampling locations are revisited at t_2, then it may be possible to discern set membership at that time providing that tree locations and sizes were recorded at t_1, or that trees tallied at t_1 were also marked. We are interested in an estimator of the component of growth, C, associated with the set of trees **C**, based on a sample of trees from this set. Likewise, we are interested in estimates of M, I, and $S_2 - S_1$ associated with sets **M**, **I**, and **S**.

To avoid the notational disorder that results from the possibility that u_{1i} and u_{2i} may be different trees, let **W** be the set $\{w_i, i = 1, \ldots, N\}$ of distinct trees in $\{u_{1i}, i = 1, \ldots, N_1\} \cup \{u_{2i}, i = 1, \ldots, N_2\}$. Furthermore, let v_{o1i}, v_{o2i} denote the value of the characteristic of interest associated with the ith distinct tree, w_i, at t_1, t_2, respectively. If w_i is a member of the **C** or **M** sets, then $v_{o2i} = 0$, $\pi_{2i} = 0$, and $\delta_{2ki} = 0$, where δ_{2ki} is the indicator of inclusion of w_i at the kth sample location at t_2, and where $\pi_{2i} = E[\delta_{2ki}]$ denotes its inclusion probability at the kth sample location at t_2. If w_i is a member of the **I** set, $v_{o1i} = 0$, but π_{1i} and δ_{1ki} both may be nonzero,[2] where δ_{1ki} is the indicator of inclusion of w_i at the kth sample location at t_1 and $\pi_{1i} = E[\delta_{1ki}]$ denotes the corresponding inclusion probability.

Writing down the estimators in a meaningful sample-based format requires definitions of sample sets similar to the population sets **S**, **M**, **C**, and **I**. We call these **s**, **m**, **c**, **i**, respectively, where **s** is the subset of trees in **S** that are tallied at t_1 and t_2; **i** is the subset of trees in **I** that are tallied at t_2; **c** and **m** are subsets of trees in **C** and **M**, respectively, that are tallied at t_1.

4.10.1 Estimating Growth with Fixed-Area Plot Designs

We consider first the case where the second survey is a remeasurement of the plots established at t_1. Let m denote the common number of plots. We assume that the plot size a and the plot shape are the same on both occasions, and that the area being surveyed, A, is likewise held constant. Assuming that there is no boundary bias, we consider the difference between

[2] If the threshold size is zero DBH, then **I** consists exclusively of trees that did not exist at breast height at t_1, as in Grosenbaugh (1958). If the threshold size is some nonzero value of DBH, then a tree may be selected into the sample with positive probability even though its DBH is less than the threshold DBH; because it has a corresponding zero-valued characteristic of interest, it does not contribute to any estimator of the aggregate value of the characteristic at t_1.

$$\hat{V}_{o1} = \frac{A}{am} \sum_{k=1}^{m} \sum_{i=1}^{N} v_{o1i} \delta_{1ki}$$

$$= \frac{A}{am} \sum_{k=1}^{m} \sum_{i=1}^{N_k} v_{o1ki} , \tag{4.38}$$

where v_{o1ki} is volume of the ith sample tree on plot k at t_1, and

$$\hat{V}_{o2} = \frac{A}{am} \sum_{k=1}^{m} \sum_{i=1}^{N} v_{o2i} \delta_{2ki}$$

$$= \frac{A}{am} \sum_{k=1}^{m} \sum_{i=1}^{N_k} v_{o2ki} , \tag{4.39}$$

where v_{o2ki} is volume of the ith sample tree on plot k at t_2. This difference, namely, $\widehat{\Delta V_o} = \hat{V}_{o2} - \hat{V}_{o1}$, unbiasedly estimates the net increase ΔV_o, but does not allow estimation of the components of growth. If tree locations and sizes were marked and recorded at t_1 such that it is possible at t_2 to distinguish the set to which w_i belongs, then it is possible to estimate $\Delta S = S_2 - S_1$ by

$$\widehat{\Delta S} = \frac{A}{am} \sum_{k=1}^{m} \sum_{i \in \mathbf{S}} \Delta v_{oi} \delta_{2ki}$$

$$= \frac{A}{m} \sum_{k=1}^{m} \sum_{i \in \mathbf{S}} \Delta v_{oki} \tag{4.40}$$

where $\Delta v_{oi} = v_{o2i} - v_{o1i}$ is the observed growth on w_i, and Δv_{oki} is the observed growth on the ith tree sampled at plot k. Similarly, M, C, and I can be estimated by

$$\hat{M} = \frac{A}{am} \sum_{k=1}^{m} \sum_{i \in \mathbf{M}} v_{o1i} \delta_{1ki}$$

$$= \frac{A}{am} \sum_{k=1}^{m} \sum_{i \in \mathbf{M}} v_{o1ki} \tag{4.41}$$

$$\hat{C} = \frac{A}{am} \sum_{k=1}^{m} \sum_{i \in \mathbf{C}} v_{o1i} \delta_{1ki}$$

$$= \frac{A}{am} \sum_{k=1}^{m} \sum_{i \in \mathbf{C}} v_{o1ki} . \tag{4.42}$$

$$\hat{I} = \frac{A}{am} \sum_{k=1}^{m} \sum_{i \in \mathbf{I}} v_{o2i} \delta_{2ki}$$

$$= \frac{A}{am} \sum_{k=1}^{m} \sum_{i \in \mathbf{I}} v_{o2ki} \tag{4.43}$$

These formulas ignore the partial periodic growth exhibited by trees in **C** and **M**. It is evident that $\widehat{\Delta V_o} = \widehat{\Delta S} - \hat{M} - \hat{C} + \hat{I}$, that is, the estimators of the components of net growth sum to the overall estimator of net growth.

For the case where the survey at t_2 establishes different sampling locations from those used at t_1, let the number of plots established at t_1 and t_2 be denoted by m_1 and m_2, respectively, and let

$$\hat{V}_{o1} = \frac{A}{am_1} \sum_{k=1}^{m_1} \sum_{i=1}^{N} v_{o1i}\delta_{1ki} = \frac{A}{am_1} \sum_{k=1}^{m_1} \sum_{i=1}^{N_k} v_{o1ki} \tag{4.44}$$

and

$$\hat{V}_{o2} = \frac{A}{am_2} \sum_{k=1}^{m_2} \sum_{i=1}^{N} v_{o2i}\delta_{2ki} = \frac{A}{am_2} \sum_{k=1}^{m_2} \sum_{i=1}^{N_k} v_{o2ki} \,. \tag{4.45}$$

With V_{o1} and V_{o2} so estimated, $\Delta\hat{V}_o = \hat{V}_{o2} - \hat{V}_{o1}$ is an unbiased estimator of ΔV_o. The survivor-growth-plus-ingrowth portion of ΔV_o, namely $\Delta S + I$, can be unbiasedly estimated by measuring the growth on trees tallied at t_2 and then expanded, as given by

$$\widehat{\Delta V'_o} = \frac{A}{am_2} \sum_{k=1}^{m_2} \sum_{i=1}^{N} \Delta v_{oi}\, \delta_{2ki} = \frac{A}{am_2} \sum_{k=1}^{m_2} \sum_{i=1}^{N_k} \Delta v_{oki} \,. \tag{4.46}$$

Because $\widehat{\Delta V'_o}$ does not account for the M and C components, it is obviously an incompatible estimator of growth in the sense that $\hat{V}_{o2} \neq \hat{V}_{o1} + \widehat{\Delta V'_o}$.

In the foregoing discussion we assumed that plot size and shape did not change between surveys. If this is not the case, the estimators of the components of growth will be influenced by sampling error that results from the changed inclusion probabilities. Separating the sampling or recalibration effect, as it has been called, from the estimate of actual growth may be difficult for reasons that are detailed in Section 4.10.2. As a simple illustration of this recalibration effect, consider a concurrent survey where two separate tallies are made from the same set of sampling locations, one tally for trees within a plot of size a_1 and another tally for trees within a plot of size a_2, where $a_1 \neq a_2$. Because different sets of trees are sampled, the respective estimates of V_o will differ, generally speaking. The expected value of this difference, which is due solely to sampling error, is zero, and its variance will depend on both the relative size difference between a_1 and a_2 and on the spatial arrangement of the trees. When the two surveys are not concurrent, this sampling error will be confounded with the difference between estimates, which is due to the actual growth of the forest.

4.10.2 Estimating Growth with VRP Designs

Consider first the case where the second survey is a remeasurement of the sample points established at t_1, where m denotes the common number of point locations. We assume that F and A are unchanged from t_1 to t_2. Barring edge effects, net growth, ΔV_o, can be unbiasedly estimated by $\widehat{\Delta V_o} = \hat{V}_{o2} - \hat{V}_{o1}$,

where

$$\begin{aligned}\hat{V}_{o1} &= \frac{FA}{m} \sum_{k=1}^{m} \sum_{i=1}^{N} \frac{v_{o1i}\delta_{1ki}}{g_{1i}} \\ &= \frac{FA}{m} \sum_{k=1}^{m} \sum_{i=1}^{N_k} \frac{v_{o1ki}}{g_{1ki}} \end{aligned} \tag{4.47}$$

and

$$\begin{aligned}\hat{V}_{o2} &= \frac{FA}{m} \sum_{k=1}^{m} \sum_{i=1}^{N} \frac{v_{o2i}\delta_{2ki}}{g_{2i}} \\ &= \frac{FA}{m} \sum_{k=1}^{m} \sum_{i=1}^{N_k} \frac{v_{o2ki}}{g_{2ki}} \end{aligned} \tag{4.48}$$

Grosenbaugh (1958) suggested that ΔS, M, and C be estimated by

$$\begin{aligned}\widehat{\Delta S_1} &= \frac{FA}{m} \sum_{k=1}^{m} \sum_{i \in \mathbf{S}} \frac{\Delta v_{oi}\delta_{1ki}}{g_{1i}} \\ &= \frac{FA}{m} \sum_{k=1}^{m} \sum_{i \in \mathbf{S}} \frac{\Delta v_{oki}}{g_{1ki}} \end{aligned} \tag{4.49}$$

$$\begin{aligned}\hat{M} &= \frac{FA}{m} \sum_{k=1}^{m} \sum_{i \in \mathbf{M}} \frac{v_{o1i}\delta_{1ki}}{g_{1i}} \\ &= \frac{FA}{m} \sum_{k=1}^{m} \sum_{i \in \mathbf{M}} \frac{v_{o1ki}}{g_{1ki}} \end{aligned} \tag{4.50}$$

$$\begin{aligned}\hat{C} &= \frac{FA}{m} \sum_{k=1}^{m} \sum_{i \in \mathbf{C}} \frac{v_{o1i}\delta_{1ki}}{g_{1i}} \\ &= \frac{FA}{m} \sum_{k=1}^{m} \sum_{i \in \mathbf{C}} \frac{v_{o1ki}}{g_{1ki}} . \end{aligned} \tag{4.51}$$

These estimators are based on the initial selection probabilities $\pi_{1i} = a_{1i}/A = g_{1i}/(FA)$, and hence are free of the recalibration effect. Grosenbaugh suggested that ingrowth be estimated by expanding the volume on trees that are sampled from the **I** domain

$$\hat{I}_1 = \frac{FA}{m} \sum_{k=1}^{m} \sum_{i \in \mathbf{I}} \frac{v_{o2i}\delta_{2ki}}{g_{2i}}$$

$$= \frac{FA}{m} \sum_{k=1}^{m} \sum_{i \in \mathbf{I}} \frac{v_{o2ki}}{g_{2ki}} . \tag{4.52}$$

$\hat{I}_1$ is obviously an unbiased estimator of I. Using it presumes that one can recognize whether $w_i \in \mathbf{I}$. Grosenbaugh (1958) regarded zero as the threshold DBH, in which case sample trees shorter than breast height at t_i but taller than breast height at t_2 are in **I**. He suggested that a newly qualifying tree be cored to determine its breast-height age, because when using a zero-DBH threshold, only those sample trees whose age at breast-height is less than the sampling interval, $t_2 - t_1$, are members of the set **I**. When the threshold size is based on a larger minimum DBH, similar steps can be taken to distinguish **I** trees from the subset of newly qualifying trees in **S** that exceeded the threshold size at t_1. Combining $\hat{I}_1$ with $\Delta\hat{S}_1$, $\hat{C}$, and $\hat{M}$ yields a composite estimator of ΔV_o as

$$\Delta\tilde{V}_o = \widehat{\Delta S_1} - \hat{C} - \hat{M} + \hat{I}_1 . \tag{4.53}$$

It is evident that $\Delta\hat{V}_o \neq \Delta\tilde{V}_o$, implying that $\hat{V}_{o2} \neq \hat{V}_{o1} + \Delta\tilde{V}_o$, that is, $\Delta\tilde{V}_o$ is not a compatible estimator.

In an undated errata sheet to Beers and Miller (1964), those authors suggested a "revised Purdue" estimator of I that is identical to $\hat{I}_1$, and another estimator (erroneously) credited to G. M. Furnival,[3] which is based on a slightly smaller set of possible sample trees $\mathbf{I}_2$, which consists of only those trees in **I** that are sampled at both t_1 and t_2:

$$\hat{I}_2 = \frac{FA}{m} \sum_{k=1}^{m} \sum_{i \in \mathbf{I}} \frac{v_{o2i}\delta_{1ki}\delta_{2ki}}{g_{1i}}$$

$$= \frac{FA}{m} \sum_{k=1}^{m} \sum_{i \in \mathbf{I}_2} v_{o2ki}/g_{1ki} . \tag{4.54}$$

Whereas newly qualifying trees that exceed the threshold size at t_1 are excluded from $\hat{I}_1$, $\hat{I}_2$ excludes all newly qualifying trees. If the sampling

[3] Trees in **I**, when the threshold size is $dbh > 0$, are members of the **S** domain when the threshold is established at $dbh = 0$. Furnival's estimator is identical to Grosenbaugh's (1958) estimator of growth for these trees, and he feels that the credit appropriately belongs to Grosenbaugh (personal communication).

interval is short enough that all $w_i \in \mathbf{I}$ existed at t_1, $\hat{I}_2$ is unbiased. Its unbiasedness in this situation is proven by recognizing that $E[\delta_{2ki}|\delta_{1ki}=1]=1$ for all $w_i \in \mathbf{I}$, hence $E[\delta_{1ki}\delta_{2ki}]=g_{1i}/(FA)$. To use $\hat{I}_2$, one must be able to distinguish sample trees in $\mathbf{i}_2$ at t_2 from the larger subset of sample trees in $\mathbf{i}$. In the situation where the sample interval is so long that some $w_i \in \mathbf{I}$ did not exist at t_1, $\hat{I}_2$ is a biased estimator of I, as is explained by Martin (1982).

Like $\hat{I}_1$, $\hat{I}_2$ can be combined with $\widehat{\Delta S}_1$, $\hat{C}$, and $\hat{M}$ to yield a composite estimator

$$\widehat{\Delta V}_o = \widehat{\Delta S}_1 = \hat{C} - \hat{M} + \hat{I}_2 , \tag{4.55}$$

which when added to $\hat{V}_{o1}$, is not compatible with $\hat{V}_{o2}$.

Van Deusen *et al.* (1986) proposed an alternative estimator of ΔS based on a slightly larger subset of $\mathbf{S}$, namely $\mathbf{s}_2$, which consists of those trees in $\mathbf{S}$ that are sampled at t_1 or t_2. Van Deusen's estimator expands the value of each $w_i \in \mathbf{s}_2$ by its unconditional inclusion probability at the occasion it is tallied.

$$\begin{aligned}\widehat{\Delta S}_2 &= \frac{FA}{m}\left[\sum_{k=1}^{m}\sum_{i\in\mathbf{S}}\frac{v_{o2i}\delta_{2ki}}{g_{2i}} - \frac{v_{o1i}\delta_{1ki}}{g_{1i}}\right] \\ &= \frac{FA}{m}\left[\sum_{k=1}^{m}\sum_{i\in\mathbf{s}_2}\frac{v_{o2ki}}{g_{2ki}} - \frac{v_{o1ki}}{g_{1ki}}\right]. \end{aligned}\tag{4.56}$$

With $\widehat{\Delta S}_2$, a compatible decomposition of $\widehat{\Delta V}_o$ is

$$\hat{V}_{o2} = \hat{V}_{o1} + \Delta\hat{S}_2 - \hat{M} - \hat{C} + \hat{I}_1 .$$

This estimator of survivor growth, $\widehat{\Delta S}_2$, makes use of precisely that set of newly qualifying trees that must be distinguished and excluded from $\hat{I}_1$. In other words, no new practical difficulties in distinguishing set membership are introduced by using $\widehat{\Delta S}_2$ instead of $\widehat{\Delta S}_1$ in conjunction with $\hat{I}_1$. Van Deusen *et al.* (1986) found $\widehat{\Delta S}_2$ to be slightly more precise than $\widehat{\Delta S}_1$ as an estimator of ΔS.

Roesch *et al.* (1989) proposed yet another alternative estimator of ΔS, namely,

$$\begin{aligned}\widehat{\Delta S}_3 &= \frac{AF}{m}\sum_{k=1}^{m}\sum_{i\in\mathbf{S}}\frac{\Delta v_{oi}\delta_{2ki}}{g_{2i}} \\ &= \frac{FA}{m}\sum_{k=1}^{m}\sum_{i\in\mathbf{S}}\frac{\Delta v_{oki}}{g_{2ki}} , \end{aligned}\tag{4.57}$$

which requires knowledge of v_{o1i} in the sample set $\mathbf{s}_2$. This estimator of ΔS was superior to $\widehat{\Delta S}_2$ and $\widehat{\Delta S}_1$ in a small empirical study on volume growth conducted by Roesch *et al.* (1989) with known values of v_{o1i} for this subset of trees. In the field, if a newly qualifying tree is not cored to determine its value at t_1, v_{o1i} will have to be estimated for these trees, thereby augmenting the MSE of $\widehat{\Delta S_3}$ by an amount that may be difficult to determine. It is unclear whether regression prediction of v_{o1i} in $\mathbf{s}_2$ will negate possible gains in precision promised by $\widehat{\Delta S_3}$.

Roesch *et al.* (1989) suggested that compatibility can be achieved by altering $\hat{V}_{o1}$. Of course, this tactic can be employed with any combination of estimators of the components of growth, yet as the authors realized, it may conflict with the reason for seeking a compatible estimator in the first place. This tactic had been mentioned earlier by Iles (1988).

For estimation of basal area and volume, the influence on the variance caused by the recalibration effect (changing the inclusion areas of trees) is greatly reduced with critical height sampling, as illustrated well by Iles (1988) and Iles and Beers (1983) and with the "variable basal-area-factor" estimator proposed by Flewelling (1981).

Considering next the case where the survey at t_2 establishes differing sampling locations from those of t_1, the situation is completely analogous to that which holds when fixed-area plots are used, that is, the simple difference estimator $\Delta V_o = \hat{V}_{o2} - \hat{V}_{o1}$ unbiasedly estimates ΔV_o, where

$$\hat{V}_{o1} = \frac{AF}{m_1} \sum_{k=1}^{m_1} \sum_{i=1}^{N} \frac{v_{o1i}\delta_{1ki}}{g_{1i}} = \frac{AF}{m_1} \sum_{k=1}^{m_1} \sum_{i=1}^{N_k} \frac{v_{o1ki}}{g_{1ki}} \tag{4.58}$$

$$\hat{V}_{o2} = \frac{AF}{m_2} \sum_{k=1}^{m_2} \sum_{k=1}^{N} \frac{v_{o2i}\delta_{2ki}}{g_{2i}} = \frac{AF}{m_2} \sum_{k=1}^{m_2} \sum_{i=1}^{N_k} \frac{v_{o2ki}}{g_{2ki}} \tag{4.59}$$

Alternatively, one can measure Δv_{oi} for each tree sampled at t_2 and jointly estimate $\Delta S + I$ as

$$\Delta\hat{V}'_o = \frac{AF}{m} \sum_{k=1}^{m_2} \sum_{i=1}^{N} \frac{\Delta v_{oi}\delta_{2ki}}{g_{2i}} = \frac{AF}{m} \sum_{k=1}^{m_2} \sum_{i=1}^{N_k} \frac{\Delta v_{oki}}{g_{2ki}} \tag{4.60}$$

4.10.3 The Importance of Compatibility

A great deal of effort has been expended by forest biometricians in order to achieve compatible estimators. Van Deusen *et al.* (1986) stated that compatibility has been a matter of concern because a lack of additivity leads to a lack of confidence in the estimation of change, especially by users of change information who themselves may not be survey specialists. However, in some surveys, it is pointless being overly concerned about it. For example, in large regional forest inventories, there may be substantial change in land use during the sampling interval (due perhaps to urbanization or other societal forces), which affects the amount of land area under forest cover. These changes will obviously confound any attempted reconciliation of past and present inventory statistics to a much greater extent than will the recalibration effect due to size-related changes in tree inclusion probabilities.

There is an acute need for an in-depth analysis of the advantages of VRP sampling and fixed area plot sampling, or of clusters of points and fixed-area plots, in estimating growth or change. Such an analysis should address the advantages and disadvantages of additive estimates. In our opinion, the reduced efficiency and increased complexity of estimation that ensures additivity may not be justifiable. Estimates of growth derived by remeasuring the same trees at both t_1 and t_2 need to be evaluated in terms of efficiency, simplicity, and the need for few assumptions. Such an analysis is particularly relevant now in the USA and other countries that have made a commitment to monitoring forest health. This involves estimating change and requires frequent visits to the sample sites, time consuming and accurate measurements, and a certain amount of destructive sampling (Schreuder and McClure, 1991).

4.11 COMPARISON OF SOME SELECTED SAMPLING STRATEGIES

Some selected sampling procedures are compared in the following sections to illustrate advantages and disadvantages in certain situations. These comparisons include fixed-area plot sampling versus VRP sampling, centroid versus importance sampling, control-variate versus importance sampling and pps versus Poisson sampling. The results of some recent simulation studies are also reviewed.

4.11.1 Fixed-Area Plots Versus Variable Radius Plot Samples

Whyte and Tennent (1975) concluded that horizontal VRP sampling provides more efficient estimates of stand basal area (G) than fixed-area plot samples and is also more efficient in practice (see Ch. 7 also). Oderwald (1981) found that horizontal VRP sampling was more precise for estimating G in clumped and random spatial patterns (also found by Matérn 1972) and plot sampling was more precise in square lattice stands. Oderwald assumed a general, discrete probability distribution of tree diameters for each pattern.

In a similar study, Martin (1983) assumed both a completely random and a clustered random spatial distribution of trees. The latter is rather unrealistic in that the clusters were located at random within the forest, the number of trees in a cluster was independent of the size and location of other clusters, the trees in a cluster were assumed to be located at the same variable radius plot, and the growth of a tree in both diameter and volume was independent of its location. This last assumption was also made for the random spatial distribution. Given these assumptions, however, Martin concluded that for any spatial distribution, the horizontal VRP sampling estimator of volume growth is more efficient if the correlation between volume growth and basal area exceeds a critical positive value determined by the diameter distribution and the spatial distribution of the trees. This critical value is largest for a completely random distribution and approaches zero as the distribution becomes more clustered.

Ranneby (1980) analytically compared the precision of horizontal VRP sampling and fixed area plots for estimating change in basal area over time. He assumed that trees are located randomly and that growth in basal area is independent of initial basal area. Under these conditions, VRP sampling was less efficient than fixed-area plot sampling for estimating change in basal area (this is not surprising given that pps sampling is efficient when the variable of interest is positively correlated with the probability of selection).

Schreuder *et al.* (1987a) compared the use of fixed-area plots, horizontal line samples, and horizontal VRP samples for estimating total number of trees, number of trees by diameter class, total basal area, sum of tree diameters (D), and average tree diameter ($\bar{D}$) in a 60-ha tropical forest in Surinam that had been mapped. Drawing samples by computer simulation, they found that horizontal line sampling was most efficient for estimating average tree diameter and VRP sampling was most efficient for estimating number of trees by diameter class. As one would expect, VRP sampling was most efficient for G and fixed-area plot sampling for N. They concluded that a combination of sampling methods was generally most efficient for estimating all parameters provided that sampling costs and sample sizes did not vary among methods. In a follow-up study, Schreuder *et al.* (1992a) estimated the number of trees, average D^2H, and sum of D^2H values by diameter class (a good surrogate for total volume by diameter class) for the three methods. When estimated costs of implementation were considered, VRP sampling was easily the most efficient for all three parameters in all but the two smallest diameter classes.

Scott (1990a) reviewed estimation using fixed area plots versus horizontal VRP for repeated inventories. He noted that originally, VRP sampling was used only for quick, temporary surveys, and when the need for estimates of change between successive surveys became important, sometimes only VRP samples were available for remeasurement. Scott compared the two procedures with respect to practicality, sample designs, estimating change and components of change, and estimating current conditions. He noted that

VRP sampling applied on one occasion does not require that all trees be visited to compute precise estimates of basal area and volume, but to estimate the components of change (ingrowth, accretion, mortality, and removals), each tree must be identified. Estimates from VRP sampling are also affected more by incorrectly sampled borderline trees. Using a larger BAF and checking all borderline trees by direct measurement are recommended to minimize this problem. Two other complicating factors in VRP sampling are that the effective plot size for a tree changes as it grows so that its frequency or expansion factor (number of trees per unit area that it represents) changes over time and the components of change do not necessarily sum to the difference between the two survey estimates (see Sect. 4.10.3). In comparison, estimating the components of change with fixed-area plots is straightforward.

Scott (1990a) concluded from his review that fixed-area plots are more efficient for estimating number of trees and change in number of trees and growth in volume over time, whereas VRP sampling is more efficient for estimating current basal area and volume. In a field study in Maine based on 58 forest sites, Scott and Algeria (1990) compared 0.08-ha (0.2-ac) fixed-area plots with horizontal VRP samples with BAF 6.1 m^2/ha (26.5 ft^2/acre) for estimating change in stand parameters over time. After allowing for differences in the costs of sampling, they concluded:

1. Fixed area plots are more efficient for estimating:
   (a) Number of trees/unit area (all attributes)
   (b) Ingrowth (number of trees, basal area and, volume).
   (c) Mortality (number of trees, basal area, and volume).
   (d) Net growth and net change in basal area.
2. VRP samples are more efficient for estimating:
   (a) Current basal area.
   (b) Net volume (accretion, removals, net growth, net change, and current values—but not ingrowth and mortality).
   (c) Removals (number of trees, basal area and volume).

Generally, attributes more closely associated with the large diameter classes or correlated with current basal area, such as volume, were better estimated with VRP samples, whereas attributes associated with the small diameter classes, such as mortality and ingrowth, were more efficiently estimated with fixed-area plots.

4.11.2 Centroid Versus Importance Sampling

Wood *et al.* (1990) found that when comparing volumes from IS with those computed from intensive dendrometry, minimum sampling error occurred in the estimate of bole volume when half the estimated volume occurred on either side of the point of importance measurement, that is, at the centroid or 50% volume position. They established that this position occurs at about 30% of total tree height. This unique point of measurement was also found useful for volume estimation by Pollanschütz (1965), Ueno (1978), Schmid-Haas and Winzeler (1981), and Forslund (1982). Based on their findings, Wood *et al.* (1990) developed a variant of IS they called centroid sampling. Simulation studies conducted on sample tree data of American and Australian hardwoods have shown promising results for the centroid method in deriving bole volume (Wood and Wiant 1990; Wiant *et al.* 1991).

Centroid sampling is applied at a fixed position on the stem, the centroid, the height of which is the expected mean height at which diameter measurements would be made in importance sampling if the sample were repeated many times (Van Deusen and Lynch 1987). The estimator of tree volume is

$$\hat{v}_o = v_{o1} c(h_c)/t(h_c) \tag{4.61}$$

where h_c is the height of the centroid and $t(h_c)$ and $c(h_c)$ are, respectively, the measured and proxy-function cross-sectional areas at h_c, and v_{o1} = proxy function volume.

Evidently, $\hat{v}_o$ is identical to $\hat{v}_{oIS}$ except that only one measurement at a nonstochastic height is taken. Wiant *et al.* (1992b) noted that with centroid sampling, measurements in the upper crown are not required, as they are occasionally with IS.

An inappropriate proxy function in IS lowers precision but does not alter the unbiasedness of the estimate. Centroid sampling, in contrast, is inherently biased unless the centroid determined by the proxy function coincides with the centroid of the bole. However, results so far indicate that this bias is negligible even with a generic type of proxy function. Thus in terms of error, regardless of whether it is systematic or random in origin, centroid sampling performs well (Wiant *et al.* 1992b). As noted in Chapter 6 and Chapter 10, the credibility of purposive sampling designs such as centroid sampling may be undermined when a single reasonable example is found where the bias is serious.

Importance and centroid sampling should prove especially useful in the tropical mixed-moist forest where reliable volume tables or equations are not available. In many cases, these methods preclude the necessity of large investments in time and money to develop species-specific volume and taper functions.

4.11.3 Control-Variate Versus Importance Sampling

Whereas the IS estimator (Section 4.7) involves a multiplicative adjustment to the integrated proxy function, the control-variate estimator makes an additive adjustment. In the context of bole volume estimation, if v_{o1}

represents the proxy volume, then a control-variate estimator of tree volume v_o is

$$\hat{v}_o = \hat{v}_{o1} + \frac{1}{m} \sum_{i=1}^{m} \left[\frac{c(\theta_i) - t(\theta_i)}{f(\theta_i)} \right]. \tag{4.62}$$

This estimator was presented by Gregoire *et al.* (1987) and is an alternative to IS for bole volume estimation. Valentine *et al.* (1991) later described both as horizontal methods because they entail a measurement in the horizontal plane. If the points of measurement are selected according to a uniform $U[h_l, h_u]$ density, then

$$\hat{v}_o = v_{op} + \frac{h_u - h_l}{m} \sum_{i=1}^{m} [c(\theta_i) - t(\theta_i)] . \tag{4.63}$$

Van Deusen (1990) proposed control-variate sampling in a vertical mode, that is, choosing a diameter at random and then measuring bole height at that diameter. It is apparent that vertical IS likewise is possible (Valentine *et al.* 1992).

Simulation studies conducted by Van Deusen (1990) and Valentine *et al.* (1992) suggest that control-variate estimation with a uniform density is more precise than horizontal IS for estimating bole volume, the relative precision of the former increasing as h_u approaches total tree height. The magnitude of difference depends, too, on the departure of the proxy function from the actual tree profile.

Field implementation of horizontal IS is apt to be easier than vertical control-variate sampling, especially since the former tends to locate points of measurement closer to the ground. Wood *et al.* (1990) reported that 40-50% or more of the horizontal IS locations were within easy reach of the ground in a field test applied to radiata pine.

If the difficulty of measurement does not depend on location, for example, if trees are felled prior to sampling, then a considerable increase in precision may be realized by taking measurements in pairs that are negatively correlated, regardless of whether IS or control-variate sampling is used. Hammersley and Morton (1956) first recognized the value of generating antithetic variates, as the procedure has come to be known, and Van Deusen and Lynch (1987) suggested its use with IS.

4.11.4 Probability Proportional to Size Versus Poisson Sampling.

Comparing the efficiency of pps and Poisson sampling is appropriate to illustrate that a random sample of fixed size n_e is usually more efficient than one with random sample size $n(s)$ with expected sample size n_e.

Pps sampling without replacement using the Horvitz-Thompson estimator is most efficient for model (2.70) with $\alpha = 0$ and $k = 2$. As noted earlier, the

Horvitz-Thompson estimator reaches the Godambi-Joshi bound (eq. 3.93) for $1 \leq k \leq 2$, so that for pps sampling (with $k = 2$)

$$V(\hat{Y}_{HT}) = \sigma^2\left(X^2/n_e - \sum_{i=1}^{N} x_i^2\right). \tag{4.64}$$

Under this same model for Poisson sampling, and using the adjusted estimator $\hat{Y}_a$ in eqn. (3.76), the approximate variance is

$$V(\hat{Y}_a) \doteq \sigma^2 X^2/n_e\left[1 + 1/n_e - \sum_{i=1}^{N} x_i^2/X^2\right] \tag{4.65}$$

(Schreuder 1990).

Pps sampling schemes with a fixed sample size n_e have an approximate efficiency of $(1 + 1/n_e)$ relative to Poisson sampling with expected sample size n_e (Schreuder 1990) so that fixed sample size pps procedures should be preferred if both schemes are practically feasible and selection probabilities are the same.

4.11.5 Simulation Studies

Ek (1971), in a simulation study, compared the simple expansion, classical ratio of means ($\hat{Y}_{rm}$), Mickey's unbiased ratio, and linear ($\hat{Y}_{lr}$) and quadratic regression ($\hat{Y}_{cl}$) estimators for equal and unequal probability sampling schemes. Estimates of total merchantable volume, height, and crown area were generated for three test populations for samples of size 4, 12, 24, and 40. Diameter, height, and crown area, plus several transformations and combinations of these variables, were used as independent variables. Ek compared the variances and biases of the estimated totals from repeated sampling of the test populations. For larger sample sizes, $\hat{Y}_{lr}$, $\hat{Y}_{cl}$, and $\hat{Y}_{HT}$ (pps sampling) were usually best. For the smaller sample sizes, $\hat{Y}_{lr}$, $\hat{Y}_{HT}$ (pps sampling), and $\hat{Y}_{rm}$ were best. Ek mentioned that sample size and population characteristics affect estimator performance, but he discussed this only briefly.

Schreuder *et al.* (1987b) used pps and SRS with several ratio and regression estimators in a simulation study of four forest populations from which samples of size 30 were drawn. Ouyang *et al.* (1992b) used the same data sets and added to the comparison a general regression estimator $\hat{Y}_{gr}$ due to Särndal (1980) (equation 3.134) and two Srivastava-type estimators (Ouyang 1990) which are special cases of

$$\hat{Y}_{gre} = \left[\left(\sum_{i=1}^{n} y_i/\pi_i\right)\Big/\left(\sum_{i=1}^{n} y_i^*/\pi_i\right)\right]Y^* \tag{4.66}$$

with

$$y_i^* = \hat{\alpha} + \hat{\beta} x_i\,, \qquad i = 1, \ldots, N \tag{4.67}$$

and

$$Y^* = \sum_{i=1}^{n} y_i^*\,,$$

where $\hat{Y}_{gre} = \hat{Y}_{grl}$ if $\hat{\alpha}$ and $\hat{\beta}$ are simple least square estimates of α and β and $\hat{Y}_{gre} = \hat{Y}_{grw}$ if $\hat{\alpha}$ and $\hat{\beta}$ are best linear unbiased (BLU) estimators of α and β in model (2.70), assuming $k = 1.5$.

The data sets used by both Schreuder *et al.* (1987b) and Ouyang *et al.* (1992b) comprised real data selected with the expectation that they would show strong linear, weak linear, or well defined curvilinear relationships between the variable of interest and covariate.

Schreuder *et al.* (1987b) concluded that:

1. The weighted regression estimator $\hat{Y}_{wr}$ (eq. 3.139) should be used for pps sampling rather than $\hat{Y}_{HT}$.
2. The linear regression estimators should be used in preference to the ratio estimator for both pps and random sampling.
3. Curvilinear or nonlinear regression estimators do not provide additional gains in efficiency to warrant their use unless well-defined curvilinear or nonlinear relationships are known to exist for the data.

Ouyang *et al.* (1992b) concluded that:

1. The Srivastava estimator $\hat{Y}_{grw}$ is generally the most efficient estimator because it uses prior information effectively in pps sampling and posterior information through model (2.70) with $k = 1.5$, which was apparently sufficiently reliable for most of the data sets. The efficiency of $\hat{Y}_{grw}$ decreased as the validity of the model became questionable.
2. Särndal's estimator $\hat{Y}_{gr}$ was essentially as efficient as $\hat{Y}_{grw}$ for all populations.

Schreuder *et al.* (1990b) compared seven model-based and design-based sampling strategies with ratio and regression estimators using eight forestry populations, seven real and one artificial, for estimating the population total Y for a variable y. Samples of size 20 were used. The most efficient strategies under a wide variety of conditions were found to be those relying heavily on an underlying linear model between the variable of interest y and a covariate x, where y is usually tree or plot volume and x is D^2H or previous plot volume, namely, stratified sampling with two units selected per stratum by pps sampling (spps); stratified sampling from the cumulated x with one unit randomly selected per strata (spsc$x$); and restricted simple random sampling (balanced sampling) eliminating sample means $\bar{x}$ more than 5% from the population mean $\bar{X}$ and sample variances of x more than 50% from the population variance of x (RSRS) when used with the weighted regression estimator ($\hat{Y}_{wr}$) and the BLU estimator ($\hat{Y}_{blu}$).

In a follow-up simulation study, Schreuder and Ouyang (1992) focused on weighted regression estimators $\hat{Y}_{gr}$ (eq. 3.134), $\hat{Y}_{piwr}$ (eq. 3.142), $\hat{Y}_{pi}$ (eq. 3.138), and $\hat{Y}_{wr}$ (eq. 3.139), assessing the efficiency of the estimators and the reliability of the variance estimators. A set of populations was used in a factorial-type arrangement consisting of:

1. Three artificial populations with a known linear model and variance structure as described in model (2.70) with $k = 2.0$. The y-variable is tree volume, $x = D^2H$. This is also true in sets 2–4.
2. Three populations with a very strong linear model and variance structure approximated well by model (2.70) with $k = 1.5$.
3. Three populations with a very strong linear model and variance structure approximated well by model (2.70) with $k = 1.0$.
4. Three populations with a very strong linear model and variance structure approximated well by model (2.70) with k unknown but probably between 1 and 2.
5. Three populations with a strong linear model but with an unknown variance structure in model (2.70). The y variable is plot volume at time t, x = corresponding volume at $t - 5$ or $t - 10$.
6. Three populations with a weak linear model and unknown variance structure. The y-variables were plot volume at time t, tree radial diameter growth, and tree height with, respectively, plot volume at time $t - 15$, tree diameter, and tree age as independent variables.
7. Three populations with a definite quadratic relationship and unknown variance structure. The y-variables were tree bole length, height, and height with DBH as x-variable. Three different data sets were involved.

The sampling designs consisted of three stratification designs based on sorted x-values selecting one unit per stratum, one stratified sampling and a U-distribution design selecting small and large x-values with high probabilities, and a restricted random sampling design (RSRS).

Schreuder and Ouyang (1992) concluded that:

1. The stratified sampling designs selecting one unit per stratum are consistently the most efficient, regardless of the estimator used.
2. None of the four estimators is superior to the others across all designs and populations.
3. Bootstrap variance estimation using SRS with replacement is often reliable, whereas classical variance estimates are not.

4. Jackknife variance estimates consistently overestimate the actual variances of the estimates (approximately 10% on average).

4.12 VARIANCE ESTIMATION USING JACKKNIFING AND BOOTSTRAPPING

In most forest sampling situations classical variances and estimators of these variances can be derived and are exact or good approximations. For some sampling strategies only rough approximations are available, and there jackknifing and bootstrapping may be appropriate. A brief summary of the limited literature on these techniques in forestry is given. We anticipate that additional applications of these techniques will be forthcoming.

Schreuder and Brink (1983) and Gregoire (1984) discuss some applications of jackknifing in forestry. For estimating the variance of $\hat{Y}_{rm}$ based on small samples ($n = 4, 6, 8$), Gregoire (1984) found the jackknife estimator v_J (eq. 3.109) to be generally more accurate and, because of its tendency to overestimate, a safer estimator of variance than the traditional one v_1 (eq. 3.106).

Schreuder and Anderson (1984) estimated the total volume of loblolly pine for two realistic populations using D^2H as covariate. They found in weighted regression that a jackknife variance estimator, and confidence intervals based on it, are much more reliable than the traditional variance estimator and two other robust estimators proposed by Royall and Cumberland (1978).

Schreuder *et al.* (1987b), in a simulation study of 16 populations, compared jackknife and classical variance estimates using ratio, linear, and nonlinear regression estimators. The populations were selected on the basis of having underlying linear and curvilinear relationships between y and x. Samples of size 30 were drawn from the populations which contained from 1794 to 3153 units. Schreuder *et al.* (1987b) found that jackknife variance estimates are not consistently better than the classical variance estimates for estimating the true variances.

Schreuder and Ouyang (1992) compared several sampling designs on a series of populations with varying degrees of linearity and heterogeneity of variance using various weighted regression estimators. Variance estimates of the estimated totals based on SRS with-replacement bootstrapping, ignoring the original designs, proved to be essentially unbiased for all sampling strategies using the bootstrap version of the regression coefficients used in the regression estimators.

4.13 FOREST SAMPLING SIMULATORS

The ideal way to develop sampling expertise is to become involved in actual forest surveys and inventories but computer simulated sampling is less costly and often a more feasible option. Several simulation programs are available that can generate a variety of (simulated) field conditions and/or offer a choice of several sampling strategies for comparing sampling efficiency.

Newnham (1968) and Wensel (1975) developed programs that generate artificial populations of points on a plane. Newnham's program allows the user to generate a number of spatial patterns ranging from clumped (contagious) through random to a uniform pattern. Wensel's program can generate populations that are relatively free from the problem of edge effect in sampling. He used the program to study the effects of degree of contagion, randomness, or uniformity on the sampling efficiency of sampling designs. Dress (1970) generated randomly distributed populations by combining a Poisson birth with a binomial death process for grid cells of equal area. Tree coordinates within the cells were selected using randomly selected azimuths and distances from the center of the cell. Wensel (1975) noted that Newnham's (1968) method is especially good for generating special distributions such as stands characterized as plantations with infection centers (e.g., seed trees) or density gradients, whereas Dress's method generates only even-aged stands with either uniform or random spatial patterns.

Arvanitis and Reich (1989) developed an interactive computer program called FOSS (Forest Sampling Simulator) for IBM compatible personal computers. FOSS can be used to assess the relative merits of several sampling designs. It allows imputation of several types of tree populations in terms of spatial pattern (random, aggregated, or uniform), diameter distribution (normal or Weibull distribution), and density gradient. This last option is used to simulate differences in field conditions such as regeneration patterns. For each tree the program generates tree diameter, height, and volume. The user can also select elementary sampling units classified as fixed-area plots, VRP samples, lines, and strips. Various shapes and lengths are feasible and edge effect is accounted for in VRP and line sampling. Sampling designs considered are SRS, systematic sampling, stratified random sampling, Poisson sampling, distance sampling, and multiphase and multistage sampling. Output consists of parameter values, VRP estimates, standard errors of estimates, and 95% confidence limits for average tree d, h, and v and for mean stand basal area and volume.

Mackisack and Wood (1990) described a computer package written in Fortran 77 that simulates VRP sampling in a simulated 1000-ha forest. The program requires specification of a number of forest parameters including stocking rate (trees/ha), spatial pattern of the trees, range of values of *dbh* outside bark (cm) and basal area (m^2/ha). Several BAFs can be used at any specified location in the forest. The program generates estimates of stand density (m^2/ha) and its coefficient of variation. Results of simulations are in agreement with field experience and show the utility of the simulator for comparing several VRP sampling strategies.

Brink and Schreuder (1992) developed a Fortran simulation program that selects samples of size n from known populations by various sampling schemes including SRS, RSRS (samples with sample mean and variance of x that deviate more than a prespecified amount from the corresponding population parameters are thrown out), pps sampling, stratified pps sampling, stratified random sampling, and a procedure that selects the largest and smallest units (in terms of x) with higher probabilities than units intermediate in size. The Horvitz-Thompson estimator $\hat{Y}_{HT}$ (eq. 3.7) and weighted regression estimators $\hat{Y}_{gr}$ (eq. 3.134), $\hat{Y}_{piwr}$ (eq. 3.142), $\hat{Y}_{pi}$ (eq. 3.138), and $\hat{Y}_{wr}$ (eq. 3.139), as well as the standard errors of the estimates (expressed as percent of the true total, standard classical, bootstrap, and jackknife standard errors), and coverage rates (percentage of confidence intervals containing the true total) are computed for each sample selection method. A write-up of 21 populations from which samples can be drawn is included.

4.14 PROBLEMS AND EXERCISES

1. A survey is made of the survival rate of tree planting in a region. Results derived from a random sample involving 12 1-ac plots are as follows:

   0, 20, 50, 60, 100, 90, 90, 60, 40, 100, 0, 70%

   What is the average survival rate?
2. The following are three simple random samples of size $n = 20$ drawn from three different populations. Graph each of the samples and determine which population is best suited for linear regression, ranked set sampling, and stratified sampling estimators.

Data Set 1		Data Set 2		Data Set 3	
x	y	x	y	x	y
1	2.3	1	3.2	1	14.0
13	25.0	13	21.3	13	24.7
5	10.0	5	5.7	5	11.0
12	23.4	12	22.1	12	9.1
16	34.1	16	34.7	16	27.6
3	7.2	3	6.2	3	17.2
20	46.0	20	40.2	20	3.1
7	12.7	7	23.4	7	20.4
11	20.3	11	20.6	11	6.8
4	9.1	4	4.5	4	14.5
15	32.3	15	35.8	15	22.7
19	37.6	19	36.3	19	31.4
2	4.3	2	7.5	2	16.1

Data Set 1		Data Set 2		Data Set 3	
x	y	x	y	x	y
9	16.8	9	25.9	9	24.9
6	11.2	6	21.2	6	10.0
14	31.0	14	20.0	14	8.8
10	22.7	10	25.4	10	17.9
8	12.7	8	23.3	8	6.3
17	34.4	17	35.0	17	15.4
18	39.2	18	34.8	18	18.5

3. If you have unlimited free computing power available and you want to assess the "randomness" of two random number generators, how would you compare their relative merits using some sampling scheme?

4. A professional timber cruiser frequently uses Poisson sampling. She tells you that she has used it five times on five different timber sales and each time the actual number of sample trees (20) was the same as the target (n_e) she was aiming for. What might you say to her with respect to her credibility from a scientific point of view?

5 Multi-Information Sources for Sampling

5.1 INTRODUCTION

In Chapters 3 and 4 we discussed single-level surveys where either the variables of interest, or these variables plus covariates on the same sampling units, were measured. We assumed that the covariate values for all population units (or at least their sum) were known. Often the covariates are correlated with the variables of interest but the covariate information is unknown too and thus also has to be collected by a survey. Collecting covariate information on a large sample and information on the variable of interest for a smaller, usually a subsample of the large sample, is done by multilevel sampling. An example is a survey for timber volume where D^2H values are obtained from a large sample of trees, some of which are measured for volume. Multilevel sampling designs can be separated into multiphase and multistage designs. Often the two types are used in combination. Such designs with appropriate estimators are discussed in Section 5.2.

Measurement over time for the purpose of deriving estimates of growth, mortality, and other parameters of change is critical in forestry and is discussed in Section 5.3. Often, too, there is a need to combine information from two or more surveys on one or more areas (populations) of interest. Methods of doing this are discussed in Section 5.4. Finally, in Section 5.5, we discuss specific applications of multilevel sampling in forestry.

5.2 MULTILEVEL SAMPLING

Multistage (or nested sampling) refers to sampling designs where the ultimate sampling units, called elements, are selected in stages and samples at each stage are taken from the sampling units comprising clusters of units selected in the previous stage. In contrast, multiphase sampling retains the same size of sampling unit at each phase (level) but samples fewer units at each succeeding phase.

5.2.1 Multistage Sampling

There are two primary reasons for adopting multistage as opposed to single-level sampling. Although one may want to draw a set of elements from a population, it may be quite expensive to obtain a list of the population elements. For example, for any reasonably large forest, it would be difficult to obtain a list of all the trees in the population; even if such a list were available, economics might dictate that groups of elements (clusters) rather than single elements be chosen. For example, it is less time consuming and much cheaper to sample 20 randomly located clusters of 30 trees in a forest than to sample 600 randomly located individual trees. As indicated in Chapter 3, there is a definite trade-off in efficiency between cluster sampling and random sampling of elements. Generally, cluster sampling is less precise for a given number of sample elements because elements close together are often less variable than those farther apart. Hence there is some repetition in the information collected if elements are within a cluster.

Estimation for Multistage Sampling. Following Murthy (1967), consider two-stage sampling applied to a population where the elements are grouped into N groups or clusters with M_i elements ($i = 1, \ldots, N$) in the ith cluster. If y_{ij} is the value of interest for the jth element in the ith cluster, then

$$Y_{i\cdot} = \sum_{j=1}^{M_i} y_{ij} \tag{5.1}$$

is the total for the ith cluster and the population total Y is

$$Y = \sum_{i=1}^{N} Y_{i\cdot} \,. \tag{5.2}$$

If the N clusters are primary sampling units (psu's) and the elements are secondary sampling units (ssu's), then select n psu's and m_i ssu's from the ith selected psu with some arbitrary probability scheme. If y_{ij} is the value of the jth selected ssu in the ith selected psu ($j = 1, 2, \ldots, m_i$, $i = 1, 2, \ldots, n$), and if the total values of each selected psu are known (the $Y_{i\cdot}$), then an estimator of Y can be obtained through the probability sampling scheme as in cluster sampling without subsampling. But in two-stage sampling, the totals $Y_{i\cdot}$ of the selected psu's are estimated on the basis of the probability scheme adopted in selecting the ssu's. Hence, the estimator in cluster sampling is of the form

$$\hat{Y}_c = \sum_{i=1}^{n} a_i Y_{i\cdot} \tag{5.3}$$

where $Y_{i\cdot}$ is the known total of the ith sample cluster and a_i is the proper expansion factor. The value of $Y_{i\cdot}$ for the ith cluster is estimated by

$$\hat{Y}_{i\cdot} = \sum_{j=1}^{m_i} a_{ij} y_{ij} \tag{5.4}$$

where a_{ij} is the expansion factor for the jth ssu within the ith psu at the second-stage selection so that

$$\hat{Y} = \sum_{i=1}^{n} a_i \hat{Y}_{i\cdot} = \sum_{i=n}^{n} a_i \sum_{j=1}^{m_i} a_{ij} y_{ij} . \tag{5.5}$$

(Murthy 1967).

The following theorem (Murthy 1967, Ch. 2) is extremely useful in deriving results for multistage sampling as well as for other situations where conditional estimates and variances are needed.

Theorem 5.1 (Conditionality Theorem). Let $\hat{Y}$ be the estimator of Y based on a sample consisting of r stages of random sampling. Then the expected value $E(\hat{Y})$ and the variance $V(\hat{Y})$ are

$$E(\hat{Y}) = E_1 E_2 \cdots E_r(\hat{Y}) \tag{5.6}$$

and

$$V(\hat{Y}) = V_1 E_2 E_3 \cdots E_r(\hat{Y}) + E_1 V_2 E_3 \cdots E_r(\hat{Y}) + \cdots + E_1 E_2 E_3 \cdots E_{r-1} V_r(\hat{Y}) \tag{5.7}$$

where r is the number of stages in the multistage sampling design, E_1 and V_1 are the expected value and variance for the first stage of randomization, and E_i and V_i $(i = 2, \ldots, r)$ are conditional expected value and conditional variance at the ith stage given that particular samples have been selected at the earlier stages, for example, for two-stage sampling,

$$E(\hat{Y}) = E_1 E_2(\hat{Y}) \tag{5.8}$$

and

$$V(\hat{Y}) = V_1 E_2(\hat{Y}) + E_1 V_2(\hat{Y}) . \tag{5.9}$$

Note that $V_1 E_2(\hat{Y})$ is a measure of the variation among psu's and $E_1 V_2(\hat{Y})$ is a measure of the variation among ssu's within psu's. In an r-stage design, the total variance consists of r parts, each part giving the variation among units of a particular stage within the units of the previous stage. The result is that the variability in the last stage can usually be ignored because the within-stage variation at this stage is divided by the product of the sample sizes at all stages and so is often negligible, as can be seen in the formula for $V(\hat{Y})$ in two-stage sampling below.

Use of eq. (5.7) makes it relatively simple to derive the variance of $\hat{Y}$ for any r-stage sampling design with specified probabilities of selection at each stage. Formulas are given below only for two-stage sampling since the extension to r-stage sampling is straightforward.

Simple Random Sampling Without Replacement at Both Stages. Estimation of the population total and the population variance are as follows:

$$\hat{Y} = \frac{N}{n} \sum_{i=1}^{n} \frac{M_i}{m_i} \sum_{j=1}^{m_i} y_{ij} , \tag{5.10}$$

$$V(\hat{Y}) = N^2 M_a^2 (1-f) \frac{\sigma_b^2}{n} + \frac{N}{n} \sum_{i=1}^{N} M_i^2 (1-f_i) \frac{\sigma_{wi}^2}{m_i} \tag{5.11}$$

with

$$\sigma_b^2 = \frac{1}{N-1} \sum_{i=1}^{N} \left(\frac{M_i}{M_a} \bar{Y}_i - \bar{Y} \right)^2 \qquad \text{the between-cluster variance ,}$$

$$\sigma_{wi}^2 = \frac{1}{M_i - 1} \sum_{j=1}^{M_i} (y_{ij} - \bar{Y}_{i\cdot})^2, \qquad \text{the within} - \text{cluster variance ,}$$

where $f = n/N$, $f_i = m_i/M_i$, $\bar{Y}_{i\cdot} = Y_{i\cdot}/M_i$, $\bar{Y} = \Sigma_{i=1}^N Y_{i\cdot}/N$, and $M_a = \Sigma_{i=1}^N M/N$ is the average number of ssu's per psu.

An unbiased estimator of $V(\hat{Y})$ is obtained by substituting unbiased variance estimators s_b^2 and s_{wi}^2 for σ_b^2 and σ_{wi}^2:

$$v(\hat{Y}) = N^2 M_a^2 (1-f) \frac{s_b^2}{n} + \frac{N}{n} \sum_{i=1}^{n} M_i^2 (1-f_i) \frac{s_{wi}^2}{m_i} \tag{5.12}$$

where

$$s_b^2 = \frac{1}{n-1} \sum_{i=1}^{n} \left(\frac{M_i}{M_a} \bar{y}_{i\cdot} - \bar{y} \right)^2 ,$$

$$\bar{y}_{i\cdot} = \frac{1}{m_i} \sum_{j=1}^{m_i} y_{ij} ,$$

$$\bar{y} = \hat{Y}/(NM_a) ,$$

and

$$s_{wi}^2 = \frac{1}{m_i - 1} \sum_{j=1}^{m_i} (y_{ij} - \bar{y}_{i\cdot})^2 .$$

Sampling Proportional to Cluster Sizes at the First Stage and Simple Random Sampling Without Replacement at the Second Stage. The efficiency of multistage sampling can be increased using the covariate information obtained in earlier stages at later stages. For example, in two-stage sampling, it may be desirable to select larger primary units with higher probability than smaller ones. This can be done by either sampling proportional to size or by stratifying the psu's by size and using two-stage sampling in each stratum.

We consider only pps sampling with replacement in order to simplify the sample draw and estimator of variance.

An unbiased estimator of Y is

$$\hat{Y} = \frac{1}{n} \sum_{i=1}^{n} M_i \bar{y}_{i\cdot}/p_i \,, \tag{5.13}$$

where p_i is the probability of selecting the ith psu at each draw, with variance

$$V(\hat{Y}) = \frac{1}{n} \sum_{i=1}^{N} p_i \left(\frac{Y_{i\cdot}}{p_i} - Y \right)^2 + \frac{1}{n} \sum_{i=1}^{N} \frac{M_i^2}{p_i} (1 - f_i) \frac{\sigma_{wi}^2}{m_i} \,, \tag{5.14}$$

and an unbiased variance estimator of $\hat{Y}$ is

$$v(\hat{Y}) = \frac{1}{n(n-1)} \sum_{i=1}^{n} \left(\frac{M_i \bar{y}_{i\cdot}}{p_i} - \hat{Y} \right)^2 . \tag{5.15}$$

Cost and Variance Trade-Offs. As mentioned earlier, there are two parts to the variance in two-stage sampling, the between-psu and within-psu variances. The first decreases as a function of n and the second decreases as a function of both n and the number of elements m_i. The sampling variance of an estimator of Y in a two-stage design can be written as

$$V(\hat{Y}) = \frac{1}{n} \left\{ A_1 + \sum_{i=1}^{N} \frac{A_{2i}}{m_i} \right\} + A_3 \,, \tag{5.16}$$

where A_1 and A_{2i} are, respectively, the contributions to the variance of the first and second stages, and A_3 is a constant. If sampling is with replacement at the first stage, $A_3 = 0$ (Murthy 1967).

An increase in n reduces both the components A_1/n and $\Sigma_{i=1}^{N} A_{2i}/(nm_i)$, whereas an increase in m_i only reduces the latter term. Hence, for increased statistical efficiency, it makes more sense generally to put extra effort into establishing additional psu's rather than ssu's. The opposite is true when costs are considered.

A simple cost function for collecting information for two-stage sampling is

$$C = C_0 + nC_1 + nm_a C_2 \,, \tag{5.17}$$

where

- C is the total cost of the survey,
- C_0 is overhead cost,
- C_1 is the cost incurred due to the location of the psu, planning the best route to it in the office, actual travel time to psu location and

establishment of the psu in the field, preliminary tabulation of psu information for such purposes as checking for error and missing data, and subsequent cost of psu revisitation if errors, missing data or lost data are noted,

C_2 is the cost of locating and establishing the ssu in the field, collecting the necessary complete information, preserving it satisfactorily, and processing the information collected,

m_a is the average number of ssu's sampled per psu, which depends on the procedure used to allocate all the sample units.

Clearly there is some ambiguity about whether some particular costs should be assigned to C_1 and C_2, and more reasonable cost functions may readily come to mind. The great advantages of eq. (5.17) are that it is simple, reasonable cost estimates C_1 and C_2 can often be obtained, and often better cost functions cannot readily be postulated.

In practice, C_1 is almost always larger than C_2, so that n increases total cost more than m. But to reduce the variance, we prefer an increase in n over one in m; hence some optimum balance between variance and cost is desired in the selection of n and m. To determine optimum n and m, we can either fix total cost C at C_F or fix $V(\hat{Y})$ at a fixed value V_F.

In the following discussion, simple random sampling (SRS) is assumed, and the optimum values

$$m_i = Nm\sqrt{A_{2i}} \Big/ \sum_{i=1}^{N} \sqrt{A_{2i}} \tag{5.18}$$

are used so that

$$\sum_{i=1}^{N} A_{2i}/m_i = A_2/m ,$$

where A_2 is independent of m. In practice the values A_{2i} $(i = 1, \cdots, N)$ are usually not available for the variable of interest, so that it has to be calculated for a suitable auxiliary variable from data in a previous survey or based on a pilot study conducted for this purpose.

For fixed total cost $C_F = C$, the optimum values of n and m that minimize the variance are

$$m_0 = \sqrt{A_2 C_1/(A_1 C_2)} , \tag{5.19}$$

$$n_0 = (C_F - C_0)(\sqrt{A_1/C_1})/(\sqrt{A_1 C_1} + \sqrt{A_2 C_2}) \tag{5.20}$$

and the minimum variance is

$$V_{\text{opt}}(\hat{Y}) = \frac{1}{C_F - C_0} (\sqrt{A_1 C_1} + \sqrt{A_2 C_2})^2 + A_3 . \tag{5.21}$$

If the variance is fixed at V_F, then the optimum values of n and m are m_0 given by eq. (5.19) and

$$n_0 = \frac{\sqrt{A_1C_1} + \sqrt{A_2C_2}}{V_F - A_3} \sqrt{A_1/C_1}\,, \tag{5.22}$$

and the minimum cost is

$$C_{\text{opt}} = C_0 + (\sqrt{A_1C_1} + \sqrt{A_2C_2})^2/(V_F - A_3) \tag{5.23}$$

(Murthy 1967).

Note that these results depend on the simple cost function in eq. (5.17). Other cost functions that may be more realistic can be selected. For example, when the cost of travel between primary units is substantial, a better cost function may be

$$C = C_0 + C_1 n + C_t\sqrt{n} + C_2 nm \tag{5.24}$$

since travel cost C_t is often proportional to $\sqrt{n}$. For fixed cost C_F, Hansen *et al.* (1953) gave an approach to determine the optimum n_0 and m_0 for minimizing the variance, as well as a table for facilitating the choice. Generally, only rough estimates of C_0, C_1, C_2, A_1, and A_2 are available and complex cost functions resulting in complex calculations are not worthwhile.

Three-Stage Sampling with Probability Proportional to Size Sampling and Replacement at Each Stage. Extension of estimation for two-stage sampling to three-stage sampling is straightforward using the conditionality theorem 5.1. If n psu's, m ssu's per psu, and l elements per ssu are selected with probabilities of selection at each draw for each stage $\{p_i\}$, $\{p_{ij}\}$, and $\{p_{ijk}\}$ ($i = 1, 2, \ldots, N$, $j = 1, 2, \ldots, M_i$, $k = 1, 2, \ldots, L_{ij}$), then if y_{ijk} is the value of the kth element in the jth ssu of the ith psu, the estimator of the total Y is

$$\hat{Y} = \frac{1}{nml} \sum_{i=1}^{n} \frac{1}{p_i} \sum_{j=1}^{m} \frac{1}{p_{ij}} \sum_{k=1}^{l} \frac{y_{ijk}}{p_{ijk}}\,, \tag{5.25}$$

and

$$V(\hat{Y}) = \frac{1}{n}\left(\sum_{i=1}^{N} Y_{i\cdot\cdot}^2/p_i - Y^2\right) + \frac{1}{mn}\sum_{i=1}^{N}\frac{1}{p_i}\left(\sum_{j=1}^{M_i}\frac{Y_{ij\cdot}^2}{p_{ij}} - Y_{i\cdot\cdot}^2\right)$$

$$+ \frac{1}{nml}\sum_{i=1}^{N}\frac{1}{p_i}\sum_{j=1}^{M_i}\frac{1}{p_{ij}}\left(\sum_{k=1}^{L_{ij}}\frac{y_{ijk}^2}{p_{ijk}} - Y_{ij\cdot}^2\right) \tag{5.26}$$

where

$$Y_{i\cdot\cdot} = \sum_{j=1}^{M_i} Y_{ij\cdot} \qquad \text{and} \qquad Y_{ij\cdot} = \sum_{k=1}^{L_{ij}} y_{ijk}\,.$$

In addition an unbiased variance estimator for $\hat{Y}$ is

$$v(\hat{Y}) = \frac{1}{n(n-1)} \left(\sum_{i=1}^{n} t_i^2 - n\hat{Y}^2 \right), \tag{5.27}$$

where

$$t_i = \frac{1}{p_i} \frac{1}{m} \sum_{j=1}^{m} \frac{1}{p_{ij}} \left(\frac{1}{l} \sum_{k=1}^{l} \frac{y_{ijk}}{p_{ijk}} \right) \quad \text{(Murthy 1967)}.$$

The simple cost function

$$C = C_0 + nC_1 + nmC_2 + nmlC_3, \tag{5.28}$$

is an extension of eq. (5.17) where

- C_0 is the overhead cost,
- C_1 is basically the same as described for cost function (5.17),
- C_2 is the cost of locating and establishing the ssu in the field, and
- C_3 is the cost of locating and establishing the elements in the ssu, collecting the necessary information on the elements, preserving the information satisfactorily, and processing the information collected.

The variance function (5.26) can be written in the form

$$V(\hat{Y}) = \frac{A_1}{n} + \frac{A_2}{nm} + \frac{A_3}{nml}. \tag{5.29}$$

Then for fixed $C = C_F$,

$$n_0 = \frac{(C_F - C_0)\sqrt{A_1/C_1}}{\sqrt{A_1C_1} + \sqrt{A_2C_2} + \sqrt{A_3C_3}}, \tag{5.30}$$

$$m_0 = \sqrt{A_2C_1/(A_1C_2)}, \tag{5.31}$$

$$l_0 = \sqrt{A_3C_2/(A_2C_3)} \tag{5.32}$$

with minimum variance

$$V_{\text{opt}}(\hat{Y}) = \frac{1}{C_F - C_0} (\sqrt{A_1C_1} + \sqrt{A_2C_2} + \sqrt{A_3C_3})^2. \tag{5.33}$$

With variance fixed at V_F, the corresponding sample sizes to minimize cost are

$$n_0 = \frac{\sqrt{A_1C_1} + \sqrt{A_2C_2} + \sqrt{A_3C_3}}{V_F\sqrt{C_1/A_1}}, \tag{5.34}$$

and m_0 an l_0 are as in eq. (5.31) and (5.32) with minimum cost

$$C_{\text{opt}} = C_0 + \frac{1}{V_F} (\sqrt{A_1 C_1} + \sqrt{A_2 C_2} + \sqrt{A_3 C_3})^2 \tag{5.35}$$

(Murthy 1967).

5.2.2 Multiphase Sampling

In multiphase sampling, data are collected on suitable covariates in the early phases and a subsample is selected of the units for the main survey, the last phase, by utilizing the covariate information from the early phases either in design (stratification or pps sampling) or estimation (regression or ratio estimation). In multiphase sampling, a complete frame of elements is required since a sample of elements is selected at each phase. In other words, definition of the sampling unit is the same for each phase of sampling. This design is used primarily to reduce the cost of sampling by collecting a relatively large amount of data on covariates that are easier and cheaper to obtain than the variable of interest and then measuring the variable of interest (and the covariates if needed) on an efficiently selected subsample at the last phase.

The discussion that follows concentrates on double (two-phase) sampling. In multiphase sampling with more than two phases, information from the different phases can be combined in quite different ways. This will be discussed briefly later and illustrated with an example from the Alaska Forest Survey.

Double Sampling for Stratification. Assume a population is to be divided into K strata and the first-phase sample of n' units is to be used to estimate the size of each stratum. For example, the survey objective is to estimate the timber volume on a large, well-defined forested area containing three strata: pine, hardwood, and mixed pine-hardwood. On the basis of low-altitude photographs, n' sample points are located in the population to estimate areas covered by the strata. Then a subsample of n points are located on the ground within the strata and sampled for timber volume.

Based on the n' first-phase samples, estimate either the proportion of the population falling in stratum $h(W_h = N_h/N)$ or the number of population units falling in stratum h, $N_h(h = 1, \ldots, K)$. The latter requires knowledge of the number of units in the population, N. Then an estimator of W_h is $w_h = n'_h/n'$, where n'_h = number of sample units falling in stratum h ($h = 1, \ldots, K$).

The n sample units in the second-phase sample can be selected in one of two ways, namely, at random from the n' units by selecting $w_h.n$ units from each stratum or by using some unequal probability sampling scheme. The latter is an unlikely option unless some continuous covariate of interest is available on the n' units.

The following theorem assumes the simple situation of random selection of the n sample units with a fixed percentage (proportional allocation) from each stratum. The results can easily be generalized to more complex sample designs.

Theorem 5.2. If n' sample units are selected to estimate the proportion of the population falling in each of K strata and n sample units are selected at random (with fixed proportions $v_h = n_h/n \leq 1$, $h = 1, \ldots, K$) from n' and measured for the variable of interest y, then

$$\hat{Y}_{dst} = N \sum_{h=1}^{K} w_h \sum_{i=1}^{n_h} y_{hi}/n_h = \sum_{h=1}^{K} \hat{N}_h \bar{y}_h = N\bar{y}_{st} \tag{5.36}$$

is an unbiased estimator of Y with variance

$$V(\hat{Y}_{dst}) = N^2 S^2\left(\frac{1}{n'} - \frac{1}{N}\right) + N^2 \sum_{h=1}^{K} \frac{W_h S_h^2}{n'}\left(\frac{1}{v_h} - 1\right), \tag{5.37}$$

where S^2 is the population variance of y, S_h^2 is the variance of y in stratum h, and $\bar{y}_h$ and $\bar{y}_{st}$ are the sample mean for stratum h and overall mean defined in Ch. 3 for stratified sampling.

An almost unbiased sample estimator of $V(\hat{Y}_{dst})$, if both $1/n'$ and $1/N$ are negligible (Cochran 1977 recommends < 0.02), is

$$v_1(\hat{Y}_{dst}) = N^2 \sum_{h=1}^{K} \frac{w_h^2 s_h^2}{n_h} - N \sum_{h=1}^{K} w_h s_h^2 + N^2 \frac{g_1}{n'} \sum_{h=1}^{K} w_h(\bar{y}_h - \bar{y}_{st})^2 \tag{5.38}$$

where $g_1 = (N - n')/(N - 1)$.

To show the utility of Theorem 5.1 we derive $V(\hat{Y}_{dst})$ in eq. (5.37).

Proof: Since $V(\hat{Y}_{dst}) = V_1 E_2(\hat{Y}_{dst}) + E_2 V_1(\hat{Y}_{dst})$ by Theorem 5.1, then

$$\begin{aligned}
V_1 E_2(\hat{Y}_{dst}) &= V_1\left(\sum_{h=1}^{K} w_h N_h \bar{Y}_h\right) \\
&= \sum_{h=1}^{K} \bar{Y}_h^2 N_h^2 V_1(w_h) + \sum_{h \neq h'}^{K} N_h N_{h'} \operatorname{Cov}(w_h, w_{h'}) \\
&= \sum_{h=1}^{K} N_h^2 \bar{Y}_h^2 \frac{W_h(1 - W_h)}{n'} \left(\frac{N - n'}{N - 1}\right) \\
&\quad - \sum_{h \neq h'}^{K} N_h N_{h'} \bar{Y}_h \bar{Y}_{h'} \frac{W_h W_{h'}}{n'} \left(\frac{N - n'}{N - 1}\right) \\
&= \frac{1}{n'}\left(\frac{N - n'}{N - 1}\right)\left[\sum_{h=1}^{K} W_h Y_h^2 - \sum_{h=1}^{K} W_h^2 Y_h^2 - \left(Y^2 - \sum_{h=1}^{K} W_h Y_h^2\right)\right] \\
&= \frac{1}{n'}\left(\frac{N - n'}{N - 1}\right)\left[\sum_{h=1}^{K} W_h Y_h^2 - Y^2\right].
\end{aligned}$$

Also,

$$
\begin{aligned}
E_2V_1(\hat{Y}_{dst}) &= E_2\left[\sum_{h=1}^{K} w_h^2 N_h^2 V(\bar{y}_h)\right] \\
&= E_2\left[\sum_{h=1}^{K} w_h^2 N_h^2 \left(\frac{N_h - n_h}{n_h N_h}\right) S_h^2\right] \\
&= \sum_{h=1}^{K} E(w_h^2)\left(\frac{N_h - n_h}{n_h}\right) N_h S_h^2 \\
&= \sum_{h=1}^{K} \{V(w_h) + W_h^2\}\left(\frac{N_h - n_h}{n_h}\right) N_h S_h^2 \\
&= \left(\frac{N - n'}{N - 1}\right)\frac{1}{n'}\sum_{h=1}^{K} W_h(1 - W_h)\left(\frac{N_h - n_h}{n_h}\right) N_h S_h^2 + V(\hat{Y}_{st})
\end{aligned}
$$

The last term is the variance of $\hat{Y}$ with single-phase stratified sampling, that is, known W_h.

Combining $V_1E_2(\hat{Y}_{dst})$ and $E_2V_1(\hat{Y}_{dst})$ gives

$$
V(\hat{Y}_{dst}) = V(\hat{Y}_{st}) + \left(\frac{N - n'}{N - 1}\right)\frac{1}{n'}\left[\sum_{h=1}^{K} W_h(Y_h - Y)^2 + \sum_{h=1}^{K} W_h(1 - W_h)\left(\frac{N_h - n_h}{n_h N_h}\right) N_h^2 S_h^2\right].
$$

The second term on the right represents the increase in variance due to double sampling for stratification relative to simple stratified sampling. This term consists of two components: the term $\Sigma_{h=1}^{K} W_h(Y_h - Y)^2$ represents variation among strata totals, and this term is large if stratification has been effective; the term

$$
\sum_{h=1}^{K} W_h(1 - W_h)\left(\frac{N_h - n_h}{n_h N_h}\right) N_h^2 S_h^2
$$

is the within-strata variation, which should be small if stratification has been effective. Note that if the within-strata variation can be ignored, we obtain

$$
V(\hat{Y}_{dst}) \approx V(\hat{Y}_{st}) + \left(\frac{N - n'}{N - 1}\right)\frac{1}{n'}\sum_{h=1}^{K} W_h(Y_h - Y)^2 ,
$$

which is the form shown for $V(\hat{Y}_{dst})$ in Sukhatme *et al.* (1984) and Cochran (1977).

If $1/n'$ and $1/N$ are not negligible, the following complex expression gives an unbiased estimator of $V(\hat{Y}_{st})$:

$$v_2(\hat{Y}_{dst}) = \frac{Nn'(N-1)}{n'-1}\left[\sum_{h=1}^{K} w_h s_h^2\left(\frac{1}{n'v_h} - \frac{1}{N}\right) + \frac{g_1}{n'}\sum_{h=1}^{K} s_h^2\left(\frac{w_h}{N} - \frac{1}{n'v_h}\right) + \frac{g_1}{n'}\sum_{h=1}^{K} w_h(\bar{y}_h - \bar{y}_{st})^2\right]. \tag{5.39}$$

In practice, it is often desirable to ensure that the sampling rates in the second phase differ among strata. This requires prior knowledge of the variability within strata for allocating the n sample units. Let x be a variable closely correlated with our variable of interest y and assume that information on x can be collected quickly prior to the first-phase sample (e.g., x may be a visual estimate of timber volume associated with each of n_0 preliminary sample points in our earlier example). Then for the simple cost function

$$C = C'n' + \sum_{h=1}^{K} c_h n_h , \tag{5.40}$$

where C' is the cost of classifying a unit for the first phase and C_h is the cost of measuring a unit in stratum h, the expected cost for selected n' and v_h is

$$E(C) = C'n' + n'\sum_{h=1}^{K} C_h v_h W_h \tag{5.41}$$

and substitute

$$\hat{v}_h = s_{xh}\left[\frac{C'}{C_h}\left(s_x^2 - \sum_{h=1}^{K} w_{0h} s_{xh}^2\right)\right]^{1/2} , \tag{5.42}$$

for v_h where s_x^2 and s_{xh}^2 are the estimated variance for variable x in the population and stratum h, respectively, and w_{0h} is the estimate of W_h for stratum h based on the preliminary sample. Then n' can be computed from the expected cost eq. (5.41).

Cochran (1977) dealt with the situation where estimates are desired only for the overall total, the purpose of stratification being primarily to increase the efficiency of sampling. But there are situations where interest is in parameters of both the strata and the overall population.

For example, the FIA unit of the Intermountain Forest Experiment Station in Ogden, Utah classifies a large number of n' systematically located photo points into K strata. A subsample of $n_h (h = 1, \ldots, K)$ plots is visited on the ground. Each of the $n = \Sigma_{h=1}^{K} n_h$ ground plots is classified into one of K strata based on ground information, and various tree and stand characteristics are sampled on the ground plots. Classification into the K strata from the photos and on the ground is not always the same; classification based on the ground information is usually considered more accurate. Hence the photo information is corrected by ground-based information to obtain reliable estimators of stratum acreage and estimators of tree and stand information for these strata. Li *et al.* (1992) addressed the situation where the first-phase estimators are corrected by information collected at the second-phase.

The estimated strata weights are then

$$w_{\cdot j} = \sum_{h=1}^{K} w_{h\cdot} n_{hj} / n_{h\cdot} \qquad j = 1, \ldots, K \tag{5.43}$$

where

$w_{h\cdot} = n'_h / n'$,
$n_h{}'$ = the number of sample units in stratum h based on the first-phase sample,
n_{hj} = the number of units classified in first-phase stratum h and in second-phase stratum j, and
$n_{h\cdot}$ = the number of second-phase samples selected in first-phase stratum h.

Li *et al.* (1992) give a reliable classical variance estimator for the $w_{\cdot j}$:

$$\begin{aligned} v(w_{\cdot j}) = \sum_{h=1}^{K} \Bigg\{ \frac{n_{hj}}{v_h n'} \Bigg[\frac{N-n'}{(n'-1)N} \left(1 - \frac{n_{hj}}{n' v_h}\right) \\ + \frac{n_{h\cdot} - n_{hj}}{v_h n' (n_{h\cdot} - 1)} \left(\frac{N-n'}{(n'-1)N} + 1 - v_h \right) \Bigg] \Bigg\} \\ - \frac{N-n'}{N} \sum_{h \neq j}^{K} \frac{1}{n'^2 (n'-1)} \frac{n_{hj} n_{lj}}{v_h v_l} \end{aligned} \tag{5.44}$$

where $n_{h\cdot}$ is the number of second-phase units selected from first-phase stratum h and $v_h = n_{h\cdot} / n'_h$.

An approximately unbiased estimator of the means for the second-phase strata is

$$\bar{y}_{\cdot j} = \left[\sum_{h=1}^{K} w_{h\cdot} \bar{y}_{hj} n_{hj} / n_{h\cdot} \right] \Big/ w_{\cdot j} \tag{5.45}$$

where $\bar{y}_{hj}$ is the mean of the second phase sample units falling in first-phase stratum h and second-phase stratum j.

A classical estimator of $V(\hat{Y}_{\cdot j})$ derived by Li *et al.* (1992) was shown to be less reliable than a bootstrap variance computed as follows: draw n_B bootstrap samples from the two-phase sample consisting of n' first-phase and n second-phase units. There were $n' - n$ units not sampled at the second phase. Bootstrapping at the first phase was done by choosing $n' - n$ units with replacement from those $n' - n$ units. Bootstrapping also was applied to each second phase sample of size n_h within the first-phase stratum h. Together there were n' resampled first-phase and n resampled second-phase samples.

Two bootstrap procedures were used for the second-phase sample. In the first procedure, the with-replacement bootstrap ground sample size $n_{h\cdot}$ was fixed for every stratum h; in the second, the bootstrap sample size varied within strata, but sampling fractions v_h within strata were fixed for each of the n_B samples. The latter implies that sometimes a with-replacement sample taken in some particular stratum may be larger or smaller than the original second-phase sample from that stratum since the value of $n'_{B.h}$, the number of bootstrap units falling into first-phase stratum h, varies between bootstrap samples. To calculate the resulting variance estimator, a bootstrap second-phase sample of size $n'_{B.h}(n_{h\cdot}/n'_{h\cdot})$ was chosen from $n_{h\cdot}$ second-phase units with replacement. Both bootstrapping variance estimators were about equally reliable.

Double Sampling with Ratio and Regression Estimators. Covariate information can also be used in ratio or regression estimation. This means that the covariate x and the variable of interest y are both continuous variables and some underlying relationship between them is assumed although often not stated explicitly.

Let n' units be selected in the first phase with any probabilistic sampling design and let a subsample n of these n' units be selected by any probabilistic design that only uses the first-phase information in estimation. Let x be a variable measured on the n' first-phase units and let y and x be variables measured on the second-phase sample. If $\hat{X}_1$ is an unbiased estimator of X based on the first-phase sample and $\hat{Y}_2$ and $\hat{X}_2$ are unbiased estimators of Y and X based on the second-phase sample, then

$$\hat{Y}_{dr} = \frac{\hat{Y}_2}{\hat{X}_2}\hat{X}_1 = \hat{R}\hat{X}_1 \tag{5.46}$$

is generally a biased but consistent ratio estimator of Y with approximate variance

$$\begin{aligned} V(\hat{Y}_{dr}) = {} & V(\hat{Y}_2) - 2R\,\mathrm{Cov}(\hat{Y}_2, \hat{X}_2) + R^2V(\hat{X}_2) + R^2V(\hat{X}_1) \\ & + 2R\,\mathrm{Cov}(\hat{Y}_2, \hat{X}_1) - 2R^2\,\mathrm{Cov}(\hat{X}_1, \hat{X}_2) \end{aligned} \tag{5.47}$$

where $R = Y/X$ (Murthy 1967).

An estimator of $V(\hat{Y}_{dr})$ can be obtained by substituting suitable sample estimates for the various terms in eq. (5.47). If SRS with replacement is used at both phases, then the components on the right-hand side in eq. (5.47) are given by

$$V(\hat{Y}_2) = N^2\sigma_y^2/n\,, \qquad V(\hat{X}_2) = N^2\sigma_x^2/n\,,$$

$$V(\hat{X}_1) = N^2\sigma_x^2/n'\,, \qquad \mathrm{Cov}(\hat{Y}_2, \hat{X}_2) = N^2\rho\sigma_x\sigma_y/n\,,$$

$$\mathrm{Cov}(\hat{Y}_2, \hat{X}_1) = N^2 \rho \sigma_x \sigma_y / n', \quad \text{and}$$

$$\mathrm{Cov}(\hat{X}_1, \hat{X}_2) = N^2 \sigma_x^2 / n'$$

where ρ is the correlation coefficient between x and y.

Then,

$$V(\hat{Y}_{dr}) = \frac{N^2}{n} \{\sigma_y^2 - 2R\rho\sigma_x\sigma_y + R^2\sigma_x^2\} + \frac{N^2}{n'} \{2R\rho\sigma_x\sigma_y - R^2\sigma_x^2\} . \tag{5.48}$$

If the first- and second-phase samples are drawn independently from the population, then $\mathrm{Cov}(\hat{Y}_2, \hat{X}_1)$ and $\mathrm{Cov}(\hat{X}_1, \hat{X}_2)$ are zero and

$$V(\hat{Y}_{dr}) \doteq \frac{N^2}{n} \{\sigma_y^2 - 2R\rho\sigma_x\sigma_y + R^2\sigma_x^2\} + \frac{N^2}{n'} R^2\sigma_x^2 . \tag{5.49}$$

If n' first-phase and n second-phase units are selected by any probabilistic design, then the regression estimator is

$$\hat{Y}_{dlr} = \hat{Y}_2 + \hat{\beta}_2(\hat{X}_1 - \hat{X}_2) \tag{5.50}$$

where $\hat{\beta}_2$ is an estimator of the regression coefficient between y and x based on the second-phase information, $\hat{X}_1$ is an unbiased estimator of X based on the first-phase sample, and $\hat{X}_2$ and $\hat{Y}_2$ are unbiased estimators of X and Y based on the second-phase sample. To derive the variance of $\hat{Y}_{dlr}$, the assumption is generally made that n is sufficiently large so that $\hat{\beta}_2 \doteq \beta$. Then $\hat{Y}_{dlr}$ becomes the much simpler estimator

$$\hat{Y}_{dlr1} = \hat{Y}_2 + \beta(\hat{X}_1 - \hat{X}_2) \tag{5.51}$$

and, using the conditionality theorem 5.1,

$$\begin{aligned} V(\hat{Y}_{dlr}) \doteq V(\hat{Y}_{dlr1}) &= V_1E_2(\hat{Y}_{dlr1}) + E_1V_2(\hat{Y}_{dlr1}) \\ &\doteq V(\hat{Y}_1) + E_1V_2(\hat{Y}_2)(1 - \rho^2) \end{aligned} \tag{5.52}$$

where $\hat{Y}_1$ is an unbiased estimator of Y that could be used if x were observed for all the units in the first-phase sample.

If replacement sampling is used at the first phase and if the second-phase sample is drawn with SRS with replacement from the first-phase sample, then $V(\hat{Y}) = \sigma_y^2/n_a$ and $V(\hat{X}) = \sigma_x^2/n_a$ where $n_a = n'$ or n depending on whether the first- or second-phase sample is involved. Then

$$E_1V_2(\hat{Y}_{dlr1}) \doteq \left(1 - \frac{n}{n'}\right) V(\hat{Y}_2)(1 - \rho^2)$$

and

$$V(\hat{Y}_{dlr}) \doteq \left[1 - \left(1 - \frac{n}{n'}\right)\rho^2\right]V(\hat{Y}_2) = \left[\frac{\rho^2}{n'} + \frac{1-\rho^2}{n}\right]\sigma_y^2 , \qquad (5.53)$$

which is clearly larger than for single-phase regression estimation. However, it is usually costly to collect information on x for all n' units so that double sampling is likely to be less costly than single-phase sampling. In double sampling, increasing the number of first-phase samples reduces the number of second-phase samples that can be selected for a given cost. Hence, some judicious allocation of sampling effort between the two phases is needed. Assume that the cost function is

$$C = C_0 + n'C_1 + nC_2 , \qquad (5.54)$$

where C is total cost, C_0 is the overhead cost, and C_1 and C_2 are the costs of surveying a unit for x and y, respectively. Then the optimum values of n' and n to minimize variance for a fixed cost C_F are

$$n' = \frac{C_F - C_0}{C_1}\left\{1 + \left(\frac{C_2}{C_1}\frac{1-\rho^2}{\rho^2}\right)^{1/2}\right\}^{-1} \qquad (5.55)$$

and

$$n = n'\left(\frac{1-\rho^2}{\rho^2}\frac{C_1}{C_2}\right)^{1/2} \qquad (5.56)$$

with minimum variance

$$V_{\min}(\hat{Y}_{dlr}) = \frac{\sigma_y^2}{C_F - C_0}\{(C_1\rho^2)^{1/2} + [C_2(1-\rho^2)]^{1/2}\}^2 . \qquad (5.57)$$

If the variance is fixed at V_F, optimum n' and n are

$$n' = \frac{\sigma_y^2}{V_F}\rho^2\left\{1 + \left(\frac{1-\rho^2}{\rho^2}\frac{C_2}{C_1}\right)^{1/2}\right\} \qquad (5.58)$$

and

$$n = n'\left(\frac{1-\rho^2}{\rho^2}\frac{C_1}{C_2}\right)^{1/2} \qquad (5.59)$$

with minimum cost

$$C_{\min} = C_0 + \frac{V_y}{V_F}\{(C_1\rho^2)^{1/2} + [C_2(1-\rho^2)]^{1/2}\}^2 \qquad (5.60)$$

(Murthy 1967).

5.3 MONITORING OVER TIME

Duncan and Kalton (1987) discussed alternative sampling designs to assess a range of objectives in sampling populations over time. They called these designs repeated (or longitudinal) panel, rotating panel and split panel surveys. We call them complete replacement (CR_p), complete remeasurement (CR_m), sampling with partial replacement (SPR), and a combination of CR_m with either CR_p or SPR sampling. The time dimension is important because the characteristics of units in the population may change over time (for example, a tree may be a pulpwood tree at time 1 and a sawtimber tree at time 2), as may the population (e.g., an area may convert from being part of the forest population to urban land, or agricultural land may revert to forest and so become part of the forest population.)

When monitoring change, the following objectives may be of interest:

- To estimate population parameters at a given time, for example, total volume of trees (V_t) and percent of pathogen-infested trees (PI_t) at times t_1 and t_2.
- To estimate a population parameter averaged over two time periods t_1 and t_2, for example,

$$\frac{PI_{t_1} + PI_{t_2}}{2} .$$

- To estimate net change in volume over time as discussed in Section 4.10. There may be interest in attempting to explain such changes, for example, relating change in growth to climate, pollution, or stand conditions (Schreuder and Thomas 1991).
- To estimate components of change over time. These may include:
  - Change in units over time (where unit denotes a tree or sample plot), for example, we may be interested in the change in vigor and degree of pathogen infestation of a tree (if these characteristics can be measured objectively for a tree).
  - Average change for each unit over several time periods; for example, we may be interested in "typical" annual growth of a tree over its first 20 years.
  - Variability in a characteristic of a unit over time, for example, the variability in annual growth rate of a tree over a 20-year period.
- To aggregate data for units over time; for example, we may be interested in the silvicultural history of a plot and how it relates to such characteristics as mortality, growth, and vigor of trees on the plot over time.
- To measure the frequency, timing, and duration of events over a given time period such as the number, month of occurrence, and duration of ice storms in a defined area.

- To cumulate samples over time, especially for rare subpopulations; for example, a single survey may not contain a sufficient number of trees of a rare species that is chronically infested with a specific pathogen or a sufficient number of trees that recently became a preferred habitat for some species of wildlife. A key difference between these two examples is that infestation with the pathogen is a fairly static condition but becoming a preferred habitat is a new event that has consequences for the sample design to be considered.

Duncan and Kalton (1987) summarized the properties of various sampling designs in relation to the sampling objectives (Table 5.1). Descriptions of the designs proposed to meet the several sampling objectives are discussed in Sections 5.3.1–5.3.4.

5.3.1 Complete Replacement Sampling

In complete replacement (CR_p) sampling, measurements are made from basically the same population at different times but no attempt is made to sample any unit more than once. Such designs are generally simple and cheap to implement since the locations of units do not have to be maintained, and in each time period a sample of the population existing at that time is routinely selected. These designs are particularly good for estimating parameters at a given time, for cumulating samples over time for rare subpopulations and for estimating population parameters averaged over several time periods. Their most serious limitations are that they cannot estimate components of change or aggregate data for units over time, and net change is estimated inefficiently.

5.3.2 Complete Remeasurement Sampling

Complete remeasurement (CR_m) sampling employs measurements that are repeated on the same units at different times. Designs of this type are highly efficient in estimating net change and components of change and are useful for limited aggregation of data for units over time (e.g., habitat trees for wildlife). The method is not efficient for cumulating samples of units that occur infrequently or for chronic conditions such as fungal infections on trees. Net change derived from CR_m sampling may differ from that derived from CR_p sampling because the former reflects changes in values only, whereas the latter also reflects changes in the composition of the population, that is, additions to or subtractions from the population. CR_m sampling requires that the same sampling units be relocated at each measurement and that they remain representative of the population of interest (which includes ensuring that sample units are not visited excessively or managed differently from the rest of the population and often that sample unit locations be kept secret).

TABLE 5.1 Properties of Alternative Survey Designs in Relation to Various Objectives (adapted from Duncan and Kalton 1987)

Sampling Objective	Complete replacement sampling	Complete remeasurement sampling	Sampling with partial replacement sampling	Combination of CR with CRp or SPR
a) estimate population parameters at distinct times b) Estimate average values of population parameters	Automatically takes into account population changes	Needs mechanism for taking population changes into account	Needs mechanism for taking into account population changes during life of replacement group. Composite estimation can be used to produce efficient estimates	Remeasurement component needs mechanism for taking population changes into account
c) Estimates net changes	Estimates combined effects of changing values and changing population	Needs mechanism for taking population changes into account. Variance of change reduced by positive correlation of values between surveys	Needs mechanism for taking into account population changes during life of remeasurement. Composite estimation can be used to produce efficient estimates	Remeasurement component needs mechanism for taking population changes into account. Variance of change in remeasurement component reduced by positive correlation of values between surveys
d) Estimate components of change over time e) Aggregate data for individuals over time	Not possible	Well-suited for these purposes	Can be used for change estimation or aggregate information over time periods shorter than the time a sample is to be replaced in sample. Only the sample to be replaced can be used	Complete remeasurement component is well-suited for these purposes. Not possible for complete replacement surveys component
f) collecting data on events occurring in specified time periods	Not possible	Can construct long-term event history data by combining data from several surveys	Can construct long-term history data but on a more limited basis than complete remeasurement surveys	Can construct limited long-term history data
g) Cumulate samples over time	Excellent for static characteristics and for new events	Not useful for static characteristics, but useful for new events	Of some use for static characteristics and useful for new events	Complete replacement survey component excellent. Complete remeasurement survey component useful for new events but not for static characteristics

Problems with CR_m surveys include the loss of sample units that leave the population (e.g., forest to urban conversion, failure to relocate the unit, destruction of the unit due to wildfire, windblow, etc.) and the entry of new units into the population (e.g., reversion of agricultural lands to forest). These problems can be handled by adding new sample units located at random to replace the lost ones or by selecting new units at random in the new part of the population. Another problem is that no new units in the existing population enter at successive points in time so maintaining representativeness is critical.

5.3.3 Sampling with Partial Replacement

In sampling with partial replacement (SPR), the life of the sample units varies, some units being replaced by new ones at each remeasurement. Hence, during the period of three surveys, some units will be visited thrice, some twice, and others once only. Jessen (1942) was probably the first to use SPR, applying it to a problem in agriculture.

The continued introduction of new sample units helps to maintain a current sample of a changing population and reduces problems due to loss of units and excessive revisitation. Such designs are well suited for estimating population parameters at a given time and average parameters and net change over time. The precision of these estimates can be improved by combining regression-type estimates with simple mean estimates (Ware and Cunia 1962, Hazard and Promnitz 1974, Hazard 1977, Kilpatrick 1981, Kumar and Lee 1983). A disadvantage is that estimation can become quite complicated if maximum efficiency is desired and many remeasurements are involved. In addition the assumption of linearity required for regression estimation may not be valid, especially if the population is changing rapidly, as in the southern USA where timber harvesting is common and frequent.

SPR designs can be used to measure the components of change, to aggregate data for sample units across time, to assess the frequency, timing, and duration of events (such as the number, month of occurrence, and duration of ice storms), to estimate net change, and to cumulate sample units for rare subpopulations. Basically SPR can assess all goals but it is a compromise, except for current estimation where it does well.

Because of its importance in forestry, we address SPR sampling for two occasions and multiple occasions.

SPR Sampling—Two Occasions. The basic concept of SPR sampling on two occasions is that on the first occasion the population total is estimated from the total of all sample units measured at that time, while on the second occasion the sample consists of some new as well as some remeasured units. When estimating the population total on the second occasion, two independent estimates are combined. One of the estimates is the total estimated from the new units measured only on the second occasion and the other is a regression estimate where the coefficients are estimated using those units measured on both occasions, and the estimated total for the independent variable is based on all first-occasion units.

Formally, let n units be drawn at each of two occasions from N units in the population by some sampling scheme such that m units are common between the two samples and let $u = n - m$ be the number of units not shared by the two samples. (This restriction is not required as pointed out in Scott 1981; we make it for simplicity in understanding.) Let $\hat{Y}_{im}$, $\hat{Y}_{iu}$, and $\hat{Y}_{in}$ be the estimates of Y_i, the population total, on the ith occasion $(i = 1, 2)$ based on the common units (m), units not in common (u) and all (n) sample units, respectively. In addition let $\hat{\beta}$ be the regression coefficient estimator based on the m common units; let σ_1^2, and σ_2^2, respectively, be the variances of y at times 1 and 2; let σ_{12} be the covariance of y at times 1 and 2; and let ρ be the correlation between observations at times 1 and 2. Then an unbiased estimator of Y_2 based on the u new units at time 2 is

$$\hat{Y}_{2u} = N \sum_{i=1}^{u} y_{2i}/u \tag{5.61}$$

with variance

$$V(\hat{Y}_{2u}) = N^2\sigma_2^2/u$$

and variance estimator $v(\hat{Y}_{2u}) = s_2^2(1)/u$ where $s_2^2(1)$ is the within-sample variance of the u units on which the y_{2i} were measured. A regression-based estimator of Y_2 using the m common units at times 1 and 2 is

$$\hat{Y}_{2mr} = \hat{Y}_{2m} + \hat{\beta}(\hat{Y}_{1n} - \hat{Y}_{1m}) \tag{5.62}$$

with variance

$$V(\hat{Y}_{2mr}) = N^2 \frac{\sigma_2^2}{m}\left(1 - \frac{u}{n}\rho^2\right)$$

To arrive at a combined estimator of Y_2, one usually weights the two independent estimators in inverse proportion to their sample variances, that is,

$$\hat{Y}_2 = (\hat{w}_1\hat{Y}_{2u} + \hat{w}_2\hat{Y}_{2mr})/\hat{w} \tag{5.63}$$

with approximate variance

$$V(\hat{Y}_2) \doteq 1/w \tag{5.64}$$

and variance estimator

$$v(\hat{Y}_2) = \left[1 + \frac{4}{\hat{w}^2} \sum_{i=1}^{2} \hat{w}_i(\hat{w} - \hat{w}_i)/d_i\right]\hat{w} \tag{5.65}$$

where $\hat{w}_1 = 1/v(\hat{Y}_{2u})$, $\hat{w}_2 = 1/v(\hat{Y}_{2mr})$, $\hat{w} = \hat{w}_1 + \hat{w}_2$, w is estimated by $\hat{w}$, $d_1 = m - 1$, and $d_2 = u - 1$.

If $\sigma_1^2 = \sigma_2^2 = \sigma^2$ and $V(\hat{Y}_{2u})$ and $V(\hat{Y}_{2mr})$ are known, so that known weights w_1 and w_2 can be used in eq. (5.63), then

$$V(\hat{Y}_2) \doteq \left(\frac{n - u\rho^2}{n^2 - u^2\rho^2}\right)\sigma^2 . \tag{5.66}$$

This reduces to σ^2/n for complete matching ($m = n$) and no matching ($m = 0$). By differentiating $V(\hat{Y}_2)$ in eq. (5.66) with respect to the ratio m/n, equating the derivative to zero and then solving for m/n, the optimum matching ratio is

$$\left(\frac{m}{n}\right)_{\text{opt}} = \frac{\sqrt{1 - \rho^2}}{1 + \sqrt{1 - \rho^2}} \tag{5.67}$$

with minimum variance

$$V(\hat{Y}_2) = \{1 + \sqrt{1 - \rho^2}\}\sigma^2/(2n) . \tag{5.68}$$

If $\rho = 0$, $(m/n)_{\text{opt}} = 1/2$ and if $\rho = \pm 1$, $(m/n)_{\text{opt}} = 0$. Thus, the matched portion should not exceed 1/2 when estimating Y_2 (Murthy 1967).

Sampling over time usually requires reliable estimates of Y at the beginning and end of the period (Y_1, Y_2), as well as change over time $\Delta Y = Y_2 - Y_1$. Scott (1981) gave two estimators of change. The more efficient one is

$$\widehat{\Delta Y} = (\hat{w}_1\widehat{\Delta Y_1} + \hat{w}_2\widehat{\Delta Y_2})/\hat{w} \tag{5.69}$$

where

$\widehat{\Delta Y_1}$ = the estimator of change from the m remeasured plots,
$\widehat{\Delta Y_2}$ = the difference estimator based on the $n - m = u$ unmatched plots,
$\hat{w}_1 = s_1^2/m + s_2^2/m - 2s_{12}/m$ is an unbiased estimator of the population weight $w_1 = \sigma_1^2/m + \sigma_2^2/m - 2\sigma_{12}/m$,
$\hat{w}_2 = s_1^2/u + s_2^2/u$ is an unbiased estimator of $w_2 = \sigma_1^2/u + \sigma_2^2/u$, and
$\hat{w} = \hat{w}_1 + \hat{w}_2$ estimates $w = w_1 + w_2$.

Then

$$V(\widehat{\Delta Y}) = 1/w \tag{5.70}$$

and

$$v(\widehat{\Delta Y}) = \left[1 + \frac{4}{w^2} \sum_{i=1}^{2} \hat{w}_i(\hat{w} - \hat{w}_i)/d_i\right] \Big/ w \tag{5.71}$$

where $d_1 = m - 1$ and $d_2 = u - 1$.

Scott (1981) showed that $\widehat{\Delta Y}$ is more efficient than the alternate unbiased estimator of change

$$\widehat{\Delta Y_1} = \hat{Y}_2 - \hat{Y}_1 \tag{5.72}$$

where $\hat{Y}_2$ and $\hat{Y}_1$ are the best estimators of Y at occasions 1 and 2 defined earlier.

Clearly the optimum scheme for detecting change over a period is zero replacement, that is, remeasure all sample units, whereas our earlier result showed that the matching proportion should not exceed 1/2 for optimizing the estimator $\hat{Y}_2$.

Instead of change we might be interested in the rate of change of R_y. The estimator of R_y is

$$\hat{R}_Y = (\hat{Y}_2 - \hat{Y}_1)/\hat{Y}_1 \tag{5.73}$$

with approximate variance

$$V(\hat{R}_Y) = \{\sigma^2 + (1 + R_Y^2)\sigma^2 - 2(1 + R_Y)f\sigma_{12}\}/(nY_1^2)\,, \tag{5.74}$$

where $f = m/n$ and it is assumed that $|(\hat{Y}_1 - Y_1)/Y_1| < 1$ (Murthy 1967).

Tikkiwal (1967) discussed the efficiencies of several estimators for double sampling and the optimum matching of units. Raj (1965), Ghangurde and Rao (1969), and Chotai (1974) modified double sampling to allow for unequal probability sampling. Chakrabarty and Rana (1977) derived results for joint estimation of change and means in three-stage sampling on two occasions. Tam (1987) and references therein cast the SPR estimation problem in the form of a dynamic linear model with close ties to the Kalman filter. The relationship between SPR and the Kalman filter was discussed in the forestry literature by Dixon and Howitt (1979).

Sampling with Partial Replacement—More than Two Occasions. When a population is sampled on more than two occasions, data from different occasions can be combined in several ways but estimation becomes more complex. For a comprehensive discussion of the problem, see Tikkiwal (1967) and Newton *et al.* (1974). At time t, there may be parts of the sample matched with time $t - 1$, others matched with both times $t - 1$ and $t - 2$, and so on. To obtain the most efficient estimate of Y for the current period, multiple regression might be used involving matchings of all previous occasions. Estimates for previous periods could also be revised based on both more current and earlier information if suitably matched portions of the sample are available. Arnab (1978) generalized the two-phase varying probability sampling schemes of Raj (1965), Ghangurde and Rao (1969), and Chotai (1974) to more than two occasions.

SPR over time has several advantages and disadvantages relative to remeasurement of all units at each occasion, namely:

- Advantages:
  - It can be highly efficient in estimating population means and totals at a given time by utilizing information from earlier surveys combined with new information.
  - New sampling units can enter the sample, which helps in ensuring that the current sample is representative of the current population. This may be a considerable advantage if, for example, the locations of permanent plots on public lands are to be public knowledge, which could imperil the representativeness of the plots.

- Disadvantages:
  - SPR sampling can be a statistician's paradise but a practitioner's dilemma. Even for the simpler sampling strategies, variance estimation theory is complex and requires various assumptions such as known correlations between variables. However, techniques such as jackknifing (Miller 1974, Schreuder and Brink 1983, Gregoire 1984) make it considerably easier to deal with these problems.
  - SPR may not be efficient when estimating for subpopulations. Sample sizes are likely to be inadequate for relatively small subpopulations such as counties or groups of counties.
  - Visualizing and implementing the total design may be difficult for multilevel sampling designs such as SPR. Errors in implementation are more likely to occur with such designs than with simpler sampling strategies.
  - The estimation of multiple resource characteristics, many of which may be uncorrelated or poorly correlated over time, may reduce the gains in efficiency from SPR designs.

5.3.4 Combination Sampling

This involves combining remeasurement sampling with either CR_p or SPR. The remeasurement sampling component can be used to estimate components of change, aggregate data for units over time, estimate net change between any two sample periods, and estimate frequency, timing, and duration of events in a given time period. The CR_p or partial replacement sample can be used to check for potential biases due to remeasurement sampling and, having accounted for that, can be used with that sample to provide estimates of parameters at a given time, averaged over time, of net change, and to cumulate units with chronic or rare conditions. No example of this is available in forestry but Lievesley and Waterton (1985) discuss one for the social sciences.

5.4 COMBINING ESTIMATES FROM SEVERAL SURVEYS

Often estimates are made based on information from different surveys. We may be interested in linear combinations of estimators, product estimators, or ratios with products.

5.4.1 Combining Linear Combinations of Estimators

If estimates for the same parameter are available from several surveys, then we would normally be interested in combining these estimates into a single estimate with greater precision. Two situations are possible and these are discussed below.

All Estimates Are Made on the Same Area. This situation is fully discussed by Rao (1980) and references therein. The estimators that are considered most efficient are of the basic type

$$\hat{Y}_w = \sum_{i=1}^{k} w_i \hat{Y}_i / w \tag{5.75}$$

where $\hat{Y}_w$ = the combined estimate,
k = number of estimates to be combined,
$\hat{Y}_i$ = estimator of Y from the ith survey,
w_i = the weight attached to $\hat{Y}_i$ when the sample size $n_i > 1$,
w = the sum of the weights $w_i (i = 1, \ldots, k)$.

Olkin (1958) gave the optimum weights $w_i (i = 1, \ldots, n)$, which are a function of the variances σ_i^2 and covariances σ_{ij} as well as the minimum variance achievable. But as noted in Li *et al.* (1984), this results in a complex estimator and in the four-phase sampling simulation they considered it was only slightly more efficient than an estimator with weights $w_i = 1/\sigma_i^2 (i = 1, \ldots, k)$. Clearly they are the same when the estimates are independent so that all the $\sigma_{ij} (i \neq j, i = 1, \ldots, n; j = 1, \ldots, n)$ are zero. We ignore the σ_{ij} in the following discussion for the sake of simplicity. If the weights are not known, several alternate estimators of w_i are available (see Rao 1980 and references therein).

A highly efficient estimator that appears to use about all the available information in the samples is

$$\hat{Y}_{w_1} = \sum_{i=1}^{k} w_{1i}\hat{Y}_i / w_1 \tag{5.76}$$

where

$$w_{1i} = 1/v(\hat{Y}_i)\,, \qquad i = 1, \ldots, k\,,$$

$$y_{ij} = \text{observation } j \text{ for sample } i\,,$$

$$\bar{y} = \sum_{i=1}^{k} n_i\bar{y}_i \Big/ \sum_{i=1}^{k} n_i\,,$$

$$w_1 = \sum_{i=1}^{k} w_{1i}\,,$$

$$v(\hat{Y}_i) = \sum_{j=1}^{n_i} (y_{ij} - \bar{y})^2/n_i \qquad \text{for SRS for } n_i \geq 1\,,$$

and n_i is the size of sample i on which the estimate is based (Rao 1980).

A computationally simpler, but usually less efficient, estimator that is frequently used is

$$\hat{Y}_{w_2} = \sum_{i=1}^{k} w_{2i}\hat{Y}_i / w_2\,, \tag{5.77}$$

where

$$w_{2i} = 1/v(\hat{Y}_i)\,, \qquad i = 1, \ldots, k$$

$$w_2 = \sum_{i=1}^{k} w_{2i}\,,$$

and

$$v(\hat{Y}_i) = \sum_{j=1}^{n_i} (y_{ij} - \bar{y}_i)^2/(n_i - 1) \qquad \text{for SRS for } n_i > 1$$

with

$$\bar{y}_i = \sum_{j=1}^{n_i} y_{ij}/n_i\,.$$

Cochran and Carroll (1953) gave an approximate variance estimator for eq. (5.77), which may be adequate for values of $n_i > 4$ ($i = 1, \ldots, k$):

$$v(\hat{Y}_{w_2}) \doteq \frac{1}{w_2}\left[1 + 4\sum_{i=1}^{k} \frac{1}{n_i'}\, w_{2i}(w_2 - w_{2i})/w_2^2\right] \tag{5.78}$$

where

$$n'_i = \begin{cases} n_i - 1 & \text{if } n_i \le 4 \\ (n_i - 1) - 4(k-2)/(k-1) & \text{if } n_i > 4 \end{cases} .$$

An approximate variance estimator for eq. (5.76) is

$$v(\hat{Y}_{w_1}) \doteq \frac{1}{w_1}\left[1 + \frac{4}{w_1^2}\sum_{i=1}^{k}\frac{1}{n'_i} w_{1i}(w_1 - w_{1i})\right] \tag{5.79}$$

(Rao 1980). Both approximations ignore terms of $1/n_i^2$ and smaller.

Distribution-free jackknife variance estimators can also be used to estimate the variance of $\hat{Y}_{w_1}$ or $\hat{Y}_{w_2}$. Preliminary results in Rao (1980) showed that these are likely to be overestimates. If the survey estimates are correlated, the jackknife variance estimators may be preferable. Bootstrap variance estimators may be appropriate too.

Estimates Are Made from Different Areas and an Overall Estimate for the Total Area is Desired. This is stratified random sampling, which is covered in Section 3.2.3. The combined estimate generally used is

$$\hat{Y} = \sum_{h=1}^{k} A_h \bar{y}_h \tag{5.80}$$

where A_h is the area of stratum h, and $\bar{y}_h$ is the estimator of the characteristic of interest on a per unit area basis. The variance of this estimator is

$$V(\hat{Y}) = \sum_{h=1}^{k} A_h^2 S_h^2 \tag{5.81}$$

where S_h^2 is the variance of estimator $\bar{y}_h$.

5.4.2 Combining Products and/or Ratios of Estimators

In cases where different parameters are estimated by different surveys and interest is in their product or ratio, it is customary to assume that the estimates come from the same area and are independent. These two situations are handled as follows:

Interest is in the Product of Estimates. In this case

$$\hat{Y} = \prod_{h=1}^{k} \hat{Y}_h , \tag{5.82}$$

where $\hat{Y}$ is the combined estimate and $\hat{Y}_1, \ldots, \hat{Y}_k$ are the independent estimators from surveys 1, . . . , k. Then the variance of $\hat{Y}$ is

$$V(\hat{Y}) = \sum_{i \neq j}^{k} [Y_i^2 V(\hat{Y}_j) + Y_j^2 V(\hat{Y}_i) + V(\hat{Y}_i)V(\hat{Y}_j)] \tag{5.83}$$

(Goodman 1960), where $V(\hat{Y}_i)$ $(i = 1, \ldots, k)$ and $V(\hat{Y}_j)$ $(j = 1, \ldots, k)$ are the variances of the independent estimators. Then $V(\hat{Y})$ can be estimated by

$$v(\hat{Y}) = \left[\sum_{i \neq j}^{k} \hat{Y}_i^2 v(\hat{Y}_i) + \hat{Y}_j^2 v(\hat{Y}_j) - v(\hat{Y}_i)v(\hat{Y}_j) \right]. \tag{5.84}$$

If the estimates are correlated, jackknifing or bootstrapping techniques should be used to obtain estimators of the variance of $\hat{Y}$.

Interest is in the Ratios and Products of Several Estimators. In this case

$$\hat{Y} = (\hat{Y}_1)(1/\hat{Y}_2)(\hat{Y}_3)(1/\hat{Y}_4)(\hat{Y}_5) \cdots (1/\hat{Y}_k) \tag{5.85}$$

where $\hat{Y}$ is the combined estimator and $\hat{Y}_1, \ldots, \hat{Y}_k$ are the independent estimators from surveys 1, . . . , k $(k = k_1 + k_2)$. Again, the estimators $\hat{Y}_i$ are from the same area or population. If we redefine some of the estimators such that $\hat{Z}_1 = 1/\hat{Y}_2$, $\hat{Z}_2 = 1/\hat{Y}_4, \ldots, Z_{k_2} = 1/\hat{Y}_k$, then

$$\hat{Y} = \prod_{i=1}^{k_1} \prod_{j=1}^{k_2} \hat{Y}_i \hat{Z}_j \tag{5.86}$$

with variance (obtained from eq. 5.83):

$$\begin{aligned} V(\hat{Y}) = {} & \sum_{i \neq i'}^{k_1} [\hat{Y}_i^2 V(\hat{Y}_{i'}) + \hat{Y}_{i'}^2 V(\hat{Y}_{i'}) + V(\hat{Y}_i)V(\hat{Y}_{i'})] \left(\prod_{j=1}^{k_2} \hat{Z}_j^2 \right) \\ & + \sum_{j \neq j'}^{k_2} [\hat{Z}_j^2 V(\hat{Y}_{j'}) + \hat{Z}_{j'}^2 V(\hat{Y}_{j'}) + V(\hat{Y}_j)V(\hat{Y}_{j'})] \left(\prod_{i=1}^{k_1} \hat{Y}_j^2 \right) \\ & + \sum_{i \neq i'}^{k_1} \sum_{j \neq j'}^{k_2} [\hat{Y}_i^2 V(\hat{Y}_{i'}) + \hat{Y}_{i'}^2 V(\hat{Y}_{i'}) + V(\hat{Y}_i)V(\hat{Y}_{i'})] \\ & \times [\hat{Z}_j^2 V(\hat{Y}_{j'}) + \hat{Z}_{j'}^2 V(\hat{Y}_j) + V(\hat{Y}_j)V(\hat{Y}_{j'})] \end{aligned} \tag{5.87}$$

where the $V(\hat{Y}_i)$ and $V(\hat{Y}_j)$ are as defined as for eq. (5.83). Eq. (5.87) is an approximation since the variance of the $\hat{Z}_i (= 1/\hat{Y}_{i+1})$ is only approximately equal to $1/V(\hat{Y}_{i+1})$. This is another good example where jackknifing or bootstrapping should be used for variance estimation.

5.4.3 Adjusting Estimates Based on Information from Independent Surveys.

Schreuder *et al.* (1988) and Li *et al.* (1990) dealt with situations where estimated proportions in two-way tables are adjusted by incorporating outside information, with or without error. The cell values of two-way tables are to be estimated. Because of the adjustments, constraints are required, so that the estimates sum to the original marginal or overall totals. The variables are: F for original estimates, O for outside estimates, and C for combined or adjusted estimates. Subscripts i and j represent the row and column number of the cell within the table. Subscripts I and J represent the subset of i and j for which outside information is available. For the $n \times 2$ case the marginal totals $F_{\cdot j}$ and $F_{i\cdot}$ are fixed; adjusted cell values C_{ij} are a function of both original cell values F_{ij} and marginal values ($F_{i\cdot}$ and $F_{\cdot j}$). The solution for the $n \times 2$ case with fixed marginals cannot be readily extended to the $n \times m$ case, so the latter is treated separately. For the $n \times 2$ case, for $i \in I$, the combined estimator weights the independent estimators inversely to their sample variances:

$$C_{iJ} = W^{-1}\{v(F_{iJ})^{-1}F_{iJ} + v(O_{iJ})^{-1}O_{iJ}\} \tag{5.88}$$

with

$$v(F_{iJ}) = F_{iJ}(1 - F_{iJ})/(m_1 - 1)\,,$$
$$v(O_{iJ}) = O_{iJ}(1 - O_{iJ})/(n_0 - 1)\,,$$
$$W = v(F_{iJ})^{-1} + v(O_{iJ})^{-1}\,,$$

m_1 is the sample size of the original sample, and n_0 is the sample size of the outside sample. The cells with row index $i \notin I$ are adjusted so that $\Sigma_{i=1}^{n} C_{iJ} = C_{\cdot J} = F_{\cdot J}$. For the cells in the column containing the cells with outside information,

$$C_{iJ} = \frac{(F_{\cdot J} - C_{IJ})}{F_{\cdot J} - F_{IJ}}\, F_{iJ} \qquad \text{for } i \notin I\,, \tag{5.89}$$

where $C_{IJ} = \Sigma_{i \in I}\, C_{iJ}$. The other column is computed as

$$C_{iJ} = F_{i\cdot} - C_{iJ} \qquad \text{for } j \in J\,.$$

For the $n \times m$ case, all cells (i, j) not in set (I, J) are adjusted, holding the overall total $T = \Sigma_{i,j}\, C_{ij} = \Sigma_i\, F_{i\cdot} = \Sigma_j\, F_{\cdot j}$ fixed, but allowing the margins $F_{i\cdot}$ and $F_{\cdot j}$ to change due to the adjustment. Then

$$C_{iJ} = \begin{Bmatrix} \left[\dfrac{T - C_{IJ}}{T - F_{IJ}} \right] F_{ij} & \text{for } (i, j) \not\in (I, J) \\ C_{ij} & \text{otherwise} \end{Bmatrix} . \tag{5.90}$$

For $(i, j) \in (I, J)$, C_{ij} can be computed from eq. (5.88).

For both situations, bootstrap estimates of the variance were reliable for the adjusted cell proportions in $n \times 2$ tables, but the classical variance estimator

$$v(C_{ij}) \doteq (T - C_{Ij})^2 [F_{ij}(1 - F_{ij}) + \hat{R}^2 F_{IJ}(1 - F_{IJ}) \\ + 2\hat{R} F_{ij} F_{IJ}] / \{(T - F_{IJ})^2 (m_1 - 1)\} \tag{5.91}$$

where $\hat{R} = F_{ij}/(T - F_{IJ})$, was more stable and usually had the least bias for the $n \times m$ table (where no marginal constraints were imposed).

5.5 MULTILEVEL SAMPLING IN FORESTRY—EXAMPLES

The following sections deal with specific applications of multilevel sampling in forestry.

5.5.1 Double Sampling for Stratification

Double sampling for stratification consists of two phases. In the first phase a large sample of units is used to estimate strata weights and in the second the desired information within the strata is obtained. A variation of this basic design is used by several of the Forest Inventory and Analysis (FIA) units of the United States USDA Forest Service (Forest Resources Economics Research Staff 1987). It involves interpreting a large number of sample points on aerial photographs and subsampling a proportion of the points on the ground. The aerial photo phase of sampling is used to obtain reliable estimates of the percent of the total area occurring in various strata (e.g. forest and nonforest). The ground sampling phase is comprised of clusters of five or ten VRP samples. If there are ten points, they are in one unit spaced 21.34 m (70 ft) apart, covering a total area of 0.46 ha (1.12 ac). Area attributes and vegetation data are collected at each sample point by measuring or counting the lesser vegetation and trees with diameter at breast height (DBH) <12.5 cm (5.0 inches DBH) on a fixed area around the point, and sample trees 12.5 cm (5.0 inches) DBH and larger with probability proportional to DBH squared. The theory for this sampling design is covered in Theorem 5.2 and subsequent discussion.

5.5.2 Three-Phase Sampling with Subsampling

In the early 1980's, the US Forest Service's FIA unit in Anchorage, Alaska, began testing a multiresource inventory system involving Landsat multispectral scanner imagery (LS), high-altitude color infrared photography (HAP), low-altitude color infrared photography (LAP), and ground-plots (G). Recent advancements in remote sensing (i.e., modern stereo-plotters) and aerial stand volume tables were used to efficiently gather data and information about Alaska's inaccessible timber and nontimber vegetation. This system was named the Alaska Integrated Resource Inventory System (AIRIS) (LaBau and Winterberger 1989). An intensive grid of 8-ha (20-acre) LS plots was used with a subgrid of 8-ha HAP plots with a subgrid of 8-ha LAP plots of which a subgrid of 8-ha plots was subsampled on the ground. This is three-phase sampling with subsampling. Since variability due to sampling at the ground sampling stage is generally a minor component of the total variability of the system and, in any case, is reflected in variance estimation for the regression or ratio estimators used, we refer to the general design as four-phase sampling.

AIRIS contains a number of important subdesigns, for example, census for using a single phase, two-phase sampling with or without subsampling, one-phase sampling with subsampling, and one-phase cluster sampling. There was some expectation early on that the two-phase sampling with subsampling might be a general enough design for all variables of interest, since using both LS and HAP may not be cost effective.

Vegetation class was evaluated on sample plots spaced on a 5-km square grid for LS, 10-km grid for HAP, 20-km grid for LAP, and 40-km grid for G (Figure 5.1).

Since the mid-1970's, imagery from satellite-borne sensors (i.e., Landsat multispectral scanner: MS) has proved particularly useful for classifying and inventorying vegetation (Hoffer *et al.* 1979; Walker *et al.* 1982; Häme 1987). The approach used in AIRIS differed from most previous multistage or mapping-oriented applications since it attempted to improve sampling efficiency using multiphase regression estimation.

The LAP photography was flown using both a radio-based positioning system and visual controls, and aerial stand volume tables were developed for estimating volumes from these photos (Mead and Setzer 1984; Setzer and Mead 1986).

AIRIS was first used to inventory the Tanana River basin (13.6 million ha) during the summers of 1982 and 1983. For this and other large areas, the data had to be manipulated in a piecemeal fashion based on universal transverse mercator (UTM) zones. Next, AIRIS was used in coastal southeast Alaska (11.2 million ha) during the summers of 1985 and 1986. In this rougher terrain, the large-scale nominal photo scale was reduced to 1:5000 and later to 1:7000 to facilitate photography and reduce the safety hazard for low-flying aircraft.

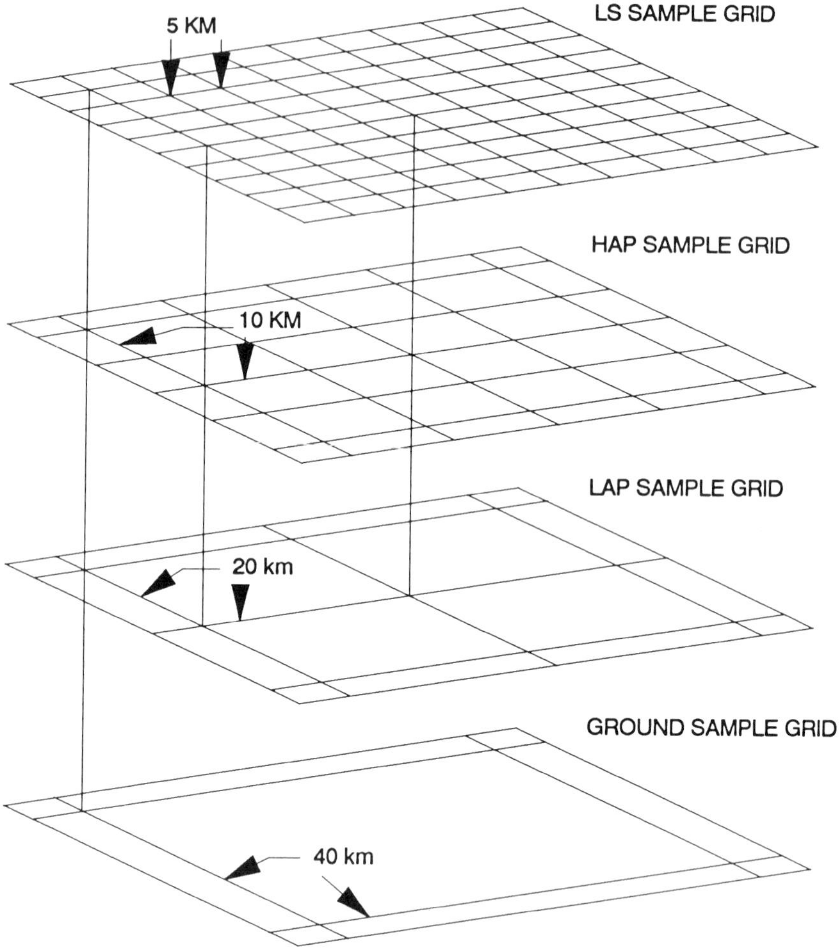

Figure 5.1 AIRIS four-phase sample grid intervals.

Estimation Theory. The basic concept underlying estimation in the four-phase sampling used in AIRIS is that covariate information obtained in the three aerial phases may be highly correlated with information of interest gathered in the fourth (ground) phase using double sampling with ratio or regression estimates. The three double sampling estimates are pooled by combining them with estimated weights $\hat{w}_i (i = 1, 2, 3)$. Schreuder (1981) developed the theory using weights based only on the inverses of the variances. Li *et al.* (1984) compared weights based on the estimated variance-covariance matrix and the estimated variances. The estimator using variance-covariance weights is more complex but more efficient, while the estimator using variance weights is simpler and preferable if computational simplicity is desirable.

Three regressions are of interest:

$$y_i = a_{\mathrm{LAP}} + b_{\mathrm{LAP}} x_{\mathrm{LAP}i} ,$$
$$y_i = a_{\mathrm{HAP}} + b_{\mathrm{HAP}} x_{\mathrm{HAP}i} ,$$

and

$$y_i = a_{\mathrm{LS}} + b_{\mathrm{LS}} x_{LSi} ,$$

where y_i is the value for the variable of interest on plot i, $x_{\mathrm{LAP}i}$, $x_{\mathrm{HAP}i}$, and $x_{\mathrm{LS}i}$ are the corresponding covariate values on plot i, and a and b are the regression coefficients for LAP with ground information (a_{LAP}, b_{LAP}), for HAP with ground information (a_{HAP}, b_{HAP}), and for LS with ground information (a_{LS}, b_{LS}).

The regressions are then used to yield estimates

$$\hat{Y}_1 = N(a_{\mathrm{LAP}} + b_{\mathrm{LAP}} \bar{X}_{\mathrm{LAP}}) ,$$
$$\hat{Y}_2 = N(a_{\mathrm{HAP}} + b_{\mathrm{HAP}} \bar{X}_{\mathrm{HAP}})$$

and

$$\hat{Y}_3 = N(a_{\mathrm{LS}} + b_{\mathrm{LS}} \bar{X}_{\mathrm{LS}}) ,$$

where N is population size (number of 8-ha plots) and $\bar{X}_{\mathrm{LAP}}$, $\bar{X}_{\mathrm{HAP}}$ and $\bar{X}_{\mathrm{LS}}$ are the large sample estimates for the covariates from LAP, HAP and LS, respectively. These are then combined with weights $\hat{w}_1$, $\hat{w}_2$, and $\hat{w}_3$ into an overall estimate $\hat{Y}$ as follows:

$$\hat{Y} = \sum_{i=1}^{3} \hat{w}_i \hat{Y}_i / \hat{w} \qquad (5.92)$$

where the weights were discussed earlier, and $\hat{w} = \Sigma_{i=1}^{3} \hat{w}_i$.

Results. The two surveys in which AIRIS was applied allowed evaluation on a wide range of vegetation types in Alaska. Schreuder *et al.* (1992a) discussed the advantages and disadvantages of implementing each of the four phases.

The results of AIRIS were disappointing. The effort was too ambitious for the available resources and objectives changed over time. The correlations between dependent variables of interest and covariates were not satisfactory (Table 5.2) and some of the better correlations were misleading. When AIRIS was first conceived, the emphasis was on multiresource as opposed to timber inventories. Subsequently the emphasis reverted to timber. This posed a problem, particularly in the Tanana Basin where only a

TABLE 5.2 R^2 Values for Key Variables of the Three Higher Phases with the Ground Sampling Phase for the Tanana ($n = 86$) and SE ($n = 65$) Regions of Alaska

		Phase		
Variable	Region	LS	HAP	LAP
Commercial forest land	SE	0.522	0.722	0.350
	Tanana	0.387	0.268	0.400
Softwoods	SE	0.914	0.759	0.529
	Tanana	0.690	0.229	0.204
Hardwoods	SE	0.966[a]	0.000[b]	0.005
	Tanana	0.345	0.516	0.079
Mixed	SE	0.962	0.856	0.588
	Tanana	0.677	0.105	0.001
Nonforest	SE	0.966	0.857	0.588
	Tanana	0.908	0.803	0.478

[a] 84 plots measurements with LS = 0, 2 plots with LS = 100.
[b] ALL HAP measurements are zero.

small portion of the plots represented the timber resource. Increased sampling intensity of forested as opposed to nonforested land would have been preferable and would have produced better results (higher R^2 values). Considerable difficulty was experienced in identifying a key variable, "commercial forest." Classifying a plot into this category is largely subjective even when inspected on the ground. Thus it was possible to classify an area as commercial forest at one phase of sampling and noncommercial forest at another: such a conflict in classification makes it difficult to establish good relationships.

In planning a major sampling effort the resources available to do the job well need to be considered carefully as well as the possibility of a change in objectives. In AIRIS, too much optimism was displayed to solve expected and unexpected problems and some shift in objectives could have been anticipated. With additional people and currently available computer power and with the experience from the first effort, a second effort using the same approach would be much more likely to be successful.

5.5.3 Point-Poisson Sampling and Alternatives

Grosenbaugh (1971, 1979) introduced the highly efficient point-Poisson (point-3P) sampling procedure for estimating total volume and different grades of volume in tracts of timber. Wood and Schreuder (1986) described in detail the following approach for implementing the procedure.

A preliminary sample of k_1 horizontal variable radius plots (VRP) is established on the tract and total or merchantable height h is measured or estimated for trees determined to be "in" using an angle count device such as a Spiegelrelaskop. This preliminary sample is used to determine the number k of first stage units (plots) and second stage units (Poisson-selected trees, m_e) required in the main survey and to estimate the sum of heights ($\hat{H}_o = \Sigma_{i=1}^{n} h_{oi}$) of the trees ($n$) selected in the preliminary point sample. A random number list for selecting the Poisson sample in the main survey is derived from this sum of heights and ancillary information.

Next, the k first-stage sample points are established on the area (either at random or on a grid) and the number of in-trees (n_i) at each point $i(i = 1, \ldots, k)$ is determined by the angle count device and their diameters are measured. This results in trees being selected proportionally to basal area. To implement the second-stage Poisson sample, h is estimated (by h_o) for each tree selected in the first-stage sample. The estimated height is then compared with a paired number from the random number list. If the random number is less than or equal to h_o, that tree becomes a Poisson sample and is measured by dendrometer (pentaprism caliper, Spiegelrelaskop, Telerelaskop, etc.) to compute its volume. These measurements include h. Because these trees are selected proportional to d^2h_o, i.e., almost proportional to their volume (if height is estimated in a reliable fashion), the procedure can be expected to be highly precise. It is also cost effective.

The estimator for total stand volume generally used for this procedure is

$$\hat{Y}_{pp} = \frac{AFH_{o\cdot}}{km} \sum_{i=1}^{m} y_i/(g_i h_{oi}) \tag{5.93}$$

(Grosenbaugh 1971, Yandle and White 1977) where

F = basal area factor (BAF) used in horizontal VRP sampling,
A = total area being sampled,
$H_{o\cdot} = \Sigma_{i=1}^{n} h_{oi}$ = sum of the estimated heights of the n first-stage trees,
k = number of first-stage point samples,
m = actual number of Poisson-selected trees, and
y_i and g_i = volume and basal area of tree i.

Eq. (5.93) is similar to Grosenbaugh's adjusted estimator in simple Poisson sampling (Schreuder *et al.* 1971).

Estimator $\hat{Y}_{pp}$ in eq. (5.93) does not appear to use all the information available for the first-stage sample trees (Schreuder *et al.* 1984a). Point-Poisson sampling can be viewed as double sampling where a covariate (either g or d^2h_o) is measured on the large sample (the n first-stage trees), and the variable of interest and the covariates are measured on the small sample (the m second stage (Poisson) trees). Thus the ratio estimator

$$\hat{Y}_{ppa} = \hat{Y}_{pp} * D_0^2 H_{02} / D_0^2 H_{01} = \hat{Y}_{pp} * R_2 \tag{5.94}$$

appears to be an alternative estimator. But substituting

$$D_o^2 H_{o1} = \sum_{i=1}^{m} \left(\frac{1}{km} \right) d_0^2 h_{oi} \frac{AH_o F}{g_i h_{oi}}$$

and

$$D_o^2 H_{o2} = \sum_{i=1}^{n} \left(\frac{AFd_0^2 h_{oi}}{kg_i n} \right)$$

into eq. (5.94) yields $\hat{Y}_{pp}$ in eq. (5.93), so the latter does use all the available information.

Wood and Wiant (1992a) and Schreuder *et al.* (1992c) introduced an alternative to point-Poisson sampling, called modified point-list sampling, in which one or more trees are selected proportional to height at each point. This guarantees a fixed number of sample trees. Wood and Wiant (1992a) found in a field study that the modified point-list-procedure using $\hat{Y}_{HT}$ was slightly less efficient than point-Poisson sampling using $\hat{Y}_{pp}$. Schreuder *et al.* (1992c) also considered point-pps sampling as a theoretical alternative to point-Poisson and modified point-list sampling in a simulation study using $\hat{Y}_{HT}$ ($\hat{Y}_{pp}$ for point-Poisson) and regression estimators $\hat{Y}_{gr}$ (eq. 3.134), $\hat{Y}_{piwr}$ (eq. 3.142), $\hat{Y}_{wr}$ (eq. 3.139), or $\hat{Y}_{pi}$ (eqn. 3.138) to estimate total volume and board foot volume. They concluded that point-Poisson sampling was always somewhat more efficient than the modified point-list procedure with all estimators. The reason was that modified point-list trees are selected for each point so that they are not always selected proportional to height in the list phase whereas they are selected proportional to height in the Poisson phase of point-Poisson sampling. Point-pps was somewhat more efficient than point-Poisson sampling for all estimators except $\hat{Y}_{HT}$ versus $\hat{Y}_{pp}$ (for board foot volume only). $\hat{Y}_{gr}$ was generally the most efficient estimator for all three procedures.

Ouyang and Schreuder (1992) suggested a possible application of Srivastava estimation in point-Poisson sampling. A general ratio estimator can be generated by estimating the regression coefficients for

$$y_i = \alpha + \beta g_i h_{oi}, \quad i = 1, \ldots, m,$$

computing the predicted values y_l^* for the n VRP sample trees and using those as follows:

$$y_{sr}^{(pp)} = \frac{FA}{k} \left[\frac{\sum_{l=1}^{m} y_l / (h_{ol} g_l)}{\sum_{l=1}^{m} y_l^* / (h_{ol} g_l)} \right] Y^*, \tag{5.95}$$

where $y_l^* = \hat{\alpha} + \hat{\beta} x_l^*$, $l = 1, \ldots, n$, and $Y^* = \Sigma_{l=1}^{n} y_l^*$. This is an approximate Srivastava estimator. For point-Poisson sampling we could obtain an estimator similar to $\hat{Y}_{s1}$ in eq. (4.23) but the π_i' for point-Poisson would be difficult to compute. Estimator $\hat{Y}_{sr}^{(pp)}$ was slightly more efficient than $\hat{Y}_{pp}$ for point-Poisson sampling (Ouyang and Schreuder 1992).

It seems unlikely that the variance and an unbiased estimator of the variance of the adjusted Poisson estimator can be obtained in a readily computable form (Schreuder *et al.* 1971). This is likely to be true also for estimator $\hat{Y}_{pp}$ in point-Poisson sampling. The difficulty in obtaining readily computable variances results from the variable sample size achieved in Poisson sampling. To obtain approximate variances and variance estimates, a logical approach is to treat Poisson sampling as sampling with probability proportional to size (pps) with fixed sample size either with replacement (ppswr) or without replacement (ppswor).

Grosenbaugh (personal communication) suggests an approximate variance estimator for $\hat{Y}_{pp}$ using point-Poisson sampling, namely,

$$v(\hat{Y}_{pp}) = \hat{Y}_{pp}^2 \frac{k'}{(k'-1)} \frac{\left[\left(k' \sum_{i=1}^{\acute{k}} \sum_{j=1}^{m_i} y_{ij}/(g_{ij} h_{oij})\right)^2 \Big/ \left(\sum_{i=1}^{m} y_i/(g_i h_{oi})\right)^2 - 1\right]}{\left(\acute{k} \sum_{i=1}^{k} m_i^2/m^2 - 1\right)}$$

$$\cdot \frac{1}{k}\left[k \sum_{i=1}^{k} \left(\sum_{j=1}^{n_i} h_{oij}\right)^2 \Big/ H_{o\cdot}^2 - 1\right] \tag{5.96}$$

$$+ \hat{Y}_{pp}^2 \frac{k'}{(k'-1)} \left[\frac{m \sum_{i=1}^{\acute{k}} \left(\sum_{i=1}^{n_i} y_{ij}/g_{ij} h_{oij}\right)^2 \Big/ m_i}{\left(\sum_{i=1}^{m} y_i/g_i h_{oi}\right)^2} - 1\right]\left(\frac{1}{k_a} - \frac{1}{k}\right)$$

where y_{ij}, h_{oij}, g_{ij} are the volume, visually estimated height, and basal area, respectively, of Poisson-selected tree j on point i. In addition,

k' = number of points with Poisson trees,
k = number of points sampled,
$k_a = k\{1 - \exp(-m/k)\}$,
m = number of Poisson trees, and
m_i = number of Poisson selected trees at point i.

The jackknife procedure can also be used as an alternative to $v(\hat{Y}_{pp})$ (Schreuder *et al.* 1984a). However, Schreuder *et al.* (1992c) found that simple bootstrapping, mimicking the actual sampling procedures used, gives reliable variance estimates for all regression estimators as well as for $\hat{Y}_{HT}$ and $\hat{Y}_{pp}$. There appears to be little justification for using anything but the bootstrap to estimate the variance for any of the sampling strategies. To illustrate the care needed in the mimicking process, consider the bootstrapping technique for modified point-list assuming, say, one tree/point is selected from k_1 points and two trees/point from k_2 points, $k = k_1 + k_2$:

1. Divide the k selected points into the two groups of points, k_1 and k_2.
2. Select two trees per point with replacement from each of the k_2 points so that a set of k_1 points with one tree and k_2 points with two trees, which may or may not be the same tree, is obtained.
3. Select k_1 points with one tree and k_2 points with two trees/point with replacement from the set in step 2 above.

5.5.4 Poisson-Poisson and Binomial-Poisson Sampling

For stands of timber of high value, an accurate estimate of true usable volume may be desired. Actual harvesting of some of the timber and destructive sampling (in a sawmill) may be necessary to assess reliably the extent of defect. In such cases Poisson-Poisson sampling may be a useful sampling design to apply (Ouyang *et al.* 1992a).

1. Select n_1 out of N trees in the population by Poisson sampling, selecting the trees proportional to some estimate of gross volume, say $x_1 = d^2$ or d^2h.
2. Select n_2 out of the n_1 sample trees proportional to some measure of defect, x_2, by Poisson sampling. Here x_2 could be an ocular estimate of defect by a highly skilled cruiser for example. (Experience has shown that even highly skilled cruisers may have trouble estimating defect on a few trees and it may be desirable to have the cruiser identify these during sampling and measure them fully, treating them as a separate stratum for which no estimate of defect is made).

The n_2 sample trees are then cut and sampled destructively to measure usable volume. Clearly, both sampling phases need to be implemented at the same time except that determining usable volume should probably be done at the sawmill.

The most promising ratio estimator for total usable volume is a modification of $\hat{Y}_a$ (eqn. 3.76) in Poisson sampling, that is,

$$\hat{Y} = (n_{e1}/n_1)(n_{e2}/n_2) \sum_{i=1}^{n_2} y_i/\pi_i \tag{5.97}$$

where n_{e1} = expected sample size at the first phase, n_{e2} the expected sample size at the second phase, and

$$\pi_i = \pi_{1i}\pi_{2i} = \left(n_{e1}x_{1i}\Big/\sum_{i=1}^{N} x_{1i}\right)\left(n_{n2}x_{2i}\Big/\sum_{i=1}^{n_1} x_{2i}\right).$$

An approximate variance estimator of $\hat{Y}$ is

$$v(\hat{Y}) = \frac{n_{e1}n_{e2}}{n_1} \sum_{i=1}^{n_2} (1 - \pi_{1i})\left(\frac{y_i}{\pi_{1i}} - \frac{\hat{Y}}{n_{e1}}\right)^2 \Big/ \sum_{i=1}^{n_2} \pi_{2i} + \frac{n_{e2}}{n^2} \sum_{i=1}^{n_2} (1 - \pi_{2i})\left(\frac{y_i}{\pi_{1i}\pi_{2i}} - \frac{n_1 \hat{Y}}{n_{e1}n_{e2}}\right)^2 \qquad (5.98)$$

(Ouyang *et al.* 1992a).

A special case of Poisson-Poisson sampling occurs if units are selected with equal probabilities at the first phase. This is binomial-Poisson sampling. This may have application in sampling truckloads of timber (Ouyang *et al.* 1992a).

5.5.5 Sampling with Partial Replacement

Bickford (1956) first proposed applying the basic, two-occasion SPR model to forest inventory and later reported results of a test he made using SPR on the Allegheny National Forest (Bickford 1959). Ware and Cunia (1962) discussed the procedure that links plot volumes at two different times by a simple linear regression function. They extended the theory by suggesting an optimization approach to the sampling design to include varying costs of plot establishment. Cunia (1965) advocated use of multiple instead of simple linear regression to estimate the second occasion parameter and Frayer (1966) proposed using the weighted least squares method of estimation whenever the conditional variance is not homogeneous. Frayer and Furnival (1967) considered SPR sampling for area estimation rather than for volume. Cunia and Chevrou (1969) extended the theory of SPR sampling to cover forest inventories with more than two successive measurements. Schreuder *et al.* (1987c) compared classical, jackknife, and bootstrap variance estimators for SPR on two occasions for three large samples of FIA plots in the northeastern USA which were treated as populations. The jackknife estimators were generally least biased for estimating the variance but the classical estimators were more stable for variables with highly skewed distributions except when sample size was small. In this situation, the jackknife estimators would probably be preferred. Generally, the jackknife confidence intervals were most reliable for all the populations, variables, and sample sizes that were considered. All these papers dealt with estimating a single parameter.

Newton *et al.*'s (1974) formulation of multivariate SPR sampling differed somewhat from that of Tikkiwal (1967), who assumed that the variables have the following correlations:

1. For any variable i at times r and s, the correlation between observations at times t and s is the product of the year-to-year correlations

$$\rho_{r,s}(i) = \rho_{r,s} = \prod_{t=r}^{s-1} \rho_{t,t+1}$$

where $\rho_{t,t+1}$ = correlation between variables $X_{t,i}$ and $X_{t+1,i}$ for $t = 1, 2, \ldots, h-1$ with $|\rho_{t,t+1}| \neq 1$.

2. For any variables (i, j) at any occasion (time) r,

$$\rho_{r,r}(i, j) = \rho'_{ij} = \prod_{t=1}^{j-1} \rho'_{t,t+1}$$

where $\rho'_{t,t+1}$ is the correlation between variables $X_{r,t}$ and $X_{r,t+1}$ for $t+1$ for $t = 1, 2, \ldots, k-1$ with $|\rho'_{t,t+1}| \neq 1$ and k = the number of variables.

3. For any variable (i, j) at occasions r, s,

$$\rho_{r,s}(i, j) = \rho_{r,s}\rho'_{ij} .$$

Tikkiwal dealt primarily with two variables where variable y is of more interest than x, so that on any occasion when double sampling is used, x is measured on a large sample and y on a subset of that sample. This procedure is extended easily to more variables and it also handles cases where the variables are of equal interest (i.e., no double sampling).

Newton *et al.* (1974) assumed that all p variables are of equal interest and their variance-covariance matrices over time are known. They considered only two occasions, used SRS to locate plots on each occasion, and examined the change in variables over time, whereas Tikkiwal did not.

Newton *et al.* (1974) listed three practical disadvantages of SPR estimates for the situation they discussed:

1. The assumption that the variance-covariance matrices are known is never fulfilled.
2. The SPR estimates depend on the existence of the inverse of the variance-covariance matrix of the vector of average differences between plots measured on the two occasions.
3. Even if the SPR estimators exist, it is not clear that they have minimum variance.

Matis *et al.* (1984) reviewed the literature of SPR sampling emphasizing forestry. Van Deusen (1989) proposed a generalized least squares (GLS) approach that can be used for multiple-occasion SPR. He provided an iterative GLS approach that simultaneously estimates population parameters and their variances and avoids incompatible estimates. Ware and Cunia's (1962) estimator for Y_2 is shown to be a special case of Van Deusen's GLS estimator.

SPR is known to be very efficient in estimating both current values and change in those values (Scott 1981). Despite this and its availability over several decades, few planners of resource surveys utilize it. For example, of the six Forest Service FIA units in the USA, only the unit covering the Northeastern states uses it and that unit is discarding it. But Sweden (Ranneby *et al.* 1987) and Switzerland (Michael Koehl, personal communication) are planning to use it in their national forest inventories.

Barnard (1974) noted, based on experience with SPR during the entire second cycle of forest surveys of 100 + million acres of forest land in the 14 Northeastern states of the USA, that SPR is flexible in meeting the objectives of the surveys. For example, the manager has flexibility in targeting different precision levels at successive occasions since the precision attained at time 1 can be improved upon or relaxed as needs and costs dictate; the mix of new and remeasured plots can be changed to emphasize either current or trend estimates, and different types of sample plots based on new sampling concepts and data collection techniques can be introduced; a check of the estimates can be made (the new and remeasured samples are independent and can provide a mutual check whether or not the parameter of interest is within the confidence interval for either estimate). Of these "advantages" the last is probably the most important when compared with the usual alternative of CR_m on each occasion. It is useful to be able to document whether or not the remeasured sample remains representative over time. Decreased or increased precision levels can be attained by reducing the number of plots remeasured or adding new (presumably temporary) plots. New sampling concepts and new data collection techniques can indeed be more easily implemented with SPR (on the new plots) than when all plots are remeasured.

Scott (1986a) noted several disadvantages of SPR arising from the flexibility of the design. The designer has more decisions to make regarding sample size and complexity increases as the number of remeasurements increases. For example, at the second remeasurement there are only three types of plots, whereas at the third there are seven. Thus the designer may be reluctant to add other methods of increasing sampling efficiency such as stratification. SPR also causes complexity in estimation. Regression equations have to be developed and carefully monitored, for example, strata may need to be grouped if the number of remeasured plots in a stratum is too small, and counties may need to be grouped to ensure an adequate sample size. Grouping has to be done carefully in both cases to avoid putting dissimilar plots together. In estimation, current volume is regressed on previous volume. For undisturbed plots, the current volume is usually greater than the previous volume but for disturbed (say harvested) plots current volume can be much less than previous volume. Clearly, for efficiency it would be desirable to separate the two in the regression process. However, this requires being able to identify disturbance for unremeasured plots without visiting them. Another problem arises because strata means are estimated separately for remeasured and new plots. These are then combined with weights inversely proportional to their variances. Maximum efficiency in estimation requires that this be done for each cell in a table and for row and column totals but the sum of the cell estimates will differ from that of the row and column estimates. Resource analysts feel that this additivity in the estimates is needed for their results to be credible. Thus adjustments are needed to ensure additivity but this increases the complexity of estimation even more.

Scott (1986a) suggested some ways around these disadvantages, for example, limit the estimation to information derived from the last two measurements only because experience indicates that plots remeasured after more than 15 years do not contribute much to the combined estimate of current volume, and place disturbed plots in a separate stratum and assign a new stratum to all plots (this is difficult to implement since the strata sizes have to be estimated from aerial photography for all strata including the disturbed stratum). To eliminate the problem of small sample size, he suggested grouping the remeasured plots in adjacent counties and developing a single set of equations that is applied to each county. Similarly group the strata. His objective was to have information from a minimum of 20 plots contributing to each regression. Sometimes, however, one may have to settle for fewer plots. The Northeast FIA unit uses a ratio adjustment procedure to ensure the additivity of cells using computer program FINSYS (Born and Barnard 1983).

Scott (1986b) noted that in response to a perceived change in user's needs, the NE FIA unit has abandoned SPR sampling. Rather than simply knowing the current situation well, the needs today are for a better understanding of trends and changes resulting from various forest management practices and policy decisions and for publishing state resource statistics promptly after data collection is completed.

SPR sampling may be an example where the potential gains from a highly efficient sampling strategy may be outweighed by the need for practical simplicity. (This same argument can be made regarding the four-phase AIRIS design discussed earlier.) In the case of the Northeast FIA unit, emphasis over time has shifted from primarily estimating timber volume (for which SPR sampling on a statewide basis is quite efficient) to estimating a multitude of parameters, several of which are not correlated with timber volume. The need for a quick and accurate response to requests for estimates for smaller areas, such as counties or groups of counties within a state, also requires a simpler design. Simple designs facilitate post-stratification for subpopulation estimation.

5.6 PROBLEMS

1. Prove that $v(\hat{Y})$ in eq. (5.15) for two-stage sampling is an unbiased estimator of $V(\hat{Y})$ when SRS is used in both stages.

2. For two-stage sampling, with SRS used in both stages, derive the variance and variance estimators of cluster sampling where clusters are selected by SRS using eq. (5.14) and (5.15).

3. Provide an unbiased estimator of Y for two-stage sampling where the primary sample units are selected without replacement proportional to some variable x_1 (which may be related to a cluster size M) and where the secondary sample units are selected without replacement proportional to some other variable x_2.

4. Prove the results for optimum n' and n for fixed variance in eqs. (5.58) and (5.59) (use method of Lagrangian multipliers).

5. Assuming SRS without replacement at both phases, derive a variance estimator for double sampling with ratio estimation by substituting suitable expressions into eq. (5.47).

6. Identify four sampling strategies that are special cases of the Alaskan (AIRIS) sampling strategy and that may have practical utility.

7. Using point-Poisson sampling in a simulation study, 30 trees are selected ultimately (on average). The Horvitz-Thompson estimator $\hat{Y}_{HT}$ (eq. 3.7) is used with variance estimator $v(\hat{Y}_{pps}(wr))$ in eq. (3.13). Results indicate that the variance estimator is negatively biased and inappropriate. Why is this variance estimator inappropriate?

6 Model-Based Inference

6.1 INTRODUCTION

In the design-based approach to inference the population values y_i ($i = 1, \ldots, N$) are regarded as fixed constants. Inference about a population parameter is reckoned with respect to the distribution of the estimator of the parameter, where the estimator is a function of the data $D = \{(i, y_i) : i \in S\}$. One can consider y_i as conforming to a rudimentary model

$$y_i = \bar{Y} + e_i \tag{6.1}$$

where $e_i = y_i - \bar{Y}$ by definition, and therefore $\Sigma_{i=1}^{N} e_i = 0$. When inference is design-based, nothing needs to be assumed about the distribution of e_i for any of the estimators considered.

If the population is stratified by the design, a model for the ith observation in the hth stratum can be written as

$$y_{hi} = \bar{Y}_h + e_{hi} \tag{6.2}$$

where $e_{hi} = y_{hi} - \bar{Y}_h$ and $\Sigma_{i=1}^{N_h} e_{hi} = 0$. As discussed in Chapter 3, stratified random sampling is an effective tool to reduce the design-based variance of an estimator of the population total. In this sense model (6.2) is used in the design stage of sampling in order to improve design-based inference.

In addition to using a model at the design stage, one can also use a model at the estimation stage, as with the ratio-of-means estimator. In an unstratified design one can express the model corresponding to the ratio-of-means estimator as

$$y_i = Rx_i + e_i \tag{6.3}$$

where $R = Y/X$, $e_i = y_i - Rx_i$, and $\Sigma_{i=1}^{N} e_i = 0$. Model (6.3) always holds regardless of any other aspects of the relationship between the y_i, x_i values. However, the model suggests conditions under which it would be worthwhile to use $\hat{Y}_{rm}$ instead of the usual expansion estimator $\hat{Y}$, specifically if the x_i values are positively correlated with the y_i values, and if $y_i \doteq 0$ when $x_i = 0$. When both conditions are approximately met, the design-based MSE of $\hat{Y}_{rm}$ is smaller than the design-based variance of $\hat{Y}$. It is in this sense that a model can be used in the estimation stage to improve design-based inference. When one or both of the favorable conditions are not met, the design-based approximations to bias and mean square error (MSE) of $\hat{Y}_{rm}$ presented in Chapter 3 remain valid. That is, the use of an implied model to estimate Y does not affect the manner in which the properties of $\hat{Y}_{rm}$ are evaluated, although it may greatly affect their value.

Other examples can be given that demonstrate the use of models like eq. (6.2) and eq. (6.3) in traditional sampling strategies, either at the design stage or the estimation stage, in order to obtain a more accurate estimator of a population parameter. Little (1983a) suggested a hierarchical classification to describe survey samplers based on their use of models: D-modelers are those who use models only at the design stage; E-modelers are those who use them only at the estimation stage; DE-modelers use models at both stages (e.g., by using the ratio estimator in conjunction with a stratified design); and I-modelers use models for inference only (e.g., using autocorrelation models in variance estimation for systematic sampling).

Hansen *et al.* (1983) used the term "model-based" to pertain to sampling strategies that may rely on a model for sample selection and/or estimation. They reserved the term "model-dependent" to refer to the dependence of inference on models, too. This nuance may be useful in some settings, but it has yet to become widely accepted. We depart from their terminology and use the term "model-based" exclusively in order to refer succinctly to this mode of inference, in contrast to traditional design-based inference, in the spirit of Little (1983a). In Little's classification, samplers who employ model-based inference are I-modelers, DEI-modelers, or are somewhere intermediate depending on the role of the model in design and estimation, as well as inference.

In the model-based approach to inference both the nature of the model and its role in the inferential process is considerably different from that described above. With the model-based approach, the population values y_i $(i = 1, \ldots, N)$ are regarded as realizations of random variables, Y_i. The analog to model (6.1) under a model-based approach is

$$\begin{aligned} Y_i = \mu + e_i\,, \qquad E_m[Y_i] &= \mu \\ V_m(Y_i \mid \mu) = V_m(e_i) &= \sigma_i^2 \\ \mathrm{Cov}_m(Y_i, Y_{i'}) = \mathrm{Cov}_m(e_i, e_{i'}) &= 0\,. \end{aligned} \tag{6.4}$$

Model (6.4) describes an infinitely large population of which N values, y_i, constitute the finite population available to the sampler. The current population may itself be regarded as a sample from a "superpopulation" that has the structure stipulated in (6.4). Because of the stochastic nature of Y_i in this model the finite population mean $\bar{Y} = \Sigma_{i=1}^{N} y_i/N$ generally will not coincide with μ, nor will it be true that $\Sigma_{i=1}^{N} e_i = 0$, as it is in the algebraic model (6.1). The sampler must decide, therefore, whether the objective is to predict the mean of the finite population, $\bar{Y}$, or to estimate the expected value, μ, of the superpopulation. When N is large, the difference between the two parameters will be negligibly small.

When prediction of $\bar{Y}$, or equivalently of $Y = N\bar{Y}$, is the objective, then the type of inference has been described as enumerative or descriptive, as opposed to analytic inference when μ is the target (Smith 1981). For the remainder of this chapter we restrict our attention to enumerative model-based inference, and we refer the interested reader to the text by Skinner *et al.* (1989) for discussion of analytic model-based inference.

Regardless of the sample design that is used to select an element into the sample, its expected value under model (6.4) is μ. The population total Y is a random variable with expected value $N\mu$. If $\hat{Y}$ is an estimator of Y using sample data D, then $E(\hat{Y}) \neq Y$; however, $\hat{Y}$ is a model-unbiased estimator of Y by definition when

$$E_m[\hat{Y} - Y] = 0 \tag{6.5}$$

where expectation is with respect to model (6.4) (Cassel *et al.* 1977, p. 92). The model-based variance of $\hat{Y}$ is $V_m(\hat{Y}) = E_m[(\hat{Y} - Y)^2]$, which is a function of the σ_i^2 for those units of the population that appear in D. Usually some simplifying assumption about the behavior of the σ_i^2 is made in order to estimate $V(\hat{Y})$. Two common assumptions are either that $\sigma_i^2 = \sigma^2$, $\forall i$, or that $\sigma_i^2 = \sigma^2 v_i$, as in eq. (2.69) and eq. (2.70).

Models more elaborate than eq. (6.4) are possible and are used in this chapter. This rudimentary model is suitable, however, to establish the basic ideas of enumerative model-based inference. We emphasize that the most fundamental idea is that the model and its specification of the behavior of the Y_i provide the basis for determining the properties of an estimator, not the probability weighted distribution of all possible estimates. The model-based bias and MSE of an estimator depend upon how well the actual superpopulation model corresponds to the assumed model. Likewise, the validity of model-based estimates of bias, variance, and MSE depend on the validity of the assumed model.

The model-based properties of an estimator are reckoned conditionally upon the observed sample. Model-unbiasedness, because it depends on the validity of the model and not the manner of sample selection, therefore does not require that the sample be probabilistic. As an example where a nonprobabilistic selection may be worthwhile, consider the superpopulation model

$$Y_i = \alpha + \beta X_i + e_i\,, \qquad V_m(e_i) = \sigma^2\,. \tag{6.6}$$

Least-squares estimators of α and β generally will be more precise when the sample includes predominantly large and small X_i values. Since the precision of $\hat{Y}$ will depend jointly on the precision of α and β, it may be worthwhile to purposely choose a sample so as to ensure a large spread in X_i sample values. As long as the sample is not chosen in a manner that changes the sample distribution of Y_i conditionally upon X_i from that specified in the superpopulation model, then there are persuasive arguments in favor of purposive sampling. This is especially true when estimation must proceed with data collected from small samples. When the validity of the model is uncertain and when parameters of multiple characteristics are to be estimated, the cogency of purposive sampling is diminished. Likewise if the data are likely to be used for secondary estimation and analyses, purposive sampling may be inadvisable.

The distinctions and comparative advantages of design-based and model-based inference will strike many as rather obscure at first. Hansen *et al.* (1983) defined a model-based sampling strategy as consisting of: (1) sample selection that may involve randomization, (2) estimators that are unbiased or consistent if the model is valid, and (3) procedures for inference such that the correctness of the inference depends on an assumed model. In contrast, a design-based strategy requires probabilistic selection, and neither the properties of the estimators nor the correctness of inference depend on the correctness of an assumed model. The important issue to be resolved by the survey analyst has less to do with which mode of inference is judged correct or incorrect, than with which is the more relevant to the analytical or enumerative purposes of the survey. At present there is little consensus, indeed there is broad disagreement as the discussion following Hansen *et al.* (1983) attests.

Särndal (1978) emphasized that every survey sampling statistician mixes elements of randomization and modeling, but with varying emphases. Basu (1971) gave an interesting (extreme) example where complete reliance on a probability sampling strategy is ridiculous.

To those who are accustomed to inference in survey sampling in the traditional mode established by Neyman (1934), model-based precepts may be anathema. Yet there are important examples within the traditional framework, too, where the presumed validity of an underlying model has enabled or enhanced design-based inference. For example, Osborne (1942) estimated the precision of systematic sampling for assessing vegetation cover type by fitting models to estimate the correlation between observations. Similarly, Matérn (1947) postulated a model for the autocorrelation between units to deal with variance estimation for systematic sampling in forestry and land-use applications.

6.2 CLASSIFICATION SCHEME OF ENUMERATIVE SURVEY SAMPLERS

Brewer and Särndal (1983) described six approaches to sample selection, estimation, and enumerative inference, which are progressively less dependent on model assumptions.

1. *Model-Based Bayesian Approach.* Inference is based on the presumption of a model and on the specification of the prior distributions of the model parameters. Estimators are developed and sampling is conducted so as to minimize Bayes risk.
2. *Model-Based Non-Bayesian Approach.* Inference is based on the model without stipulating priors for model parameters. Estimators are developed so as to have minimum variance in the class of best linear unbiased (BLUE) estimators, and sampling is conducted so as to minimize the variance under the model.

For neither of these approaches is there a logical requirement for random selection of the sample. Both are liable to lead to erroneous inferences if the chosen model is not an accurate description of the state of nature. On the other hand, such approaches should be reliable and cost-effective if the assumed model is correct.

3. *Robust Model-Based Non-Bayesian Approach.* This differs from the preceding approach because the sample is selected in order to achieve balance. Estimators are developed so as to be BLUE under the assumed model, and, given a balanced sample, to remain unbiased under closely related alternative models. This approach is less efficient than approach 2 if the model is correct but is not as disastrous if a closely related alternative model is valid.

Randomization is not required with approach 3 but it may nonetheless be used because simple random sampling (SRS) assures a balanced sample on average. A balanced sampling of order J is defined in Royall and Herson (1973a) as a sample for which $\bar{x}_s^{(j)} = \bar{X}^{(j)}$ for $j = 1, 2, \ldots, J$, where J is any positive integer and $\bar{x}_s^{(j)}$ and $\bar{X}^{(j)}$ are the jth sample and population moments. Regardless of the sample selection criteria, inference under approach 3 remains model-based.

4. *Probability Sampling with Modeling.* With this approach, inference may be either design-based or model-based. The design is chosen to minimize the model expectation of the design-based MSE.

Under a correctly assumed model, the model-based variance of an estimator should be smaller than the design-based variance when an unbalanced sample has been selected. Yet under an incorrectly specified model, the design-based variance is still interpretable. With approach 4, when the two measures differ greatly, the statistician is alerted to the fact that either an atypical sample has been selected or that the model has broken down.

5. *Classical Probability Sampling.* Selection is probabilistic and inference is design-based as explained in Chapter 3 and in Section 6.1.

Estimators are developed so as to be Cochran-consistent with small MSE, using models to guide either the design or the estimation.

6. *Inference Without Exchangeability.* This approach attempts to make inferences without ignoring population labels and without assumptions about exchangeability.

The rationale behind this approach is that random sampling cannot be implemented without giving each population unit a label and this label might have useful information about the unit. The approach has yielded some useful theorems on nonexistence of optimal estimators but has had little practical value.

Brewer and Särndal (1983) suggested that the viable choices for enumerative inference are approaches 3, 4, and 5, but they preferred approach 4. Discussions of Brewer and Särndal's paper offered a wide variety of different classification schemes and agreements and disagreements with Brewer and Särndal's choices and schemes.

6.3 THE ARGUMENTS: MODEL-BASED VERSUS DESIGN-BASED INFERENCE

The recent statistical literature contains a number of instructive discussions among advocates of one or the other modes of inference outlined in the preceding sections. Arguments advanced in this literature for model-based versus design-based inference, and the appropriate sampling strategies, focus on both philosophical and practical issues.

6.3.1 Philosophical Issues

Hansen *et al.* (1983) cited two advantages for design-based inference:

(i) A minimum number of assumptions is made for the size and nature of the sample, that is, the approach is robust against invalidity of assumptions.

(ii) The results are generally acceptable without the underlying assumptions having to be defended, and personal bias in sample selection and model selection is eliminated. This is quite critical for surveys used in making decisions of concern to opposing factions with different interests.

In addition, large sample confidence intervals are valid in that they contain the target parameter at approximately the nominal probability level, independent of the distribution of the population from which the sample is drawn. The central limit theorem (Theorem 2.6) guarantees the dependability of these confidence intervals on average. Saddle point methods

(Reid 1988) may allow reliable construction of confidence intervals even for small sample sizes.

Proponents of model-based inference argue that after a sample has been drawn, the soundness of inference depends on the sample actually obtained and not the process used to obtain it. The model should be used, therefore, to guide the selection of the sample in a manner which is beneficial to the inference to be drawn. Usually one seeks unbiasedness under the model, robustness against bias for small departures from the model form, and an estimator with a small variance under the model. As mentioned earlier, model-based variance is reckoned with respect to the error structure of the superpopulation model for those elements of the superpopulation that are included in the sample. In other words, the variance of an estimator of $\bar{Y}$ is with respect to the possible realizations of the Y_i that might have been observed for those units in the sample, in contrast to the design-based approach, in which the variance is averaged over the possible samples of realized values. Advocates of model-based inference feel that it is misleading to use variability over all possible samples as a measure of precision when the sample is atypical, especially when the extent of that variability is estimated from the same atypical sample. Tacit acceptance of this belief is evidenced by the practice of some survey samplers of rejecting certain random samples when one that is obviously imbalanced has been selected.

If the model is incorrect, overly optimistic estimates of the precision of the estimates will occur frequently. The design-based variance is the probability-weighted squared error of the estimates over all possible samples that can be obtained from the finite population under a specified sampling design. Royall and Cumberland (1981a,b) argued that this can yield unreliable estimates of precision too, but they noted that improved estimates of precision can be derived more readily for design-based estimation.

Hansen *et al.* (1983) strongly advocated design-based inference coupled with the use of models whenever appropriate at the design and estimation stage. They argued that one appeal of design-based inference is its dependence for validity on only two criteria: consistency and asymptotic normality. With the design-based approach, estimators can be chosen whose bias diminishes more rapidly with increasing sample size than does the standard error of estimate. For large samples, therefore, the design-based MSE is essentially the variance of the estimator of the population parameter. In contrast, if an assumed model is incorrect, then the model-based bias does not diminish with increasing sample size, and since this bias is not reflected in the estimated variance, a spurious level of accuracy is implied.

Neither consistency nor normality of an estimator can be defined for a given finite population. To establish both conditions requires that one postulate a hypothetical sequence of finite populations of increasing size, and as demonstrated by Hansen *et al.* (1983), an estimator can be either consistent or inconsistent depending on the sequence chosen; a similar problem arises when establishing asymptotic normality. Godambe (1978) questioned the practical utility of these properties, therefore, because of the impossibility of interpreting them for a given finite population. As an advocate of model-based inference, Smith (1983a) voiced a similar concern and he countered the arguments of Hansen *et al.* with the argument that these sequences of populations effectively constitute a model, too. He contended that compared to a regression model describing a superpopulation relationship between two variables, say y and x, a model that consists of a sequence of hypothetical populations is purely arbitrary and cannot even be tested against real data.

In their advocacy of the design-based approach to inference, most proponents will agree that model-based strategies and inference may possess substantial advantages in some situations. For example, useful inferences can be based on quite small samples and, in specific cases, it may be possible to adopt unequivocally a best model-based strategy for estimation.

Analytical studies based on underlying models and follow-up simulation studies have highlighted some weaknesses in design-based inference. "Large" sample sizes are assumed in design-based inference (large never being defined, although samples larger than 100 units are generally accepted to be large) because of the reliance on asymptotic properties. Not surprisingly, for example, classical survey sampling estimation of standard error is sensitive to misspecification of the error structure of the models. The standard design-based variance estimators for both the ratio and regression estimators can be quite unreliable as revealed by recent model-based analyses (Royall and Cumberland 1978) and subsequently verified empirically (Royall and Cumberland 1981a,b). These papers showed that more reliable model-based estimators of precision are available. However, Joshi (1988) argued that Royall and Cumberland (1981a) should have used stratified random sampling with proportional allocation rather than SRS in their simulations, which would have yielded more reliable estimates and (presumably) more reliable estimates of precision. Hansen *et al.* (1983) noted that design- and model-based samplers use different ratio estimators, the former use weighted ratio estimators and the latter unweighted estimators since probabilities of selection can be ignored.

DE-modelers who advocate design-based inference assert that such inference is robust to departures from the model. If the model that has been used in the sample design or the choice of estimator is incorrect, the resulting inference may be poor but it will remain valid. Royall, a discussant of Hansen *et al.* (1983), argued that they err when crediting randomization with producing the desired robustness. Royall argued that the balance of the selected sample and stratification of the population are the essential guarantors of robust estimation, not random selection of samples. Balanced sampling and its pps analogue π-balanced sampling are discussed more fully later in this chapter.

6.3.2 Practical Issues

What, then, are the practical issues relating to model-based or design-based sampling? Royall and Cumberland (1981a) posed the following basic questions (couched in terms of the ratio estimator for their purpose):

Question 1. Which of the two approaches is more appropriate for inference from an observed sample and a calculated estimate? The key question is: How much better can we expect model-based inference to be relative to design-based inference? This translates into questions of what are good model-based strategies; how much closer are estimates of a parameter based on such strategies than on design-based strategies; and which of the definitions of bias and variance are more useful to judge the reliability of the estimates based on a given sample?

Question 2. Is the model-based (prediction) theory seriously misleading when the superpopulation model fails while probability sampling theory continues to provide reliable inferences, or is it true, as prediction theory suggests, that the probability sampling distribution is irrelevant for inference from a given sample?

These broad issues are addressed by discussing research that focuses on one or both of them in specific circumstances and then by citing some results for efficient model-based sampling strategies.

Royall and Cumberland (198la) examined these issues by contrasting the performance of design-based inference with model-based inference, focusing on bias and variance estimation. Question 1 (appropriateness or reliability) and question 2 (validity of inference under model failure) are considered below in the context of ratio estimation with SRS and restricted SRS(RSRS).

Question 1. Royall and Cumberland (1981a) focused on bias and reliability of variance estimators for the following model:

$$E(Y_i) = \beta x_i\,, \qquad V(Y_i) = \sigma^2 x_i\,, \qquad \mathrm{Cov}(Y_i, Y_j) = 0\,, \qquad i \neq j\,. \tag{6.7}$$

If $\bar{x}_r$ and $\bar{y}_r$, $\bar{x}$ and $\bar{y}$ and $\bar{X}$ and $\bar{Y}$ denote average x and y values in the nonsample units, sample units, and the population, respectively, then the weighted least-squares predictor of the population total (Y) is the ratio-of-means estimator

$$\hat{Y}_{rm} = \Sigma_s y_i + (\bar{y}/\bar{x})\Sigma_r x_i = N\bar{X}\bar{y}/\bar{x}\,. \tag{6.8}$$

The estimator is unbiased under model (6.7) with error variance

$$V_m(\hat{Y}_{rm} - Y) = E_m[(\hat{Y}_{rm} - Y)^2]$$
$$= \frac{N}{n}(N - n)\sigma^2\bar{x}_r\bar{X}/\bar{x}\,, \tag{6.9}$$

which decreases as $\bar{x}$ increases. If model (6.7) is correct, $\hat{Y}_{rm}$ is optimal among unbiased estimators and the best sample consists of the n largest units in terms of x. Departures from model (6.7) can badly bias both $\hat{Y}_{rm}$ and the estimator of variance from standard least squares theory:

$$v_2 = \frac{N}{n}(N - n)\{(\Sigma_s d_i^2/x_i)/(n-1)\}\bar{x}_r\bar{X}/\bar{x} \tag{6.10}$$

where $d_i = y_i - (\bar{y}/\bar{x})x_i$.

The variance estimator,

$$v_1 = \frac{N}{n}(N - n)\Sigma_s d_i^2/(n-1)\,, \tag{6.11}$$

given in eq. (3.106) for the ratio estimator, can have serious bias as an estimator of $V_m(\hat{Y}_m - \hat{Y})$. The relative bias is approximately equal to $N[\{\bar{x}^2/(\bar{X}\bar{x}_r)\} - 1]$, which increases with $\bar{x}$ and vanishes only in samples balanced on x, namely, when $\bar{x} = \bar{X}$.

A variance estimator that removes the bias in v_1 under model (6.7) is (eq. 3.110)

$$v_3 = v_1(\bar{x}_r\bar{X}/\bar{x}^2)/(1 - c^2/n) \tag{6.12}$$

where

$$c^2 = [\Sigma_s(x_i - \bar{x})^2/\bar{x}^2]/(n-1)\,.$$

Royall and Eberhardt (1975) showed that v_3 is asymptotically unbiased even when the error variance of the model is incorrectly specified. It is also asymptotically equivalent to the jackknife variance estimator (eq. 3.109)

$$v_J = N(N - n)\bar{X}^2(n-1)\Sigma_s D_{(j)}^2/n \tag{6.13}$$

where, for every j in s, $D_{(j)}$ is the difference between the ratio $(n\bar{y} - y_j)/(n\bar{x} - x_j)$ and the average of these n ratios." The variance estimator (eq. 3.111),

$$v_4 = \frac{N}{n^2}(N - n)(\bar{x}_r\bar{X}/\bar{x}^2)\Sigma_s d_i^2/[1 - (x_i/n\bar{x})]\,, \tag{6.14}$$

is unbiased under model (6.7), approximately unbiased for more general variance models, and asymptotically equivalent to v_J.

If $E(Y_i) \neq \beta x_i$ but model (6.7) is assumed, then the expected values of each of the variance estimators and the MSE increase. If the sample is balanced so that the ratio estimator is unbiased, then v_3 and v_4 tend to overestimate the actual MSE.

Contrasting $V_m(\hat{Y}_{rm} - Y)$ to the usual linearization approximation $V(\hat{Y}_{rm})$ (eq. 3.105),

$$V = \frac{N}{n}(N - n) \sum_{i=1}^{N} [y_i - (\bar{Y}/\bar{X})x_i]^2/(N - 1) \tag{6.15}$$

it is evident that the observed sample plays an explicit role through $\bar{x}$ in the model-based variance, whereas its role is obscured in the design-based approach (Royall and Cumberland 1981a,b). With the design-based approach $V(\hat{Y}_{rm})$ is constant for a fixed sample size. Using the standard error as a measure of the reliability of an estimator, no sample is any better or worse than any other. This will strike many as intuitively false, and among advocates of model-based inference this issue is one basis for complaint with design-based inference. The observed sample provides the current estimate of Y, and evaluating the behavior of an estimator over all possible samples is not informative about the quality of the particular estimate at hand. If a badly unbalanced sample was selected, it is small comfort and deceptive to rationalize that when it is averaged with other possible estimates that could have been obtained, the adverse effect of the current estimate is minimal. On the other hand, by evaluating the performance of an estimator conditionally upon the observed sample, one hopes to learn the characteristics of a sample that contributes to a good estimate, and then, perhaps, to select a sample in such a way as to assure a good sample. As seen above, a balanced sample is desirable when model (6.7) holds and when the performance of $\hat{Y}_{rm}$ is assessed under it instead of under the randomization distribution.

Question 2. Royall and Cumberland (1981a) noted that if both a superpopulation model and a probability sampling distribution are available, the bias, variance, and other parameters of interest can be defined with respect to the combined model. For example, bias $= \Sigma_s p(s) E_m(\hat{Y}_{rm} - Y) = E_m[\Sigma_s p(s)(\hat{Y}_{rm} - Y)]$. Royall and Cumberland (1981a) argued that the combined model approach may be useful for planning purposes, to study the properties of $\hat{Y}_{rm}$ with respect to various probability distributions, but not for inference.

For each of Royall and Cumberland's (1981a) six populations, five sets of samples were selected using SRS, RSRS, and three purposive plans respectively. Sample sizes $n = 16$, 32, and 64 were used. The six populations represented essentially linear or slightly curvilinear relationships with increasing variance of y as x increases.

Royall and Cumberland (1981a) concluded that:

1. It can be dangerous to examine one's sample to see what model fits

best and then make inferences based on that model. Curvilinearity in one population was not detected often and serious bias occurred with both the ratio estimator and v_2 in poorly-balanced samples.

2. v_2 in eq. (6.10) is clearly inferior to v_3, v_J, and v_4 in eq. (6.12)–(6.14).

T. M. F. Smith, a discussant of Royall and Cumberland (1981a), concluded that properties of the sample, not of the sampling plan, provide robust inferences and that model-based estimates of variance are inadequate. Smith advocated estimating variances by procedures such as the jackknife since they are based on the internal structure of the data and are not model dependent.

M. E. Thompson, another discussant, suggested that the "protective" role of randomization adds a feeling of confidence or safety to an expression of inference because, although the inference for the specific sample may be misleading, the expression has desirable and known frequency properties for the design. The utility of this protective role may be less in the case of populations with a strong relationship between y and x. Thompson suggested that Royall and Cumberland should have bolstered their arguments for the use of good-fitting models by examining either less regular departures from the model than the polynomial trend considered or weaker relationships. He argued that balanced samples are demonstrably representative with respect to certain functions of the x characteristic but that to assume that balanced samples will then yield reliable inference about the population total requires a "leap of faith." He concluded that with an appropriate randomized design, a reliable and defensible estimate of the confidence interval for the population total will be obtained.

Royall and Cumberland (1981b) addressed the same two issues for regression estimation. They concluded that for the same six populations as described in Royall and Cumberland (1981a), their results illustrate the fallacy of the randomization principle for regression estimation and that samples should be well balanced.

Under model (6.7) the ratio-of-means estimator $\hat{Y}_{rm} = (\bar{y}/\bar{x})X$ is an unbiased estimator of the population total Y. Cumberland and Royall (1988) concluded that the variance estimators v_1, v_4, and v_J are all asymptotically unbiased and that the confidence intervals are generally asymptotically correct. But if the model is not correct, then samples and estimators should be chosen with robustness in mind by balancing sample selection through restricted randomization, stratification, or systematic sampling. Choosing the sample at random does not produce robust confidence intervals because robustness requires that the sample mean not fall in the tails of the distribution. This is not guaranteed with SRS even for large n.

Hansen *et al.* (1983) gave some guiding principles on the two issues in choosing between model-based and design-based inference. These principles are partially based on numerical comparisons made for some generated populations:

- In general, the search for a "best" estimator is unrealistic because such an estimator cannot be found except by unduly restricting the class of estimators. A best estimator obtained under a model depends on the validity of the model and may yield confidence intervals that are seriously misleading.
- It is advantageous, as compared with the use of best model-based estimators, and sufficient to have a "good" estimator based on a reasonably large probability sample that provides a valid and acceptably small confidence interval.
- Model-based designs, even "robust" ones, face the risk of substantially understating the MSE, even when the model appears satisfactory. Approaches in which the model is adjusted to be consistent with survey results may substantially understate the length of the confidence intervals, whether the confidence interval is viewed as conditional on a single sample or for all possible samples. This is especially true as sample size increases.
- Probability sampling with reasonably large samples provides protection against failures of assumed models and robustness for all estimates, including estimates for any subdomains for which the samples are reasonably large. Models are useful for determining design-based samples. For example, the (perceived) lack of assumptions in this approach may be vital when sample estimates are for finite populations and results are to be used for important public-policy actions or used by opposing factions with different interests.
- Ordinarily, with reasonably large samples, sample selection methods and estimators based on good design-based sampling methods are almost as efficient as model-based methods even when the models are valid.
- When inferences are to be made about a causal system, rather than a finite population, there is no choice but to use a model-based sample. Design-based methods are not available for drawing samples from a causal system, but only from some finite realization of that system. However, in many situations, design-based samples of finite realizations of the causal system may be quite useful.
- Proper use of models has much to contribute to survey design. Models should be used to the fullest, but preferably by using designs and estimates that are not model-based.

C. E. Särndal, in comments on (an earlier version of) the Hansen *et al.* (1983) paper, made the following points:

- Most statisticians feel more at home with model-based inference.
- Model-based thinking has more direct appeal and is preferred on intuitive grounds. It is generally more comprehensible to people.

- The model-based framework should be more seriously considered in data analysis for surveys rather than in the traditional situations of estimating totals where old methods generally work well.
- The numerical example used by Hansen *et al.* (1983) shows the traditional approach in a favorable light because:
  - Model-based estimators are used with models known to be false; hence, they can yield biased estimates.
  - Estimators based on probability sampling benefit because the distribution of the population is known, which is not the case for model-based sampling. For example, the conditions considered are ideal for the weighted mean estimator used in probability sampling.

We close this section on practical issues by citing some theoretical developments that have practical implications in showing how model-based sampling strategies can be quite efficient under certain idealized circumstances. Considerable theoretical and simulation work has been performed in order to discern the advantages and liabilities of this mode of inference. Two interesting concepts resulting from Royall's work are balanced sampling and π-balanced sampling as described below.

Consider the polynomial model, denoted by $\xi[\delta_0, \delta_1, \ldots, \delta_J : v(x)]$ of the form

$$Y_k = \delta_0\beta_0 + \delta_1\beta_1 x_k + \cdots + \delta_J\beta_J x_k^J + e_k v(x_k)^{1/2}$$

where e_k is a random variable with zero mean and constant variance; and where $\delta_j = 1$ if the term $\beta_j x_k^j$ appears in the model, and $\delta_j = 0$ otherwise. Let $\hat{Y}[\delta_0, \delta_1, \ldots, \delta_J : v(x)]$ denote the BLUE estimator of Y under the model. Royall and Herson (1973a) proved the following theorem.

Theorem 6.1. If a sample is a balanced sample of order J as described in Section 6.2, then for any sequence $\delta_0, \delta_1, \ldots, \delta_J$ of zeros and ones in which $\delta_j = 1$, $\hat{Y}[\delta_0, \delta_1, \ldots, \delta_J : x_k^j] = N\bar{y}$.

In other words, when a balanced sample is selected, the expansion estimator ($N\bar{y}$) is the BLUE estimator under any Jth degree polynomial regression model in which the variance function is proportional to x^j for some $j = 0, 1, \ldots, J$ as long as the regression function contains the jth degree term $\beta_j x^j$.

This theorem can be extended to:

Theorem 6.2. For any sequence $\delta_0, \delta_1, \ldots, \delta_J$ of zeros and ones and any variance function $v(x) = \Sigma_{j=0}^{J} \delta_j a_j x^j$(with $a_0, a_1, \ldots, a_J$ any positive constants), then $\hat{Y}[\delta_0, \delta_1, \ldots, \delta_J : v(x)] = N\bar{y}$, if the sample is balanced to order J.

Random sampling is fundamentally justified in that it provides samples that are approximately balanced with respect to various characteristics.

Experimenters often reject random samples that appear to be "extreme" or "unrepresentative"; and after the sample is chosen and the data are gathered, they frequently make adjustments in calculations if the randomization process has failed to divide the units into "sample" and "nonsample" groups that are similar with respect to relevant characteristics. One familiar technique of this type is post-stratification. This is also the basic rationale behind the ratio-of-means estimator, which adjusts the sample mean $\bar{y}$ by the ratio adjustment $\bar{X}/\bar{x}$.

If not enough is known about the relationship between y and x to correct for imbalance in the sample, then one should aim for a sample that is balanced, or representative, with respect to x. This is one justification, for example, for stratified sampling and ranked set sampling (RSS, discussed in Ch. 3).

Royall and Herson (1973b) considered stratification on a size variable as a further technique (besides balancing) to protect against model failure. They showed that stratification and balanced sampling provide more protection for efficient estimation than balanced sampling alone using ratio-of-means estimators. Let the population total be

$$Y = \sum_{h=1}^{H} \sum_{k=1}^{N_h} y_{hk} = \sum_{h=1}^{H} Y_h \tag{6.16}$$

where strata are defined so that stratum 1 contains the N_1 smallest units, stratum 2 the next N_2 smallest units, and so on, for h strata. Then a sample s_h of n_h units is selected from stratum h and the total Y_h for the stratum is estimated by the ratio-of-means estimator

$$\hat{Y}_h(0, 1 : x) = \left(\sum_{s_h} y_{hk} \Big/ \sum_{s_h} x_{hk}\right) \sum_{k=1}^{N_h} x_{hk} \,. \tag{6.17}$$

The overall total Y is estimated by the separate ratio-of-means estimator

$$\hat{Y}^*_{rm} = \sum_{h=1}^{H} \hat{Y}_h(0, 1 : x) \,. \tag{6.18}$$

This estimator can be generated by the stratified balanced sample of degree J denoted by $s^*(J)$ where $\bar{x}^j_h = \bar{X}^j_h$ for $j = 1, 2, \ldots, J$.

Royall and Herson's main results are stated in the following two theorems.

Theorem 6.3. If n_h is proportional to $N_h/\sqrt{\bar{x}_h}$, then the strategy $[s^*(J), \hat{Y}^*]$ is more efficient than $[s(J), \hat{Y}(0, 1 : x)]$ under $\xi = \xi[\delta_0, \delta_1, \ldots, \delta_J : x]$. That is, $E_m[\hat{Y}(0, 1 : x) - Y]^2 \geq E_m(\hat{Y}^* - Y)^2$. The model ξ is $y = \delta_0\, \beta_0 + \delta_1 x + \delta_2 \beta_2 x^2 + \cdots + \delta_J\, \beta_J x^J + e$ where $V_m(e) = \sigma^2 x$.

Theorem 6.4. When the size measures $x_1, \ldots, x_N$ all have different values and the stratified balanced sampling strategy $[s^*(J), Y^*]$ with equal allocation is employed, then a necessary condition for optimal stratification under all models of the form $\xi[\delta_0, \delta_1, \ldots, \delta_J : x]$ is

$$1.\ N_1 \geq N_2 \geq \cdots \geq N_h \tag{6.19}$$

and

$$2.\ N_1\bar{X}_1 \leq N_2\bar{X}_2 \leq \cdots \leq N_h\bar{X}_h \tag{6.20}$$

if compatible with eq. (6.19) above.

There are three special cases for which both eq. (6.19) and eq. (6.20) in Theorem 6.4 are satisfied:

$$1.\ N_1 = N_2 = \cdots = N_H \tag{6.21}$$

$$2.\ N_1\bar{X}_1 = N_2\bar{X}_2 = \cdots = N_H\bar{X}_H \tag{6.22}$$

$$3.\ N_1^2\bar{X}_1 = N_2^2\bar{X}_2 = \cdots = N_H^2\bar{X}_H\,. \tag{6.23}$$

Royall and Herson suggested that equal allocation in eq. (6.23) above is usually close to optimal.

Kott (1986) showed that the Horvitz–Thompson estimator $\hat{Y}_{HT} = \Sigma_{i=1}^{N} (y_i/nx_i)X$ is the BLUE estimator for samples satisfying the condition (called π-balanced)

$$\sum_{i=1}^{n} x_i/(nX) = \sum_{i=1}^{N} x_i^2/X^2 \tag{6.24}$$

under the model

$$y_i = bx_i + cx_i^2 + \eta_i$$

where $E_m(\eta_i) = 0$, $V_m(\eta_i) = \sigma^2 x_i^2$, and the η_i are independently distributed.

Under the model

$$y_i = \alpha + \beta x_i + \gamma x_i^2 + \eta_i$$

with $E_m(\eta_i) = 0$, and $V_m(\eta_i) = \sigma^2 x_i^2$, the Horvitz–Thompson estimator is on average the optimal, linear, robust estimator if π-balanced samples are selected satisfying eq. (6.23) and

$$\frac{X}{n} \sum_{i=1}^{n} 1/x_i = N\,. \tag{6.25}$$

Iachan (1985) noted that the natural superpopulation model for which the ratio estimator $\hat{Y}_{rm}$ is optimal is model (6.7). A purposive sample is optimal generally under eq. (6.7) because

$$E_m(\hat{Y}_{rm} - Y)^2 \le E(\hat{Y}' - Y)^2$$

for all linear ξ-unbiased estimators $\hat{Y}'$ for any sampling design p', where p' denotes any possible distribution over the set of all possible samples. The estimator $\hat{Y}_{rm}$ can be badly biased however. For example, when

$$E_m(y_i) = \beta_0 + \beta_1 x_i \qquad \text{for } \beta_0 \ne 0 \tag{6.26}$$

and

$$\begin{aligned} \text{Cov}_m(y_i, y_j) &= \sigma^2 x_j && \text{if } i = j \\ &= 0 && \text{if } i \ne j \,, \end{aligned}$$

the model-bias of $\hat{Y}_{rm}$ becomes

$$E_m(\hat{Y}_{rm} - Y) = N\beta_0(\bar{X} - \bar{x})/\bar{x}$$

which is small only if $|(\bar{x} - \bar{X})/|\bar{x}|$ is small. This leads naturally to the idea of balanced sampling such that $\bar{x} \doteq \bar{X}$ (Royall and Herson 1983a).

A criterion for robustness of the estimator $\hat{Y}_{rm}$ against model departure is the mean square expected error (MSEE)

$$\text{MSEE} = E_p(\xi - \text{bias})^2 = E_p\{[E(\hat{Y}_{rm} - \bar{Y})]^2\} \,, \tag{6.27}$$

which is a term in the model-expected MSE.

For model (6.7) and $\hat{Y}_{rm}$, this becomes

$$\text{MSEE} = E_p\left[\frac{(N\beta_0)(\bar{x} - \bar{X})}{\bar{x}}\right]^2 = (N\beta_0)^2 E_p\left(\frac{I_1}{1 + I_1}\right)^2 \tag{6.28}$$

where $I_1 = (\bar{x}\text{-}\bar{X})/\bar{X}$. Note that MSEE vanishes if either model (6.7) holds exactly or the samples are balanced.

The generalized regression estimator $\hat{Y}_{gr}$ (eq. 3.135) can be viewed as a compromise between model-based and design-based inference (Little 1983b).

Little wrote $\hat{Y}_{gr}$ as

$$\hat{Y}_{gr} = \sum_{i=1}^{n} (y_i - \hat{y}_i)/\pi_i + \sum_{i=1}^{N-n} \hat{y}_i \tag{6.29}$$

where the $\hat{y}_i$ are the predicted values of the y_i from regression. He noted that the first component on the right-hand side can be viewed as the

Horvitz–Thompson estimator applied to the residuals from the regression that protects the estimator from model misspecification by ensuring asymptotic design unbiasedness. The second component predicts Y from the regression model. Robinson and Särndal (1983) established that this estimator is asymptotically design unbiased for conditions where the superpopulation model is not valid. As noted earlier, this estimator appears to be quite efficient in simulation studies (Schreuder and Ouyang 1992) and may be quite efficient even for small sample sizes.

6.4 COMPARISON OF ESTIMATORS ASSUMING SUPERPOPULATIONS

Considerable work has been done comparing the utility of sampling strategies for estimating population totals. As indicated earlier, covariate information can be used in either or both the design and estimation stages. Basically, criteria for selecting a sampling strategy should include ease of execution, efficiency of the sampling strategy under an assumed model, and sensitivity to deviations from this model.

Comparing the efficiency of sampling strategies is often worthwhile to gain a better understanding of the strategies and to determine how much improvement can be obtained by incorporating covariate information in the design stage. However, expressions of relative efficiencies are often so complex that their evaluation does not lead to meaningful insights. Assuming appropriate superpopulations may lead to simplified expressions of relative efficiencies.

Most of the results in either theoretical or empirical comparisons are based on special cases of the multiple regression model, G_{MR}, which assumes that the values $Y_1, \ldots, Y_N$ in the population are independently distributed with

$$\mu_i = E_m(Y_i) = \beta_1 + \sum_{j=2}^{q} \beta_j x_{ij} , \qquad \sigma_i^2 = E_m(Y_i - \mu_i)^2 = \sigma^2 v_i , \tag{6.30}$$

with $\beta_1, \ldots, \beta_q$ and σ^2 unknown and $x_{i2}, \ldots, x_{iq}$ and v_i a set of known numbers for every i ($i = 1, \ldots, N$) (see Definition 2.21).

Special cases considered are models G_R, where the regression is linear through the origin ($\beta_1 = 0$) with only one x-variable, and G_{Rk}, which is G_R with the added specification that $\sigma_i^2 = \sigma^2 x_i^k$ ($i = 1, \ldots, N$) with known or unknown k.

Cassel *et al.* (1977, p. 151) pointed out the interrelationship between many sampling strategies. Let

$$t = \hat{Y}_{HT} + h(X - \hat{X}_{HT}) \tag{6.31}$$

be a predictor where $\hat{Y}_{HT} = \Sigma_{i=1}^{n} y_i/\pi_i$, $\hat{X}_{HT} = \Sigma_{i=1}^{n} x_i/\pi_i$, $\hat{R} = \Sigma_{i=1}^{n} y_i/(nx_i)$, and $\pi_i > 0$ $(i = 1, \ldots, N)$ are functions of the x_i $(i = 1, \ldots, N)$ and are the inclusion probabilities of any given fixed effective sample size design p, and $h = h[\{(i, y_i, x_i, \pi_i) : i \in s\}]$ is a specified function. This includes many predictors plus designs as special cases. For example:

1. If $\pi_i = f = n/N$, then one can reasonably say that the auxiliary information has been ignored at the design stage.
2. The auxiliary variable is only used in estimation when $h \neq 0$. The ratio-of-means estimator with SRS is an example where x_i plays no role in the design but does in estimation. In contrast, if $h = 0$, but the π_i are functions of the x_i, then the dependence of $\hat{Y}_{HT}$ on the x_i is introduced by the choice of design but not at the estimation stage.
3. $h = 0$, $\pi_i =$ constant $(i = 1, \ldots, N)$ implies ignorance of the auxiliary information at both the design and estimation stage. Hence $t = \hat{Y}$ in this case. Special cases of t are summarized in Table 6.1 (Cassel *et al.* 1977).

Note that if $\pi_i = nx_i/X$, $t_{GR} = \hat{Y}_{mr}$ is the mean-of-ratios estimator in eq. (3.113). These two estimators may be interesting possibilities for multipurpose surveys where the means of several y-variables have to be estimated using one fixed set of unequal inclusion probabilities $\pi_1, \ldots, \pi_N$, where $v(x_i) =$ variance of x at i, usually $v(x_i) = x_i^k$.

An estimator of β is the weighted least squares estimator

$$\hat{\beta} = \frac{\sum_s (x_i - \bar{x})(y_i - \bar{y})/v(x_i)}{\sum_s (x_i - \bar{x})^2/v(x_i)} . \tag{6.32}$$

Cassel *et al.* (1977) compared unequal probability sampling procedures using mean-of-ratios type estimators (called M-estimators) with equal probability sampling using ratio-of-means estimators (called R-estimators). Their

TABLE 6.1. Estimators of the Type $\hat{Y}_{HT} + h(X - \hat{X}_{HT})$[a].

Estimation Stage:		Design stage: Choice of π_i	
Choice of h	$\pi_i = f$	$\pi_i \propto x_i$	Arbitrary π_i
$h = 0$	$\hat{Y}$	XR_{yx}	$\hat{Y}_{HT}$
$h = c$	$t_D = \hat{Y} + c(X\text{-}\hat{X})$	XR_{yx}	$t_{GD} = \hat{Y}_{HT} + c(X\text{-}\hat{X}_{HT})$
$h = \hat{Y}_{HT}/\hat{X}_{HT}$	$t_R = X\hat{Y}/\hat{X}$	XR_{yx}	$t_{GR} = X\hat{Y}_{HT}/\hat{X}_{HT}$
$h = \hat{\beta}$	$t_{REG} = \hat{Y} + \hat{\beta}(X - \hat{X})$	XR_{yx}	$t_{GREG} = \hat{Y}_{HT} + \hat{\beta}(X - \hat{X}_{HT})$

[a]Adapted from Cassel *et al.* 1977.

most useful conclusions concern comparisons of pps sampling with the Horvitz–Thompson estimator $\hat{Y}_{HT}$ (eq. 3.80) called M_{HT}, of SRS with the ratio-of-means estimator $\hat{Y}_{rm}$ (eq. 3.97) called R_c, and of SRS with the Lahiri-strategy called R_{La} ($\hat{Y}_{mr1}$ (eq. 3.100)). This last strategy was devised by Hájek (1949) but is generally attributed to Lahiri (1951). R_{La} is implemented by drawing the first unit pps and using SRS for drawing the remaining $n - 1$ units. Cassel *et al.* concluded that for moderately large to very large populations, the M-strategies can be treated as one group of estimators and the R-strategies as another. The pertinent question then becomes: should one use pps sampling on which the M-strategies are based or the much simpler SRS on which the R-strategies are based?

Cassel *et al.* (1977) made the following recommendations when k can only be specified in a range, N is large, n is not too small and the population of X-values is approximately gamma shaped:

- If $1 \leq k \leq 2$, an M-strategy (such as M_{HT}) is likely to be a good choice and should be preferred to R-strategy competitors, particularly if the population is highly skewed.
- If $0 \leq k \leq 1$, M_{HT} should be avoided especially for skewed populations, and R_{La} or R_c should be used.
- If $0 \leq k \leq 2$, an R-strategy may be more prudent than an M-strategy, that is, if the population is highly skewed, an R-strategy at least avoids a highly inefficient procedure.
- If k is approximately 0, both M_{HT} and R_{La} can probably be improved considerably.
- If skewness is negligible, then both the M- and R-strategies are equally efficient but the R-strategy should probably be preferred because of its simple design.

Other types of strategies shown in Table 6.1 are of interest too, but their performances, particularly those of the generalized ratio and regression estimators, are not yet known (Cassel *et al.* 1977).

6.5. MODEL-BASED INVESTIGATIONS IN FORESTRY

An understanding of the relative efficiency of sampling strategies can be obtained from analytical comparisons assuming superpopulations and also from simulation experiments where samples are drawn from completely enumerated populations or from artificial populations.

Schreuder and Wood (1986) discussed the merits of probability and purposive sampling in estimating population totals in forestry. They concluded that probabilistic techniques should be used if there is doubt about the validity of the underlying model but that the non-random procedures are preferable when the model is acceptable. Non-random sampling techniques may be well suited to sampling for volume in timber sales (Schreuder, 1984) since it is known that model (2.70) is an excellent approximation of reality if y_i = volume of tree i and x_i = tree basal area times total height (McClure *et al.* 1983).

Model-based procedures were usually found to be as efficient as, and sometimes considerably more efficient than, competing probabilistic approaches (e.g., point-Poisson and pps or stratified pps sampling) for timber inventories and for updating estimates of statewide forest volume using current plot volume as the dependent variable and past plot volume as the independent variable (Schreuder *et al.* 1984b; Schreuder and Thomas 1985; Wood *et al.* 1985, Biggs *et al.* 1985). Particularly efficient were sampling from the cumulated x-values (pscx) and stratified pscx (spscx) sampling.

The pscx procedure is as follows: Sort the population of x-values and divide the sum of the x, X, by the desired sample size n. Calculate the number of trees N_i in each of the n classes $[(X/n)i, (X/n)(i+1)]$ $(i = 0, 1, \ldots, (n-1))$ and select those n trees for the sample that have x-values close to $i\,(X/n)/N_i$ $(i = 1, \ldots, n)$.

The spscx procedure is a stratified random sampling procedure selecting one unit at random in each stratum set up by the pscx procedure.

Wood and Schreuder (1986) described how to implement point-model-based (PMB) sampling in forest inventory. PMB sampling is a model-based procedure in which trees are selected by variable radius plot (VRP) sampling at the first stage and a subsample of these trees is selected by a purposive scheme for the second stage. Wood and Schreuder recommended a procedure such as spscx if interest is in total volume. If interest is in volume by diameter classes, point-Poisson or point-pps may be preferable (Schreuder *et al.* 1992a).

Schreuder *et al.* (1990b) conducted extensive simulation studies on eight forestry populations using restricted simple random sampling (RSRS), sampling with probability proportional to size (pps), stratified pps (spps), pscx, spscx, and sampling based on the range of x (psR sampling) with seven ratio and regression estimators. All populations had linear relationships between the variable of interest and the covariates but simple linear R^2 values varied from 0.22 to 0.98. Spscx, pscx, and spps were generally the best sample selection schemes over all populations using weighted regression estimators $\hat{Y}_{wr}$ (eq. 3.139) but the purposive sampling scheme, pscx, showed evidence of estimation bias in several instances. No reliable variance estimator was found for most of the sampling strategies tested, although bootstrap variance estimation with spps tended to have the least bias ($\leq$8.6% bias for all populations). The jackknife variance estimator always overestimated the variance. Schreuder *et al.* (1990b) concluded that more reliable variance estimators are needed for many sampling strategies.

In a follow-up simulation study on seven sets of three populations, varying in degree of linearity and variance structure, the weighted linear regression estimator $\hat{Y}_{gr}$ (eq. 3.134) was one of two estimators (the other being $\hat{Y}_{pi}$ in eq. (3.138)) that was quite efficient across the range of populations and showed less than 0.5% bias for all populations using the most efficient sample selection procedures giving one unit per stratum for samples of size 20 (Schreuder and Ouyang 1992). A simple bootstrap variance estimator generally had small bias, tended to be conservative, and yielded reliable coverage rates with 95% confidence intervals for most populations. Bootstrapping was done differently from the procedure used in Schreuder *et al.* (1990b), which accounts for the difference in reliability of the bootstrap variance estimates. In Schreuder and Ouyang (1992), a SRS of size n with replacement was drawn from the n sample units, regression coefficients were generated for each bootstrap sample, and the corresponding bootstrap estimated totals and their variances were computed.

6.6. RECOMMENDATIONS

What circumstances dictate the use of design-based or model-based strategies? The answer depends on objectives. If one is interested only in an efficient estimate, and is little concerned about the error of that estimate, a model-based strategy can be a sensible approach, particularly for small samples.

But when one is definitely interested in a scientifically defensible estimate, which implies that a valid estimate of precision can also be computed, and when there is doubt about the validity of the underlying model, then design-based inference should be used. If there is no doubt that the underlying model is quite reasonable, our recommendation still is to use a probabilistic selection procedure but to rely heavily on the model to make the design as efficient as possible. For example, in the simulations of Schreuder and Ouyang (1992), selecting one unit per stratum using stratified random sampling was consistently very efficient with weighted regression estimators. The strata were set up based on the covariate x. Such stratified sampling methods are probabilistic in that every unit in the population has a positive probability of selection but the joint probabilities of selection for units in the same stratum are of course zero. Rennolls (1981) made similar recommendations. Consideration of an underlying model should impact the method used to select the design-based sample selection method, too, that is, one may want to examine the sample selected to see how representative, well balanced, or π-balanced it is. As noted by Royall and Cumberland (1981a,b), variance estimators should be used that rely more on the actual sample obtained than on properties based on all possible samples (e.g., in the case of $\hat{Y}_{rm}$, $v_2(\hat{Y}_{rm})$ should be better than $v_1(\hat{Y}_{rm})$ since v_2 adjusts for differences between sample mean, $\bar{x}$, and population mean, $\bar{X}$, in the sample obtained). We recommend bootstrap variance estimates in general for this purpose.

Model-based inference and related sampling strategies deserve serious consideration for estimation from small samples when a valid model can be assumed. The argument becomes less persuasive for larger sample sizes, although one can argue that a smaller sample size may be required for model-based inference than for design-based inference. At the same time, efficient estimators, such as the generalized regression estimator (Särndal 1982), may make design-based inference satisfactory for smaller sample sizes too.

A primary contribution of model-based inferential procedures with their reliance on infinite population and superpopulation models is to identify more clearly the strengths and limitations of traditional design-based inference, resulting in sampling strategies that are still essentially design-based but more robust. Superpopulation models may also provide a coherent, natural framework for both design-based and model-based approaches to inference (Mandallaz 1990, 1991).

Keeping in mind the concept introduced in Section 6.1, that survey samplers only differ in their emphasis on what mixture of design-based and model-based precepts to use, the following recommendations are made:

1. Randomization in sample selection should generally be used whenever possible. Randomization provides a feeling of security in inference even though inference from a specific sample may be misleading. The need for this feeling of security may be less in the case of populations with a strong relationship between y and x.
2. Stratified random sampling or systematic sampling with a random start should frequently be used to minimize the risk of a specific sample being misleading in inference. Such procedures do not sacrifice the feeling of security engendered by randomization and are often attractive from a practical point of view as well. Purposive balanced sampling is an attractive concept to help us understand why a particular sample may be better than another, but in practice it should only be used when an efficient estimate is wanted with no concern about the error of the estimate. This results in samples that are demonstrably representative with respect to certain functions of the population but they are not necessarily representative of all characteristics of the population, and reliable inference about population totals Y is not assured.
3. If only a small number of sample units can be selected, say five to ten, purposive sampling guided by a credible model should be considered. Balanced sampling may be attractive in this context. The estimate of precision is likely to be poor for any design in this case. One might seek "good" estimates but sacrifice one's ability to state with confidence how good the estimate is.

4. More use needs to be made of asymptotically designed unbiased regression estimators such as Särndal's generalized regression estimator $\hat{Y}_{gr}$. This particular estimator and others like it, which have appeared in the model-based sampling literature, may help to blend the efficiency of model-based sampling strategies with the inferential robustness of design-based sampling strategies, even for small sample sizes.
5. Variance estimators based on the internal structure of the data, for example, the jackknife and bootstrap, should be used for design-based estimators such as ratio and regression estimators. Classical ratio and regression variance estimators are not based sufficiently on the properties of the specific sample, as is well illustrated in Royall and Cumberland (1981a,b).

6.7 PROBLEMS

1. You work for an international timber company and are asked to cruise a tract of timber for both total volume $V_{01}(m^3)$ and total merchantable volume $V_{02}(m^3)$. At the end of sampling, you will have accurately measured D^2H values for every tree in the population as well as volumes v_{01} and v_{02} for n randomly selected sample trees. Would you recommend an estimator based on model (2.70)?

2. The tract of land you cruised in Problem 1 is owned by Ms. Brown. Based on your cruise, you tell her that your estimates of V_{01} and V_{02} are $2M\ m^3$ and $1M\ m^3$, respectively. It turns out that she had an independent cruise made for which the estimates yielded $V^*_{01} = 4M\ m^3$ and $V^*_{02} = 3M\ m^3$. She is furious and you worry about a lawsuit against your company for undervaluing the timber. What would you do to minimize the chances of a lawsuit occurring and to protect yourself and your company if one did occur?

3. You have a boss who is not an inventory specialist but thinks he is. He has total faith in model (2.70) with $k = 2$ and always wants you to use an estimator based on that model. You are asked to cruise a tract of timber for total volume V_0.

(a) If you have complete faith in cruiser A and know that model (2.70) holds with $k = 2$ with y = tree volume and x = ocularly estimated tree volume (by cruiser A), what sampling strategy would you use if α can be assumed to be negligible?

(b) What sampling strategy would you recommend if α is substantially different from zero?

(c) Cruiser A is involved in a car accident and will be out of action for a year. Your boss introduces you to cruisers B, C, and D and gives you the following facts about them. Cruiser B is totally unreliable, that is, any correlation between his ocular estimate x and volume y would be pure coincidence. Cruiser C is much more reliable on small trees but always underestimates the volume of large trees. Cruiser D is extremely reliable but true volume and ocular volume for her are best related by the function $y = \alpha + \beta x + cx^2$. Which cruiser would you use?

7 Mensurational Aspects of Forest Inventory

7.1 INTRODUCTION

Forest mensuration or measurement is a very broad subject covering principles applicable to measurement in all aspects of forestry including forest management, wildlife management, watershed management, recreation, and other aspects of multiple-use forestry. This chapter covers only that section of forest mensuration dealing with the more common techniques of measuring or assessing single trees or parts thereof, groups of trees, the understory vegetation, regeneration, and forest health. It begins by defining measurement and discussing significant figures and rounding off, and concludes with a general discussion of sample plots in forest inventory and data recording. For comprehensive coverage of the field of forest mensuration, the reader should refer to such texts as Loetsch *et al.* (1973), Husch *et al.* (1982), Avery and Burkhart (1983), and Philip (1983).

Measuring variables of the forest and its components provides the forest manager with the quantitative information essential for sound planning and management of the resource, but measurement of extensive forests is costly. Thus the great challenge for the forest mensurationist is to develop simpler, cheaper and yet more accurate methods of taking measurements and deriving population estimates.

The principles of measurement used in forestry apply to any field of specialization where objectives are defined, needs are established, and technology and methodology have been evolved but, understandably, the methods of measurement vary from place to place depending on the purpose of measurement, the nature, quantity, and value of the resource, local conditions, time, labor, money, and market factors. Whatever the situation, however, the value of the data gathered depends on the dedication and skill of the technical personnel whose responsibility it is to take and record the measurements. Thus, it is vital that the technician be thoroughly familiar with the tools of measurement, be skilled in their use, and be aware of the sources of error likely to affect the measurements because, as discussed in Chapter 9, it is difficult to deal with measurement errors and missing data in a statistically valid manner. For this reason, particular emphasis is given in this chapter to recognizing sources of error in the types of measurement discussed and describing the action necessary to minimize their occurrence.

Lord Kelvin once remarked that,

> When you measure what you are speaking about and express it in numbers, you know something about it, but when you cannot (or do not) measure it, when you cannot (or do not) express it in numbers, your knowledge is of a meagre and unsatisfactory kind.

This remark highlights the essential difference between quantitative and qualitative assessment. The latter involves subjective judgment and is much more prone to personal bias and error. For this reason, sound management of forest resources relies heavily on data derived from measurement.

We may define measurement as the determination of size in relation to some observed standard, such as meter, kilogram, second, ampere, degree Kelvin, candela, mole, or some unit derived from these seven basic units (only the first two of which affect forest measurements). For measurements to be reliable, they should be taken carefully and in statistically acceptable ways. Thus before beginning any measurement exercise, it is wise to:

- Define the problem and state specifically what needs to be measured.
- Explore possible approaches to measurement.
- Design suitable sampling and measuring techniques.
- Test the techniques in the field, checking that they are efficient and yield sensible and sufficiently representative answers.

Then, as measurement proceeds, continually scan the data for errors so that corrections can be made before leaving the field site because, in a majority of forest measurement situations, cost as well as other factors prohibit the measurer from returning to the site to check suspect data.

7.2 SIGNIFICANT DIGITS AND ROUNDING OFF

Significant digits and rounding off frequently pose problems when data derived from measurement are processed and presented in scientific publications. It is appropriate to consider the general rules that apply to each.

7.2.1 Significant Digits

Because all measurements are subject to error, one is never able to quote an exact value for a measured physical quantity. Thus if the height of a tree is estimated to lie somewhere between 27.5 and 28.5 m, the result should be quoted as 28 m (2.8×10). Such a measurement is said to have been made to two significant digits.

It is incorrect to record more significant digits than are actually observed; for example, a length measurement of 15 taken to the nearest meter should not be written 15.0 m. Scientists particularly must always be on guard against writing down a long string of numbers that have no significance, because it is misleading. Spurious digits are particularly liable to arise when a series of calculations is performed. To avoid confusion, it is often preferable to write a result using an appropriate power of 10. For example, consider the following calculation:

$$\begin{aligned} \text{forest volume} &= (\text{total area} \times \text{mean volume per plot})/\text{plot area} \\ &= (2274.6\ \text{ha} \times 6.36\ \text{m}^3)/0.15\ \text{ha} \\ &= 96\,443.04\ \text{m}^3 \end{aligned}$$

The result could be written $96\,000\ \text{m}^3$ but 9.6×10^4 is better because it indicates clearly that only two digits are significant in the result (see rule 2 below). Because the great majority of calculations today are done by machine, there is no need to address the question of significant digits until all calculations have been completed, but it is essential that the operator know or record the measurement or factor embodied in the series of calculations that carries the least number of significant digits.

The rules that apply are:

Rule 1. The significant digits in a number are those reading from left to right, beginning with the first nonzero digit and ending with the last digit written, which may be zero but, if zero, it must not result from rounding off. We have, for example:

2 Significant Digits	3 Significant Digits
$0.\underline{87}$	$0.\underline{876}$
$0.0\underline{87}$	$\underline{8.76}$
$\underline{8.7}$	$0.0\underline{876}$
$0.000\underline{90}$	$0.0\underline{900}$

Rule 2. In multiplication and division, the factor with the least number of significant digits limits the number of significant digits in the product or quotient. For example, $306.17 \times 8.14 = 2492.2238$. The result should be written $249. \times 10$. However, $21.556 \times 40 = 862.24$ and may be written as 862.24 because the number of significant digits in an integer (40) is infinite. Thus the factor with the least number of significant digits in the expression is 21.556 with five.

A good rule in a series of multiplications or divisions is to note the factor with the least number of significant digits in the set and round off to that number of digits at the end. For example:

Set 1	$A(7\text{ digits}) \times B(3) \times C(4) = D$
Set 2	$E(8\text{ digits}) \times F(2) \times G(7) = H$
Then suppose	$D \times H = I$.

Round off I to two significant digits because the factor with the least number of significant digits is F with two.

Determining the number of significant digits in addition and subtraction requires that the numbers in the set be first aligned according to their decimal places. The number of significant digits in the result can never be greater than that of the largest of the numbers in the set and is often less. The following rule applies:

Rule 3. Align the numbers according to their decimal places. Identify the largest number in the set and locate the right-most of its digits, which aligns with at least the right-most digit of all the other numbers in the set. Counting the number of digits to the left of and including that digit gives the number of significant digits in the result.

Consider the following examples:

Example 1	Example 2	Example 3
123.623	1786.413	1671.53
+1.1	−6.2	+1547.97
4.72	1780.213	2000.34
129.443		8619.29
		13 839.13

Record the results as 129.4, 1780.2, and 13 839.1, respectively.

These rules should be rigorously applied to processed data, such as computer printouts, because the precision implied in the output is almost invariably spurious.

7.2.2 Rounding Off:

Normal practice with field records is to round them off to the nearest smallest unit attainable by measurement. The accepted convention for rounding off may be illustrated using a two-digit number $n.5$, where n is the ones digit.

Consider a number $<n.5$, say 7.3. Round off to n, that is, 7.

For a number $>n.5$, say 7.7, increase n by 1, that is, record as 8.

For a number $= n.5$,

1. *If* n is even, round off to n; for example, record 12.5 as 12.
2. *If* n is odd, increase n by 1, for example, record 13.5 as 14.

This convention avoids bias.

7.3 MEASURING SINGLE TREES

A tree comprises three main components: main bole (including the stump), crown, and roots, each of which may be measured in greater or lesser detail depending on the purpose of measurement. In this section dealing with measurement of the characteristics of single trees, we confine our attention to those characteristics most commonly measured or derived from measurement in resource inventory, namely, diameter, height, bark thickness, taper, bole or surface area, volume, weight, crown parameters, and age. We do not consider measurement of the roots, which is both difficult and costly and only feasible in research projects. Readers interested in measuring roots should refer to Hermann (1977) and Böhm (1979).

7.3.1 Diameter

Trees taper from ground to tip, that is, the diameter of the bole decreases with height. Despite this, a standard point of measurement on the bole is necessary as an aid in tree identification and to sort trees into groups of similar size for description and management. The most convenient point on the bole for measuring diameter is near the ground but as trees vary in height, it is desirable to measure at a fixed relative height. This presents difficulties however, for example, the 5% point would be out of reach for trees exceeding about 30 m. If a relative height less than 5% was accepted, measurement of diameter, particularly on smaller trees, would be adversely affected by buttswell. Thus the universal convention is to measure at a fixed height above ground called *breast height*. This fixed height is 1.4 m or 4′6″ above ground in the USA, New Zealand, Burma, India, Malaysia, South Africa, and some other countries, and 1.3 m or 4′3″ above ground in continental Europe, the United Kingdom, Australia, Canada, and some other former members of the British Commonwealth. The international symbol for diameter at breast height outside bark is d (IUFRO 1959).

Measuring the diameter of tree boles at 1.3 or 1.4 m is convenient but it is not the best place to take a measurement. Due to tree to tree variation in buttswell, cross-sectional shape, and bark thickness near the base of the bole, diameter measurements at breast height may not be typical of the solid content of the tree as a whole. Recent work (Wood *et al.* 1990, Wiant *et al.* 1991) indicates clearly that, for deriving an estimate of the total volume of a tree bole, a measurement of diameter at the centroid position (approximately one-third of total tree height) is desirable. In most situations, however, this is impractical as the point of measurement is out-of-reach from the ground. Thus the convention of measuring at 1.3 or 1.4 m is unlikely to change.

Rules have been developed to improve the consistency of measurements representing DBH and these should always be applied with common sense. The most general of these rules are well covered in the mensurational texts referred to earlier, but special comments on unrepresentative points are justified.

Unrepresentative points occur along the boles of trees due to the presence of branches, nodal swellings, wounds, and malformities. Adjusting for them is essential if the purpose of measurement is to derive the volume of sawlogs in a specified area of forest but may be ignored if the trees are to be used solely for wood fiber or pulp. Two methods are commonly used to make the adjustment:

Method 1. Take two measurements equidistant above and below the nominal point of measurement and average the two diameters. However, if the two diameters differ appreciably in size, average their sectional areas and convert the result back to its diameter equivalent.

Method 2. Choose for measurement a point on the bole that is judged by eye to be representative of the diameter at the initial point of measurement had the swelling, branches, and so on, not been present. There is a danger of bias with this approach but it is unlikely to be serious, particularly if the purpose of the data is to derive information on growth and increment. The bias can be avoided entirely in periodic measurement by using Method 1 initially to derive the diameter and then, by trial and error, finding a point of equivalent diameter on the stem and marking it permanently with a scribe mark, soft nail, or paint.

In forests where buttressing and fluting are common or where many trees have suffered butt damage due to fire or other causes, special instructions may need to be issued to the field crews detailing acceptable procedures for measurement. Usually in these cases, measurements are taken above the buttressed, fluted or damaged section but if the condition extends well up the bole, an arbitrary height is usually specified, for example, 3 m above ground.

d is the most important dimension assessed on trees. It is easily measured and is one of the dimensions from which basal area, surface area, and volume can be determined. The cross-sectional area at breast height, derived from d assuming a circular cross section, is termed *basal area* and is designated by the internationally recognized symbol g (IUFRO 1959). Sectional area at any other point on the stem is called *sectional area*.

The diameter of a standing tree at breast height can be measured directly by calipers and indirectly by linear tape graduated in π units (indirectly, because with tapes, girth is measured and diameter is estimated from the girth measurement assuming a circular cross section). A calipered diameter is the distance d_v between parallel tangents to the region of convex closure (Gregoire *et al.* 1990, Matérn 1990), whereas a taped diameter is computed from a parametric measurement of the convex closure (the region bounded by the tape wrapped firmly around the cross section). The diameter d of a tree is the average d_v over all directions v (Matérn 1990).

Sometimes, however, diameters are required at points along the bole that are out of reach from the ground, for example, for assessing stem taper (defined as the decrease in diameter per unit length of stem), form quotient (the ratio of the diameter at some point above breast height to that at breast height), or for deriving volume from a series of measurements along the stem. These diameters can be measured directly by climbing if precision and accuracy of measurement are considered to be more important than cost, as in some research projects. However, because climbing can be hazardous and special safety precautions must be taken to protect the climber, indirect methods of measurement are usually employed, such as pole tapes or calipers (useful for limited heights above ground), optical calipers (e.g., Wheeler pentaprism caliper, McClure mirror caliper), and optical dendrometers (e.g., Spiegelrelaskop, Telerelaskop, Barr and Stroud dendrometer). On felled trees, diameter may be measured directly by caliper or by reading from a linear scale placed at the log ends, or indirectly by tape.

Because tapes and calipers are the instruments most widely used for measuring diameter in forest trees, it is appropriate to compare the sources of error that apply to each as well as their accuracy, precision, practicability, cost, and general usage.

Sources of Error. Errors in diameter measurement arise from three sources: the tree, the instrument, and the operator.

Tree Related Errors. For a perfectly circular tree, which is rare, no error occurs with either instrument provided each is in perfect working order and is used properly.

For a tree of truly elliptical cross section, which is also rare, the caliper gives a better estimate of cross-sectional area than the tape if the major and minor axes are measured and area is derived using the formula: area = $\pi ab/4$, where a and b are the lengths of the two axes. The estimate is only slightly better however and only if the diameters measured are the true major and minor axes. Otherwise, the error may be larger and can be either positive or negative (Carron 1968a).

For trees of noncircular cross section, the tape is always positively biased. It is well known that if g_t is the basal area of a stem calculated from an exact taped measurement of its d and if g is its true basal area, then $g_t \geq g$. The excess of g_t over g depends on both the departure of the stem from convexity and the departure of the convex region encompassing the stem cross section from circularity (Gregoire *et al.* 1990). Matérn (1956) called these areas the convex deficit (CD) and the isoperimetric deficit (ID), respectively. The ID of a convex, noncircular section is always positive. Normally, this bias is small.

The error by caliper on noncircular tree boles may be larger or smaller than that of the tape and positive or negative, depending on where the caliper arms are placed; that is, it depends on the axes selected for measurement.

In practice, several procedures are used to derive diameter from caliper measurements made on irregularly shaped trees, namely measurements of:

- The maximum and minimum diameters.
- The maximum diameter and that at right angles to it.
- Any two diameters at right angles to each other.

The third option is most commonly used. Usually, the diameter recorded is accepted as being the arithmetic mean of the two measured diameters, but it is of minor importance when deriving sectional area whether it is the arithmetic, geometric, or quadratic mean (Matérn 1990).

The caliper has an advantage over the tape when diameter is required inside bark (= under bark) because bark thickness can be measured at the point of contact of the caliper arms with the tree; that is, an actual inside bark equivalent of an outside bark (= over bark) dimension can be obtained more reliably by caliper. However, this does not imply that the inside bark measurement is always more accurate. This depends on how accurately bark thickness is measured at each contact point and on the representativeness of the axes selected for measurement.

Instrumental Error. This error is negligible for correctly graduated steel tapes but can occur with cloth and fiberglass tapes due to stretching with continued use, particularly in rough or harsh environments. The only way to avoid the problem is to check the tape periodically against a standard length.

Calipers are more subject to instrumental error than tapes because they are more easily damaged. If the sliding arm is not perpendicular to the scale when read, a negative bias usually results; but if the fixed arm has been bent toward the sliding arm, which can happen if the instrument is dropped on the tip of the fixed arm, a positive bias can occur. Wear in the shoe at the base of the sliding arm has always been a problem with wooden calipers but this defect has largely been eliminated in modern steel and metal alloy calipers.

Operator error. The main sources of operator error with both instruments are:

- Incorrect location of the point of measurement on the tree bole.
- Incorrect placement of the instrument perpendicular to the long axis of the bole irrespective of whether the tree is vertical or leaning. This error is directly proportional to the angle of displacement (θ) and is always positive. Note particularly that with the caliper this error can occur on both axes (each measuring $d \sec \theta$) whereas, with the tape, it occurs on one axis only because the tape measures the perimeter of an ellipse with major axis $d \sec \theta$ and minor axis d.

- Incorrect tension (tape) and pressure on the arms (caliper) at the time of measurement.
- Incorrect reading of the scale of the instrument. This occurs much more commonly than is realized.

Most of the errors described above can be minimized and even avoided entirely by thorough training of the measuring crew and by choosing a person to record measurements who has the capacity to remain alert for extended periods under sometimes difficult conditions.

Accuracy and Precision. For many decades foresters have argued about the relative merits of the caliper and tape for accurate measurement of tree diameter (see McArdle 1928, Brickell 1970). The question may never be resolved satisfactorily because so many factors are involved. Certainly, a deterrent for using the tape is that it is positively biased on noncircular trees (most trees are noncircular), but as stated earlier this bias is usually small.

Gregoire *et al.* (1990) investigated the accuracy of tree diameter measurements at breast height by tape and caliper. They compared the estimates of basal area so derived (assuming circularity) with the actual basal areas derived by digitization of photographic images of discs excised from trees at breast height. They found that the errors, including those due to mensuration, averaged 3.1% with taped diameters, of which noncircularity accounted for about 2.6% of the bias, and −2.5% when the means of two calipered diameters taken perpendicular to each other (one randomly located) were used. On average, they observed a difference of 5% in the basal areas computed using the two approaches.

Gregoire *et al.* (1990) noted that many empirical studies have shown that measurements of DBH by caliper lead to underestimates of actual basal area, whereas analytic results show the direct opposite. Matérn (1990) suggested that the only possible explanation for the discrepancy is the influence of mensurational errors, such as the pressure of the caliper arms on the outer bark and more or less intentional adjustment of the direction of the caliper arms to avoid protruding ridges of bark.

Many foresters prefer the tape to the caliper because the tape is more portable and can be used more precisely even by inexperienced crews. The ratio of the coefficients of variation, tape to caliper, is approximately 2 to 3 : 1, the advantage for the tape increasing with increase in tree size. This lower precision of the caliper implies that it is an instrument much more prone to error in use. Unquestionably, the tape is preferable for any work related to tree growth for which consistency of measurement is important.

General Usage. Choice between the tape and caliper has been dictated mostly by tree size and custom. Calipers are used widely in Europe and Asia but infrequently in tropical countries and the southern hemisphere where many trees are too large to be measured by caliper. Some countries of the

British Commonwealth, for example, Australia, adopted tapes for measurement when forestry activity first began in the country both because of the size of trees in the virgin forest and because they inherited from Britain the Hoppus system of estimating log volume which required measurement of girth rather than diameter (Carron 1968a). Some optical instruments for measuring out-of-reach diameters such as the penta-prism caliper (Benson and Slinn 1964, Robbins and Young 1968) have the same limitation on measurement imposed by tree size as does the conventional caliper, but others based on the optical fork principle are not so affected, such as Bitterlich's Spiegelrelaskop and Telerelaskop. The Spiegelrelaskop (henceforth called Relaskop) deserves special mention here because it is a compact, robust, versatile and reasonably priced instrument that is proving indispensable in many countries for use in modern forest inventory. The Relaskop can be used to measure the diameter at any point along the bole of trees of any size, but what is not always appreciated by the user is that the precision of measurement is directly proportional to the horizontal distance from the tree and inversely proportional to the diameter under scrutiny, that is, $P(\%) = 10 \times s/d$, where s is the horizontal distance in meters from the observer's eye to the center of the tree bole and d is the diameter of the tree in centimeters. Thus the closer the observer is to the tree and the larger the tree, the better the precision of measurement.

One version of the Relaskop, the wide-angle Relaskop (Bitterlich 1962), incorporates a large number of reference angles and height scales and is invaluable for measuring diameters and heights in tropical high forest where tree boles are large and visibility is restricted (Philip 1983). Optical dendrometry is discussed further in Section 7.3.6.

This discussion would be incomplete without some mention of ocular estimation of tree diameter and length, a practice adopted in many countries earlier this century for low-cost forest resource inventory; for example, it was used to estimate the center diameter or girth and commercial log length of standing trees for deriving their volume. Today, ocular estimation is important in several types of Poisson or 3P sampling (see Section 4.5). It is a useful skill for a recorder to develop because by quick appraisal of a tree, one can often identify a suspect measurement by a crew member and have it checked before the information is recorded. Remember, any error in a measurement of diameter translates into an error in volume directly proportional to the difference between the squares of the correct and incorrect diameters.

7.3.2 Height

Height of trees and stands is a variable of fundamental importance in forest measurement because it is used to describe both the quantity and quality of tree and forest growth. For example, height is an important variable in tree and stand volume tables and height at a given age or size of tree is an index

commonly used to classify forest sites. Thus particular attention is paid in this section to identifying sources of error when measuring tree height and describing the procedures that can be taken to minimize them.

The height of a tree is defined as the vertical distance from ground level to its uppermost point (Empire Forestry Association 1953). It is commonly called *total height* to distinguish it from *merchantable height* (defined below). In forests where leaning trees are common, measuring the length of the tree axis from ground level to its tip may be more appropriate because an underestimate of standing volume will occur otherwise. Correction for lean is usually ignored in extensive surveys of forest crops but may be essential in intensive surveys.

Measurement of total height is straightforward in most young and semimature coniferous stands (these mostly have an excurrent habit, that is, a well-defined tip) but can be difficult in very dense stands and in mature (old) stands, in young plantations before pruning and thinning, and in moderately dense to dense stands of mature hardwoods, which typically have a deliquescent habit (rounded, spreading crowns). The convention for measuring the total height of trees of deliquescent habit is to sight to the uppermost point in the crown plumb above the top of the main bole. On slopes, ground level is usually taken to be on the upper side of the tree.

Merchantable height is defined as the height above ground (above stump height in some countries) to which a tree stem is salable (Ford–Robertson 1971); that is, it is the height to the highest point on the bole where diameter is not less than some specified value or where utilization of the bole is limited by branching or defect. Mostly, merchantable height rather than total height is measured in mature hardwood stands.

The total and merchantable stem length of felled trees is measured directly by linear tape and sometimes by graduated pole. For standing trees, both direct and indirect methods of measurement are available. Direct methods are more precise and accurate but they are not as widely used because of cost and other constraints discussed below.

Direct Methods of Measuring Height (Standing Trees). The two most common methods involve climbing with a tape and graduated pole or using height sticks or rods.

Climbing is accurate but practicable only with some tree species under certain conditions. It is costly and dangerous and is never used in routine forest inventory and assessment work, but it may be justified in the occasional research project where accuracy is all important.

Height sticks offer the most reliable and practical method of measuring the height of trees up to 25 m tall. Using 1.5 m length sticks made of duralumin tubing, experienced crews can measure trees up to 30 m tall in still conditions. For measuring heights up to 12–15 m, sticks are faster than any other method. Fiberglass sticks are some 25% heavier than duralumin sticks and their use is restricted to trees up to 20 m tall.

Some height sticks are manufactured with a locking device to prevent them pulling apart in use. This eliminates a safety hazard for the field crew but slows up measurement and increases cost. They are unnecessary if the measuring crew is properly trained.

The most serious error affecting measurement by height sticks is parallax error, which can be as much as ±10%. It can be eliminated by the observer standing well away from the tree and perpendicular to the vertical plane generated by the upper bole and sticks. The vertical plane is defined by the observer moving around the tree until the top of the bole is obscured by the top of the sticks, or vice versa. The observer then moves away from the tree perpendicular to the defined plane to locate a point on the ground convenient for observing the tree tip.

Indirect Methods of Measuring Height (Standing Trees). Owing to the practical limitations of height sticks, indirect methods of measuring tree height are more common in practice than direct methods. The instruments used involve two broad groups of hypsometers, one group based on geometric and the other on trigonometric principles. The former group includes the staff or pole hypsometer (Carron 1968a), the Vorkampff–Laue hypsometer (Hummel 1951), the Christen and Weise hypsometers (Jerram 1939), the Merrit hypsometer (Husch 1963), and the Jal hypsometer (FAO 1954). We do not discuss this group further because they are insufficiently precise and accurate for serious forest assessment and inventory work.

In considering methods of measuring tree height based on trigonometric principles, assume for the present that the boles of all trees are vertical and they stand on level ground. Given this situation, the total height (h) of a tree is easily determined by setting off a horizontal distance (s) from the observer to the geometric center of the tree bole and recording both the angle of elevation to its tip (ρ) and angle of depression to its base (θ). Total height is then given by: $h = s(\tan \rho + \tan \theta)$.

Prior to World War II, the instrument most widely used for measuring tree height based on trigonometric principles was the Abney level. Since the War, the Abney has been replaced by instruments of similar precision (realistically ±2.5%) that can be read directly and are more robust and quicker to use, such as the Blume Leiss altimeter, the Haga altimeter, the Relaskop and the Suunto clinometer. Although most of these instruments have a range finder attachment for setting off the horizontal distance, we advise against using the facility unless the operator is working alone, the reason being that a further subjective decision is required, which decreases the precision of the final estimate. This advice may change, however, with the increasing availability of small, precise, relatively low-cost electronic range finders, which allow the operator to move around a tree until a good sight of its tip is obtained. However, unless these instruments have a built-in facility for slope correction, it is more satisfactory, and just as quick, to set out the horizontal distance manually.

The decision on which height measuring instrument to use in a particular project depends on a number of factors including availability of instrument, personal preference and the several features listed and summarized by instrument in the comparative table below which has been compiled based on personal experience. (Note: A = Abney, B = Blume Leiss, H = Haga, S = Suunto clinometer, R = hand held Relaskop):

Steadiness:	$B = H \gg A = S = R$
Line of sight:	$B = H \gg R > S > A$
Durability:	$R > S > B > H > A$
Compactness:	$S \gg R > A = B = H$
Ease of use:	$B = H > R > S^{a} > A$
Speed of use:	$B > H > S = R > A$
Price:	$S < A = H = B \ll R$.

For highly precise and accurate measurements of tree height by indirect means (as might be required, e.g., to determine the tallest tree in a country), instruments such as the Barr and Stroud dendrometer, Todis dendrometer, Telerelaskop, theodolite, and so on would have to be used (see Optical Dendrometry in Section 7.3.6). These more sophisticated instruments, however, are never used in forestry for routine height work because of their bulk, weight, time required for measurement, and cost.

Consider now the case of a tree on sloping ground. The horizontal distance can still be set out manually either by holding the tape high at one end and low at the other (if down or up slope and one is tall enough) or by moving on the contour. If feasible, the latter is preferable because it is not uncommon for a tree on a slope to have a slight lean downhill. In such cases, setting out the horizontal distance perpendicular to the lean minimizes the error in the height estimate due to the lean. Should either of the above procedures prove impractical in a particular case, a solution is to measure slope distance in any convenient direction from the tree, record the angle of slope, and derive the equivalent horizontal distance either by pocket calculator or from prepared tables. Because the derived distance is unlikely to equal one of the fixed horizontal distances for which the instrument is calibrated, the height recorded for the tree will have to be adjusted by multiplying it by c/s, where c is the calculated horizontal distance and s is the horizontal distance for which the scale is calibrated. Alternatively, if the instrument has a percentage scale, use it rather than a "direct reading" scale.

Much of the "noise" in data bases of tree height occurs due to failure by measuring crews to account for tree lean. It is a fact that only a small minority of trees are truly vertical, that is, the tip of the tree is vertically above the geometric center of its base. Thus setting off the horizontal distance from the base of the tree will incur some error. In many cases, the

[a]The Suunto causes eyestrain in some people due to the method of projecting the line of sight.

lean and associated error will be very small and can be ignored but in many others it cannot. Consider, for example, ten trees each 30 m tall but with leans increasing from 1° to 10° in increments of 1°. Suppose now an assessment crew ignored the leans and measured the height of each tree by setting off a horizontal distance of 30 m from its base (1) in the direction of lean and (2) in the opposite direction. By simple trigonometry it can be shown that the errors incurred will be as summarized in table (7.1.) These results imply that even for a very slight lean of 2°, the error incurred in assuming the tree is vertical will range from +3.6% to −3.4% depending on the direction from which the tree is observed. To minimize this error, the horizontal distance must be set out perpendicular to the plane of lean. Thus it is imperative that field crews involved in measuring tree heights inspect each tree closely for lean and, if a lean is detected, establish its plane and move away perpendicular to it. For leans greater than about 6°, however, the tip of the tree must be projected to ground and the horizontal distance set off from the projected point. Otherwise, errors well in excess of 1% can result (Table 7.2). These errors are not uncommon and partly explain why forest modelers frequently complain about excessive "noise" in tree-height data bases.

For leans up to 8° (such a lean looks quite severe), the height of the vertical component (H_v) can be accepted as the true height (H_t) of the tree because the error will be less than 1% (<2% for an 11° lean and <5% for an 18° lean). For most types of forest assessment and inventory, leans greater than 8° require correction. This is done by measuring the displacement of the tip, that is, the horizontal distance (s) from the geometric center of the base of the tree to the plumb point and correcting the height of the vertical component by Pythagoras' theorem: $H_t = \sqrt{(H_v^2 + s^2)}$.

TABLE 7.1 Maximum Errors Incurred in Measuring the Height of Leaning Trees as if They Were Vertical[a].

	Lean Toward Observer	Lean Away from Observer
Lean in Degrees	Percentage	Error
1	+1.8	−1.7
2	+3.6	−3.4
3	+5.4	−5.1
4	+7.2	−6.7
5	+9.1	−8.4
6	+11.1	−10.0
7	+13.0	−11.5
8	+15.0	−13.1
9	+17.1	−14.6
10	+19.2	−16.1

[a]Height of tree, horizontal distance, and angle of observation are assumed to be 30 m, 30 m, and 45°, respectively.

TABLE 7.2 Errors Incurred in Measuring the Height of Leaning Trees, Having Set Off the Horizontal Distance from the Tree Base Perpendicular to the Plane of Lean[a].

Angle of Lean (degrees)	Error %
1	−0.0
2	−0.1
3	−0.3
4	−0.5
5	−0.8
6	−1.1
7	−1.5
8	−1.9
9	−2.4
10	−3.0
11	−3.6
12	−4.2
13	−4.9
14	−5.7
15	−6.5
16	−7.3
17	−8.2
18	−9.1
19	−10.1
20	−11.1

[a]Angle of observation = 45°.

It is unnecessary to measure the lean of the tree by instrument to determine whether or not a correction for lean is necessary. The maximum size of s allowable ($s_{\max}$) to restrict to 1% the error incurred by assuming that the height of the vertical component equals true height, is easily determined from the formula

$$s_{\max} = H_{vm} \times k ,$$

where H_{vm} is the measured height of the vertical component and $k = 0.1425$. Both H_{vm} and k are independent of the unit of measurement.

For precise measurement of height of leaning trees, and as an absolute check on the accuracy of measurement, readings should be taken from two independent positions on the ground roughly 90° apart. To be acceptable, the two readings should agree within the limits set by the precision of the instrument, namely, ±2.5% for the Abney, Haga, Blume Leiss, Relaskop, and Suunto clinometer.

Seldom should one need to substitute ocular estimates of height for measurement except for quick approximations and point-3P (point-Poisson) sampling. Ocular estimation of tree heights up to about 18 m is used in stripline assessment of site quality in *Pinus radiata* plantations aged 9–10 years in South Australia where experience has been that the estimates are reliable provided the assessors are highly trained. A wise precaution to adopt if ocular estimates have to be used for any reason is to make frequent checks by instrument to "calibrate the eye." If practical, stand a "calibration" pole of known length (2 or 3 m) vertically at the base of each tree being assessed. The pole is particularly useful in point-3P sampling, where it improves the precision of the volume estimate.

7.3.3 Bark Thickness

Bark thickness may be defined as the distance between the cambium and region of convex closure outside bark, but this is problematical on stems with concave regions. Bark thickness varies around the stem and a recorded measurement is usually an average of four measurements taken at points roughly equidistant around the bole.

For most species, the average volume of bark in a tree ranges from 10 to 20% of the outside bark volume. Because the thickness of bark on a tree varies with species, genetic constitution, site conditions, age, size, rate of growth, and position along the bole, the percentage of bark in truck loads of unpeeled, round timber may vary appreciably. In a given log, bark percent can be estimated from the formula

$$\text{bark } \% = 100\,((d_1^2 - d_2^2)/d_1^2)$$

where d_1, d_2 are the outside and inside bark diameters respectively, at the center of the log.

Bark thickness decreases more or less regularly from the base of a tree to its tip, the rate of taper usually but not necessarily being correlated with the rate of taper of the stem inside bark. Despite this variation, a relationship between average bark thickness at breast height, bark volume, and stem variables (d outside bark and height) can usually be established. Today, however, owing to the increasing use of high-performance optical dendrometers in forest assessment and inventory to estimate the upper stem dimensions and other characteristics of standing forest trees, attention is being directed more to developing functions permitting inside-bark diameters at any point on the tree bole to be derived from outside measurements. A simple function may be adequate for this, for example, $d_i = kd_o$, where d_i is the estimated diameter inside bark, d_o is the measured diameter outside bark, and k is a simple ratio of the measured diameters outside and inside bark at a specified point on the stem, usually breast height. More complex functions, however, may be necessary.

Grosenbaugh (1967b) included three options (equations) in his computer program STX for estimating bark thickness (BT) along a tree bole given measurements of BT and diameter outside bark (DOB) at breast height and DOB at a series of points along the bole. One advantage of these equations is that different coefficients are not required for different species. The equations were designed to cater for the three broad patterns of behavior observed in the ratio of diameter inside bark (DIB) to DOB along tree boles, as follows:

Option 1: Constant at all levels (applies to many hardwoods; Wiant and Koch 1974; Castaneda 1979; Wingerd and Wiant 1982; Biggs *et al.* 1985):

$$\mathrm{DIB} = \mathrm{DOB}(\mathrm{DBHIB}/\mathrm{d})\,. \tag{7.1}$$

Option 2: Increases curvilinearly above breast height and decreases curvilinearly below (applies to many softwoods and some hardwoods: Mesavage 1969; Wingerd and Wiant 1982):

$$\mathrm{DIB} = \mathrm{DOB}\;[1.0 - (1.0 - \mathrm{DBHIB}/\mathrm{d})\{1.0/(2.0\text{-}\mathrm{DOB}/\mathrm{d})\}]\,. \tag{7.2}$$

Option 3: Decreases curvilinearly above BH and increases curvilinearly below BH (applies to some hardwoods: Boehmer and Rennie 1976; Wingerd and Wiant 1982):

$$\mathrm{DIB} = \mathrm{DOB}(\mathrm{DBHIB}/\mathrm{d})\{9.0/(10.0 - \mathrm{DOB}/\mathrm{d})\}\,. \tag{7.3}$$

If Grosenbaugh's general functions are known not to fit a particular species as well as desired, the only option available to the prospective user is to collect appropriate sample tree data and derive a better function. Thus Johnson and Wood (1987) developed for radiata pine plantations (*Pinus radiata* D. Don) in the Australian Capital Territory a bark function for use in determining the inside bark volume of sample trees measured by dendrometry. The equation, which is independent of site and age predicts bark thickness (RB_i) at any point i along the bole, relative to bark thickness at breast height, from the fourth power of relative diameter (RD_i), RD_i being defined as the ratio of DOB at i to that at breast height. The form of the model is

$$\mathrm{RB}_i = b_0 + b_1\,\mathrm{RD}_i^4 \tag{7.4}$$

where b_0, b_1 are the simple linear regression coefficients. Using this model, Johnson and Wood estimated the inside bark volume of 109 trees with an average arithmetic error of -0.5%.

In most countries, timber is sold on an inside bark basis so there is a need to derive inside bark dimensions. These can be obtained indirectly using bark functions as described above but, sooner or later, measurements will have to be taken.

Direct measurement of bark thickness (and inside-bark diameter) is confined to felled trees where it can be applied at the log ends using a scale rule and at points along the bole either by removing the bark or by cutting out chips of wood with bark attached. With the latter, care must be taken that the surface of the bark is not compressed and that bark does not flake off during cutting.

On standing trees, bark thickness has to be measured indirectly using instruments such as a bark probe (Gill *et al.* 1982) or bark gauge, the best known of which is the Swedish gauge. On particularly thick and tough barks, a hammer and nail or brace and bit may be required. Users are often surprised by the accuracy that can be achieved with a hammer and nail and a thin, metal scale rule. If funds are limited, use of this equipment should be considered seriously. Unfortunately, no satisfactory way of ensuring exact bark penetration has yet been devised with any instrument, but with repeated and intelligent use an operator can develop a "feel" for the gauge and detect when wood penetration occurs.

Improper use of the bark gauge is one of the most common sources of error when estimating the volume inside bark of trees and logs. Even an error as small as 25 mm in radial bark thickness in a 25-cm DOB log incurs a 4% error in sectional area inside bark and hence volume. Thus sound training and extreme care is needed at all times when using a bark gauge.

7.3.4 Taper

The taper of a solid body of regular outline (i.e., the way it decreases in diameter from base to tip) is governed by the values of k and b in the expression $y^2 = kx^b$, where y is the radius of cross section, x is the distance from the apex, b determines the way the body tapers, that is, its solid shape or form, and k determines the rate of taper within that shape. Values of b for the solid shapes that tree boles approximate are 3 (neiloid), 2 (conoid), 1 (second-degree paraboloid), and 2/3 (third-degree paraboloid). Most tree boles comprise frusta of several solid shapes, such as neiloidal butt, paraboloidal main bole, and conoidal tip. Of these, the conoidal frustum is dominant in immature trees and the paraboloidal frustum in mature trees.

Many mathematical models describing the profiles of tree stems have been developed (Sterba 1980) but no single model has been found that applies generally to trees of all species and sizes in all geographic regions (Czaplewski *et al.* 1989b). However, the segmented polynomial model of Max and Burkhart (1976) has consistently performed well (Cao *et al.* 1980; Martin 1981; Gordon 1983; Amidon 1984; Alberta Forest Service 1987).

The form of this model is

$$\mathrm{RD}^2 = b_1(h/H - 1) + b_2(h^2/H^2 - 1) + b_3(a_1 - h/H)^2 I_1 + b_4(a_2 - h/H)^2 I_2 \qquad (7.5)$$

where RD = relative diameter as defined in eq. (7.4)
b_i = regression coefficients, $i = \{1, 2, 3, 4\}$
h = height above ground to estimated top diameter $\hat{d}$
H = total tree height
a_i = join point parameters estimated from sample data, $i = \{1, 2\}$
$I_i = 1$, for $h/H < a_i$, $i = \{1, 2\}$; $= 0$, for $h/H > a_i$.

Cao (1978) derived equations for estimating diameter, height, and volume using Max and Burkhart's model. Czaplewski and McClure (1988) developed a variant of this model conditioning it for two diameter measurements, one at DBH and another on the upper stem. Given DBH and total tree height, scatterplots of the empirical rate of change of stem diameter with height can indicate changes in taper that provide tentative estimates for join points in segmented profile models, and can help in identifying the desired form of the model (Czaplewski 1989; Czaplewski *et al.* 1989a).

Goulding and Murray (1976) and Gordon (1983) developed polynomial, volume-compatible taper equations to predict DIB at any point along the bole of radiata pine (*Pinus radiata* D. Don) in New Zealand, the term "volume-compatible" implying that the volume between the base and tip of the tree derived from the equation equals the known (measured) bole volume. Such equations can be integrated to estimate volume between any two points on the bole. The general form of Goulding and Murray's and Gordon's equations is

$$d(h) = [v_o/(KH)(b_1x + b_2x^2 + b_3x^3 + b_4x^4 + b_5x^5 + b_6x^6)]^{0.5} \qquad (7.6)$$

$$v_t = v_o(b_7x^2 + b_8x^3 + b_9x^4 + b_{10}x^5 + b_{11}x^6 + b_{12}x^{l+1}) \qquad (7.7)$$

where

$d(h)$ = DIB at any point along the bole (cm),
v_o = total volume inside bark (m^3),
K = a constant (in metric: 0.000 078 539 8 m^2/cm^2),
H = total tree height (m),
$x = (H - h)/H$, where h is the height (m) to $d(h)$,
v_t = volume from tip to $d(h)$, that is, volume of the section $(H - h)$ (m^3),
$b_1, b_2, \ldots, b_{12}$ = coefficients, some of which may be zero, and
l = an integer, $5 < l < 41$.

Very good fits to diameter and height data have been achieved with simpler functions than the New Zealand one described above. For example, Kozak *et al.* (1969) developed a taper function for 19 species and species groups in British Columbia from the basic parabolic function

$$d^2(h)/D^2 = b_0 + b_1 h/H + b_2 h^2/H^2 , \tag{7.8}$$

namely:

$$d^2(h)/d^2 = b_1(h/H - 1) + b_2(h^2/H^2 - 1) \tag{7.9}$$

or

$$d^2(h)/d^2 = b_2(1 - 2h/H + h^2/H^2) , \tag{7.10}$$

the particular function used depending on the species, where:

d = diameter at breast height outside bark (DBHOB),
h = height of δ above the ground (m),
H = total height of tree (m),
b_0, b_1, b_2 = regression coefficients.

Kozak *et al.* (1969) noted that the standard error of estimates of DIB at different heights using their function was comparable with errors reported by Pfeiffer (1967), who measured upper stem diameters (outside bark) by Barr and Stroud dendrometer under good sighting conditions and adjusted them for bark thickness to derive the corresponding DIB. Thus Kozak *et al.* (1969) suggested that it will be more convenient and less costly to predict upper bole DIB from reliable taper functions (if such functions are available) than from measurement by dendrometer.

Their suggestion is highly relevant today because of the interest developing in the new methods of sampling forest trees discussed in Chapter 4, namely, importance sampling (Gregoire *et al.* 1986; Wiant *et al.* 1989), centroid sampling (Wood and Wiant 1990; Wood *et al.* 1990; Wiant *et al.* 1991) and critical height sampling (McTague and Bailey 1985; Lynch 1986; Van Deusen and Meerschaert 1986; Iles 1988; Van Deusen 1990). Each of these methods either requires a taper function for successful implementation or becomes practical for forest assessment and inventory when used in combination with a reliable taper function. Taper functions for deliquescent tree species up to crown break (from above the buttress to crown break for many tropical species) are badly needed to improve assessment of the world's indigenous hardwood forests. Bruce and Max (1990) gave an excellent summary of stem taper equations and their use in tree volume estimation.

7.3.5 Bole or Surface Area

The surface area of a tree bole inside bark, called bole area, represents the base for potential increase in bole volume (Lexen 1943, Boyce 1975). Similarly, the total bole area for a forest stand represents the base for potential increase in stand volume. Wood increment over a period is superimposed on the bole area at the beginning of the period. Also, the outer surface area of a tree is the habitat and source of supply of food for various insects, fungi, lichens, and algae. Thus tree and stand bole area both inside and outside bark are variables that are of inherent interest to foresters and forest ecologists, entomologists, and pathologists. Whittaker (1966) used outer-bark surface area expressed as a proportion of ground surface area as a measure of the maturity and degree of closure of a forest stand.

Lexen (1943) argued that volume of a forest stand provides a measure only of wood capital and not of the potential of the stand to grow. He suggested that estimates of stand volume should be supplemented by estimates of stand bole area, and he developed bole area tables for *Pinus ponderosa* based on DBHOB and total tree height. He found that the area of an individual tree bole could be estimated by applying the formula

$$\text{bole area } (\text{m}^2) = kdH \tag{7.11}$$

where $d = \text{DBHOB}$ (cm), $H = \text{total height}$ (m), and $k = 0.0172\ \text{m}^2/\text{cm}$ ($=0.143\ \text{f}^2/\text{in}$ imperial).

The bole area of an individual tree can be derived indirectly from a reliable taper function by calculus or directly from a series of height and diameter measurements along the bole, that is, from any of the direct methods for estimating bole volume (Section 7.3.6). If, for example, a tree is measured by the standard sectional method (Section 7.3.6), then bole area is given by the formula

$$\text{bole area } (\text{m}^2) = \pi\{K\,\Sigma d_i + l_1 d_{n+1} + l_2 d_{up}/2\} \tag{7.12}$$

where

K = constant length of section (m),
d_i = DIB (m) at midlength of the n K meter sections,
d_{n+1}, d_{up} = DIB (m) at midlength and top end, respectively, of the odd log, and
l_1, l_2 = length (m) of odd log and tip, respectively.

If the length of stem to the upper diameter limit is an exact multiple of K meters, the $l_1\, d_{n+1}$ term drops out.

Carron (1968a) investigated bole area in plantation-grown *Pinus radiata* covering all sites and ages in the Australian Capital Territory and found that k approximated the value derived by Lexen for *Pinus ponderosa*, varying from 0.0164 (0.137 imperial) to 0.0185 (0.154 imperial) and was correlated with d but unaffected by height within a diameter class. Overall, Carron found variation in k was so small that he suggested a useful approximation to stand bole area could be derived from the formula

$$\text{stand bole area } (\text{m}^2) = 0.0172\bar{d}\bar{H}N \tag{7.13}$$

where $\bar{d}$ = stand mean d (cm) outside bark,
$\bar{H}$ = stand mean h (m), and
N = total number of trees.

More usually, the surface area of woody plants and forest communities is estimated by applying an appropriate regression function (Whittaker and Woodwell 1967, Swank and Schreuder 1974). Gregoire *et al.* (1992) proposed an alternative means of estimation when an appropriate regression function is unavailable or when the accuracy of prediction from regression cannot be determined. Their method allows the surface area (and bark volume) of the entire bole or any portion of it to be estimated.

Further research on the bole area of forest trees and communities (both estimation and application) is justified. We see from formula (7.13) that stand bole area is a function of diameter, height, number of trees, and tree form and is thus a measure of the amount of stand in a physical sense. Also, as it is the base for potential increase in stand volume and describes a surface that absorbs short wave radiation from the atmosphere and re-radiates long wave radiation to surrounding areas, it has particular biological and ecological significance.

7.3.6 Volume

Each year, more and more wood is being sold by weight but forest managers still use volume data extensively. Thus even though sale, processing, and marketing procedures are changing, volume and the measures used to derive it (diameter, height, form, basal area, etc.) are still important variables to be assessed during forest inventory.

When foresters speak of volume, they mean the amount of wood in a tree, stand of trees, or other specified area according to some unit of measurement and some standard of use. The unit of measurement[1] most

[1]In the USA, lumber is sold in the marketplace in board foot units. The board foot is a measure of products derived from the forest resource as opposed to the volumetric measure (cubic meters) of the resource itself. Thus a single meaningful board foot volume cannot be specified for a particular tract of forest because it depends on how the trees are sawn. In practice, trees are "scaled" assuming that 1″ boards will be cut from them, and buyers make their own adjustments for other cutting plans. As a medium of exchange, the board foot has worked well in the USA and it is unlikely to be discarded by industry until a complete switch to the metric system is made. The measure has absolutely no appeal outside the USA.

commonly used worldwide is cubic meters or cubic meters/hectare and the standard of use may be, for example, veneer timber, sawtimber, round timber (poles, piles, etc.), pulpwood, or firewood. Usually, volume is expressed outside or inside bark according to different specifications, namely, gross total volume, gross merchantable volume, and net merchantable volume.

The *gross total volume* of a tree (or stand) is the volume of the main stem (or of all stems) including its stump and all defective and decayed wood. Contrast this with *gross merchantable volume*, which excludes the stump and top but includes the defective and decayed wood, and *net merchantable volume*, which excludes the stump and top as well as the decayed and defective wood. When assessing standing forest trees, it is much more difficult to derive reliable estimates of net merchantable volume than gross total or gross merchantable volume because assessing decay and defect, and their extent, depends entirely on the experience, skill, and judgment of the assessor. Even very skilled assessors frequently make estimates that are seriously in error.

Two other terms requiring definition at this stage are biological volume and utilizable volume. *Biological volume* is defined as the volume of the stem with branches trimmed at their junction with the stem. It excludes all irregularities not part of the natural growth habit of the tree, for example, malformation due to insects, fungi, fire, and mechanical damage. *Utilizable volume* (or merchantable volume) excludes as well the normal irregularities of episodic growth, namely the branch nodes. This requires some subjective judgement by the assessor. Potential utilizable volume may closely approximate biological volume in some situations, for example, smooth stemmed species in a pulp operation. However, in sawlog operations, utilizable volume is invariably less than biological volume.

Directly or indirectly, estimates of the volume of wood in forests are based on the volumes of stems of individual trees. Thus estimation of stem volume is a most important aspect of forest mensuration.

The methods available for estimating stem volume can be grouped into two classes: direct and indirect. The direct methods involve physical contact with the tree throughout measurement whereas the indirect methods involve measurement mostly from a distance using linear tapes, hypsometers, dendrometers, and so on.

Direct Methods. These methods apply mostly to felled trees but some can be applied to standing trees in certain circumstances. Though they give more reliable estimates of stem volume than indirect methods, they are more hazardous, more time consuming, and more costly to apply. For these reasons they are seldom used in forest inventory unless internal defect is serious or the number of trees to be measured is small. They are used mostly to derive the volumes of sample trees for developing tree volume tables and functions.

The direct methods fall into three main classes: xylometry, summation of the volume of stem sections, and graphical procedures.

Xylometry. Determining the volume of a tree bole (or bole sections) in a xylometer from either the weight of water displaced or the difference in weight of the bole measured both in air and water, is the most accurate method available (Loetsch *et al.* 1973, p. 148), but it requires special equipment, is laborious and costly, and can only be applied to felled trees. Thus, xylometry has no application in routine forest inventory, but it is useful in research projects (e.g., Dargavel and Ditchburne 1971) and has been used successfully in Finland for measuring the volume of pulpwood bundles (Nylinder 1967) and also in Canada (David Bruce, personal communication).

Summation of the Volume of Stem Sections. The tree is assumed to comprise a number of sections. The dimensions of each defined section are measured and volumes are then derived and summed to give stem volume. With the exception of the tip, whose volume (v_o) is usually derived assuming a conoid shape ($v_o = l\pi d^2/12$, where l is the length of the tip and d is the diameter of its base), section volumes are mostly calculated using either Huber's or Smalian's formula, both of which assume the solid shape of the section is a second-degree paraboloid. The reliability of the estimate of stem volume then depends on the validity of the assumption, the representativeness of the points of measurement, and the accuracy of measurement.

$$\text{Huber's formula:} \qquad v_o = l\pi d_m^2/4\,, \tag{7.14}$$

$$\text{Smalian's formula:} \qquad v_o = l\pi(d_{le}^2 + d_{se}^2)/8\,, \tag{7.15}$$

where l = section length and d_{le}, d_m, d_{se} = diameters at the large end, midlength, and small end, respectively, all units in meters.

Both Huber's and Smalian's formulas are accurate if the shape of the section is a second degree paraboloid, but if it is not both will give biased estimates. For Huber, the bias is negative for shapes less than second-degree paraboloid (conoids and neiloids) and positive for the fuller shape of a third-degree paraboloid. The bias for Smalian in all cases is opposite in sign to and twice that of Huber. Thus it is preferable to use Huber's formula in all cases (all log lengths) where section length can be held constant. If this is not possible, there are two alternatives: (1) use Smalian's formula with section lengths shorter than 3 m and exercise more care in measurement particularly in the region of the lower bole affected by buttswell, or (2) derive volume by integrating a cubic spline curve fitted through data points of cross-sectional area as described by Goulding (1979).

The volume of stem sections can also be derived using Newton's formula, which generally is more accurate than Huber's and Smalian's formulas.

Newton's formula makes no assumptions about solid shape, but because it requires measurement at the middle of the log and both ends, it is much more costly to apply. Thus its use is limited to research and to serving as a check on volumes derived by other methods. Newton's formula is

$$v_o = l\pi(d_{le}^{\,2} + 4d_m^{\,2} + d_{se}^{\,2})/24 . \tag{7.16}$$

Particular care should be exercised when using Newton's formula on butt logs because, like Smalian's formula, it can give seriously biased results unless allowance is made for butt flare when measuring d_{le}. A more appropriate method of measurement in these cases is the centroid method (Wiant *et al.* 1992a).

The best known methods used for deriving stem volume by accumulating the volumes of stem sections are the standard sectional, Hohenadl, and taper step methods. The first two of these use a constant section length and Huber's formula to derive volume, whereas the taper step method uses a variable section length and Smalian's formula.

STANDARD SECTIONAL METHOD. This method is suitable for estimating total volume and volume to any merchantable limit, outside or inside bark, of both excurrent and deliquescent trees. It involves simple calculations easily programmed for the computer.

Any convenient length of section (K) can be used but 3 m is conventional. Diameter is measured at the midlength of each section and volume is derived using Huber's formula. The constant length of section can be maintained on felled stems to within K m of the tip, leaving usually a short length ($< K$) whose volume is derived using the formula for a conoid. Alternatively, an upper diameter limit (d_{up}, e.g., 0.1 m) on the bole is specified and measurements cease at this point. If the height to d_{up} is not an exact multiple of K meters, a section of stem less than K m (called the odd log) will occur. The total volume of the bole is then derived as follows:

$$v_o = (\pi/4)\{K \Sigma\, d_i^2 + l_1 d_{n+1}^2 + l_2 d_{up}^2/3\} \tag{7.17}$$

where

K = the constant length of section,
d_i = the diameter at the midlength of the n K meter sections,
d_{n+1}, d_{up} = the diameters at the midlength and top end, respectively, of the odd log, and
l_1, l_2 = the lengths of the odd log and tip, respectively.

All units are measured in meters. If the length of the stem to the upper diameter limit is an exact multiple of K meters, the $l_1 d_{n+1}^2$ term drops out.

In practice, the first or basal section may start at an arbitrary stump height but it is common to start from ground level and make a stump allowance when calculating merchantable volume.

The standard sectional method is unbiased to the extent that each K meter section and the odd log are frusta of second degree paraboloids, the tip is conoid, and each point of measurement is representative of the section. Owing to buttswell, the butt section is usually the frustum of a neiloid so there is a possibility of bias (negative), the extent depending on the flare of the butt and how far it extends up the bole. This bias can be reduced by dividing the basal section into two $K/2$ meter sections measured at 0.25 K and 0.75 K, respectively, above ground level and adjusting the formula accordingly.

Because points of measurement on the stem are fixed, the standard sectional method is usually precise, though some imprecision may arise if standard procedures are not applied when measuring at unrepresentative points (see Section 7.3.1).

The standard sectional method does not allow any simple visual check on measurements so the recorder must be alert to minimize mistakes in measurement. Errors can often be detected by checking that the measured diameters and bark thicknesses decrease up the bole in a sensible manner.

HOHENADL'S METHOD. This method is based on sectioning the stem into equal lengths, usually five or ten sections per stem. Its advantage is that stems of different length can be compared, which is important in investigations of stem form (Loetsch *et al.* 1973). Diameter outside or inside bark is measured at the midpoint of each section and volume is determined using Huber's formula. Volume of the stem is then derived by accumulating the volume of the sections as follows:

$$v_o = \pi h(d_n^2 + d_{n-1}^2 + d_{n-2}^2 + \cdots + d_2^2 + d_1^2)/(4n)\,, \tag{7.18}$$

where n = number of sections, h = total tree height (or height of the main bole in deliquescent trees), and $d_n, d_{n-1}, \ldots, d_2, d_1$ = diameters at the midpoints of the n sections. Sectioning the bole into 10 or even 20 sections of equal length yields a more accurate result than using only 5 sections because the bias arising from the butt section is more tightly controlled, but cost increases.

As with the standard sectional method, care needs to be taken that all points of measurement are representative and that the measured diameters and bark thicknesses decrease in a regular manner up the bole.

TAPER STEP METHOD. This method, first described by Whyte (1971), uses variable section lengths and Smalian's formula for deriving volume. It allows freedom in choosing points of measurement along the stem so the measurer can ensure that each point of measurement is biologically sensible. Whyte claimed that the method gives a more reproducible and realistic index of stem volume.

Measurements are made at mid-internodes, the lengths of sections being determined by taper steps (large end minus small end diameter). Whyte suggested the following taper steps for *Pinus radiata* to minimize errors in the basic assumptions regarding the solid shape of stem sections and to improve the estimate of volume:

Large End DOB	Taper Step	Tolerance
<25 cm	2.5 cm	±1 cm
25–50 cm	5.0 cm	±1 cm
50–75 cm	7.5 cm	±1 cm

The tolerance of ±1 cm allows some freedom in choosing the actual points of measurement based on the taper steps, that is, one should not attempt to have steps exactly 2.5 cm, 5.0 cm, and so forth, as this increases the cost of measurement unnecessarily.

Theoretically, volume derived using the taper step method is less than that derived using more traditional methods because measurements are confined to the mid-internodes. Whyte claimed that this slight bias is of no consequence as total stem volume is simply a standard from which recoverable and merchantable out-turns (products) are deduced. He also pointed out that precision of the volume estimate is improved and the between-tree variance is reduced.

A valid criticism of the taper step method is that the frequency of measurement is proportional to the taper of the tree and measurements tend to be concentrated along the upper bole due to changing the steps from 5 cm to 2.5 cm when the large end diameter decreases to below 25 cm. Goulding (1979) estimated that 50% of the taper step measurements on a sample of *Pinus radiata* trees in New Zealand were used to calculate less than 5% of the total volume, and he recommended that the method be modified as follows:

- Take measurements in the butt region at stump height, half-breast height (ca 0.65 m), breast height, and about 3 m.
- Over the rest of the stem, take measurements at regular intervals corresponding to the level of accuracy required.
- Where a sharp drop in diameter occurs, measure both above and below the point of sudden change.
- If true (= biological) volume is required, do not confine measurement to mid-internodes but avoid points with a severe abnormality in cross section such as deep scars and local large swellings.

These modifications suggested by Goulding are sensible and offer considerable flexibility for measuring the volumes of sample trees. Basically, the onus is on the measurer to determine the set of measurements that best suits his or her purposes, and proceed with measurement accordingly. It should not be forgotten as a general principle, however, that the center of volume (the centroid) of most tree boles occurs at a relative height of approximately 0.3 (Wood *et al.* 1990). Thus the center of value will occur well below this, probably at a relative height of 0.15 to 0.20. This implies that for efficient estimation of the volume of sample trees, half of all measurements taken should be spread over the lower one-fifth or one-sixth of the bole.

Most processing of sample tree measurements is now done by computer so little is gained by keeping section length constant. Thus in Australia and New Zealand, a modified taper step method is now preferred to the standard sectional method (which aims at simplifying manual calculation rather than adequately assessing tree volume) for routine measurement of sample trees.

HEIGHT ACCUMULATION METHOD. Like the standard sectional method, this unique method of determining the volume of a tree bole, conceived and developed by Grosenbaugh (1948, 1954), was designed to facilitate calculations both in the field and office, a feature that is relatively unimportant today given the ready availability of programmable calculators. Basically, the method requires choosing a constant taper step for the end diameters of logs within a tree bole, defining these end positions on the bole (preferably by Relaskop or similar instrument), measuring and accumulating the lengths of the resulting logs, and applying a simple formula to derive bole volume. This formula may include specific coefficients derived assuming a fixed taper step, unit length, and mean DIB/DOB ratio[2] as described by Husch *et al.* (1982, p. 103) or may be of generalized form as derived by Philip (1983, p. 73), namely,

$$v_{om} = (\pi d_t^2(h)/4) \sum_{i=1}^{n} (i^2 l_i + i l_i + l_i/k)$$

or

$$v_{om} = (\pi d_t^2(h)/2)\left(\sum_{j=1}^{n'} a_j + \sum_{i=1}^{n} l_i/(2k)\right)$$

(both formulas above exclude the cone above the top DOB (d_t)) and

$$v_o = (\pi d_t^2(h)/2)\left(\sum_{j=1}^{n} a_j + \sum_{i=1}^{n} l_i/(2k) + l_c/6\right)$$

(which includes the cone above $d_t(h)$) where

v_{om} = volume of tree to top of last log,
n = number of logs measured, excluding the cone,
n' = number of taper steps to a top DOB of $d_t(h)$,
v_o = total volume of tree,

[2]Formulas for calculating the coefficients were given in Grosenbaugh (1954). Loetsch *et al.* (1973) gave tables of constants for various taper steps and height units.

l_i = length of the ith log ($i = 1, \ldots, n$),
k = constant (= 2 for paraboloid, 3 for conoid, 4 for neiloid),
a_j = accumulated height to top of jth log ($j = 1, \ldots, n'$), and
l_c = length of cone.

The method has been applied extensively in the USA for several decades and has been shown to be efficient and accurate. It can be applied to groups of trees by accumulating the field data for all trees rather than by summing the individually calculated volumes.

Graphical Procedures. The best known methods employing graphical procedures to derive stem volume are the taper line method (Gray 1956) and the graphical or taper curve method first proposed by Reineke (1926).

TAPER LINE METHOD. Gray (1956), using data derived from many sample trees, particularly *Pinus radiata*, plotted sectional area against height. He found that excluding a variable distance from both the ground and tip, sectional area inside bark decreased linearly with height, at least within the limits to be expected with biological data and the usual standards of accuracy of measurement. Hence he hypothesized that the central length of the tree bole inside bark approximates a solid body generated by a second-degree paraboloid as closely as any other geometrical body. Based on this hypothesis, he proposed the taper line method for deriving the volume of a wide range of tree species, particularly conifers and young eucalypts (*Eucalyptus* sp.), from a limited number of representative measurements on the lower bole, namely, two in the region of buttswell and three to five along the main stem up to at least half height. From these measurements, total and merchantable volume of the stem are derived by plotting sectional area against height and making an assumption about the location of the point of inflection between the paraboloidal main bole and conoidal tip. The estimation procedure was well described by Carron (1968a). Though the original taper line procedure was graphical, the estimation algorithms are easily programmed into a computer.

The main criticisms of the method are that the volume estimate is liable to bias (because the points of measurement are chosen subjectively and linear relationships are fitted to the data ocularly in the field) and precision of the estimate is likely to be low. Despite this, the method usually gives acceptable estimates of stem volume (accuracy ±5%) for a minimum of time, effort, and cost. Thus it warrants close consideration whenever time and funds are limited. Bias and imprecision arising from fitting the linear relationships by eye are eliminated if the data are processed in the field using a portable computer.

GRAPHICAL METHOD. As originally conceived, this method involves measuring diameter at short intervals along the stem, plotting the data, and joining the points by straight lines or curves giving a diametral outline of the stem called a taper curve, or stem profile. If sectional area rather than diameter is plotted against height, volume is given directly by the area under the curve, which can be measured by planimeter or by counting squares. Today, the volume estimate is derived by computer by aggregating the volumes, derived by Smalian's formula, of all the short frusta comprising the section of stem of interest.

The graphical method has many advantages. It is flexible and applicable to all types of trees (excurrent and deliquescent); measurements can be made at any odd and convenient interval along the stem without complicating volume calculations; it makes no assumptions about overall stem shape; calculating volume from the area under the curve can be used with equal accuracy for all degrees of taper; total volume and volume to any diameter limit outside or inside bark can be determined; and it is simple to program for the computer. Its main disadvantage is that it is time consuming and costly if many measurements are taken, for example if the interval between points of measurement is set at approximately 1 m. But if the full advantages of the method are to be realized, this should be done.

There is a danger of bias with the graphical method because points of measurement are chosen subjectively, but the large number of measurements tends to control the bias. Likewise, imprecision of the method decreases, as does the correspondence between the estimate of total volume and actual biological volume, as the number of measurements along the stem is increased.

Estimates of volume derived from the graphical method are commonly used as standards for comparison with estimates derived by other methods.

Indirect Methods. Most estimation of tree volume in forest inventory involves standing sample trees for which direct methods of estimation are generally inapplicable because of cost and the hazards associated with climbing. Thus the assessor has to resort to indirect methods of estimation, the most notable being volume equations or functions and methods incorporating optical dendrometry. If reliable volume equations are available, they should be used, but if they are unavailable or if their reliability is suspect, optical dendrometry offers a viable alternative at reasonable cost (e.g., in the height accumulation method of Grosenbaugh 1948, 1954).

We do not discuss the compilation and derivation of volume equations for estimating the volume of sample trees and stands (their application for estimating stand volume is discussed later) because the subject is not an integral part of forest inventory *per se* and it is covered thoroughly in textbooks on forest mensuration (e.g. Loetsch *et al.* 1973; Clutter *et al.* 1983; Philip 1983). Optical dendrometry, however, is becoming an increasingly important component of sound and efficient forest inventory in many countries of the world and it warrants further discussion.

Optical dendrometry. Initially, the development of instruments described as optical dendrometers arose from the need of forest managers and scientists for more accurate estimates of tree and stand volume than were being provided by one- and two-way (single- and dual-entry) volume tables: large errors were often found in the estimates when compared with volumes computed from direct measurement of felled trees and stands. These errors were due mostly to differences in tree height for a given *d* and differences in stem form, that is, solid shape (Bruce 1925).

Since World War II, there has been considerable development and improvement in the volume equations available to the forest manager, some equations based on four independent variables being reliable not only for the species for which they were developed but also for other species (and even other genera) of similar habit, such as the regional volume table for *Pinus radiata* in South Australia (Lewis *et al.* 1973). Despite these advances, reliable volume (and taper) equations are not available for many commercially important tree species in both developed and underdeveloped countries of the world. To obtain volume estimates for trees of these species without felling the trees requires either estimates of the length and mid-diameter of the merchantable boles or, preferably, a series of diameter and height measurements along the length of each stem. These measurements can be made directly using ladders or climbing irons (a hazardous, costly, and mostly impractical procedure); by using conventional instruments such as a tape or caliper elevated manually (Ferree 1946; Godman 1949; Loetsch *et al.* 1973, p. 80, a procedure suitable for short length boles (6 to 10 m) or relatively small trees only); or by using optical dendrometers with built-in scales for reading diameter and height along the stem from a point on the ground remote from the tree.

In a classic paper, Grosenbaugh (1963) reviewed the theory and performance of the three main classes of optical dendrometers, namely optical forks, optical calipers, and range finder dendrometers. He concluded that for precise, accurate, and rapid estimation of upper-stem diameters, the dendrometer must have magnification and coincident or superposed alignment. It is desirable also that the instrument be simple to use, accurate at all heights, portable, relatively inexpensive, and operable independently of the distance from the point of measurement (Avery and Burkhart 1983). No available instrument meets all these requirements but some possess a sufficient number of the attributes to be valuable tools for forest research, assessment, and inventory, such as the Barr and Stroud dendrometer (FP 12 and FP 15 models), Zeiss teletop dendrometer, Relaskop, Telerelaskop, and Wheeler's penta-prism caliper with clinometer attached. Chehock (1982) described in some detail how to use each of these instruments for taking a series of diameter and height measurements along the bole of a standing tree and a method of recording and processing the data by computer to give product volumes. The best of the instruments listed is the Barr and Stroud dendrometer, which is a split-image, coincident type, magnifying, short-base range finder dendrometer; unfortunately, it is no longer manufactured. The best readily available alternative is the Telerelaskop but it is expensive. Next in line is the Relaskop, a moderately priced instrument that is increasing in popularity because of its small size, compactness, durability, and versatility. The wide-angle version of the Relaskop (Bitterlich 1962), with its large number of reference angles (for measuring diameter) and height scales giving direct height readings for horizontal distances from 4 m to 20 m in multiples of 2 m, has proven invaluable for inventory of tropical forest (Philip 1983). Methods of improving the accuracy of measurement when using optical dendrometers have been discussed by Bruce (1975) and Ferguson *et al.* (1984).

Laser Technology Inc. in the USA has developed a prototype of a portable, lightweight, laser ranging instrument that is fully integrated with an electronic vertical angle encoder and flux-gate compass, a graduated scale reticle, a key pad, and a digital two-line display. The instrument can also be interfaced with data recording equipment. It is claimed that, when operational, the instrument will be more efficient and accurate than the dendrometers conventionally used in many forest survey and inventory tasks for measuring range and the diameter and height of trees.

Achieving accurate estimates of volume inside bark by dendrometry requires a reliable bark function (refer to Section 7.3.3) and the following actions:

- Check the calibration of the instrument before work commences.
- Measure diameters of the butt section (heights ≤ breast height) by hand.
- Choose a ground position for the instrument which affords good visibility of most sections of the tree bole and from which the sighting angle to the highest point to be observed on the bole is about 45°.
- Check the tree for lean. If lean is apparent, take appropriate action, for example, locate the dendrometer at a position on the ground in a plane perpendicular to the plane of lean.
- Concentrate at least half of the intended measurements in the lower third of the bole including remeasurements at stump height, and at approximately 0.7 m, breast height, and 3 m. Scan all readings ensuring that they increase or decrease sensibly along the bole. Check deviant measurements.
- Ensure that the first diameter reading by dendrometer above breast height is consistent with that measured manually at breast height.

7.3.7 Weight

A forester speaking of tree weight (or volume) almost invariably means the weight (or volume) of the commercially marketable portion of the tree. This may comprise a variable portion of the main bole if the tree is used to produce veneer and saw timber, or the greater part of it if it is utilized for pulp, chips, and firewood.

The weight of a standing tree is estimated using linear or nonlinear regression equations developed for a specified species and population from information gathered previously from sample trees. These sample trees are selected objectively to represent the various size strata in the population. They are then felled and measured directly to provide the information needed to solve the regression model, namely usually DBHOB and total or merchantable height (independent variables) and green weight from which the dependent variable (oven-dry weight) is derived.

Gathering information on tree weight usually requires that the main bole be cut into short lengths or bolts (either of fixed length or fixed drop in diameter), which are then weighed individually or collectively, with or without bark, on a platform scale. Fixed-width disks are cut from the lower end of each bolt and these are weighed collectively either before and after or after the bark is removed. The disks are then taken to the laboratory where their oven-dry weights are measured. The dry weights of the bolts, and of the tree by aggregation, are derived from these measurements using the formulas of Clutter *et al.* (1983, pp. 19-20), for example,

$$w = l[a_{le}d_{se} + 2a_{le}d_{le} + 2a_{se}d_{se} + a_{se}d_{le}]/6 \tag{7.19}$$

where

w = dry weight of the bolt
l = bolt length
a_{le}, a_{se} = cross-sectional area of wood only at the lower and upper ends, respectively, of the bolt
d_{le}, d_{se} = dry-weight density at the lower and upper ends, respectively, of the bolt.

Proven techniques for establishing weight prediction models are also covered well by these same authors (Clutter *et al.* 1983, pp. 20–27).

Alternatively, tree weight can be estimated entirely using sample discs cut from the bolts. The sample discs with bark are weighed green in the field before significant moisture loss occurs. They are then taken to the laboratory where their cross sections are measured for diameter outside and inside bark. The bark is then removed from the disks and the disks are oven dried and weighed. Green- and dry-weight density, green bolt weight (wood and bark), and dry bolt weight (wood only) are then derived using the formulas of Clutter *et al.* (1983, pp. 19–20). This ratio method of estimation is biased but it is practical. A procedure for estimating without bias the aboveground biomass, woody volume, or mineral content of a tree and their variances is described by Valentine *et al.* (1984). Their procedure uses techniques involving randomized branch sampling and importance sampling, the selection probabilities in each case being proportional to estimated size.

7.3.8 Crown Parameters

The functions of a tree crown are to display its actively photosynthesizing leaves most efficiently to radiant energy and provide for the renewal of leaves. The size of the crown has a marked effect on, and is strongly correlated with, the growth of the tree and its various parts. Thus measurement of crown parameters is important in forestry.

The dimensions and parameters of tree crowns most commonly measured or estimated in forest surveys are diameter or width, length or depth, height, cross sectional area, surface area and volume.

Crown Diameter or Width. As with tree stems, crown diameter is an ill-defined characteristic due to lack of circularity. It is measured directly by projecting the edges of two opposite sides of the crown vertically to ground (using a plumb line, optical prism, or similar instrument capable of projecting a vertical line of sight), marking them on the ground, and measuring the horizontal distance between them using a linear tape. The measurement is applied to two or more axes depending on the irregularity of the crown, small individual branches and minor crown irregularities being ignored, and the results are averaged. Readings are taken usually to the nearest 0.1 m, but if subjective judgement of the projection is acceptable, results should be rounded off to the nearest meter.

Diameters of tree crowns can also be recorded indirectly using aerial photographs, the accuracy of measurement depending on image scale, film resolution, the ability of the assessor, and the extent to which the crowns interlock or mask each other (both from above and below). With few exceptions, estimates derived this way are less reliable than those derived from direct measurement because only those parts of the crown visible on the photo are measurable.

Estimates of crown diameter are used for several purposes, namely, to determine the crown diameter ratio (ratio of crown diameter to DBHOB), which is of silvicultural interest; to derive tree DBHOB from crown width measured on aerial photographs using established relationships; and as an independent variable in aerial tree volume tables. They are also used to derive one measure of stand density called the crown competition factor (Krajicek *et al.* 1961; Leech 1984), which is potentially useful in growth and yield studies because it is independent of age and site and applies to both even-aged and all-aged stands.

Crown Length (Depth) and Height. The length or depth of a crown and crown height, the vertical distance from ground level to the base of the crown, are usually measured on the standing tree, directly using height sticks or indirectly using a hypsometer. Crown length is sometimes expressed as the crown length ratio, l/h (not to be confused with crown ratio—see above) or green crown percent, $100(l/h)$, where l is the length of the living crown and h is total height of the tree. The length l is measured either to the lowest live whorl of branches (upper crown length) or to the lowest live branch excluding epicormics (lower crown length). A sensible judgment has to be made if this branch is isolated.

Crown Cross-Sectional Area. If the crown is regular in outline, its cross-sectional area is calculated from an average diameter assuming a circular cross section, the average diameter being derived from the maximum and minimum diameters, or from the maximum diameter and that at right angles to it, or from two diameters at right angles to each other, the direction of the first being determined at random. With irregular crowns, the edges are plumbed to ground and diameter is measured on a number of axes. The data are then plotted and the area is determined by planimeter or square counting. Repeating this process for a group of trees in a stand gives one estimate of crown closure (the proportion of the area of a forest stand covered by crowns). Alternatively, for an unbiased estimate of cross-sectional area without an assumption of circularity, measure the distance to the crown edge in some randomly chosen direction from some fixed interior point and call it $R(\theta_i)$. Repeat m times, and calculate the quadratic mean radius, say R_q. An unbiased estimate of cross-sectional area is then given by $\hat{A} = \pi R_q^2$.

Crown Surface Area and Volume. Crown surface area and volume (quite distinct from foliar surface area and volume) are derived from crown diameter and length assuming a solid shape (for the crown space) that is deemed appropriate for a particular tree, such as conoidal, paraboloidal, or hemispherical (for formulas, refer to Hamilton 1975, pp. 190–191). Usually, the crowns of conifers and young hardwoods are modeled as cones, their surface areas and volumes being derived from the formula

$$\text{surface area } (\text{m}^2, \text{ft}^2) = \pi d_b L/2 \tag{7.20}$$

$$\text{volume } (\text{m}^3, \text{ft}^3) = \pi d_b l/12 \,, \tag{7.21}$$

where d_b = diameter (m, ft) at the crown base,
L = sloping length (m, ft) from the apex of the crown to its base, and
l = crown length (m, ft).

Both crown surface area and crown volume should be useful predictors of tree growth, particularly the former because the most active photosynthetic region in a tree crown comprises the fully expanded young leaves at or near the crown surface (Kramer and Kozlowski 1979).

7.3.9 Age

Information on tree age is important when discussing the increment of a tree, that is, the quantitative increase in size in a specified interval of time due to growth, where growth is defined as the biological phenomenon of increase in size with time. The interval of time most commonly used when determining increment is one year. Although the age quoted is mostly that at the beginning of the period of increment, it is immaterial whether the beginning or end is used, but it must be specified.

Determining the age of a tree is often difficult. Age is known for certain only when the year of sowing or planting is known, for example, with plantation forests age should be available from compartment history or management records. Normally, the age of trees raised from transplanted seedlings is determined from the date of planting whereas that of trees resulting from direct sowing at the plantation site is reckoned from the observed time of germination. For forest stands originating from natural reproduction following a natural catastrophe or silvicultural treatment, the span in age of the trees may vary as much as 5–10 years or more. This presents problems when deciding the age of the stand. There are two practical solutions, namely, (1) consider the age of the stand to be the average age of a specified number of the largest dominant and codominant trees, or (2) consider the age of the stand to be the average age of a specified number of sample trees approximating average tree volume. The first option is favored by Husch *et al.* (1982) for its simplicity and other advantages.

Real difficulties arise in determining the age of trees in natural forests. Sometimes, in these forests, age can be estimated from the date of some catastrophic event such as wildfire, flood, severe storm (cyclone/typhoon), or logging. In other cases, the age of some tree species can be estimated from (1) the annual development of the shoot (e.g., in some *Pinus* and other coniferous species, the annual shoots on both the stem and branches are defined by bunched bracts (modified leaves) formed during winter at the base of the spring whorl, and these shoots can be counted), or (2) from growth rings, if annual, counted on the stump of a felled tree or on core samples extracted with an increment borer from breast height or near ground level of a standing tree. If the core is taken from breast height, the number of rings counted has to be increased by the estimated number of years taken for the tree to grow as a seedling to that height. However, increment borers are impractical for determining the age of large trees both because the maximum length of a borer bit is about 40 cm (16 inches) and because it is difficult to locate the pith. Thus the increment borer is used mostly to study the diameter growth of trees of small to moderate size and the recent growth of large trees (e.g., over the past 5 or 10 years). It is preferable when boring to take two cores, one each on the long and short axes, but if only one is taken, it should be extracted midway between the long and short axes.

True annual rings are very common in species of the northern hemisphere and in temperate, deciduous, broadleaf species generally, but they are the exception rather than the rule in many countries of the southern hemisphere, for example, Australia, New Zealand and South Africa. Tree rings are useless for determining tree age in tropical regions, which do not experience a regular change in growing conditions (Husch *et al.* 1982), but for some species growing in tropical countries with monsoon climates, e.g., teak (*Tectona grandis*), Mexican weeping pine (*Pinua patula*), and Mninga (*Pterocarpus angolensis*) in Tanzania, useful information on growth can be derived although the rings are not annual (Philip 1983).

The ease of counting annual rings on stem cores and wood disks depends on the species, age of tree, environmental and biological conditions, and the extent of ring distortion. In some tree species, the color differentiation between early and late wood is slight, which makes counting difficult. The problem experienced with old trees is that annual growth is small and diameter is large so the outer rings are crowded together, making differentiation difficult. Environmental and biological conditions such as drought, early or late frosts, and insect infestation can promote the formation of false rings, while under exceptionally good growing conditions, little differentiation into early and late wood may occur. All of these situations complicate the ring counting process. This is true also where defects, compression wood, tension wood, and branch whorls have distorted or destroyed the ring pattern.

When reconstructing the past history of growth of a tree using stem analysis (Jerram 1939; Bruce and Schumacher 1950; Husch *et al.* 1982; Philip 1983), it is essential that the annual growth rings be identified reliably. If this cannot be done macroscopically using a hand lens, microscopic examination of stem sections will be required. This is both tedious and time consuming unless expensive equipment, such as that used in dendrochronologic studies, is available.

7.4 MEASURING LOGS

Logs are formed when a tree is felled and its bole and larger branches are trimmed and crosscut into short lengths. The minimum allowed diameter at the small end is dictated by end-use. Logs are also formed hypothetically when the bole and larger branches of a standing tree are subdivided into short lengths by an assessor on the ground.

Although the trend around the world is for more and more logs to be graded before sale as sawlogs, peeler logs, and so on, and sold by weight, the majority are still sold on a volume basis. Thus methods of deriving log volume are still important.

Measuring the gross volume of logs accurately is not easy because logs are often irregular in both cross section and profile. As mentioned in Section

7.3.6, the only accurate method of measurement is by water displacement in a xylometer but this is impractical in the forest and impossible for hypothetical logs in standing trees. Thus gross volume has to be determined by other means that involve measuring log length and diameter and making assumptions about log shape.

Log length is measured in meters and decimeters, directly using graduated linear tapes (reinforced plastic, fiberglass, steel) and indirectly using a scalestick, revolving wheel, and in modern industrial plants, a photogrammetric or infrared scanning device.

Log diameter is measured directly at log ends using scale rulers and at intermediate positions using calipers. With direct measurement, two readings at right angles should be taken, but a single measurement at the midpoint of each log is sufficient if a large number of logs is involved (Philip 1983). Instruments for measuring log diameter indirectly include the diameter tape, Biltmore stick (Chapman and Meyer 1949; Loetsch *et al.* 1973) and, in modern industrial plants, photogrammetric or infrared scanning devices. Girth tapes are usually made of steel. They are wound tightly around the log in a plane at right angles to the long axis. The operator must ensure that the tape is not twisted and is in contact with the surface of the log around its full circumference.

When measuring both the length and diameter of logs, care must be taken that the measuring instrument is calibrated properly, measurements are taken in the correct plane, and the specified precision of measurement is met.

To determine log volume, sectional areas are needed. These are derived from measurements of log diameter assuming that the cross sections are circular. Errors arise because this assumption is seldom met, but they are small for most logs. When the volume of a noncircular log has to be estimated accurately, Philip (1983) recommended that two diameters be measured by caliper at the midlength of the log and the volume be derived using Huber's formula.

The formulas used to derive log volume (Huber, Smalian and Newton) are given in Section 7.3.6 (eq. 7.14, 7.15, and 7.16 respectively). All three formulas are accurate if the log is a cylinder (rare) or the frustum of a second-degree (quadratic) paraboloid. Fortunately, most logs approximate this latter shape. For logs approximating frustums of other shapes (neiloids, conoids, paraboloids of the third degree), Newton's formula is accurate in all cases but Huber's and Smalian's formulas are biased, the extent of error being proportional to the length of the log and the square of the difference between the two end diameters (Philip 1983). However, use of Newton's formula is restricted to research projects because it is time consuming and costly to apply.

Huber's formula is intermediate in accuracy between those of Newton and Smalian, its error being half that of Smalian and of opposite sign for shapes other than second-degree paraboloids (and cylinders). The biases of the three formulas for various solid shapes are summarized below:

Formula	<2nd-Degree Paraboloid	2nd-Degree Paraboloid (Excluding Cylinder)	>2nd-Degree Paraboloid
Huber	Negative	Nil	Positive
Newton	Nil	Nil	Nil
Smalian	Positive	Nil	Negative

The table above should be interpreted with care. It may give the impression that Newton's formula is accurate with all logs. This is not so with butt logs, most of which comprise frustums of two contiguous solid shapes, namely, a short neiloidal frustum at the butt end merging into a paraboloidal frustum above. Newton's formula can incur positive bias in such logs unless allowance is made for basal flare when measuring the butt diameter. Of the three formulas, Huber alone avoids this problem. With all three however, the extent of error in the volume estimate is also partly governed by the representativeness of the points of measurement. Often, insufficient attention is given to this when measuring logs, for example, the end- and mid-diameters are taken as fixed points even though these positions may coincide with nodal swellings.

The volumes of logs in many countries are derived from log volume tables compiled from sample data gathered from specified populations. These tables give volume inside bark (IB) from measurements of length and DIB at either the small end or midlength of the log. Tables using DIB at midlength (volume derived by Huber's formula) are very common despite the fact that the midpoints of logs in stacks are inaccessible and bark removal, bark measurement, or an empirical bark deduction is required to obtain the mid-DIB measurement. On the other hand, Huber's formula is more robust than the other two formulas when long log lengths are involved (Ellis and Duff 1973).

Smalian's formula, which requires measurement of diameter at both ends of a log, is the easiest to apply but it is the least accurate of the three formulas if log shape does not approximate a second-degree paraboloid. This applies particularly to butt logs with flared ends. Errors in volume estimates derived from Smalian's formula increase rapidly as the interval between measurements increases, exceeding 8% in logs 5 m and over in length (Young *et al.* 1969; Ellis and Duff 1973; Goulding 1979).

The principal advantages of Smalian's formula are that it can be applied to logs in stacks and is suitable for deriving the volume of logs in standing trees when the interval between measurements is variable (unlike Huber's and Newton's formulas, which require a constant interval). This second advantage explains why Smalian's formula has been used almost invariably for processing data gathered from dendrometry of standing trees. However, Goulding (1979) has demonstrated that integrating a cubic spline curve fitted through data points representing a stem profile gave estimates of volume with only 60% of the error of Smalian. Thus it is probable that this procedure will gradually supplant Smalian's formula in data processing packages.

A three-dimensional 'universal formula' for determining volume IB of round wood (logs) of exotic conifers in New Zealand was developed by Ellis (1982). The formula, which requires input of log length, DIB at both ends, and log taper, is

$$v_o = \exp\left[b_0 \ln(l) + b_1(d_{se}) + b_2 \ln((d_{le} - d_{se})/l) + b_3\right] \times 10^{-3} + b_4 d_{se}^2 l \qquad (7.22)$$

where

l = log length (m)
d_{le}, d_{se} = large and small end diameters, respectively, inside bark (cm),
$(d_{le} - d_{se})/l$ = a measure of log taper (cm/m), and
$b_0, b_1, \ldots, b_4$ = coefficients.

This formula has been used to compile ten log volume tables, based on average tapers from 0.4 cm/m to 2.2 cm/m, which apply to any species of exotic conifer in New Zealand (and possibly in other countries). The nominal 95% confidence intervals for a single log and for 50 logs are ±16% and ±2%, respectively (Ellis 1982).

Quantity of logs can also be recorded in other ways, such as the following:

- Number of pieces or cumulative length by size class, for example, posts, poles.
- Stacked volume (m^3, cords, cunits, etc.), for example, pulpwood, fuelwood.
- Weight, for example, sawlogs and peelers (usually includes an explicit or implicit factor to convert to volume), pulpwood, fuelwood.

As mentioned earlier, there is a worldwide trend, particularly in large scale operations, to measure saw and peeler logs by weight rather than by the traditional method, volume, because of greater efficiency. Weight scaling requires the use of either truck scales or weighbridges strategically distributed with respect to the trucking route from the logging site to the mill. Alternatively, front end loaders can be fitted with a microprocessor that uses the hydraulic pressure in the loading arms to weigh bundles of logs at the logging site. Such equipment tested in New Zealand had a capacity of 40 tonnes per lift and reduced the cost of assessing log quantity by 85%

compared to the traditional method (measurement of length and diameter and reference to log tables) (Ellis 1984). A disadvantage of weight measurement of sawlogs is that weight to volume conversion factors are needed, and these vary with locality, species, age, position of log in tree, weather, time since felling, and so on. Thus it is important to continuously monitor the conversion factor by sampling.

Log defect is a problem when estimating log quantity. Traditionally, defect refers to those imperfections that result in complete loss of wood volume when sawing a log. Scale deductions may need to be made for rots, insect flight passages, ring shakes, checks, splits, and crooks. The logical, sound and practical system developed by Grosenbaugh (1952) for calculating deductions in volume to account for defect is highly recommended.

Grosenbaugh (1967c) discussed the wisdom of using the traditional units of measure, such as cubic meters, kilograms, board feet, cubic feet, pounds, and so on, to display the end results of inventory. He suggested that foresters consider assessing the standing forest resource in terms of three relatively invariant, reproducible, primary units of measure, namely, volume, surface area, and length, which tend to reflect tree size and to be linearly related to the output of each individual sawmiller, regardless of the sawing processes used or the products made. Subsequently, in a pilot test involving 60 sample trees, Space (1974) demonstrated the methodology required to implement Grosenbaugh's idea. The sample trees were measured standing by Barr and Stroud optical dendrometer, felled, milled, and details of all products, including their value, were recorded. Space demonstrated that the primary units of measure derived from processing the data (using STX, Grosenbaugh 1967b) were related to the product information gathered in the mill. He used regression analysis to derive factors for converting the data gathered during forest inventory to predict directly the volumes, values, and products. The approach has considerable potential for increasing the efficiency of logging and milling operations and it is surprising that no major forest organization or company has tested and evaluated it more thoroughly.

7.5 MEASURING GROUPS OF TREES (STANDS)

The stand variables most commonly measured or assessed in routine forest inventory are stand basal area, stand height, and stand volume. These are discussed separately in Sections 7.5.1, 7.5.2, and 7.5.3, respectively.

7.5.1 Stand Basal Area

Many measures of stand density appear in the literature, the best known being: number of trees, stand basal area; crown closure; stand density index (Reineke 1933); tree-area ratio (Chisman and Schumacher 1940); crown competition factor (Krajicek *et al.* 1961); and height/spacing ratio (Hart

1928) which was also called spacing index by Becking (1954) and Hummel (1954) and relative spacing by Beekhuis (1966). Some or all of these measures of density are described in detail in recent texts on forest mensuration and management (e.g., Husch *et al.* 1982, Avery and Burkhart 1983; Clutter *et al.* 1983). West (1983) compared 17 different measures of density in even-aged regrowth eucalypt forest in southern Tasmania and concluded that development of the more complex of them seemed to have gained little over the simpler measures.

One of the simplest measures of forest density is stand basal area (international symbol G). G is the sum of the basal areas of all living trees in a forest stand (or of the living trees of interest to the observer, e.g., those larger than a specified minimum diameter). It is expressed on a horizontal area basis and is an important reference variable for a stand and is particularly useful for quantitative description. It is assessed in the ground phase of all types of forest resource inventory because information on it is basic to deriving estimates of timber volume. For this reason, it has been singled out for special attention in this book.

$$G = (\pi/4)\,\Sigma(d_i/100)^2 = 0.000\,078\,539\,8\,\Sigma\, d_i^2$$

where G is expressed in m^2/ha and d_i = DBH of tree in cm.

Stand basal area is used extensively by foresters to evaluate quantitatively the degree of crowding in a forest stand and the level of site utilization attained. This evaluation is important because, within limits, the more growing space available to a tree, the faster it will grow. Thus an important role of the forester is to regulate stand density through initial planting spacing, thinning, and other silvicultural practices.

Values of G commonly range up to 50–60 m^2/ha, often higher in coniferous forest, and, in rare cases on exceptionally good sites, may reach 150 m^2/ha and more. The mean basal area ($\bar{g}$), derived from G/N (where N is the number of trees in the stand), is a useful reference variable in even-aged forest stands. The equivalent diameter is called the quadratic mean diameter (international symbol d_g). d_g is preferred to the arithmetic mean DBH ($\bar{d}$) as a size parameter of trees because it gives additional weight to the larger trees and is more directly related to stand volume.

The basal area of forest stands is estimated by sampling the forest using either fixed-area plots (and the formula given above) or VRP sampling. As discussed in Chapter 4, when sampling a population comprising units of variable size, it is more efficient to select the larger units with greater probability using pps sampling. Pps sampling has been applied in various forest sampling situations but most notably for estimating G using VRP.

Instruments Used in VRP Sampling and Their Calibration. The instruments that generate the reference angles are collectively termed angle count devices or angle gauges, the best known of which are the Relaskop

(Bitterlich 1949), Spiegelrelaskop (both standard and wide scale) (Bitterlich 1952, 1962), and the optical wedge or wedge prism (first used by Müller 1953). Other types of gauge have been discussed by Finch (1957).

Procedures for calibrating angle gauges, that is, for determining the factor (called the basal area factor, BAF, viz., the ratio of tree basal area to the tree's plot area) by which the tree count is multiplied to give G, have been described by many authors, for example, Grosenbaugh (1952), Carow and Stage (1953), Bay (1960), Husch (1963), Beers and Miller (1964), and Husch *et al.* (1982).

The Relaskop and wedge prism are the most widely used angle gauges for estimating G. The advantages and disadvantages of both these instruments are summarized in Table 7.3.

Whether the Relaskop or a wedge prism is used on a particular project depends on the nature of the project, forest conditions, availability of instruments, and the personal choice of the assessor(s).

Recommended Field Procedures. Accurate results cannot be expected from VRP unless the procedures outlined below are rigorously applied:

- Use a team of two persons comprising an assessor and an assistant. The role of the assistant is to define the breast high point on trees that are doubtful (i.e., near borderline), help check the doubtful trees, tally the count (self-counting by the assessor is error prone), and watch that trees are not missed during the sweep.
- Select an appropriate band or band combination (Relaskop) or choose an instrument of appropriate BAF that will give the desired average

TABLE 7.3 Advantages and Disadvantages of the Relaskop and Wedge Prism for Estimating *G*

Instrument	Advantages	Disadvantages
Relaskop	Light and robust Choice of many BAFs Automatically corrects for slope	Requires simultaneous alignment of four lines Expensive Movement of the instrument or tree interferes with the alignment
Wedge prism	Inexpensive Only two lines to align Unaffected by movement of instrument or tree Very light and portable	One BAF only Slope correction needed Problem with "floating images" in dense stands Special care needed in orienting it correctly with respect to tree under observation

count, usually 6–12 trees per point. This may require a quick reconnaissance survey, sweeping the forest at several points with an angle gauge of BAF about 2–4 m^2/ha to derive a rough estimate of $\bar{G}$. Dividing this rough estimate by the desired average count (say 10) gives the approximate BAF required for the main survey.

- Check the calibration of the instrument before first use.
- Avoid bias when locating sample points, for example, do not consciously veer away from a specified bearing or alter the survey distance to prevent the sample point, for example, from falling inside a large clump of nettle or hard up against a rocky outcrop. This applies equally well to sampling with plots of fixed area.
- Select a distinctive tree or other object to define the starting point of each sweep.
- Sight the DBH at right angles to the long (central) axis of the bole. This applies to all trees whether vertical or leaning. Judge breast height by eye (unless the tree is doubtful—see next below).
- Temporarily mark the breast high position on doubtful trees with, for example, a thin sheet of white paper. Alternatively, and if practical, use a white *T*-piece placed behind the tree with its long axis 1.4 m (or 1.3 m) long and the cross arm wider than the largest tree expected in the survey.
- If the tree is still doubtful, check it by direct measurement as described later.
- Inexperienced assessors using a Relaskop should mount the instrument on a tripod or other support to reduce the number of doubtful trees requiring checking by direct measurement.
- For leaning trees that are doubtful, measure the slope distance from eye level at the sample point to the center of the tree at breast height.
- When sampling by wedge prism, sight through the center of the prism and hold it in a plane normal to the line of sight to the breast high point on the tree. If an organization uses wedge prisms extensively, it is advisable, and much less costly than purchasing custom made prisms, to arrange for prisms of face dimensions approximately 45 mm × 12 mm to be cut from an optical lens of appropriate dioptre

[3]The strength (O) of optical lenses is quoted in dioptres, where a dioptre is defined as the right angle displacement of an image by 1 unit per 100 units of distance. The formula used to convert dioptres to basal area factor, F (m^2/ha) is

$$F = 2500\,(O/100)^2 .$$

Thus the F (m^2/ha) of lenses of strength 1, 2, 3, 4, . . . , 7, . . . , 10 dioptres is 0.25, 1, 2.25, 4, . . . , 12.25, . . . , 25, respectively. Wedge prisms cut from an optical lens need to be calibrated individually using, preferably, a flat, fixed-width target (see the standard procedures referred to earlier). Calibration is not a complicated or difficult procedure but it needs to be done with care.

strength[3], and mounted in a PVC "sighting" tube of approximate diameter and length 45 mm and 300 mm, respectively. Doing this avoids errors arising from not holding the prism normal to the line of sight, protects the wedge from scratching, minimizes smudging of the lens surface, confines the field-of-view to the tree under observation, and makes the prism much easier to handle.

- When sighting doubtful trees by wedge prism, hold the prism vertically above the sample point. With all other instruments, the operator's eye must be above the point (the vertex of the reference angle in these instruments occurs at the eye). It is not essential to follow this rule closely when a tree is obviously "in" or "out".
- If a tree is unrepresentative at breast height, select a representative point and sight to it.
- In unthinned or dense forest stands planted in rows, sight the trees in sequence by rows rather than making the conventional clockwise or anti-clockwise 360° sweep.
- On completing a sweep, check the area for slope and record the result (unnecessary with the Relaskop). Subsequently adjust the estimate of G accordingly.
- At the margins of both small and long narrow stands, corrections need to be made for the effect of the stand boundary on the assessment. Eight methods of making the correction are available of which four are biased (see Section 7.11.3 for details). Of the unbiased methods, the most practical in our experience are the mirage method of Schmid-Haas (1982) and the well-known method of Grosenbaugh (1958) which involves accepting 180° sweeps at the sides and 90° sweeps at the corners, and weighting the estimate accordingly, that is, ×2 or ×4, respectively. Unfortunately, Grosenbaugh's method is the least precise. For intensive inventories repeated over time (e.g. monitoring forest health), we strongly recommend the less practical but unbiased tree-concentric method. By measuring the distance separating a tree from the forest boundary, one can compute and store the exact probability of its inclusion in the sample, thereby permitting unbiased estimates to be made of anything associated with the tree either now or in the future (experience has demonstrated that considerable amounts of inventory data collected in the past are of questionable value for establishing baseline trends, so it is imperative that we start recording and storing data that may be of use for future, currently unforeseen purposes). Note that Bitterlich (1984) supports use of a procedure similar to that of the biased 'direct weighting method' of Beers (1964). The perpendicular distance from the sample point to the forest boundary is recorded and the estimate of G is adjusted using published procedures and formulae (Bitterlich 1982, 1984).
- When sighting to a tree hidden or partially hidden by another tree or immovable object, move to one side, keeping the distance to the tree

constant until its DBH is visible. Make the reading and then return to the sample point. Alternatively, sight to a smaller tree that is further away. If the smaller one is "in", then so must be the hidden tree.

- Watch for merged silhouettes of tree boles (the *T*-piece is particularly useful for separating silhouettes). This problem is often more serious when using a prism.
- If ground vegetation impedes the line of sight to the breast height of a tree under observation, sight to the lowest visible point higher on the bole. If the tree is "in" at that point, it will certainly be "in" at breast height. Otherwise, move sideways as described above or cut back the vegetation. If both these options are impractical, check the status of the obscured tree by direct measurement as described later.
- Be alert for trees that may not be part of the assessment, such as, dead trees.
- Use the same strength of angle gauge throughout a sampling stratum. Changing the gauge within a stratum may bias the estimates.

Determining the Size of Sample. This topic is discussed generally in Section 3.3. On a VRP survey, the number of sample points required varies with the size and variability of the forest and the desired precision of the estimate. If the coefficient of variation of the estimator of $G(\hat{G})$ is known from a pilot survey or from past experience in that forest or similar forests, then an estimate of the number (n) required can be obtained from the formula

$$n = (Ct/p)^2 \qquad (7.23)$$

where C = the estimated coefficient of variation of $\hat{G}$ (%)
t = Student's t value (usually set at 1 for the 67% and 2 for the 95% level of confidence), and
p = the desired precision of the estimate (%).

This formula gives the sample size that will achieve a $(1-\alpha)\cdot 100\%$ confidence interval, ignoring uncertainty in the estimate of C.

Thus assume that a pilot survey of a forest based on six randomly located sample points indicated a $\hat{\hat{G}}$ of 35.5 m^2/ha with standard deviation 5.2 m^2/ha ($C = 14.6\%$), and that the desired precision of estimate is $\pm5\%$ at the 95% level of confidence; then

$$n = ((14.6 \times 2)/5)^2 = 34 \text{ plots}$$

that is, another 28 sample points need to be located and assessed for G.

Another practical way, based on the law of diminishing returns, of addressing this common problem is described by Bell and Associates (1990).

Checking Doubtful Trees. The major factor contributing to error in VRP sampling is the failure to check doubtful trees. For the Relaskop, this involves the following:

1. Measuring the DBH of the tree.
2. Measuring slope distance (S) from eye height at the sample point to the center of the tree at breast height.
3. Measuring slope angle θ (from eye height to BH).
4. Calculating the limiting slope distance (L) from eq. 7.24.
5. Checking S against L and deciding the status of the tree ("in," "out," or "true borderline"), namely:

   If $S > L$, the tree is "out".

   If $S < L$, the tree is "in".

   If $S = L$, the tree is a "true borderline" (rare occurrence).

The checking procedure is similar for other types of angle gauge (those without automatic correction for slope) except that step (3) is eliminated and eq. 7.25 is used in step (4).

Formulas for calculating the limiting slope distance are:

$$\text{Relaskop:} \qquad L = 0.5d/(\text{Cos}\ \theta\sqrt{F}) \qquad (7.24)$$

$$\text{Other angle gauges:} \qquad L = 0.5d/\sqrt{F} \qquad (7.25)$$

7.5.2 Stand Height

The height of a forest stand is described by the mean total height of some or all of its trees. It is a parameter of limited practical value in irregular (uneven-aged) forests but is of considerable value in even-aged forests. For example, it is used extensively as an independent variable in stand volume equations and to determine site productivity (Carron 1968a; Loetsch *et al.* 1973; Husch *et al.* 1982; Philip 1983. However, Monserud (1984) has a different opinion about the usefulness and advisability of using height as an index of site.

Stand mean height is expressed in one of three ways, namely:

- Mean height—unweighted.
- Mean height—weighted by basal area.
- Top and predominant height.

The first and last of these expressions are interrelated and can be predicted one from the other.

Mean Height—Unweighted. This is the arithmetic average total height of all trees in the stand, $\bar{H}$, and is given by:

$$\bar{H} = \sum_{i=1}^{N} h_i / N \tag{7.26}$$

where h_i = total height of tree i and N = number of trees in the population.

It is laborious and costly to measure the height of every tree in a stand to derive $\bar{H}$. The problem is avoided by assuming that stand mean height is the mean height of those trees with DBH equivalent to the quadratic mean diameter (d_g) of the stand. Defined this way, $\bar{H}$ can be derived in two ways:

1. Directly, by determining the mean height of a specified number of trees/ha, e.g., 40–60, each with d equal to or approximating d_g.
2. Indirectly, by compiling a stand height curve (SHC, which is discussed later) and reading off the height equivalent to d_g.

Alternatively, $\bar{H}$ can be estimated from

$$\bar{h} = \sum_{i=1}^{n} h_i / n \tag{7.27}$$

where $\bar{h}$ = estimated mean height and n = number of trees sampled in the population.

The n sample trees may be selected either systematically or randomly but should be representative so that $\bar{h}$ is an unbiased estimate of $\bar{H}$.

Mean Height—Weighted by Basal Area. Weighting the estimate of mean height by g allows the larger trees to contribute more to the mean. This expression of mean height, called Lorey's mean height, is designated by $\bar{H}_L$:

$$\bar{H}_L = \sum_{i=1}^{N} g_i h_i \Big/ \sum_{i=1}^{N} g_i\ , \tag{7.28}$$

and is commonly estimated by

$$\bar{h}_L = \sum_{i=1}^{n} g_i h_i \Big/ \sum_{i=1}^{n} g_i \tag{7.29}$$

where g_i = basal area of tree i,
and N, n, and h_i are as defined previously.

The arithmetic mean height of the trees selected by VRP sampling at a specific point in a forest is a ratio estimator of $\bar{H}_L$ at that point. Lorey's mean height is less affected than $\bar{H}$ by thinning from below (low thinning), but it is affected by thinning from above (crown thinning).

Top and Predominant Height. The effect of thinning on $\bar{H}$ and $\bar{H}_L$ prompted foresters to seek a forest parameter that was relatively unaffected by thinning. In fact, they chose two parameters, top height and predominant height (international symbol h_{dom}), which can be predicted one from the other (Burkhart and Tennent 1977) and which are defined as follows (Ford–Robertson 1971):

Top Height. The mean total height of the *n* trees/ha with the largest diameters at breast height.

Predominant Height. The mean total height of the tallest *n* trees /ha.

The value of *n* in both these definitions is predetermined and varies between countries (and even within a country) from 40 to 100 trees/ha. Only a small proportion of the *n* trees/ha are actually measured, for example, if plot size is 0.05 ha and *n* is 40, two trees would be measured. However, it is prudent when assessing predominant height to consider choosing and measuring twice the number of trees required (four in the example above) because it increases the chance of actually locating the two tallest trees.

Unfortunately, the terms "top" and "predominant height" are often used interchangeably by foresters and the definitions are not universal. For example, top height in New Zealand is defined as the height predicted by the Petterson *h*/*d* curve (Petterson 1955, Schmidt 1967) for a *d* corresponding to d_g, the quadratic mean *d* of the 100 trees of largest diameter in the stand (Goulding 1986). The form of the Petterson function used in New Zealand,

$$h = 1.4 + (b_1 + b_0/d)^{-2.5} \tag{7.30}$$

is calculated as the linear regression

$$y = b_0 + b_1 d$$

where $y = d(h - 1.4)^{0.4}$,
h = total height,
d = DBHOB, and
b_0, b_1 = coefficients.

Other measures (definitions) of stand height that are grouped under the symbol h_{dom} have been listed by Loetsch *et al.* (1973). Thus it is most important when presenting quantitative data on stand height to define the terminology used.

A "rule-of-thumb" method for determining whether to assess stand height based on predominant or top height is that if the tallest trees in the stand are easily identified from the ground, use predominant height as it is usually quicker and cheaper.

Particular care is needed when assessing the stand height (top and predominant) of sample plots of 0.05 ha and greater if (1) site variation is apparent across them, and (2) they are rectangular and located on sloping terrain with their long axes aligned parallel to the slope (and, most likely, parallel to a fertility gradient). Unless such plots are stratified before sampling, the n tallest or fattest trees/ha are most likely to occur at one end of the plot, which would positively bias the estimate. In these cases, the sample plots should be broken up into nA or $nA/2$ subplots (where A is plot area). Then, one or two, respectively, of the tallest or fattest trees should be located and measured in each subplot.

Mention was made above of the *stand height curve* (SHC), which is the curve of best fit to a series of points representing the relationship between h and d for some or all trees in a stand. The trend of the SHC in even-aged stands is usually apparent but not well defined because, within any d class, the height of individual trees varies due to genetic differences and differences in point density (affects d but has little effect on h). The trend is better defined by plotting class means of h and d against each other.

Almost invariably, the SHC is concave to the abscissa. It is steep for young crops on good sites and near flat and linear for old crops and crops on poor sites. With time, the SHC moves upwards to the right, the movement being governed by the relative growth rates of h and d over time (Figure 7.1).

The SHC is important because it partly describes the structure of a stand, and it is used in one method of estimating stand volume (best suited to species or stands in which variation of height within a diameter class is small), namely:

1. Enumerate for d the full stand or a sample of it and compile a stand table (frequency by diameter class).
2. Compile the SHC.
3. Derive from the SHC the heights equivalent to the midvalues of the diameter classes.
4. Using an appropriate volume function, derive the volumes of trees of class mid-diameter and corresponding height. Multiply these volumes by class frequency to derive class volume.
5. Derive stand volume by summing the class volumes.

A stratified random sample, using the diameter classes as strata and weighting the sample by the frequency of occurrence of trees within the strata, is usually most efficient for establishing the SHC. Nevertheless,

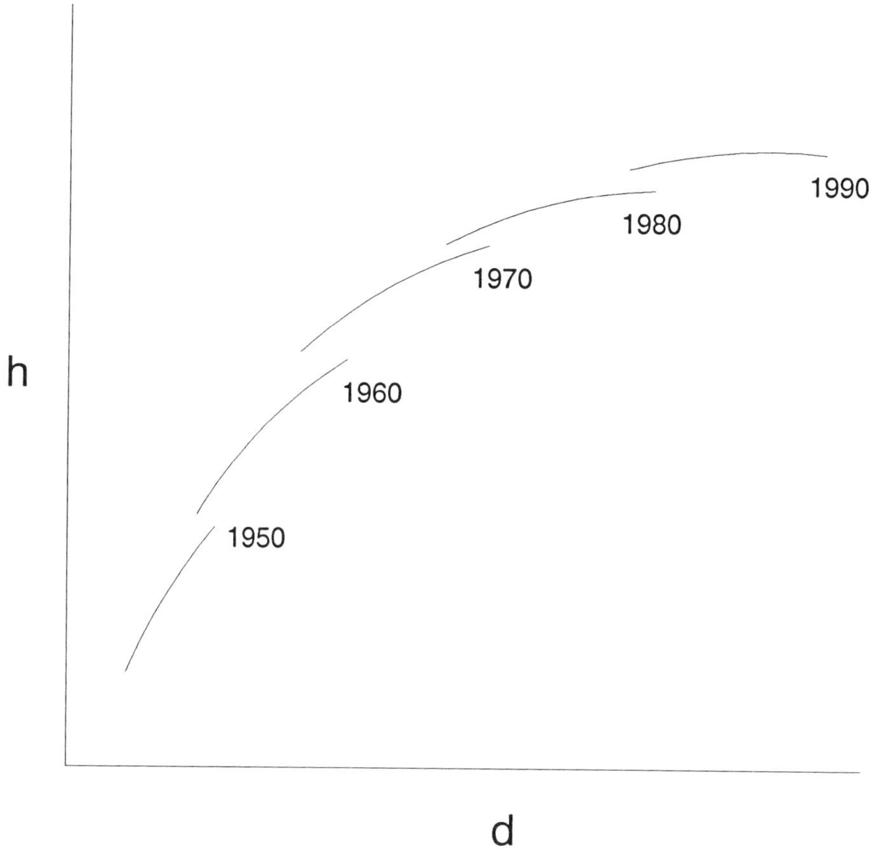

FIGURE 7.1. Movement of the stand height curve in an unthinned even-aged stand with time.

systematic sampling with the start located at random is commonly applied. Subjective sampling is sometimes used in small plantations.

If there is a need to compile the SHC periodically, say in periodic management inventory, a permanent sample may be advisable to retain uniformity in the position of the curves and in their change of slope, to maintain constant selection bias, and as a check on the measurement of tree height. However, in intensively managed forests, thinning may make it impossible to retain a permanent sample. In this case, the sample should be selected objectively on each occasion.

Fitting the SHC mathematically is desirable. Good fits are often achieved with Petterson's curve (eq. 7.30) and equations of the form

$$\ln(h) = a + bd^{-1}$$

$$\ln(h) = a + b\ln(d)$$

$$h = a + b \ln(d)$$

$$h = a + bd + cd^2 .$$

Other relationships worth testing are given in standard texts on forest mensuration and in Curtis (1967). Experience suggests that no single model is likely to be satisfactory in all cases because the SHC is dynamic, changing with time in a particular stand and from stand to stand at a particular time.

In uneven-aged stands, the relationship between h and d depends on the species composition of the stand and the treatment the stand has been given. In intensively managed selection forests of a single species on a uniform site in Europe and elsewhere, reasonable relationships can be established, but it is difficult to do this in a majority of mixed species forests, such as native hardwood forests managed extensively in Australia under a group selection system (Carron 1968a). Sometimes, however, a satisfactory relationship between merchantable height and diameter can be established that, if constant over time, can obviate the need to measure height in periodic management inventory. This has the advantage of reducing the error in the estimate of volume increment because errors in height measurement are held constant and differences of opinion on the merchantable limit are eliminated (Carron 1968a).

7.5.3 Stand Volume

When estimating the volume of timber in a forest by inventory, it is important to realize that the estimate is based on a sample that, on an area basis, is often considerably less than 1% and even 0.1%. In addition, within the sample units, there is further selection of trees and of points on those trees for detailed measurement. Thus the expansion factor in deriving the population estimate from the sample is great. When we add to this the errors arising from various sources (instruments, methods, sampling, and people), the limitations imposed on the inferences that can be drawn about the population should be obvious. In forestry an estimate of volume within ±10% of true is good, and yet a 10% error in the estimate will throw out a cutting program by 1 year in a 10-year cutting cycle.

Conventional methods of estimating the volume of forest stands include ground methods, methods based on aerial photography (e.g., aerial stand volume functions or tables), and combined ground and aerial methods. In the near future, large inventories (e.g., of states, regions, and whole countries) will undoubtedly be multistage or multiphase, combining satellite and other remotely sensed imagery with aerial photography and ground sampling methods. An example of this type of development is the Alaskan AIRIS inventory discussed in Section 5.5.2.

In this section, we concentrate on ground methods of measurement and assessment. These methods are basic in forest inventory, and the concepts and principles that apply to them are essentially the same as those that apply to aerial methods. These ground methods, which include both direct and indirect methods of estimating stand volume, are discussed in the subsections below.

Direct Estimation of Stand Volume Using Small Fixed-Area Sample Plots. Direct estimation involves measuring every tree in each sample plot by climbing (standing trees) or on the ground (felled) using techniques described in Section 7.3.6. Climbing ensures accurate results but at the expense of operator safety and cost. Felling every tree in a plot and measuring the trees directly on the ground is the most accurate way of deriving plot volume but seldom is this permissible. Thus these procedures are applied infrequently. When they are applied, volume of the forest is estimated from the formula

$$\hat{V}_o = A\left[\sum_{i=1}^{n}\sum_{j=1}^{m_i} v_{oij}/(na)\right] \tag{7.31}$$

where $\hat{V}_o$ = estimated total volume of forest (m^3),
A = total area of forest (ha),
n = number of sample plots
a = plot area (ha),
m_i = number of trees/plot, and
v_{oij} = volume (m^3) of the jth tree on the ith plot.

Indirect Estimation of Stand Volume Using Small Fixed-Area Sample Plots. If all trees in the sample plots are visited and measured for d and h, it is more practical to estimate the volume of individual trees indirectly using tree volume functions or tables, if these are available, deriving the population estimate using eq. (7.31). Alternatively, the trees in each plot can be enumerated in full for d, subsampled for h, and the SHC established as described in (Section 7.5.2). The height of each tree is then predicted from the curve using the measured DBH. This predicted height, h', and measured diameter are then used to predict volume.

If subsampling of a plot for volume is used, the mean tree method, mean form height method, volume-basal area line method, and Poisson sampling can all be used. Of these methods, the first three are restricted in use to homogeneous stands, mostly even-aged and monospecific, whereas Poisson sampling can be applied to any type of stand, even- or uneven-aged.

Mean Tree Method. This is one of the oldest methods in forestry of deriving stand volume (Carron 1968a) and is still useful for quick low-cost appraisals. It is the least precise of the four methods mentioned.

The underlying theory is to locate the tree of mean volume in each sample plot, carefully measure its volume, and multiply it by the number of trees in the plot. Alternatively, the volume of the tree is divided by its basal area (g) and multiplied by plot basal area (G in m^2/ha) to derive plot volume (m^3/ha). Volume of the forest is then estimated from the formula

$$\hat{V}_o = A \sum_{i=1}^{n} \hat{v}_{oi}/(na) \tag{7.32}$$

where $\hat{v}_{oi}$ = estimated volume of the ith sample plot (m^3/ha) and $\hat{V}_o$, A, a, and n are as defined in eq. (7.31).

In principle, the mean tree method is simple and direct but difficulty is experienced in finding the tree of mean volume. Sampling is necessary. If this is done objectively, a number of trees will be required, but subjective sampling may give as good a result with many fewer sample trees. However, this does not allow for objective assessment of the reliability of the estimate. The ultimate is to select one tree only.

Selection of the tree of mean volume is facilitated if the stand has been enumerated for d so that d_g can be derived. Trees in the stand approximating d_g are inspected and the one considered to be closest to average size (based on observed height, bark thickness, and taper) is selected for measurement of volume. To improve the estimate, additional sample trees approximating d_g can be selected and measured, or the method can be applied independently to diameter or basal area classes within each plot (see, e.g., Sture 1987), but these variations destroy the essential simplicity of the method.

Mean Form Height Method. This method is an extension of the version of the mean tree method described above using the mean v/g ratio (= mean form height) to derive plot volume. In North America, it is known as the VBAR method (Avery and Burkhart 1983, Section 10.18). Each sample plot is enumerated for d and a subsample of trees is selected and measured for volume. The mean form height of each plot is estimated from the subsample and the volume of the forest is derived using the equations given by Philip (1983):

1. Subsample selected with equal probability, e.g., systematically. Then compute

Step 1:
$$\bar{x}_i = \sum_{k=1}^{n_i} \hat{v}_{oik} \bigg/ \sum_{k=1}^{n_i} g_{ik} . \tag{7.33}$$

Step 2:
$$\hat{v}_{oi} = \bar{x}_i \sum_{j=1}^{m_i} g_{ij} . \tag{7.34}$$

Step 3:
$$\hat{V}_o = A\left(\sum_{i=1}^{n} \hat{v}_{oi}\right)/(na) \tag{7.35}$$

where $\bar{x}_i$ = the estimated mean form height of the n_i subsample trees in plot i (m),

n_i = number of trees measured for volume in the subsample of plot i,

$\hat{v}_{oi}$ = estimated volume of the trees in plot i (m^3),
v_{oik} = volume of the kth subsample tree measured for volume in plot i (m^3),
g_{ik} = basal area of the kth tree measured for volume in plot i (m^2),
g_{ij} = basal area of the jth tree in the ith plot (m^2),
m_i = number of trees in plot i,
$\hat{V}_o$ = estimated total volume of the forest (m^3),
a = area of sample plot (ha),
A = area of the forest (ha), and
n = number of sample plots.

2. Subsample selected with unequal probability, for example, proportional to g_i based on a VRP sample within each plot. This method uses in the estimation a pooled mean height from the subsample of trees from all plots.

Step 1:
$$\bar{\bar{x}}_i = \sum_{i=1}^{n} \sum_{k=1}^{n_i} (v_{oik}/g_{ik}) \tag{7.36}$$

Step 2:
$$\hat{v}_{oi} = \bar{\bar{x}}_i \sum_{j=1}^{m_i} g_{ij} \tag{7.37}$$

Step 3:
$$\hat{V}_o = A\left(\sum_{i=1}^{n} \hat{v}_{oi}\right)/(na) \tag{7.38}$$

where $\bar{\bar{x}}_i$ = the mean form height calculated from $\Sigma^n\, n_i$ subsample trees and the other terms are as defined previously.

Volume-Basal Area Line Method. This method is more precise than the mean tree and mean form height methods and is preferred, especially when a computer is available to process the data (Philip 1983).

The volume-basal area line (v_o/g line) is a simple linear regression of tree volume on tree basal area. The v_o/g line for a plot is established from sample trees selected either objectively, which is preferable to avoid bias, or subjectively. Bias from subjective selection can be kept under reasonable control in small stands if the assessor is thoroughly familiar with the population.

When the v_o/g relationship is known to be linear and the variance of v_o does not depend on g, the best estimate of the slope of the regression line is obtained when half of the observations are taken at each end of the range of the independent variable, but if there is doubt about the linearity and constancy of variance, the observations should be distributed over the whole range either uniformly or with a greater concentration toward the ends of the range (Demaerschalk and Kozak 1974). Taking more samples at the upper end of the range is favored by some workers because the larger trees contribute more to stand volume than the smaller trees and the variability of volume for a given d is greater (Klitscher 1983; Goulding 1986).

The v_o/g line takes the form

$$\hat{v}_{oik} = b_0 + b_1 g_{ik} \,. \tag{7.39}$$

Thus

$$\hat{v}_{oi} = m_i b_0 + b_1 \sum_{j=1}^{m_i} g_{ij} \tag{7.40}$$

and, if it is known that the regression constants (b_0 and b_1) apply to a larger forest area, then

$$\hat{V}_o = A\left(\sum_{i=1}^{n} m_i b_0 + b_1 \sum_{i=1}^{n} \sum_{j=1}^{m_i} g_{ij}\right) / (na) \tag{7.41}$$

where a = area (fixed) of each sample plot in the forest (ha) and $\hat{v}_{oik}$, g_{ik}, $\hat{v}_{oi}$, m_i, g_{ij}, $\hat{V}_o$, and A are as defined earlier.

The v_o/g line applies well for estimating total and merchantable volume IB of many plantation conifers and even-aged hardwoods in many countries but difficulties may be experienced on poor sites and when merchantable limits exceed about 15 cm DIB. The v_o/g line is the basis of many volume tariff systems (e.g., Hummel 1955; Turnbull and Hoyer 1965; Carron 1971).

Poisson Sampling. As mentioned in Chapter 4, Poisson sampling is a form of pps sampling that was introduced into the forestry literature by Grosenbaugh (1964) as 3P sampling (sampling with probability proportional to prediction). Poisson sampling can be applied to all or just some trees (a subsample) in a stand and is highly efficient. Trees selected in the sample are measured either directly or indirectly for volume using methods described in Section 7.3.6. Estimation formulas for stand volume and its variance are given in eq. (3.76)–(3.79).

Indirect Estimation of Stand Volume Using Stand Volume Functions. The objective of stand volume functions or tables is to derive stand volume directly and rapidly from measured stand variables, rather than by accumulating the volumes either of individual trees or of groups of trees within d classes. Stand volume functions are particularly useful when an overall estimate of volume is required, that is, an estimate not broken down into assortments (size and quality classes).

Stand volume functions usually have two and sometimes three independent variables, namely, G, some measure of stand height ($\bar{h}$ or h_{dom}) and sometimes an expression or index of tree form. Functions based on G and h_{dom}, where G is estimated by VRP sampling (Section 7.5.1) and h_{dom} is derived from the heights of a specified number of the tallest or fattest trees (Section 7.5.2), are particularly quick and easy to apply.

Models commonly used when establishing the relationship between stand volume (V_o), G and H ($\bar{h}$ or h_{dom}) include:

Combined Variable Model

$$V_{oi} = b_0 + b_1 G_i H_i \,. \tag{7.42}$$

Generalized Combined Variable Model {= Australian equation (Spurr 1952)}

$$V_{oi} = b_0 + b_1 G_i + b_2 H_i + b_3 G_i H_i \,. \tag{7.43}$$

Form Height Model

$$V_{oi}/G_i = b_0 + b_1 H_i \tag{7.44}$$

where b_0, b_1, b_2 and b_3 are regression constants.

Other models worth testing and recommended methods for fitting the relationships have been discussed by Clutter *et al.* (1983).

Volume tariffs (essentially a series of one-way tree volume tables) referenced to h_{dom} are used extensively to estimate stand volume directly at low cost. An example is the tariff compiled by Carron (1968b) for even-aged radiata pine in the Australian Capital Territory. The form of this tariff (eq. 7.45) implies that the regression constants and slopes of the v_o/g lines are related quadratically and linearly, respectively, to h_{dom}.

$$V_o = N(b_0 + b_1 h_{\text{dom}} + b_2 h_{\text{dom}}^2) + G(b_3 + b_4 h_{\text{dom}}) \tag{7.45}$$

where

V_o = stand volume to 10-cm DIB of trees ≥13-cm DOB (m^3),
N = number of trees in the stand,
h_{dom} = predominant height (tallest 40 trees/ha) (m),
G = stand basal area (of trees ≥13 cm DOB) (m^2), and
$b_0, \ldots, b_4$ = coefficients.

The Carron tariff can be applied rapidly to derive an estimate of stand volume at any point in the forest by taking a point (VRP) sample, measuring d of all trees accepted in the sweep, and assessing h_{dom} in the vicinity of the point. N is estimated by accumulating the frequencies of the included trees, namely,

$$\hat{N} = F \sum_{i=1}^{n} 1/g_i \tag{7.46}$$

where g_i = basal area of the ith tree included in the sweep (m^2), and
n = number of sample trees at the point.

Indirect Estimation of Stand Volume Using VRP Sampling. Bitterlich (1984) described several methods that have been proposed for estimating stand volume using VRP sampling, the most promising of which are critical height sampling (Kitamura 1962) and space point sampling (Ueno 1979, 1980). Both methods are described in Section 4.6 and the estimation formulas are given in Eq. (4.26) and (4.27), respectively. More extensive testing of these methods under field conditions in a variety of forest types using a taper function to estimate the critical height, is needed.

7.6 ASSESSING SITE

The term "site" is used in forestry to convey the totality of environmental influences existing at a particular location in a forest. These influences are exerted through climatic factors (air temperature, humidity, radiant energy, precipitation, wind), soil factors (physical and chemical properties, soil moisture, soil microorganisms), topographic factors (slope, elevation, aspect), and competitive factors (other trees, lesser vegetation, animals, man). The effects of the competitive factors are of minor importance in evaluating site because they are transient and can be altered by management practices, for example, thinning.

Site assessment and classification are becoming increasingly important in forestry because of the need for improved land use decisions and practices, for better delineation of homogeneous land areas, and for better estimates of forest growth and yield. The motivation behind this latter need is the commitment of the forestry profession to increasing the efficiency of assessing forest resources and to improving forest management planning, management practices, and forest production. Accurate site maps are essential for intensive forest management. They are the basis of yield regulation and forecasting, and all thinning, pruning, fertilizing, and other tending and protection strategies relate to them.

In the context of timber management, the quality of a site can be defined as its potential to produce timber given a particular species or timber type. Thus the assessed quality of a forest site has meaning only for the species or species composition occurring at the particular location. Of course, extremes exist that provide absolute limits for all species, for example, latitudinal and altitudinal range limits for trees.

Forest sites are classified either qualitatively or quantitatively into site types. Qualitative classification is based on environmental factors such as climate, altitude, topography, exposure, soil, and vegetation. Predicting site this way has two advantages: it can be applied regardless of the presence or absence of forest, and most of the factors of the environment that are measured are either stable or change slowly. Qualitative classification is not practical at the stand level, however.

Quantitative classification is applied to individual stands for the purpose of differentiating between them based on their potential to produce primary wood products. It makes use of some characteristic (or characteristics) of the trees that is considered to be sensitive to site, for example, in even-aged forest stands volume production at a specified age (called site quality), or dominant height at a specified age (called site index).

Site index is the most widely accepted quantitative measure of site in even-aged stands because it is practical and mostly consistent. However, problems arise in stands that have suffered upper stem damage affecting height growth, are of mixed species, or cover different soil types (e.g., in a region near the Arkansas–Louisiana state border, Zahner (1962) found it necessary to develop a site index system for loblolly pine comprising three separate site index equations, one for each of the soil types present). These and other problems with site index that have been documented (e.g., stands of the same site index giving grossly different volume yields (Assmann and Franz 1963)) have been discussed by Monserud (1984). Monserud suggested that individual tree-based stand simulation models have the greatest potential for avoiding many of these problems provided the models are driven by key factors (one of which is site index) that are related to site productivity.

Obviously, to assess site based on a forest characteristic, forest must be present. The causal factors in this case are not defined; rather, their net effect is inferred from the measure of stand volume or dominant height. Assessing site based on dominant height can usually be done quickly. However, as indicated above, dominant height may not be a good index of site *per se* in some forests or it may be difficult to measure in, for example, damaged stands or dense even-aged stands, when based on the tallest trees. In these cases, some or all of the following stand characteristics may also have to be assessed: stand volume, tree vigor and form, crown density, needle length and color, bark color, and height to the base of the live crown. Assessment of these additional characteristics is necessary in South Australia to classify site accurately in *Pinus radiata* plantations (Lewis *et al.* 1976).

Determining the site index of an even-aged forest stand requires that h_{dom} and age be determined as described in Sections 7.5.2 and 7.3.9, respectively. Then h_{dom} at a specified index (standard) age, that is, site index, is derived from height-age curves previously established for the forest.

An alternative to evaluating site based on dominant height is to use information on height growth over a defined short period during the life of the stand. This approach, called the growth intercept method (Wakeley 1954), is practical only for uninodal species that display annual branch whorls or for multinodal species in which the spring whorls are clearly identifiable. Its advantages are that stand age is not required and the length of the specified number of internodes is usually easier to measure than the heights of dominant trees.

To apply the method, the lengths of a specified number of successive annual internodes are measured, beginning at some well-defined point on the stem (e.g., the first whorl above breast height or above some other specified height or a whorl of known age). If the measurement is over five internodes, it is called a 5-year growth intercept.

Growth intercept values can be used directly to assess site quality (called growth intercept indices) or indirectly to estimate site index, for example,

$$\text{site index} = b_0 + b_1 \bar{I} \tag{7.47}$$

where $\bar{I}$ = the average growth intercept of sample trees selected in the stand and b_0, b_1 = regression constants.

Growth intercept indices are particularly useful for classifying site in young stands (Ferree *et al.* 1958, Wakeley and Marrero 1958, Beck 1971). In older stands, site index is considered to be more meaningful (Clutter *et al.* 1983).

In uneven-aged forests of one or more species, height in relation to age cannot be used to express site. Height growth of trees in such forests is related more to the varying stand conditions that have affected them throughout their life than to age. Sometimes, the relationship between the height and diameter of the dominant trees affords a sensitive and reliable measure of site in uneven-aged forests (McLintock and Bickford 1957). Based on this concept, site index is defined as the height attained by dominant trees at a standard DBHOB. A variant of this concept was applied successfully by Vanclay and Henry (1988) to classify site in indigenous cypress pine (*Callitris* sp.) in southern Queensland. They defined their index of site as the expected height of a 25 cm DBHOB tree predicted from the stand height curve (h vs. d).

The species composition of understory vegetation present on a given site commonly reflects the fertility of the soil and is often a good indicator of the availability of soil moisture in the upper soil horizons. Thus Cajander (1926) proposed that certain key species (indicator plants) in the forest reflect the overall quality of the site for a tree species or forest type and that they can be used to classify site. Site classification systems based on this concept have been most successful in relatively undisturbed stands or extensive forests of a few species, for example, the northern spruce-pine-birch forests of Finland, naturally regenerated forests in the more mountainous areas of Europe (Philip 1983), and the Kaingaroa plains in New Zealand (Ure 1950). Indicator plants have proved useful in Australia for assessing the suitability of a site for afforesting and reforesting an area with a particular species and as a rough guide to the potential productivity of the chosen species (see, e.g., Lewis *et al.* 1976).

For a detailed discussion of site and site assessment and methods used to construct and apply height/age curves, we refer the reader to published texts, namely, Alder (1980); Husch *et al.* (1982); Clutter *et al.* (1983); and Philip (1983).

7.7 ESTIMATING BIOMASS

Forest biomass may be defined as the amount of living matter per unit area or volume of the forest habitat. Traditionally, the portion of the forest biomass assessed in forest inventory has been restricted to trees above a certain minimum size (e.g. 15-cm DOB) and to sections of the bole above the stump usually to some upper diameter limit that is either specified (e.g. 12-cm DOB) or imposed by branching. These tight standards of merchantability are gradually being relaxed as demands increase from forest managers and scientists for information on crown biomass for studying nutrient cycling, evaluating wildlife habitat, determining site productivity, and predicting forest fire hazard (Phillips and Cost 1979). In addition, the increasing demand by society for wood and its products, the rising costs of these products, and world-wide concern about sources of energy (and global warming) have stimulated interest in utilizing the total forest biomass (Cost and McClure 1982) and minimizing forest clearing and waste.

Generally biomass sampling has involved the sampling of forest stands to develop equations for estimating the green and dry weight of wood and bark since, traditionally, these were considered the only useful biomass products (Cost and McClure 1982). However, interest in biomass extends to root biomass, and biomass in the crowns of the trees and in the lower vegetation. These parameters generally have to be estimated by different sampling and estimation procedures as discussed in the sections that follow.

7.7.1 Wood, Bark, and Foliar Biomass

These tree variables are generally linearly related to tree volume. Much of the extensive literature on volume estimation from g (tree basal area) or d^2h may be applicable here, as seen from the results in Swank and Schreuder (1974) and Geron and Ruark (1988).

7.7.2 Root Biomass

Root biomass can be measured directly from samples using soil cores. For example, Gholz *et al.* (1986) extracted cores of 5-cm diameter to an average depth of 64 cm in a 9-year-old and 108 cm in a 27-year-old *Pinus elliottii* plantation, but a standard method to quantify the fine roots in the cores is not yet available. Methods that have been applied vary considerably and few results are comparable. The technical difficulties involved have been discussed by Fogel (1983, 1985), Persson (1983), and Santantonio and Grace (1987).

It is likely that tree root biomass, like stem biomass, is allometrically related to d or d^2h but since roots from forest trees interpenetrate, root biomass cannot be measured directly for individual trees to establish the prediction equations. Indirect methods of estimating fine roots have been proposed based on balancing nitrogen mineralization budgets (Nadelhoffer *et al.* 1985) and depletion of starch reserves (Marshall and Waring 1985) but before these methods can be applied extensively, validation by direct estimates is required (Santantonio and Grace 1987).

Santantonio and Grace (1987) developed a mathematical model to integrate information on standing forest crops and decomposition of fine roots. They then applied the model to quantify fine-root processes in a 12-year-old plantation of *Pinus radiata* in New Zealand. Based on their experience, they suggest that placing emphasis on estimating processes rather than variables is a more fruitful way of developing hypotheses and generalizations about root behavior in forests.

7.8 SAMPLING LOWER VEGETATION

O'Brien and Van Hooser (1983) listed essential requirements for inventorying understory vegetation, namely:

- Methods used should be adaptable for use in extensive, sample-based inventories.
- They should be readily integratable into existing procedures and time frames of surveys.
- They should use objective measurements that people with limited taxonomic skills can make.
- They should be implementable during the seasons when sampling is done.
- They should not disturb vegetation on permanent plots.

We limit our discussion to methods satisfying these requirements. We believe that the technology has advanced sufficiently in this field that subjective and destructive methods are unnecessary, and we focus only on techniques of vegetation measurement that are totally objective. Thus we ignore methods such as described in O'Brien and Van Hooser (1983) that ocularly classify the vegetation into height classes and percent cover classes.

7.8.1 Direct Measurement Techniques

Several objective methods are available. Jukola-Sulonen and Salemaa (1985) described one such method, the point quadrat method, in which a frame with a moving needle is used to record the presence or absence of plant species. The needle, with a tip thickness of less than 1 mm, is lowered vertically onto the vegetation at each of one hundred points spaced evenly at 10 cm intervals. Touches between the needle tip and the vegetation (plant species) are counted. The proportion of points where a given species is touched gives the percentage cover for that species. All the hits in a vertical direction between the needle and parts of the plant species can also be counted, giving an approximate measurement of the species-specific percentage cover by size and vertical distribution of each species.

The authors noted that this method is purely quantitative if a sufficiently large number of points is measured but the results are affected by the thickness of the needle. An alternative procedure using an optical device was developed by Winkworth and Goodall (1962).

A related method was proposed and tested by Cline (1991) for estimating the structure of vegetation up to 9–10 m tall. In this procedure, a grid of 16 points, spaced 3.66 m (12 ft) apart, is laid out and each point is appropriately tagged. One observer holds vertical a pole (up to 10 m high) with a spirit level attached while another crew member records the substrata on which the pole rests (including mosses and lichens) and the live vegetation that touches it by 30-cm (1-ft) intervals from ground level to the top. Multiple characteristics can be listed for each height class such as species and part of the plant touched. Information is collected at all 16 points.

The following precautions should be taken when applying Cline's method:

- To minimize play in the extended pole and ensure true vertical hits at the upper heights, level the pole with the spirit level at all times.
- To record vegetation hits in the correct height class, read up to the next 30-cm (1-ft) level up to 1.5 m (5 ft) because one reads down for 1.5 m and below. Read down to the next level for data above 1.5 m (5 ft) because one usually reads up for 1.8 m (6 ft) and above.
- If the base of the pole rests on a stump, large rock, or fallen tree that raises it above ground level, this raised distance must be added to each vegetation hit recorded.

Results of field tests of this procedure are not available at present. Biased readings (due to wind) and variability (basically we are sampling a rare population, that is, we expect few hits) are major concerns.

7.8.2 Indirect Measurement Techniques

Tucker (1980) noted that nondestructive methods to estimate herbaceous biomass should be accurate and fast, should require a minimum amount of calibration, and should be little affected by environmental conditions such as clouds, wind, varying irradiational conditions, uneven microtopography, mist, and dew. Instruments used should be light, sturdy, easy to carry, reliable, and inexpensive. β-attenuation and capacitance meters meet these criteria.

β-Attenuation. Absorption of β-particles depends to a large degree on their interaction with the electrons of the absorbing material. The efficiency of the absorber made out of elements of low atomic number will be approximately proportional to the density of the absorber. The absorption of material of a given thickness (mg/cm^2) is practically independent of the atomic number and atomic structure of the material.

Plant material contains only a small proportion of heavy elements. Thus estimation of the density of plant material by β-particle attenuation between a radioactive source and detector maintained at a constant geometry gives a reliable estimate of herbage biomass. β-attenuation can also be used to measure change in water content of leaves. This requires that the leaf be placed between a weak β-emitter and a thin-window Geiger-Mulier tube.

Capacitance Meters. The principle of a capacitance meter is that there is a difference between the dielectric constant for air and that of water present in the vegetation, that is, the meter measures a variable, capacitance, that is proportional to the amount of water in the vegetation.

Advantages of such meters are the linear relationship that exists between capacitance and water in herbaceous vegetation, rapidity of measurement, and simplicity of use. Disadvantages are the inability to be used in wet weather, adverse effect of water droplets or dew on the readings, requirement for frequent recalibration, and the need to operate only on homogeneous herbaceous canopies of similar phenology.

Angelone *et al.* (1980) gave a brief review of the multiple-probe-type capacitance meter and described the principles involved. They concluded that:

- It does not take much soil moisture to effect a connection to earth so additional moisture would not affect readings much;
- plant surface moisture can inflate readings;
- the electronic readings should not drift much;
- since water is detected primarily, readings should correlate better with water content than with dry weight;
- the physical position of the herbage will influence the capacitance readings.

Tucker (1980) concluded that β-attenuation has not been tested extensively enough and that its use of radioisotopes is a serious drawback. He also concluded that capacitance measurements are significantly affected by dew, relative surface soil dampness and humidity, and that calibration relationships are specific to the species composition.

7.9 SAMPLING REGENERATION

Literature devoted specifically to sampling regeneration is sparse, the best known publications being Kaltenberg (1978) and a proceedings devoted to the topic (Society of American Foresters 1984). The explanation may be that the objectives of surveys are generally simple, and implementing a survey involves a straightforward application of general sampling techniques.

The prime concern of all forest regeneration surveys is to determine whether or not the area of interest is satisfactorily stocked with trees. Most surveys make use of equal probability sampling methods involving fixed-area plots, quadrats, or points. Statistical problems encountered when analyzing quadrat counts and points (involving nearest neighbor distances) are discussed by Greig-Smith (1964), namely, the size and shape of the quadrat affect results when the pattern of regeneration is clustered; and, because both random and systematic sampling points tend to fall more often in empty spaces of the regeneration mosaic, the distance to the nearest neighbor tends to be biased. Ward (1991) has suggested a new approach to overcome some of these problems. It involves triangular tessellation, which combines area and distance, and is easy to use in the field. The sides of a triangle formed by three trees around the sample point are measured and a hand-held computer is used to derive estimates of stems per hectare (the simple GWBASIC program required for implementing the method is listed by Ward (1991, p. 289–290)). Sampling ceases when the cumulative estimate stabilizes. The method can be applied using points located either randomly or systematically.

Given that variable probability sampling methods are often much more efficient than equal probability sampling methods, it is surprising that vertical line sampling (Beers and Miller 1976) has not been applied more often in regeneration surveys, particularly when the regeneration is well established and exceeds several meters in height. More extensive trials of this method appear to be warranted.

7.10 ASSESSING FOREST HEALTH

As physicians routinely measure body temperature, blood pressure, heart rate, and weight to monitor general fitness, comparable variables should be measured for tree and forest health (Smith 1990).

Today numerous indicators of forest health are being considered by researchers due to concern about the impact of the environment on the health of forests. These indicators range from subjective measures of visual symptoms to reliable quantitative measures such as concentrations of foliar nutrients and chemical contaminants that can be measured with high reliability.

Nilsson and Duinker (1987) discussed the extent of forest decline in Europe. They noted that variables used to quantify forest damage due to air pollutants include stress symptoms in plant cells, foliar injury at a microscopic scale, foliar concentrations of pollutants or their derivatives, reductions in diameter or volume of wood growth, color changes in foliage, and foliage loss (decrease in canopy density). Unfortunately, the cost of assessing the first three of these variables in statistically representative, large-scale surveys is very high. In addition, the usefulness of data on foliage discoloration and loss is questionable because systematically collected information on foliage discoloration is not available for forests in many European countries, and visual estimates of crown density, extensively used in these countries to assess forest damage, are suspect for four main reasons, namely (Nilsson and Duinker 1987):

1. Other variables influence foliage loss, which varies with species, temperature and precipitation events, elevation, exposure, soil moisture, soil fertility, competition, and age. In addition, some of these variables interact significantly.
2. Foliage loss is often not an early-warning indicator of forest decline. Trees may grow poorly for many years, even several decades, before exhibiting a decrease in crown density. Therefore, foliage loss, as an expression of pollution stress, greatly lags behind the concomitant growth response.
3. The relationship between foliage loss and tree growth is uncertain. For coniferous species, a loss of up to 20–25% can occur without measurable reductions in radial growth but, beyond that, the relationship is exponential. For deciduous trees, however, even a very small loss of foliage can cause a reduction in growth, and the relationship appears to be linear.
4. Tree crown density is usually estimated ocularly and is subject to observer bias (and skill). It is difficult to aggregate these estimates when specifying the area of forest in a declining state.

Nilsson and Duinker (1987) concluded that only a small loss of foliage is required in deciduous forest to affect tree growth adversely, whereas a loss of 22–25% is required in coniferous forest to have the same effect. They warned, however, that a smaller decrease in crown density may still place the coniferous forest at risk because the combined effect of other stresses, including those arising from air pollution, may carry the trees beyond the threshold.

Innes (1988a, 1988b) warned about the serious bias that can arise in surveys involving subjective assessment of a variable. However, some variables related to growth are available that can be assessed more objectively, namely basal area growth, canopy leaf area, and wood production per unit of leaf area. Of these three variables, basal area growth is easily measured and is already being monitored extensively (Bechtold *et al.* 1991); canopy leaf area is an important indicator of both forest growth and environmental stress (Waring 1985) and efficient methods of measuring it are available (Pierce and Running 1988); and wood production per unit of leaf area may be a sensitive indicator of environmental stress (Waring 1983). This latter variable may be defined more specifically as the ratio of periodic growth in stemwood volume to the capacity for growth on a unit area basis, that is, per unit of canopy leaf area (Waring 1985). Accepting that canopy leaf area can be measured quickly and reliably (Pierce and Running 1988), one might consider using growth in basal area, which also can be measured reliably and quickly, rather than stemwood volume, as the component of wood production in the numerator.

Smith (1990) challenged society to ensure that the health of our natural and artificial forest ecosystems is monitored more carefully. He emphasized that more committed support is needed for research programs directed at predicting stress processes in forests and evaluating associated risks to forest health. He contended that a comprehensive network of permanent sample plots in these ecosystems is crucial to monitoring and maintaining their health effectively and intelligently. Failure by research administrators and scientists to respond to Smith's challenge will adversely affect the health and welfare of all societies.

7.11 SAMPLE PLOTS IN FOREST INVENTORY

Sample plots are used in forest inventory to establish the present state of the forest resource and to provide information on its growth and yield. Both sets of information are vital for sound forest planning and management.

7.11.1 Plot Size and Shape

Unbiased estimates of forest populations can be obtained from any combination of size and shape of sampling unit but the optimum combination varies with forest conditions. For important surveys, it is well worth investigating the relative efficiency of different combinations by comparing their sampling errors and costs.

A guiding principle in choosing the size of unit is to have it large enough to include a representative number of trees but small enough so that the time required for measurement is not excessive, that is, the size of the unit must be related to the physical size of the population elements, the trees. For example, in native forest, one may need to use fixed-area plots of size 0.005, 0.05, and 0.5 ha in stands of regeneration, poles, and large mature trees, respectively, whereas in plantation forests, sizes of 0.02–0.05 ha are usually appropriate. Concentric plots are sometimes used in mixed native forests containing a wide range of tree sizes: a maximum of three concentric units is recommended. It is wise to remember also that small plots yield more variable information than large plots but, because of their wider distribution for a given sampling intensity, an improvement in the precision of estimate of the mean usually results—but at a cost! A compromise must be sought.

Gambill *et al.* (1985) described a method for determining the size of a plot in a particular inventory that minimizes total cruising time and provides estimates within a specified level of precision. Their approach warrants serious consideration by anyone involved in designing efficient forest inventories. Note that in VRP sampling, the size of sample unit is undefined and is controlled by the basal area factor (F) of the angle gauge. It is generally recommended that a factor be selected that results in an average count of between 6–12 trees per sweep. However, when planning large-scale costly VRP inventories, it is wise to follow Gambill *et al.*'s procedures for determining the optimum F (i.e., average tree count/point) to use in each stratum.

The shapes of fixed-area plots commonly used in forest inventory are rectangular, square, circular, and narrow-width rectangular (called striplines or transects). If a fertility gradient is recognizable in the area under survey, it is preferable to use rectangular plots with the long axis aligned parallel to the gradient, because this leads to an increase in precision of the estimate. If no gradient is recognizable, square or circular plots are appropriate. Clusters of plots are often more efficient than single plots. For example, Scott (1992) discussed a method of determining the optimal cluster design when different plot types are used to assess different forest resource variables.

Rectangular and Square Plots. These plots are usually laid out by starting with a corner point located by survey (compass and tape) using an aerial photo or map. The second corner is then located and, at both corners, right angles are established and corners three and four are located. The distance between corners three and four and the lengths of the diagonals (if possible) are usually measured for control. All distances must be corrected to horizontal.

With the square, a method that starts with the center point is less susceptible to surveying errors because shorter distances are measured. In this approach, two diagonally opposite corners are located on the survey line at a distance of one half of a diagonal ($a/\sqrt{2}$) from the plot center where a is the side length of the square. The other two points are located on a line perpendicular to the survey line crossing the center of the square, again at distance ($a/\sqrt{2}$) from the center. (Clearly a similar approach could be used for a rectangular plot.)

With the rectangular plot, if lay-out is started at a corner, an error of +8.3% can occur if the two right angles of the square are too wide by 5°.

For the square plot starting at the center point, the equivalent error is only −0.4%.

Special care should be taken when establishing plots in plantation forests where trees occur in rows that are not exactly parallel (most cases). Serious bias can occur in the plot estimates in such forests if an attempt is made to establish an exactly square or rectangular plot of a specified size. To avoid such bias, we recommend locating the plot corners mid-way between the rows. This usually results in both the plot size and shape only approximating that desired, but this is relatively unimportant.

Circular Sample Plots. A circular sample plot is defined by the plot center and radius (horizontal). A device for establishing such plots was described by Loetsch *et al.* (1973, p. 319). It consists of a metal rod driven vertically into the ground and secured by three lateral cords. A measuring tape marked with fixed radii is allowed to move freely around a guide attached to the rod 1.3 m above ground. The time required to install the guide is offset by the increased speed of determining the status of the boundary trees.

Establishing a circular plot is often simpler than for other plot types because distances from the plot center have to be checked only for those trees located within a peripheral strip of width approximately 1.5–3.0 m. The length of this strip, and hence the number of boundary trees, increases with increase in the radius of the plot. In some cases, exact measurements are required to determine whether a tree is inside or outside the plot.

With appropriate care and good instruments, the layout of circular plots can be done without bias. Loetsch *et al.* (1973) noted that optical range finders incorporated in the Blume-Leiss and Haga hypsometers can also be used to establish circular sample plots. However, any angle gauge can be used for this purpose, for example, locate a cylinder of appropriate diameter at the center of the plot (stand the cylinder vertically) and establish the boundary with the gauge by seeking the borderline position (basic principle of range finding). This procedure is sometimes used in Australia by operators working alone to establish circular plots of fixed area. Alternatively, a relatively inexpensive, electronic distance measurer can be used.

Narrow-Width Rectangular Plots (Striplines or Transects). Striplines are always used in conjunction with survey lines, which simplifies their establishment in the field. They are most convenient if information on topography and forest composition is also required as part of the survey, and if dense undergrowth or difficult terrain necessitates spending a disproportionate amount of time on plot establishment. The width of strips usually ranges from 5 to 40 m and depends on a number of factors including sampling intensity, topography, forest composition, density of undergrowth, and variability and value of the forest.

Conventionally, strips are run at right angles to the contours of the main topography of the area under survey because the fertility gradient usually runs in this direction. Thus if an area is essentially of one topographic form, the bearing of the long side (plots or strips) can be specified in the prescription for plot location. If topography varies by zones, bearing of the long side may need to be specified by zones.

For a given sampling intensity, a strip survey may be faster than a survey based on plots because the ratio of working time on the units to travelling time between them is greater for strips. However, the number of units, and therefore the number of degrees of freedom for calculating sampling error, is often far fewer for strips than plots, and this may not be balanced by the reduction in the variation between units.

Strips and plots may be combined in what are called "line plots." With these, topographical and forest-type data are gathered on the strips and quantitative information on the forest is obtained from plots located at intervals along their length.

To avoid the bias commonly found in laying out long striplines, Näslund (1941) suggested that the compass person, dragging a 100 m tape, define with the tape one of the two side boundaries, rather than the centerline, while the other boundary is defined by an assistant who stays the required horizontal distance away. Using this procedure on a 10-m wide stripline, Näslund obtained a standard error of measured strip width of 0.7% and 0.1% for 100 and 5000 observations, respectively.

7.11.2 Plot Establishment

There is always a danger of bias when locating sample units in the field. Usually, this is very slight with systematic surveys when the starting point is selected at random and located by survey method because the position of all other points is automatically fixed. With random samples, however, the sample units are usually selected on paper. This can give rise to various sources of bias when locating the units due, for example, to the maps lacking sufficient detail; the forest showing no distinguishing features; the forest composition preventing location on the ground of features apparent on aerial photographs; and the quality of the photographs preventing reliable interpretation. If an operator is unable to locate a specific point on the ground, the sensible thing to do is locate the point by objective means, for example, move a certain distance in a certain direction, the distance and direction being determined at random.

The problem of boundary trees, that is, those that occur at or near the plot boundary, always arises when establishing sample units in the field. Boundary trees can be a source of considerable error in deriving plot estimates in forest inventory. In VRP sampling involving variable-radius plots with no fixed boundary, this error is minimized by checking all the boundary trees (called borderlines) by direct measurement (refer to Section 7.5.1). With bounded fixed-area plots, the error can be reduced by selecting a plot shape that, for a given area, maximizes the ratio of area to perimeter. Thus in general, the circle is preferable to all other shapes.

7.11.3 Adjusting for Edge Effects at the Forest Boundary

Edge effects at the forest (or stand) boundary arise as a result of different microlocality conditions and may extend from 1 m to tens, and even hundreds, of meters within. In all types of forest assessment, it is important to consider these effects both when designing the sample frame and when determining the procedures for estimating the population parameters.

Depending on the sampling procedure, a tree near a forest boundary may have an inclusion probability different from that of a similar tree situated further toward the interior. Some procedures may restrict the sampling locations to certain regions of the stand in such a way that the inclusion probability of a tree is altered from what it would be under an unrestricted location scheme. The bias that results from boundary overlap, as this condition has been called, usually stems from the failure of an estimator to appropriately account for each sample tree's actual probability of inclusion. Trees near the boundary, not sample units (plots, points), are the cause of the altered inclusion probabilities and hence are the source of the problem (Gregoire 1982). The magnitude of the bias that results if neither the sampling procedure nor the estimator are modified to counter the effects of boundary overlap depends on the distribution of trees near the edge, the sample design, the characteristic to be estimated, the estimator, and the amount of edge in relation to the overall area to be sampled. This last factor can possibly be stated more aptly in terms of the numbers and sizes of trees near the edge relative to all trees in the tract and so is closely related to the first-named factor.

A number of alternative sampling procedures have been devised in an attempt to eliminate or minimize boundary bias by making the inclusion probability of an edge tree more nearly equal to its inclusion probability had it been located in the interior of the stand. Other alternatives purposely alter the inclusion probabilities of edge trees to known values that can then be used to weight the sampled observations appropriately. Yet other methods attempt to modify the weight attached to the sample observation so that it is more nearly equal to the reciprocal of the inclusion probability, since a sufficient condition for unbiased estimation with homogeneous linear estimators (Chapter 2) is that each observation be expanded by the reciprocal of its inclusion probability. A summary of eight methods is given below, using the notation established in the introductory section of Chapter 4, supplemented where needed. The summary is a less technical treatment than that given by Gregoire and Scott (1990), from which it was excerpted. All eight methods are appropriate for fixed-area circular plot or VRP designs.

The estimator corresponding to these sampling strategies is of the homogeneous linear type and, when sampling consists of a single random sampling location, may be written as

$$\hat{Y} = \sum_{l=1}^{n} w_l \, y_l = \sum_{i=1}^{N} w_i y_i \delta_i \qquad (7.48)$$

where δ_i is the usual Bernoulli indicator of inclusion and w_i is the weight associated with y_i. When there are M sampling locations and $\hat{Y}_j$ is the estimator (7.48) from the jth location ($j = 1, \ldots, M$) then instead of eq. (7.48) we consider $\hat{Y} = \Sigma \ \hat{Y}_j / M$. However, little generality is lost by restricting attention to the case $M = 1$ and hence $\hat{Y}$ as expressed in eq. (7.48). The indicator δ_i is a random variable whose value depends on the sampling location, (l_{1s}, l_{2s}), as well as on the decision rule followed to determine whether or not the unit is sampled. For circular plots and horizontal VRP sampling, the decision rule can be expressed in terms of the horizontal distance, r_i, separating the ith tree in the population, u_i, from the plot or point location (l_{1s}, l_{2s}) and a limiting distance R_i. In some of the methods to be described, the decision rule depends on the location of (l_{1s}, l_{2s}) with respect to u_i, too, so that the probability of selection may be regarded as a function of r_i and (l_{1s}, l_{2s}), both of which are random variables. In contrast, both R_i and (l_{1i}, l_{2i}) are fixed quantities. For all but the enlarged-tree-circle method to be described, $R_i = d_i/(2\sqrt{F})$ when the sample design consists of horizontal VRP sampling and R_i = plot radius under a fixed-area circular plot setup. For all methods but those introduced by Grosenbaugh (1958), Beers (1966), and Schmid-Haas (1969), w_i is fixed, not random. To provide a unified treatment of all methods, $w_i \delta_i$ must be treated as a product random variable as detailed in Gregoire and Scott (1990). Finally we emphasize that a_i denotes the actual inclusion area of u_i. We introduce a_i^o as the nominal inclusion area; if R_i is expressed in meters, then $a_i^o = \pi R_i^2 / 10\,000$ in ha.

No-Correction Method. The sampling location (l_{1s}, l_{2s}) is chosen randomly within A. If $r_i \leq R_i$ then u_i is included in the sample. Irrespective of the location of (l_{1s}, l_{2s}), $a_i < a_i^o$ if u_i is within a distance R_i of any stand boundary; $a_i = a_i^o$ otherwise. The inclusion probability is $\pi_i = a_i / A$, and the weight attached to y_i is $w_i = A / a_i^o$. Hence, in any stand where there exist trees closer to the stand edge than their respective limiting distances, this estimator of Y will yield underestimates.

Rejection-to-R Method. The sampling location is chosen randomly within A but exclusive of a peripheral zone of width R within the tract boundaries. Typically R would be fixed at the maximum R_i value expected to be encountered in the inventory, so that for circular plot sample designs, the width of the exclusion zone is one plot radius. If $R > R_i$, then u_i has zero probability of being included in the sample if it is within a distance $R - R_i$ of the stand edge. Otherwise its inclusion area a_i is the portion of the circular area concentric with u_i that is within the interior zone of the tract where (l_{1s}, l_{2s}) may be located. If this interior zone has area $A_L < A$, the inclusion probability of u_i is $\pi_i = a_i/A_L$. The weight associated with y_i is $w_i = A/a_i^o$. The feature that distinguishes this method from the no-correction method is the restricted location of plots or points, since evidently the estimator is the same. The statistical properties of the estimator with this sampling scheme differ because of the effect of the restricted sample locations on the inclusion probabilities π_i for trees near the edge.

Move-to-R Method. This is similar to the rejection-to-R method. With this method, too, sample locations are excluded from a peripheral zone of width R but the restriction is accomplished by relocating a plot or point orthogonally back from the stand boundary until it is a distance R from the edge(s); sampling proceeds from this location, (l_{1s}, l_{2s}). Plots or points that are initially situated further than R from the edge(s) are not relocated. The inclusion area of u_i is the possibly noncircularly shaped area within A where a plot or point can be situated initially (prior to possible relocation to the edge of the peripheral zone) and yet lead to the eventual location of (l_{1s}, l_{2s}) that is a distance $r_i \leq R_i$ of u_i. The inclusion area a_i may be more or less than the nominal inclusion area a_i^o. For example, if $R > R_i$ and yet u_i is within a distance $R - R_i$ from the stand boundary, then $a_i = 0$. Conversely, if u_i is within the interior zone of the stand and within a distance $R + R_i$ of the stand boundary, then a plot or point that is initially in the peripheral zone and further than R_i from u_i can be relocated to the position (l_{1s}, l_{2s}) that is within a distance R_i of u_i and hence lead to its inclusion in the sample. The weight attached to y_i is $w_i = A/a_i^o$ as with the preceding two methods. The estimator of Y is biased, and as with the rejection-to-R method, the bias may be positive or negative, depending on the factors listed earlier.

Direct Weighting Method. This method was suggested by Beers (1966). The sampling location is situated randomly within A. The usual decision rule is implemented to determine if u_i is sampled, namely that u_i is included in the sample if r_i is less than the limiting distance R_i. To this point the direct-weighting method is identical to the no-correction method. The distinguishing feature is that the weight attached to y_i is $w_i = A/a_i^*$, where a_i^* is the portion of the circular area concentric with (l_{1s}, l_{2s}) that is within the stand. The resulting estimator of Y is biased.

Grosenbaugh's Method. With this method proposed by Grosenbaugh (1958), trees within a peripheral zone of width R are sampled differently from trees that are in the interior zone of the stand. Typically R is set equal to the largest R_i expected to be encountered. The method leads to bias otherwise. We defer to Grosenbaugh's monograph for details about this method and its implementation. Suffice it to say that the inclusion area a_i, of u_i is equal to a_i^o, $a_i^o/2$ or $a_i^o/4$, depending on whether it occurs in the interior zone, side peripheral zone, or corner peripheral zone of the (rectangular) tract. Correspondingly, the weight w_i attached to y_i is set equal to A/a_i^o, $2A/a_i^o$ or $4A/a_i^o$, so that the product $w_i\pi_i$ is maintained at unity in all instances.

For irregularly shaped stands with curved boundaries, the estimator of Y may be biased. The bias is negative if the boundary is convex to the center of the tract and positive if it is concave.

Enlarged Tree Circle Method. This method was suggested by Barrett (1964). The sampling location (l_{1s}, l_{2s}) is situated randomly within A. The weight attached to y_i is $w_i = A/a_i^o$. The inclusion area, a_i of u_i is altered so as to be equal to a_i^o regardless of the proximity of u_i to the stand boundary. To do this, a circular area concentric with u_i is established such that the portion of this area within the stand is equal to a_i^o. Let R_i be the radius of this tree concentric area. Then u_i is included in the sample if $r_i \leq R_i$, thereby ensuring that the inclusion probability of u_i is $\pi_i = a_i/A = a_i^o/A$. Notice that R_i will equal $DBH_i/(2\sqrt{F})$ when sampling with horizontal variable radius plots if u_i is at least $DBH_i/(2\sqrt{F})$ from the nearest stand edge; otherwise $R_i > DBH_i/(2\sqrt{F})$. Similarly, if sampling is conducted with fixed-area circular plots of radius R, then $R_i > R$ if u_i is closer than R to the nearest stand edge.

Mirage Method. This method was introduced by Schmid-Haas (1969). A sampling location (l_{1s}, l_{2s}) is situated randomly within A. For $R_i = DBH_i/(2\sqrt{F})$ or $R_i = R$, depending on whether VRP points or fixed area circular plots are used, u_i is included in the sample if $r_i \leq R_i$. The inclusion area, a_i, is a circular area concentric with u_i but truncated by the stand boundary if it is within R_i of the tree. The weight attached to y_i is an integer multiple of A/a_i^o, where the multiplier depends on whether u_i can also be tallied from points that are projections of (l_{1s}, l_{2s}) orthogonally across the stand boundaries that truncate a_i. See Gregoire (1982), Schmid-Haas (1982), and Gregoire and Scott (1990) for details. The estimator of Y is unbiased.

Tree Concentric Method. This method is identical to the no-correction method except that measurements must be taken to permit one to determine the actual inclusion area, a_i, of tree u_i, when it is within a distance R_i of both the sampling location and one or more stand boundaries. The weight attached to y_i is A/a_i, which is the reciprocal of π_i so that the estimator of Y is unbiased.

In a simulation study of a mapped 5.2-ha stand, Gregoire and Scott (1990) compared the percent bias and root mean square error (MSE) of the estimators of the tree frequency, basal area, and volume corresponding to the eight methods described above. Sampling was conducted with 0.04-ha plots. Since bias is unaffected by the number of plots used in sampling and root MSE decreases proportionally to the square root of this number, results were reported on the basis of $m = 1$ plot. Among the unbiased methods, Grosenbaugh's method produced much more variable estimates than the other three, all of which were within $\pm 1\%$ of each other in terms of efficiency. Almost all of the biased methods had a smaller root MSE. The no-correction method showed biases ranging from -4 to -9% whereas the reject-to-R method showed biases ranging from 2 to -11%, varying with target characteristic and stand. Both the move-to-R and direct-weighting methods generally had biases within $\pm 1\%$.

Based on these preliminary results, either the enlarged tree circle, the tree concentric, or the miraging technique should be used. The difference in efficiency between these three unbiased methods is slight and the choice between them should probably be based on nonstatistical considerations such as ease of use in the field.

7.11.4 Controlling Errors in Field Procedures

Any errors in determining the horizontal ground area of a forest population, and of the sampling units *per se*, will affect the estimates of population means and totals and should be rigidly controlled. One cannot overemphasize the need for accurate determination of total forest area. Too often, area information is taken from erroneous land records, rough uncorrected maps, or aerial photographs. Errors when determining the area of the forest may be such as to completely counter the considerable care taken during the survey in establishment of sample units, tree and stand measurement, species identification, and so on, to derive accurate volume estimates.

Sources of error in forest assessment and inventory can be minimized by organizing special training programs for measure crews and by frequent checking of the methods and procedures actually being applied in the field. This latter check is essential to any serious inventory project.

7.12 RECORDING FIELD DATA

Most information gathered in the field two decades ago by forest scientists, consultants, and measurement crews was recorded on specially prepared forms that were subsequently taken or sent to the office for entering the data into a computer and processing. The time that elapsed between collecting the data and viewing the results varied from weeks to months and even years for some large projects. For a variety of reasons, this procedure is still followed in many countries, organizations, and situations.

In the early 1970's a significant development took place that revolutionized methods of recording and processing field data. It involved portable electronic devices that allowed data to be keyed into memory via a keyboard at the time of measurement, viewed on a small screen, and transferred later to a host computer for processing and analysis. Development of these devices in the periods 1978–1985 and 1986–1989 has been summarized by Nieman (1990a). The devices are microprocessor-controlled, battery-powered, hand-held computer terminals designed primarily for automating data collection at locations remote from the primary data-processing equipment (Cooney 1987). These new machines have drastically changed the habits of people collecting data, especially those involved in large organizations and large projects. In fact, large organizations, such as governments, universities, and forest companies, have to use electronic recording devices if they are to remain efficient (and competitive) because hand recording of data and subsequent data punching are slow, expensive, error-prone, and often inconvenient under adverse weather conditions (Fins and Rust 1988). The substantial savings to be made from using these machines are discussed by Nieman (1990a).

It may not be economic for small users to invest in the new devices because the establishment cost covering both hardware and software can be large and it may not be recouped by savings in time and improved accuracy. To minimize the cost of software to small users, Scott (1990b) suggests that they either investigate the utility of generalized data entry programs or use forms entry systems available from the data base management utilities already resident on their personal computers (PCs). The data recorder can then be programmed quickly to suit their specific needs.

Data recorders are now appearing on the market with more variety, more memory, and more peripherals; they use either standard disk operating systems (DOS) or systems that can process software written on DOS personal computers. These newer machines are capable not only of recording data but also of being programmed to monitor data input, detect measuring and recording errors and process the results in the field so that management and other decisions can be made on the spot (Scott 1990b). Owing to this change in function from a device with a single use to one that is programmable and thus useful for a wide variety of applications, the collective name for the machines now varies from portable data recorders, data loggers, data collectors, and electronic field data recorders (specific-use devices) to hand-held computers and field computers (multiple-use devices). However, for convenience, we shall refer to the instruments collectively in this discussion as portable data recorders (PDRs).

Forestry does not constitute a big market for PDRs (Nieman 1990a) but the manufacturers recognize that it is demanding and sophisticated and the users are becoming much more discriminating and adept at using software. PDRs are used in forestry in regeneration and stand assessment, forest inventory, seed orchard inventory, cruising for timber sales, log scaling, monitoring research projects, survey traverses, and many other field operations (Nieman 1990a; Scott 1990b). However, it is estimated that present use is only 15–20% of potential use (Nieman 1990b).

The ever increasing miniaturization of electronic components and ever broadening selection and lower prices of PDRs for collecting, recording, and processing data of almost any type are important to all sections of commerce and industry, not the least being forestry. However, selecting from the array of available machines with their impressive capabilities and multitude of features to find the product that best suits the user's needs is a daunting and

TABLE 7.4. Factors to Consider when Evaluating PDR Hardware[a]

1. Equipment	Available locally? Carry case—type, style, material, options
2. External features	Harness (neck strap, safety strap), hand strap, hand or finger grooves in instrument casing
3. Physical features	Weight, dimensions, balance. One-hand held?
4. Ruggedness	Resistance to water, dust, shock, extreme temperatures, and humidity
5. Power supply	Type (alkaline, NiCad) and life of batteries. Availability of batteries (or substitutes) and ease of replacement. Capability for checking charge of main and backup batteries. Type of display—external or internal? Battery charger. Can batteries be charged in the field? Protection against improper insertion of batteries
6. Keyboard	Acceptance by field crews and technical staff. Layout—location of numeric and alphabetic pads. Keys—raised or flat, size, color coded. Rate of key response. Function keys available? Specialized keys available (fast cursor, scroll, duplicate, insert, delete, etc.)? Durability of keys and keyboard. Arrow keys (to scan data quickly)
7. Display screen	Protection from scratching. Size, readability, background color, back lighting. Ability to tilt screen for better visibility. Visibility of cursor
8. Memory	Amount available in RAM (program execution), on disk (data storage), cassette, etc. Add-on capability (denser chips, memory cards) and cost
9. Operating system	Type—DOS or non-DOS, or menu driven processor able to execute software written on DOS PCs. Graphics capability
10. Software	Availability of application software
11. Communication system	Ports available for disks, printers, bar-code readers, etc. Telephone modem capability. Data transfer rate
12. PDR availability	Delay in supply likely?
13. Interference	Affected by shortwave radios, etc.?
14. Part replacement	Ease of replacing keyboard, display screen, battery compartment, etc.
15. Warranty	Type and period
16. Maintenance	Contract available? Type, period, and cost per unit

[a]Adapted from Hart (1990) and Wood (1990b).

TABLE 7.5. Factors to Consider When Evaluating the Manufacturer/Vendor[a]

1. Location, size, and financial stability of the company
2. Years of operation
3. Forestry expertise
4. Expertise in hardware engineering and ability to design and manufacture a PDR
5. Expertise in writing software; ability to design and write application software; willingness to write future application software; associated costs
6. Willingness to work closely with users when developing the desired hardware and software
7. Response time in repairing PDRs and debugging application software
8. Record of support for hardware and software
9. Reputation with current users
10. Support for consultant advice via telephone, and associated costs
11. Training support

[a]Adapted from Hart (1990) and Wood (1990b).

TABLE 7.6. Factors to Consider When Evaluating Software[a]

1. PDR operating system: type and ease of use (menu driven system is preferable)
2. System facility available to check charge of main battery and backup battery?
3. Can the PDR be programmed for the application? In what languages?
4. Can a compiler be used or is the machine confined to an interpreter (much slower)?
5. Amount of memory available for storing software and data
6. System's ability to check memory space left for storing software and data
7. Speed of data entry. Does software include an input data buffer? How many characters?
8. What communication software comes with the PDR? Is it compatible with the rest of the organization's hardware and software with which it will interface? If not, can it be made compatible?
9. Ability to run IBM-compatible PC software (e.g., programming languages). Any limitations?
10. Application program generator available allowing user to design own form? Ease of use?
11. Ease of transferring software and data to and from the host computer.
12. Application software: Easy to maintain, fast and well supported, constantly being upgraded?
    Speed of operation, simplicity, and user friendliness.
    Edit capabilities. Can data be edited in the same data screens as used for original input?
    User guide available?
13. Help screen facility available?
14. Willingness of company to modify application software both during the original development and subsequently
15. Warranty on the software
16. Company support for software user: training and consultant phone service available?
17. Calculator mode available?

[a]Adapted from Fins and Rust (1988), Hart (1990) and Wood (1990b).

challenging task. This task is made easier if the potential purchaser gives careful consideration to a number of questions including the following: Is a PDR really needed for the job? What reliability can be expected from a PDR? Which PDR (hardware) is right for the job? Which software is right for the job? How should these considerations be balanced to make the final decision? Can the purchase of a PDR be justified to management? How can the needed training program for field and office staff be best handled?

Readers contemplating purchase of a PDR are referred to Tables 7.4–7.6 and to Wood (1990b) who discusses each of the items listed above in considerable detail.

7.13 PROBLEMS

1. The data listed below derive from measurements taken in the laboratory and field. How many digits are significant in each?

1.000 25
0.029 40
17 007
40
40
4.000
4.550

2. A signpost gives the area of Koala State Forest as 16 800 ha. How many digits are likely to be significant in this number? Justify your answer.

3. How many digits are significant in the results of the following calculations?

(a) $17.16 \times 15 = 257.4$

(b) $20/6 = 3.33°$

(c) $1/4 = 0.25$

(d) $(2.3 \times 1774 \times 16 \times 0.1)/593.584 = 11.$

(e) How should the result of (d) be written?

4. Briefly explain why forest scientists and managers should be aware of and apply the rules relating to significant digits in a number.

5. Progressively round off 0.999 853 675 to 8, 7, 6, . . . , 1 decimal place(s).

6. Assuming you have a diameter tape and a graduated stick, describe how you would locate the point of measurement for recording d on each of the following stems:

(a) Vertical tree on a 10° slope.

(b) Leaning tree on level ground but with a 10-cm layer of litter above mineral soil level.

(c) Tree forked at 1.1 m (3′ 7″).

(d) Tree coppiced at 1.6 m (5′ 3″) with three coppice stems.

7. (a) List the main sources of error encountered when using: (i) a caliper and (ii) a tape to measure DBH.

(b) You find that measurements of DBHOB by diameter tape are negatively biased. How might this have occurred?

(c) When will a caliper give a less biased estimate of diameter than a tape?

(d) Under what conditions will a tape and caliper give the same result without bias?

8. (a) The DOBs of a tree with a malformation at breast height are measured an equal distance either side of the nominal breast high point. If the diameters recorded are 30.3 and 20.4 cm, respectively, what value would you record for *d*?

(b) Suppose the tree in (a) is located in a permanent sample plot that is remeasured every 5 years. What procedure would you adopt for measuring this stem?

9. Assume you have to choose between a tape and caliper for measuring the DBH of trees in a large measuring project. What factors would you consider before making your choice?

10. When measuring permanent sample plots, what are the relative advantages and disadvantages of:

(a) Using a graduated stick to define the breast high point on each stem at each measure?

(b) Permanently marking the point on each tree (e.g., paint, scribe mark) at the time of the first measurement?

11. List the action you would take to improve the precision of measuring tree DBHOB in a large measuring program.

12. List points relevant to a discussion of upper stem diameters and their measurement.

13. (a) What angle (minutes and seconds) is generated by two narrow bands of a Relaskop?

(b) An operator standing at a horizontal distance of 20 m from a tree determines that Band 1 and $2\frac{1}{5}$ narrow bands perfectly cover the bole at breast height. If the operator moved back from the tree a further 5 m, what bands and fractions of a narrow band would now cover the bole?

(c) Assume the d of a tree in virgin forest is 154 cm. What minimum horizontal distance would an operator need to be from the tree to measure d by a standard metric Relaskop?

14. A technician measured the d of four trees by Relaskop. Assuming he had laid out horizontal distances of 20, 15, 10, and 20 m, respectively, and recorded results of 60, 50, 35, and 75 cm, which tree would have been measured most precisely and which least precisely? Justify your answer.

15. Describe how you would check the calibration of the diameter scales on a Relaskop.

16. The height of a tree measured by Haga altimeter is recorded as 22.1 m. What precision of measurement is implied by this result? Is it realistic? Explain your answer.

17. Four operators independently estimate the height of a tree by Relaskop with the following results:

Operator	A	B	C	D
Height (m)	$39\frac{1}{2}$	40	$39\frac{1}{2}$	41

Comment on the results (precision, accuracy, bias) given that the true height of the tree is 40.3 m.

18. Determine the height of the four trees below given the information listed:

Tree Number	Instrument	Horizontal Distance (m)	Reading to Tree Base	Reading to Tree Tip	Reading to Plumb Point	Horizontal Displacement	Comments
1	Abney	20	−4°	42°	N/A	N/A	Vertical tree
2	Suunto	20	N/A	38%	+2%	5 m	Leaning tree
3	Haga	25	+4 m	26 m	N/A	N/A	Vertical tree
4	Relaskop	30	N/A	27 m	−5 m	2 m	Leaning tree

19. A technician uses four different hypsometers to estimate the height of a tree. Her raw data, are as follows:

Tree Number	Instrument	Horizontal Distance (m)	Datum (m)	Tree Base	Tree Tip
1	Relaskop	20	N/A	1 m	24 m
2	Abney level	25	N/A	−5° 30′	42° 0′
3	Haga	20	N/A	−1 m	25 m
4	Suunto	22	1 m	10%[a]	106%

[a]Sighted to the datum (a fixed point on the tree bole of known height above ground).

Assuming all instruments are in perfect working order and the measurements have been taken properly, which is the shortest tree and which is the tallest?

20. The height of a tree is measured independently by each of four operators using the same instrument, a Blume Leiss. Their results in meters are:

A	B	C	D	Average
$22\frac{1}{2}$	23	$22\frac{1}{2}$	22	22

Assuming the actual height of the tree is 25.0 m,

(a) Is the average height of 22 m acceptable? If not, why not?
(b) Is the average height recorded by operators A, B, and C acceptable?
(c) Is there any clear evidence of instrument bias in the readings?
(d) What factors could cause the mean error of 3 m?
(e) Comment on the precision of measurement.

Now suppose the same four operators measured two more trees, the full results being:

	Tree 1	Tree 2	Tree 3
	Actual Height (m)		
	25	$27\frac{1}{2}$	24
Operator	Measured Height (m)		
A	$22\frac{1}{2}$	27	24
B	23	28	24
C	$22\frac{1}{2}$	$27\frac{1}{2}$	$24\frac{1}{2}$
D	20	30	28
Average	22	28	25

What are your comments now on:

(f) Instrument bias?
(g) Operator bias?
(h) Operator error and possible causes?
(i) Operator competence?

21. A confident but inexperienced student complains that the Haga altimeter issued to his group of three students is negatively biased by 10–15%. When asked how he arrived at this conclusion, he replies, "We independently measured the same two trees and our precision of measurement was very good. We checked that the horizontal distances were correctly set out and we had sighted to the correct tips and read the correct scales." You advise the student that these are insufficient grounds for his conclusion and instruct him to think further about it. Justify your action.

22. **(a)** What readings on the degree scale of a Suunto clinometer would correspond with the readings of +105 and +8 on the percentage scale?

(b) Sighting from the ground to the top of a tree from a horizontal distance of 30 m gives a reading of 27 m on the 30 m scale of a Blume Leiss altimeter. What are the corresponding readings on the 20 m and 40 m scales?

(c) The reading to the tip of a vertical tree using the 30 m scale of a Blume Leiss altimeter is 32 m. What would be the readings on the same scale if you positioned yourself at horizontal distances of 20 m and 40 m, respectively, from the tree?

23. A member of a crew you are training sighted from a fixed point on the ground to the tip of a designated tree and obtained a reading of 47 using the percentage scale of a Suunto clinometer and 28° 36′ using an Abney level. What would be your advice to the crew member and why?

24. "What are we doing wrong?" asks a frustrated leader of an inexperienced measuring crew. "Our measurements of total tree height using height sticks are so imprecise and inaccurate." What would be the substance of your reply?

25. An observer on level ground marked positions on the north, east, south, and west sides of a tree at horizontal distances of 40 m from the geometric center of its butt. Then, using a Suunto clinometer at each position, she sighted carefully to the tip and base of the tree, recording readings as follows:

	Reading to:	
Position	Tip	Base
North	+ 89%	−2%
East	+ 94%	−3%
South	+100%	−2%
West	+ 95%	−2%

Explain the results.

26. What advice would you give to the leader of a measuring crew whom you have instructed to record the heights of the 50 tallest trees in a section of *Pinus* plantation that suffered damage in a windstorm following two days of heavy rain? Assume the trees are: (i) <15 m tall; (ii) >30 m tall.

27. You, the leader of a measure crew, have been asked to undertake a stand height survey in a specified section of *Pinus* plantation. List the matters you would consider in planning the survey.

28. Describe how you would check the calibration on the height scales of a Relaskop.

29. Describe the procedure you would adopt to gather the raw data necessary to develop for a 20 000-ha plantation resource of *Pinus* ranging in age from 10–40 years either **a**) a bark thickness function for use in deriving diameter inside bark along the tree boles, or **b**) a tree volume function.

30. Attempts to develop a general stem profile equation have proved futile. Why is this?

31. You and an assistant are given one day to estimate the bole area of a 0.15-ha sample plot of radiata pine located in a stand carrying an average of 420 trees/ha stocking. Outline your plan of action.

32. Which of the direct methods available for measuring the volume of felled sample trees will incur little or no bias if the assumed shape in sections of the bole below the merchantable limit is not approximated? Explain your answer.

33. You are instructed to measure directly the under bark volume of 50 plantation-grown pine trees ranging in age from 15 to 40 years. The data are required to test a recently compiled tree volume function that estimates the sawlog volume of standing trees given DBHOB and total tree height. Nominate the method of measurement you would use and explain why you would choose it in preference to the other methods available.

34. The tip of a tree is 6 m long and it has a basal DIB of 12 cm. What are the probable maximum and minimum values for its inside-bark volume in cubic meters.

35. You now wish to apply your function to estimate the total volume of wood in standing trees on a 0.12-ha sample plot (500 trees/ha) located within the plantation. Describe the procedure you would follow to derive the estimate given that you have one day in which to complete the work.

36. Assume you have located a regrowth tree (DBHOB = 18 cm, total height = 15 m) of excurrent habit with a crown of regular outline. Describe how you would estimate the crown ratio, crown length ratio, crown surface area, and green crown percent.

37. Discuss difficulties likely to be faced when attempting to age a tree by counting its annual growth rings.

38. "VRP sampling is useless for estimating stand basal area because it is so prone to error." How would you react to this statement made by a member of a field crew?

39. Describe how you would check the calibration of Band 4 of a Relaskop on both level and on sloping ground.

40. Assuming the desirable count per sweep in VRP is 6-12 trees, briefly describe how you would select the appropriate band or band combination of a Relaskop to give the desired count at a specified point in a forest.

41. What difference in procedure is involved in checking whether a doubtful tree is "in," "out," or "true borderline," when using a wedge prism and Relaskop?

42. What action would you take to instruct a VRP sampling crew experienced in using wedge prisms but not Relaskop, how to estimate stand basal area by Relaskop?

43. The measurement is doubtful for a tree 16.75 m from the eye of the observer (slope distance) when sighted by Relaskop with the brake freed. If the slope angle is 10° and the DBHOB of the tree is 66 cm, what band or combination of bands is being used?

44. An operator using a wedge prism achieves perfect displacement of the image of a tree of DBHOB 45.7 cm and 11.3 m distant. In sweeping a forest stand with the prism at sample point #7, the same operator counts 10 trees "in" and 2 trees "doubtful." Her check of the doubtful trees yields the following data:

Tree Number	DBHOB (cm)	Slope Distance (m)
1	43.2	10.7
2	58.4	14.9

What is the estimated stand basal area (m^2/ha) at point #7?

45. You have been instructed to estimate $\bar{G}$ of a square 400-ha block of slash pine to within ±5% at the 95% level of confidence. Outline your plan of action assuming that a survey 2 years earlier estimated $\bar{G} = 36\ m^2$/ha with a standard deviation of $6\ m^2$/ha.

46. Sweeping by Relaskop at a given point in an even-aged forest using bands 1, 2, and 4 gave counts as follows:

Band:		1	2	4
Count:	"In"	23	14	4
	"True Borderline"	1	0	0

Interpret the results in terms of the variation of G about the sample point.

47. During a sweep of an area by wedge prism (F 4.35 m^2/ha), a tree of 62.3-cm DBHOB is counted "in." What frequency (stems/ha) does this tree represent?

48. The precision of estimate of G in a compartment of mixed hardwoods varied from ± 4 m^2/ha to ± 8 m^2/ha when angle count instruments of F 1 and 5 m^2/ha respectively, were used at 20 randomly located sample points. Interpret the findings.

49. Assume 20 sample points have been located in a 10-ha plantation of *Pinus*. Under what conditions would you expect to achieve a similar precision of estimate from 2° and 4° wedge prisms when used to estimate $\bar{G}$? What would be your answer if no mention was made of the number of sample points?

50. An experienced forest technician tests four different sampling frames based on VRP to estimate $\bar{G}$ of a 20 ha forest, namely:

Sampling Frame	Number of Sample Units	F Used (m^2/ha)
A	10	1
B	10	3
C	20	2
D	20	1

List the sampling frames in descending order of the anticipated precision of estimate of $\bar{G}$ and justify your answer.

51. Describe how, single-handed, you could use Band 1 of a Relaskop and equipment selected from the items listed below to establish a temporary circular plot of area 0.050 ha at a randomly located point in a forest:

1 × survey marker
1 × diameter tape
1 × scientific calculator
1 × pressurized can of orange paint
3 × 1.5-m long cylinders of rigid plastic tubing of outer diameters 11.3, 17.8, and 25.2 cm, respectively.

52. Describe how you would estimate:

(a) The predominant height (40 trees/ha) of a rectangular sample plot (50 m × 20 m) in a mature, even-aged coniferous forest of a single species. The plot is located on a 15° slope and its long axis is parallel to the slope.

(b) The mean predominant height of a square 16 ha compartment of 14-year-old *Pinus* plantation on level terrain.

53. It is laborious and costly to measure the height of every tree in an even-aged forest stand to derive stand mean height. Nominate two much less costly approximation methods and describe in detail how you would apply one of them to estimate the mean height of a 0.5-ha block of plantation forest (250 trees/ha) whose predominant height estimated by eye is 34 m.

54. Outline a plan for establishing the equation representing the SHC of a 2-ha, 20-year-old forest stand of *Pinus* on a uniform site of stocking 300 trees/ha. Cover in the plan the sample frame, sample size, and field and office (data processing) procedures.

55. Describe how you and two assistants would estimate the standing merchantable volume of a 1.5 ha even-aged stand of a commercial pine given that you have one day in which to complete the work. You may assume that a reliable two-variable (d and h) volume function is available for the species.

56. Enumeration of a forest plot for d yields the following data:

d Class (cm)	40	42	44	46	48
Number of Trees	5	15	20	14	6

Ten trees, selected from the plot on a stratified random basis, are then measured accurately for total height (h). Data for these trees are:

Tree Number	d (cm)	h (m)
2	41.5	27.8
11	39.7	25.9
19	46.7	31.5
25	43.4	29.4
27	44.1	29.8
34	42.8	29.1
43	41.2	27.2
50	48.4	32.8
56	44.7	30.1
58	45.8	31.0

If the tree volume function applicable to the forest is

$$v_o = 0.124 + 0.000\,025\, d^2 h$$

where v_o = tree volume (m^3) and d and h are in cm and m, respectively, what is the volume of the plot?

57. Enumeration of a 0.7 ha stand of even-aged *Eucalyptus* regrowth indicates that it has 275 stems/ha and a basal area of 30 m^2/ha. A stratified random sample of ten trees selected from the stand is measured accurately for volume with the following result:

Tree Number	g (m^2)	Volume (m^3)
5	0.030	0.151
25	0.116	0.953
37	0.058	0.425
71	0.134	1.050
79	0.090	0.696
95	0.068	0.453
106	0.150	1.225
131	0.106	0.801
153	0.074	0.554
187	0.046	0.248

(a) What is your estimate of the volume of the stand?

(b) If the same stand was remeasured five years later, what changes would you expect to find in slope and position of the volume (v_0/g) line? Explain these changes.

58. Measurements taken over a period of 12 years on a series of sample plots in a 40-ha plantation of *Pinus radiata* yielded the following data:

	Estimated Parameters of the v_0/g Lines	
Mean h_{dom} (m)	b_0	b_1
15	−0.038	5.9
18	−0.048	6.7
21	−0.060	7.6
24	−0.070	8.6
27	−0.078	9.5

(a) Derive an appropriate stand volume function from these data.

(b) Estimate the quantity of timber available for sale from the plantation had it been clear felled at age 15 years when it had 800 stems/ha, a basal area of 25 m^2/ha and h_{dom} of 22 m.

59. A volume tariff compiled for an even-aged forest is of the form:

$$v_o = -0.0382 + 0.0022h_{dom} - 0.000\,141h_{dom}^2 + 0.4145g + 0.3390gh_{dom}$$

where v_o = tree volume (m^3), and h_{dom} and g are in m and m^2, respectively.

(a) What does this formula indicate about the relationship between h_{dom} and the parameters of the many v_0/g lines embodied in the tariff?

(b) Assuming the h^2_{dom} term can safely be dropped from the regression, what change does this suggest in the nature of the relationship?

60. Illustrate with an example the value to a forest manager of a well established relationship between v_o/g and h_{dom}.

61. Assume you have measured d, h, and v_0 of all 50 trees in a sample plot of even-aged *Pinus*. You now plot the data on separate graphs, namely, v_0 versus d^2 and d^2h, respectively.

(a) What similarities and differences would you expect to find in the two graphs?

(b) Three years later, you remeasure the same 50 trees and plot the data on the original graphs. Describe and explain the changes you would expect to find in the relationships.

8 Related Sampling Topics

The topics discussed in this chapter (small area estimation, sampling wildlife and fish populations, sampling rare populations, spatial distributions in forests, and remote sensing in resource inventories) are diverse but, potentially, they are of considerable interest to people engaged in forest resource inventory. An introductory review of each topic is given and key references are indicated for those who wish to study the topics in more detail.

8.1 SMALL AREA ESTIMATION

Small area estimation is applied in situations where estimates are desired or required for subdivisions of the population and design-consistent estimators have unacceptably large errors for the subdivisions. The estimation problem involves a probability sample that can be used to construct estimates for some of the subdivisions, the presence of auxiliary information for all subdivisions, and finally, a model that specifies the relationship among the sources of information to construct estimators for the subdivisions (Fuller 1986). Small area estimation is a good example where the probabilistic theory of Chapters 3–5 and the model-based concepts discussed in Chapter 6 should be integrated to give "good" estimates. Such estimates should often be better than strictly probability-based estimates. Because they are at least partially model-based, there will often be concern about the bias of the estimates and their reliability.

The problem typically arises when a large population is sampled (for example, a state in the USA) but some user groups may be more interested in reliable estimates for parts of the state such as a county. Although the state-wide estimates are often quite reliable, only a few sample units may fall in a given county (small area) and hence other sources of information (such as the same information for nearby or similar counties) are used to strengthen the estimates. Information can often be borrowed from various sources and can be used in different ways so there can be many alternative estimators.

Interest in small area data has grown markedly in recent years. Brackstone (1987) attributed this to both demand for such data and availability of data sources. For example, in several countries there is increasing government concern with issues of income distribution. Existence of regions or other subgroups of the population that are well below average in income are causes for concern and remedial action. Identifying such subpopulations and developing, administering, and evaluating the effects of programs on income distribution all imply a need for statistical data. Industry and local governments have similar requirements (for supplies of raw material, labor requirements, etc.). In forestry an example is the removal of timber volume by county (Green *et al.* 1987).

Recent advances in statistical methods and technical developments have assisted the development of small area data bases. At the same time, many administrative and technical systems are now completely automated and often include a detailed geographic code (postal code or latitude and longitude coordinates). An example is the east-wide Forest Inventory and Analysis (FIA) data base developed by the US Forest Service (Hansen 1990). This has facilitated the storage, processing and statistical analysis of large data sets.

Generally, small area estimation methods are predictive in that the estimates are derived under an appropriate population model that relates the variable of interest to a set of associated variables. Identification and evaluation of the model is the major analytical problem in small area estimation. The associated practical problem is to obtain the auxiliary data needed for both the model and the selected estimation method. Researchers in this field are now examining composite methods that combine the strengths of several population models (Brackstone 1987).

Not all small area statistics need to be statistically reliable. The analytic inference needed may be met by good measures of relative relationships between small area units. Small area estimation is most useful when objectivity is introduced by using the results of an agreed upon measurement process as the official values, rather than seeking some unobtainable "true" values (Heeringa 1981).

Besides requiring different estimation techniques, small area statistics require more caution in the use of estimates generated, due to the unfamiliar model-based methodology and questions about their reliability. These issues were raised in Chapter 6 in a somewhat different context. These small-area estimates are evaluated and criticized more frequently than conventional sample estimates of large populations. A clear distinction should be made between the two types of estimates. For example, the National Center for Health Statistics (NCHS) in the USA published one type of small area estimate called synthetic estimates of disability and health characteristics. The term "synthetic estimates" derives in part from NCHS's early attempts to separate the small area estimates of disability and the conventional estimates published for major census regions and the nation at large (Brackstone 1987). What then are synthetic estimators?

8.1.1 Introduction to Synthetic Estimators

Rao (1986) introduced the basic ideas used in small area estimation. He assumed simple random sampling (SRS) for purposes of illustration. He distinguished between situations where concomitant data is or is not available.

No Concomitant Data Is Available. If the small area means $\bar{Y}_i$ are the parameters of interest and the area sizes N_i $(i = 1, \ldots, k)$ are known, the classical estimator of $\bar{Y}_i$ is

$$\bar{y}_i = \sum_{j \in s_i} y_j / n_i \tag{8.1}$$

where s_i denotes the sample from area i, n_i is the sample size in s_i, and $n = \Sigma_{i=1}^k n_i$. The variance of $\bar{y}_i$ is of order n_i^{-1} so that $\bar{y}_i$ is unreliable for small sample sizes n_i.

A simple synthetic estimator of $\bar{Y}_i$ with variance of order n^{-1} is

$$\bar{y}_i(1) = \bar{y} = \sum_{j \in s} y_j / n \tag{8.2}$$

where s is the total sample of size n. Clearly this estimator is badly biased unless

$$\bar{Y}_i \doteq \bar{Y} \qquad \text{for} \quad i = 1, \ldots, k\,. \tag{8.3}$$

This assumption can be explicitly stated as a model on the y_j,

$$y_j = \mu + e_j \qquad \text{for} \quad j = 1, \ldots, N \tag{8.4}$$

where the e_j are uncorrelated with zero mean and constant variance σ^2. $\hat{\mu} = \bar{y}$ is the least squares estimator of μ and the best linear unbiased predictor (BLUP) of $\bar{Y}_i$ is

$$\bar{y}_i(2) = \frac{n_i}{N_i}\,\bar{y}_i + \left(1 - \frac{n_i}{N_i}\right)\bar{y}\,. \tag{8.5}$$

Note that $\bar{Y}_i$ is a random variable under the model, and $\bar{y}_i(2)$ is consistent whereas $\bar{y}_i(1)$ is not since $\bar{y}_i(2) = \bar{Y}_i$ and $\bar{y}_i(1) = \bar{Y}$ for $n_i = N_i$, and $\bar{y}_i(2) \doteq \bar{y}_i(1)$ for $n_i \ll N_i$.

The assumption $\bar{Y}_i \doteq \bar{Y}$ $(i = 1, \ldots, k)$ can be improved by assuming homogeneity only within strata set up after sampling (post-stratification),

$$\bar{Y}_{hi} = \bar{Y}_{h\cdot} \qquad \text{for} \quad h = 1, \ldots, L \tag{8.6}$$

where $\bar{Y}_{hi}$ is the mean of N_{hi} population units in post-stratum h belonging to small area i, $\bar{Y}_{h\cdot} = \Sigma_{i=1}^k W_{hi}\bar{Y}_{hi}$, $W_{hi} = N_{hi}/N_i$, and $\Sigma_{i=1}^k N_{hi} = N_i$. One synthetic estimator is obtained in standard post-stratification estimation by replacing the sample mean $\bar{y}_{hi}$ in the estimator

$$\bar{y}_i^* = \sum_{h=1}^{L} W_{hi}\bar{y}_{hi} \qquad (n_{hi} > 0 \text{ for all } h)$$

by the post-stratum sample mean $\bar{y}_{h\cdot}$

$$\bar{y}_i^*(1) = \sum_{h=1}^{L} W_{hi}\bar{y}_{h\cdot} \,. \tag{8.7}$$

The assumption $\bar{Y}_{hi} = \bar{Y}_{h\cdot}$ can be explicitly stated as a model on the y_{hi}:

$$y_{hi} = \mu_h + e_{hi} \qquad (i = 1, \ldots, N_h;\ h = 1, \ldots, L) \tag{8.8}$$

where the e_{hi}'s are uncorrelated with mean zero and variance σ_h^2. For this model the BLUP is

$$\bar{y}_i^*(2) = \sum_{h=1}^{L} W_{hi}\left\{ \frac{n_{hi}}{N_{hi}}\, \bar{y}_{hi} + \left(1 - \frac{n_{hi}}{N_{hi}}\right) \bar{y}_{h\cdot} \right\}. \tag{8.9}$$

Note that $\bar{y}_i^*(2)$ is Cochran-consistent but not consistent whereas $\bar{y}_i^*(1)$ is not, but for $n_{hi} \ll N_{hi}$, $\bar{y}_i^*(2) \doteq \bar{y}_i^*(1)$ (Rao 1986).

Concomitant Data is Available. If concomitant data x is available, with known population means $\bar{X}_i$, the usual ratio estimator of $\bar{Y}_i$ is

$$\bar{y}_{ir} = (\bar{y}_i / \bar{x}_i)\bar{X}_i \,. \tag{8.10}$$

We can then obtain improved synthetic ratio estimators analogous to the simple synthetic estimators discussed earlier.

If $R_i = \bar{y}_i/\bar{x}_i \doteq R = \bar{Y}/\bar{X}$ is assumed, that is, the ratios are homogeneous, then a synthetic ratio estimator can be used by replacing $r_i = \bar{y}_i/\bar{x}_i$ in eq. (8.10) above by $r = \bar{y}/\bar{x}$:

$$\bar{y}_{ir}(1) = (\bar{y}/\bar{x})\bar{X}_i \tag{8.11}$$

where $\bar{y}_{ir}(1)$ has properties similar to $\bar{y}_i(1)$. Under the frequently used model

$$y_j = \beta x_j + e_j \tag{8.12}$$

with mean zero and $V(e_j) = \sigma^2 x_j$ $(j = 1, \ldots, N)$, the BLUP is

$$\bar{y}_{ir}(2) = \frac{n_i}{N_i}\, \bar{y}_i + \left(\bar{X}_i - \frac{n_i}{N_i}\, \bar{x}_i\right) \frac{\bar{y}}{\bar{x}} \tag{8.13}$$

which is approximately $\bar{y}_{ir}(1)$ if $n_i \ll N_i$. $\bar{y}_{ir}(1)$ and $\bar{y}_{ir}(2)$ are more precise than $\bar{y}_i(1)$ and $\bar{y}_i(2)$ if model (8.12) holds and are also less biased if $R_i \doteq R$ is more reasonable than $\bar{Y}_i \doteq \bar{Y}$. Clearly, linear and nonlinear regression estimators can also be used instead of the simple and ratio-based synthetic estimators, depending on the relationship between y and x one is willing to assume (Rao 1986).

Purcell and Kish (1979) and Heeringa (1981) reviewed the literature on small area estimations. Heeringa, in addition, assessed the utility of various techniques used in the Survey of Income and Program Participation (SIPP) program in the USA. Several of the small area estimators used in SIPP are discussed below.

8.1.2 Synthetic Estimators

Early work on synthetic estimators was done primarily by the NCHS and the Bureau of Census in the USA, the Central Bureau of Statistics in Norway, Statistics Canada, and in Australia (Purcell and Linacre 1976; Heeringa 1981). Synthetic estimation methods assume that small areas have characteristics similar to those of the larger areas of which they are a part. If this assumption is wrong, synthetic estimators are biased (Särndal and Hidiroglou 1989).

A general synthetic estimator is (Heeringa 1981)

$$\hat{Y}_h^* = \sum_{g=1}^{G} W_{hg} \bar{y}'_{\cdot g} \tag{8.14}$$

where $\hat{Y}_h^*$ = the synthetic estimator for small area h, $\bar{y}_{\cdot g}$ = the sample mean for the gth subpopulation, W_{hg} = the weight given to the mean of small area h, subpopulation g, and G = the number of subpopulations.

The accuracy of synthetic estimates depends on the correspondence between the subpopulation means, $\bar{y}_{\cdot g}$, and the corresponding small domain values, $\bar{y}_{hg}$. Improvements may be obtained by increasing the homogeneity within the subpopulations through more detailed classification, or by reducing the heterogeneity within the large domain from which the mean $\bar{y}_{\cdot g}$ is estimated.

There are two forms of the basic synthetic estimator. The first is

$$\hat{Y}_h^* = \sum_{g=1}^{G} (X_{hg}/X_{\cdot g})\hat{Y}'_{\cdot g} \tag{8.15}$$

where $\hat{Y}'_{\cdot g}$ = the large domain estimate for the total of Y in subpopulation g, and $X_{\cdot g}$ = the large domain total for associated variable x in subpopulation g.

The second estimator, for means, rates, and proportions, is the original synthetic estimator introduced by the NCHS, namely,

$$\bar{y}_h^* = \sum_{g=1}^{G} (X_{hg}/X_{h\cdot})\bar{y}'_{\cdot g} \tag{8.16}$$

where

$\bar{y}_h^*$ = the synthetic estimator of the mean value of y for small domain h,
$\bar{y}'_{\cdot g}$ = the large domain estimator of the mean value of y for small domain h,
X_{hg} = the total for associated variable x in subpopulation g and small domain h, and
$X_{h\cdot}$ = the total for associated variable x in small domain h (Heeringa 1981).

8.1.3 Regression-Adjusted and Synthetic Regression Estimators

The regression adjustment models the relationships between the relative percentage error of the synthetic estimate and an ancillary variable x_h measured for each small area. The estimator is

$$\hat{Y}_{ra;h}^* = \hat{Y}_h^*\left(1 + \frac{\hat{\alpha} + \hat{\beta}X_h}{100}\right) \tag{8.17}$$

where $\hat{Y}_{ra;h}^*$ is the regression-adjusted estimator of the parameter of interest for small area h, X_h is the value of the covariate x for small area h, and $\hat{\alpha}$ and $\hat{\beta}$ are estimated regression coefficients, the estimation of which requires knowledge of the true values Y_h. Approximations are obtained by collapsing the sampling strata, calculating unbiased estimators $\hat{Y}'_c$ and $\hat{X}'_c$ for the collapsed strata, and fitting the regression equation to those values. The computed $\hat{\alpha}$, $\hat{\beta}$ are then used in eq. (8.17) (Heeringa 1981).

8.1.4. Composite Estimators

The bias in synthetic estimation can be reduced by combining a synthetic estimator with another estimator, for example, by combining a low variance synthetic estimator and a low bias direct estimator. Numerous approaches can be used to arrive at the weights of the two components. We basically limit ourselves to the ones discussed by Heeringa (1981) because their simplicity is attractive as an introduction to the general topic.

Simple Composite Estimator. For the larger domains of interest, fairly precise estimates may be available directly from the sample. For this situation,

$$\hat{Y}_{c,h} = c_h\hat{Y}'_h + (1 - c_h)\hat{Y}_h^* \tag{8.18}$$

where

$\hat{Y}_{c,h}$ = the composite estimator of the statistic of interest for small area h,
$\hat{Y}_{h'}$ = the direct sample estimator for small area h,
$\hat{Y}_h^*$ = the synthetic estimator for small area h, and
c_h = a weight factor chosen to minimize the mean square error (MSE) of $\hat{Y}_{c,h}$.

An approximation to the optimal value of c_h is $c_h = 1/(1 + R_h)$ where $R_h = \text{MSE}(\hat{Y}_h')/\text{MSE}(\hat{Y}_h^*)$ so that only knowledge of the ratio of mean square errors of the direct and synthetic estimators is required (Heeringa 1981).

Base Unit Method Estimator. This method is designed to be assumption free and is restricted only by the availability of predictor variable (x) data for units of small area. The small area of interest and the remaining areas covered by the sample design are subdivided into small geographic areas (base units), such as counties for which data on x may be available. The base units are grouped on the basis of the information on x into $k = 1, \ldots, K$ homogeneous clusters (Cohen 1979). From data for a sample of units in each of the K clusters, a weighted estimate of the statistic of interest is calculated,

$$\bar{y}_k' = \sum_{i=1}^{N_k} W_{ki}\bar{y}_{ki}' \qquad \text{for} \quad k = 1, \ldots, K \tag{8.19}$$

where

$\bar{y}_k'$ = the weighted mean of y for the kth base unit cluster
$\bar{y}_{ki}'$ = the estimated mean of y for the ith base unit in the kth cluster, and
W_{ki} = the weight value assigned to the ith base unit in the kth cluster.

Then the estimator for a small area is a weighted average of the sample estimators for each of the $k = 1, \ldots, K$ clusters of base units:

$$\bar{y}_{bu,h}^* = \sum_{k=1}^{K} P_{hk}\bar{y}_k' \tag{8.20}$$

where $\bar{y}_{bu,h}^*$ is the base unit estimate of the mean of y for small area h, and $P_{hk} = N_{hk}/N_{h\cdot}$ is the proportion of elements in small area h that belong to a base unit in the kth cluster.

The base unit method assumes no underlying relationship between the variable of interest and the independent variables but this is desirable for the estimator to be efficient. This method requires the same algorithm for final calculations as the synthetic method. Preprocessing of input can be considerably more complex and time consuming. One advantage of the base unit method is that estimates for the homogeneous clusters of geographical areas are a byproduct of the calculation procedure (Heeringa 1981).

Regression-Sample Estimator. If reliable data on independent variables are available for small domains, the regression-sample method may be preferable. A functional relationship between the variable of interest and the independent variable is assumed. If the relationship is linear, the overall accuracy of the estimator is better than for the base unit method. For nonlinear settings, the advantage is not as clear and may disappear completely for extreme nonlinearity (Heeringa 1981).

Ericksen (1973, 1974) tested the regression method on estimators of population growth based on the linear model

$$y_h = \beta X_h + \epsilon_h \, . \tag{8.21}$$

using estimator

$$\hat{Y}_{rs,h} = \hat{\beta}_0 + \hat{\beta}_1 X_{1h} + \cdots + \hat{\beta}_p X_{ph} \tag{8.22}$$

where

$\hat{Y}_{rs,h}$ = the regression-sample estimator of the parameter of interest for small area h

X_{ih} = the small area values of the $i = 1, \ldots, p$ independent variables, and

$\hat{\beta}_i$ = the estimated regression coefficients ($i = 0, 1, \ldots, p$).

The β_i are estimated by fitting the regression equation to unbiased primary sampling unit (psu) estimates of the dependent variable Y_{psu} and to current psu values of the independent variables. Then estimates for small areas are obtained by substituting current values of the independent variables for the small area of interest into the fitted equation.

Särndal (1984) and Särndal and Hidiroglou (1989) generalized $\hat{Y}_{rs,h}$ and noted that their generalized regression estimators were approximately unbiased (even for small n_h) and allowed for simple calculations of design-based confidence intervals.

The regression-sample method is appealing since it incorporates a current measure of the association between the variable of interest and related independent variables. It can be difficult though to obtain data on independent variables that are highly correlated with the variable of interest and to identify the appropriate model. The regression-sample method is not difficult to implement but sample size, model identification, and the availability of good data on independent variables limit its application.

James–Stein Estimator. The James–Stein (JS) estimator (Stein 1955; James and Stein 1961; Fay and Herriott 1979) assumes that two estimators exist for a small area characteristic, namely, a direct sample estimator Y'_h and another estimator Y^*_h derived using information for all small areas in a designated set. The prior distribution of the second set of estimators is assumed known. Fay and Herriott (1979) used regression-sample estimators but the prior estimators can take a number of forms. Prior estimators could be an unbiased mean for a large domain or a synthetic estimator for a small area (the latter is a composite estimator). Through the prior estimator, the JS estimator for a small area borrows information from the group at large. If the prior estimator is a mean for a larger domain, the estimator produces a shrinkage toward that average value; if it is a regression estimator, the shrinkage will be to the regression line, the amount of shrinkage depending on the variances of the two component estimators.

One form of the JS estimator is

$$\hat{Y}_{js,h} = c_h Y'_h + (1 - c_h) Y^*_h \tag{8.23}$$

where

$\hat{Y}_{js,h}$ = the JS type estimator of the statistic of interest for small area h,
Y'_h = the direct estimator for small area h,
Y^*_h = the prior estimator for small area h, and
c_h = a weight factor chosen to minimize the Bayes risk of $\hat{Y}_{js,h}$.

The risk function for $\hat{Y}_{js,h}$ is minimized by choosing $c_h = A^*/(A^* + D_h)$ where A^* is the variance of the prior estimate and D_h is the sampling variance of the direct estimator for small area h. If A^* is known exactly, $\hat{Y}_{js,h}$ is the classical Bayes estimator. If A^* is approximated, an empirical Bayes procedure can be used. The JS estimator assumes that the variance of the prior estimates is equal across the units in the group under consideration but usually it differs for subsets of units within the larger group. The problems of unequal variances and group composition require further research.

The data requirements for each component of the JS estimator are about the same as for other methods discussed. The main problem is estimating the weight factor $c_h = A^*/(A^* + D_h)$ to combine the two estimators into the JS form. Research on the James–Stein estimator provides a logical framework for addressing the problems of combining estimators.

Research on small area estimation techniques is advancing rapidly. Battese *et al.* (1988) used an error-component prediction model to estimate county crop areas and Rao (1986) explained structure preserving estimators (SPREE) for small area census estimators that use iterative procedures to improve estimates. Prasad and Rao (1990) addressed the issue of mean square estimation more fully. Hidiroglou and Särndal (1985) and Lundström (1987) conducted simulation studies to assess the reliability and estimation bias of various estimators. Dagum *et al.* (1987) in another simulation study examined the sensitivity of some small area estimators to misclassification errors in variables and to possible conceptual differences in classifying variables.

8.1.5 A Forestry Application

Green *et al.* (1987) discussed JS estimation of timber removals by county using forest survey data (USDA Forest Service 1992) (for an earlier application of the same idea, see Thomas 1986.) They noted that JS estimation is logically appealing for this application because, in addition to the typical small sample size, the counties within a survey unit are often similar in forest type and condition. County boundaries rarely reflect physiographical or biological transitions, so that samples taken in one county are not much different from samples in another. Hence it makes intuitive sense to combine information from several counties to improve the estimates for each. Timber removals for some counties were estimated based on as few as 4 survey plots and by as many as 53 in others.

Green *et al.* (1987) JS estimation procedure is as follows: Assume there are k independent unbiased estimates $\bar{y}_1, \bar{y}_2, \ldots, \bar{y}_k$ (the k sample means), of parameters $\theta_1, \theta_2, \ldots, \theta_k$ $(k > 3)$, which are normally (i.e., $\bar{y}_i|\theta_i \sim N(\theta_i, v_i)$) and independently distributed. For small v_i, one may use the estimates $\bar{y}_i$, but otherwise use

$$\hat{\theta}_i = (1 - \hat{\beta}_i)\bar{y}_i + \hat{\beta}_i \bar{y}_w \tag{8.24}$$

where

$$\bar{y}_w = \sum_{i=1}^{k} \bar{y}_i W_i \Big/ \sum_{i=1}^{k} W_i$$

$$W_i = 1/(v_i + \hat{A})$$

$$\hat{A} = \sum_{i=1}^{k} W_i^2\{(k/(k-1))(\bar{y}_i - \bar{y}_w)^2 - v_i\} \Big/ \sum_{i=1}^{k} W_i^2$$

and

$$\hat{\beta}_i = \{(k-3)/(k-2)\} v_i/(v_i + \hat{A})$$

where $\hat{A}$ represents an empirical Bayes estimate of the unconditional (prior) variance of θ_i. Then the posterior variance of $\hat{\theta}_i$ is estimated by

$$s_i^2 = v_i[1 - \{(k - r_i)/k\}\hat{\beta}_i] + v_{1i}(\bar{y}_i - \bar{y}_w)^2$$

with

$$r_i = kW_i \Big/ \sum_{i=1}^{k} 1/W_i$$

$$v_{1i} = \{2/(k-3)\}\hat{\beta}_i^2(\bar{V} + \hat{A})/(v_i + \hat{A})$$

where

$$\bar{V} = \sum_{i=1}^{k} W_i v_i \Big/ \sum W_i \, .$$

Then $\hat{\theta}_i \pm z_\alpha s_i$ or $\hat{\theta}_i \pm t_\alpha \hat{s}_i$ can be used as the approximate 95% confidence interval for $\alpha = .025$. When $\hat{A}$ is negative, it is set to 0.

The results show that because the standard errors of estimates for the counties were unequal, the county means were shrunk toward the weighted mean at different rates. Green *et al.* (1987) used bootstrap samples to compare the JS estimates with the standard estimates. They recommended use of JS estimators instead of the standard estimators because they tend to be closer to the "true" values and often have smaller confidence intervals. But JS estimators are biased. Conventional techniques should be used if individual means are well estimated. Implicit in JS sampling is the idea that all means are of equal interest.

8.2 SAMPLING WILDLIFE AND FISH POPULATIONS

Much of the previous theory on sampling finite populations is not applicable to sampling wildlife populations even though the populations are finite. Animals are mobile, they can and will hide, which makes detection difficult, and sampling may itself affect their location. The result of this is that there is usually no sampling frame. Probabilities of selection cannot be assigned in advance and have to be estimated even after the sample is drawn. The existence of a specific selection probability for each individual in the population is often only conceptual. The net effect is that sampling techniques tend to be more expensive than surveys of vegetation, and more statistical assumptions have to be made to make estimation possible. Hence, the likelihood of errors is greater (Burnham 1980).

Four recent technological advances have facilitated the inventory of animal populations (Seber 1986), namely, the development of very small radio transmitter tags (Amlaner and MacDonald 1980), remote heat sensing using thermography (Best and Fowler 1981), use of underwater acoustics (Kieser and Mulligen 1984), and statistical techniques such as simulation, Monte Carlo inference, jackknifing, and bootstrapping. Seber (1992) added a fifth technological advance, photography. He also noted that radio transmitters are now widely used even for subterranean rodents and marine mammals, and cited several references that use acoustic methods for estimating abundance of fish.

The primary parameters of interest in wildlife sampling are usually population size, birth rate, immigration rate, mortality rate, and emigration rate (Pollock *et al.* 1990). Because of the complexity of estimating some of these parameters, populations are classified as closed and open populations.

A closed population is one that has a constant size (and retains the same members during the period of closure except for known removals) during a study, while an open population is one where births, immigrations, deaths, and emigrations can occur. Pickands and Raghavachari (1987) pointed out that estimating the number of animals in a closed population is the same as estimating the number of errors in a manuscript.

The discussion in the following sections focuses primarily on estimating population size.

8.2.1 Sampling with Sample Units in Closed Populations

Closed populations are much simpler to sample than open populations and sample units such as plots, strips, lines, and points can be used. Sampling methods include those discussed in the following sections.

Census. A census can be implemented for territorial species such as birds, animal herds, migrating fish, and fish in ponds that can be drained. A total count is made of the population.

Quadrats and Strip Transects. The usual method of estimating population size is to count the numbers of animals on a random sample of plots. Quadrats are frequently used although other plot shapes are possible. Such methods are useful for ant colonies, termite mounds, animals living in the soil, some species of insects, birds, and small mammals (Burnham *et al.* (1980).

Transformation of the data can improve the statistical analysis because a large proportion of zeros often occurs (Seber 1986). Pennington (1983) suggested, with data from marine surveys, that the zeros be treated separately.

When the population is very patchy and the area covered is large, Thompson (1982, 1990) suggested an adaptive sampling scheme in which the sampling intensity in a given area depends on the density of animals found in the previous area(s). This results in a higher sampling intensity in the higher density areas. The technique is further discussed in Section 8.3.

Because animal populations are often clustered, statistical distributions are used in wildlife studies to describe the probabilities of finding a certain number of animals or signs of animals (y) in a plot. The negative binomial distribution

$$f(y) = \{k(k+1)\cdots(k+y-1)/k!\}(u/k)^y(1+u/k)^{-y-k}\,, \quad y = 0, 1, 2, \ldots \tag{8.25}$$

is often used for this purpose (Seber 1982) where u is the population mean and the population variance is $\sigma^2 = u + u^2/k$. The smaller k is, the greater the clustering.

Strip transects are usually easier to locate and sample than quadrats but are more affected by animals moving on the boundaries (Seber 1986). Such transects are used extensively in counting animal pellets and in aerial censuses described below.

Aerial Censuses. Strips are preferred to quadrats when sampling from the air since they are easier to locate and to use in computing adjustments in ratio or regression estimation. Such adjustments in wildlife sampling are called visibility factors (Seber 1982). Important research in aerial censusing has been done in Australia by the National Parks and Wildlife Service, for example, Ridpath *et al.* (1983). Various ways of adjusting the aerial counts are discussed by Seber (1986). Radiotelemetry will probably become useful in aerial censuses (Packard *et al.* 1985).

Line Transect Sampling. Strip transects are often not appropriate for studying mobile animals such as birds, but line transect or variable-width transect sampling can be appropriate (Burnham *et al.* 1980).

If an area has known boundaries and size A, and one wants to estimate the abundance of some biological population in the area, then line transect sampling establishes one or more lines of travel of known length L in the area to determine a sample of target objects (say deer). The following measurements can be taken:

x = perpendicular distance from the line to an observed object,
r = distance from the observer to the object at the moment of detection,
θ = angle between observer line of travel and the line of sight to the object at the moment of detection.

Usually x or r and θ are recorded for each object. Clearly not all objects in area A will be recorded, and those closer to the lines of travel are more likely to be recorded than those farther away. Four critical assumptions are made to obtain reliable estimates of population size. In order of importance, these are:

1. Objects on the line will never be missed.
2. The objects do not move before being detected, and none are counted twice.
3. Distances and angles are measured exactly.
4. Sightings are independent events.

In this method, objects to be sampled are considered dimensionless points (Eberhardt 1978). To make inferences about animal abundance from the set of distances, angles, and sample size, a conceptual model is required that relates the data to the abundance parameter to be estimated. The basic model is based on the idea that the probability of detecting an animal decreases as its distance from the transect line increases. This idea is mathematically represented by a function $g(x)$ called the detection function, which defines the detection probability of objects as a function of their distance from the transect line.

One does not need to assume that there is one single function that is identical for all objects over the duration of the study. Numerous factors such as weather conditions, alertness, interests and training of observers, habitat conditions, time of year and day, group size for clustered populations, and species, sex, and age of population units affect the probability of detecting objects. Mathematically the detection function can be a mixture of many simple functions $g(x) = g_1(x) + \cdots + g_k(x)$ where each of the $g_i(x)$ applies to a particular population segment or set of detection conditions.

An estimator of the number of objects in an area is $\hat{D} = n_0/(2L\hat{a})$ where n_0 = number of objects counted, L = length of the transect, and $\hat{a}$ estimates $a = \int_o^w g(x)dx$ where w is the maximum perpendicular distance at which observations are made, that is, the width of the transect. In this regard, a can be interpreted to be one-half the effective strip width of the transect. Note that the value of w is not in the formula for $\hat{D}$, so that it does not matter whether w is finite or infinite. Also, the parameter a is related to but not identical to the average detection probability over the area searched, $p = a/w$.

The basic problem, then, to obtain an estimate $\hat{N}$ of the number of objects N is to estimate the parameter a. This means one has to obtain the probability density function of the observed data on perpendicular distance x, which is related to the reciprocal of a. Burnham *et al.* (1980) rewrote $\hat{D} = n_0/(2L\hat{a})$ as $\hat{D} = n_0\hat{f}(0)/(2L) = n_0/(2Lw\hat{p})$ and devoted much of their monograph to estimating appropriate functions $f(x)$ where $f(x) = g(x)/a$. An example of such a density estimator is the Hayne estimator

$$\hat{D}_H = \frac{1}{2L}\sum_{i=1}^{n} 1/r_i\,, \tag{8.26}$$

where a fixed circular flushing radius r_i (distance between observer and animal) is assumed for each animal.

Burnham *et al.* (1980) noted that variance estimators for $\hat{D}$ (and hence $\hat{D}_H$) should be based on replicate transect lines. If there are r replicate lines with lengths l_i and sample sizes n_i and if individual sample sizes are large enough, then we have r independent estimates of density

$$\hat{D}_i = n_i\hat{f}_i(0)/(2l_i) \tag{8.27}$$

with an overall, weighted estimate of density

$$\hat{D} = \sum_{i=1}^{r} l_i D_i/L \tag{8.28}$$

where $L = \Sigma_{i=1}^{r} l_i$ with variance estimator

$$v(D) = \sum_{i=1}^{r} l_i(\hat{D}_i - \hat{D})^2/\{L(r-1)\}. \tag{8.29}$$

This estimator is often not useful because the individual $\hat{D}_i$ are often too small. Hence, Burnham *et al.* (1980) suggested the alternative jackknife variance estimator:

$$v_J(D) = \sum_{i=1}^{r} l_i(\hat{D}^{(i)} - \hat{D}_J)^2/\{L(r-1)\} \tag{8.30}$$

where

$$\hat{D}_{(i)} = \frac{L\hat{D} - (L - l_i)\hat{D}_{(i)}}{l_i}$$

and

$$\hat{D}_J = \sum_{i=1}^{r} l_i\hat{D}^{(i)}/L .$$

Transect methods do not always work well because the terrain and vegetation may make it difficult for an observer to walk quietly while, for example, looking for birds. The technique of a variable circular plot in which stations are located along the transect and the observer spends time at each station observing bird distances (Reynolds *et al.* 1980) is promising. This method is analogous to line transects (Burnham *et al.* 1980, p. 195) and does not require the assumption of a random distribution of birds. De Vries (1986, Ch. 7) gave a very complete discussion of line intersect sampling with specific orientation to forestry. Kaiser (1983) reviewed the closely related technique of line intercept sampling where the length of intercept of each object that covers the line transect is recorded. This is used primarily for habitat sampling but can also be used for sampling large animal signs, such as den sites (Eberhardt 1978). Kimura and Lemberg (1981) called this technique line intercept density estimation when estimating fish populations using hydro-acoustics.

Distance Methods. When a population is randomly distributed over an area, the distances between objects from randomly chosen points (closest-individual techniques) or distances between objects (nearest-neighbor techniques) can be used to estimate the population density. These techniques have been developed primarily for stationary populations of plants or trees but can be applied to conspicuous and relatively stationary animal populations (snails) or to well-marked colonies of ants and termite mounds. They have also been applied to more mobile populations such as grasshoppers, cricket frogs, and quail coveys (Seber 1982).

Several types of distances can be measured, such as the distance from a randomly selected point P to the rth nearest object Q_r, the distance from Q_1 to the rth nearest object in a direction away from P, and the distance from Q_1 to the nearest object (Byth 1982). Byth and Ripley (1980) preferred semi-systematic sampling to random sampling to avoid oversampling and Byth (1982) discussed some new robust density estimators for sampling spatial patterns by distance methods.

Removal Method. Samples are removed sequentially from the population and the decline in catch size is used to estimate the population size. Usually at least three catches are required and the best estimators are still not established (Seber 1986).

Relative Density Method. The density of signs of animals, such as faeces pellets or tracks, is used to provide an index of population density for situations where such an index can be expected to be fairly stable for a variety of environmental conditions (Seber 1982).

8.2.2 Capture-Recapture Sampling in Closed Populations

In addition to estimating population size, one may be interested in births, mortality, immigration, and emigration. Capture-recapture sampling methods can be used to estimate these, but are difficult theoretically. Hence, we introduce them first for simple (closed) populations where interest is only in population size.

Single Recaptures. The simplest method of sampling is to capture M out of the N animals (objects) in the population, mark or tag them for future identification, and return them to the population. Then, after allowing time for the marked and unmarked animals to mix, a second sample of n animals is taken of which m are found to be marked and $u = (n - m)$ unmarked. By assuming that the proportion of marked animals in the second sample is a reasonable estimate of the unknown population proportion, an estimate of the number of animals in the population is

$$\hat{N} = \mathrm{Mn/m} \,. \tag{8.31}$$

This estimator is called the Petersen estimator and the idea behind it underlies many of the more complex capture-recapture models (Seber 1986). Various probability functions have been used to describe the distribution of m. For example, if $\pi = M/N$ is the proportion of marked animals, then three such functions are:

1. The hypergeometric density:

$$f(m|M, N) = \binom{M}{N}\binom{N-M}{n-m} \Big/ \binom{N}{n} , \tag{8.32}$$

2. The binomial density:

$$f(m|\pi) = \binom{n}{m} \pi^m (1-\pi)^{n-m} , \tag{8.33}$$

3. The Poisson density:

$$f(m|\pi) = \exp(-n\pi)(n\pi)^m/m! . \tag{8.34}$$

If the sample marked is a random sample, and if p_i is the probability of capture in the ith sample ($i = 1, 2$), then the multinomial distribution

$$\begin{aligned} f(\{a_w\}|N) &= \frac{N!}{\Pi a_w!(N-r)!} \{p_1(1-p_2)\}^{a_{10}} \\ &\quad \times \{(1-p_1)p_2\}^{a_{01}}(p_1p_2)^{a_{11}}\{(1-p_1)(1-p_2)\}^{N-r} \\ &\propto \frac{N!}{(N-r)!} \prod_{i=1}^{2} p_i^{n_i}(1-p_i)^{N-n_i} \end{aligned} \tag{8.35}$$

can be assumed to hold where a_w represents the number with recapture history, that is, $a_w = a_{10}$ is number caught in sample 1 but not in 2, a_{11} number caught at both times, and a_{01} number caught only at time 2, n_i is number caught in the ith sample, and r is the number of all animals caught.

Then $\hat{N}$ in eq. (8.31) is the maximum likelihood estimator (mle) of N for the binomial (eq. 8.33) and Poisson (eq. 8.34) distributions. The integer value $[\hat{N}]$ is the mle for the hypergeometric distribution (eq. 8.32) and approximates $\hat{N}$ for the multinomial distribution (eq. 8.35) (Seber 1986).

The Petersen method has been applied frequently in wildlife sampling, a good example being Bowen and Sergeant (1983) because the authors carefully discussed the underlying assumptions.

Multiple Recaptures. A series of samples is taken from the closed population and the unmarked animals in each sample are tagged and are returned to the population. A series of models is used to cope with variations in the catchability of animals. Seber (1986) and Pollock *et al.* (1990) discussed the relative merits of the models in some detail.

8.2.3 Capture-Recapture Sampling in Open Populations

Multiple Recaptures. Jolly (1965) and Seber (1965) developed what is called the Jolly-Seber model for populations in which death, permanent migration, and recruitment occur. The model assumes the following:

- Every animal in the population at time i ($i = 1, 2, \ldots, k$) has the same probability of capture (p_i).
- Every marked animal present in the population after the ith sample

has the same probability of survival ϕ_i until the $(i + 1)$th sampling time $(i = 1, 2, \ldots, k - 1)$

- Marks are not lost or overlooked.
- All samples are instantaneously taken and each release after sampling is immediate (Pollock *et al.* 1990).

If M_i, the number of marked animals in the population just before the ith sample, is known for $i = 2, 3, \ldots, k - 1$, then

$$\hat{N}_i = n_i \hat{M}_i / m_i \tag{8.36}$$

is analogous to the Petersen estimator (eq. 8.31) where m_i and n_i are the marked and total number of animals captured in the ith sample, and $\hat{M}_i$ is defined later.

A reasonable survival rate estimator $\hat{\Phi}$ should be

$$\hat{\Phi}_i = \hat{M}_{i+1} / (\hat{M}_i - m_i + R_i) \tag{8.37}$$

where R_i is the number of animals caught at time i and released back into the population with marks. A logical recruitment estimator between times i and $(i + 1)$ is

$$\hat{B}_i = \hat{N}_{i+1} - \hat{\Phi}_i(\hat{N}_i - n_i + R_i) \tag{8.38}$$

and capture probabilities can be estimated by

$$\hat{p}_i = m_i / \hat{M}_i = n_i / \hat{N}_i \tag{8.39}$$

(Pollock *et al.* 1990).

The M_i usually need to be estimated, and a reasonable estimator is

$$\hat{M}_i = m_i + R_i z_i / r_i \tag{8.40}$$

defined only for $i = 2, \ldots, k - 1$, where z_i and r_i are the animals in $(M_i - m_i)$ and R_i, respectively, captured again at least once. Pollock *et al.* (1990) gave approximate variance and covariance estimates for the biased estimators in eq. (8.38–8.40).

Seber (1982) gave alternate approximately unbiased estimators if the underlying model is correct:

$$\tilde{M}_i = m_i + (R_i + 1) z_i / (r_i + 1) \tag{8.41}$$

$$\tilde{N} = (n_i + 1) \tilde{M}_i / (m_i + 1) \tag{8.42}$$

$$\tilde{\Phi}_i = \tilde{M}_{i+1} / (\tilde{M}_i - m_i + R_i) \tag{8.43}$$

and

$$\tilde{B}_i = \tilde{N}_{i+1} - \tilde{\Phi}_i(\tilde{N}_i - n_i + R_i)\,. \tag{8.44}$$

Jolly (1982) suggested the estimator

$$\tilde{p}_i = m_i / \tilde{M}_i \, . \tag{8.45}$$

m_i and r_i should exceed 10 to get satisfactory results with these estimators (Seber 1982).

Sophisticated models, methods, and software for capture-recapture data are summarized in a host of publications (Pollock *et al.* 1990; Lebreton *et al.* 1992; Seber 1992). Work has progressed to develop flexible capture–recapture methods for open populations (Lebreton *et al.* 1992). However, a major problem is still the heterogeneity of capture. Pollock *et al.* (1990) cited an application of capture-recapture models to fossil data (Nichols and Pollock 1983) as an interesting example in an entirely different field.

Recovery from Dead Animals. A special case of the Jolly–Seber model uses recoveries from dead animals. A cohort of tagged individuals is released each year and the tags are recovered from dead animals, mostly from hunting and fishing (Brownie *et al.* 1985). Survival and band recovery rates may be age- and year-dependent. As inexpensive, light, robust radio transmitters with low failure rates become more readily available, problems of identifiability, incomplete reporting, and so on, will be minimized, estimated survival rates will become more reliable, and better models will be feasible (White 1983).

Seber (1986) also discussed "catch-effort methods" and "demographic and other methods," but these methods are not well developed at this time.

There seems to be a correspondence between parameters of interest in wildlife sampling (survival, death, migration, and recruitment) and in forestry (survivor growth, mortality, cut, and ingrowth), which perhaps could lead to improved estimation techniques in both fields by taking advantage of potential cross-fertilization between the two areas of research.

8.3 SAMPLING RARE POPULATIONS

Sampling of rare (or elusive) populations is difficult because the costs of locating such populations are considerable and often exceed measurement costs (Sudman *et al.* 1988).

The fundamental problem with sampling rare populations is that many contacts may be needed to identify sample members with the rare trait in the overall population (Kalton and Anderson 1986). Most work in this field has been done on human populations. We confine our terminology to that used for human populations to retain the flavor of the complexities and advantages of sampling these populations, which in several cases, are very different from those encountered when sampling forest populations. We rely heavily on material from Kalton and Anderson (1986) in the following discussion. Three distinct objectives that may lead to different strategies in surveying rare populations are:

1. To estimate the number of individuals M in the rare population, the prevalence of the rare population in the total population ($P = M/N$ where N is the size of the total population), and the prevalence in various subgroups of the population.
2. To estimate the means of certain variables for individuals in the rare population, or proportions of the rare population with certain characteristics, such as

$$\bar{Y} = \sum_{i=1}^{M} y_i/M$$

in the case of proportions, where $y_i = 1$ if the individual has the characteristic, $= 0$ otherwise.
3. To estimate the mean or total of the y-variable for the entire population with $y_i = 0$ for those members that are not part of the rare population. The population parameters are

$$\bar{Y}_t = \sum_{i=1}^{N} y_i/N = P\bar{Y} \qquad \text{and} \qquad Y_t = N\bar{Y}_t = M\bar{Y}\ .$$

An example would be to estimate the mean cost of treatment for diseased trees in the whole population of a particular tree species, where the rare population consists of all trees with the disease.

Kalton and Anderson (1986) discussed six methods for sampling rare populations, namely, screening; disproportionate sampling; multiplicity sampling; multiple frames; snowball, or reputational, sampling; sequential sampling; we include adaptive cluster sampling.

8.3.1 Screening

A large sample from the total population is examined to identify members of the rare population. Screening costs increase rapidly as the degree of rarity rises. In screening human populations, telephone interviews and mail questionnaires are usually used. Both can be susceptible to significant bias because of nonresponse (rare population members may not have phones or may not reply to or receive questionnaires). In an analogous forestry situation, diseased trees may be missed because the disease is not always obvious.

When a rare population is geographically clustered, it can be efficient to sample the clusters in which the rare population occurs more frequently and at higher rates. The first step (e.g., in a telephone survey) is to select a cluster proportional to its size in the population and to select one element at random in that cluster. If the element is a member of the rare population, it is included in the sample and further screening of the cluster is done until a set number of K units of the rare population is selected. However, if the initial element is not a member of the rare population, the cluster is rejected. The process is repeated until the required number of sample clusters is selected. This procedure samples clusters with probability proportional to the number of members in the rare population that are part of the cluster. Sudman (1985) gave a formula for optimum choice of K and discussed various applications. If some of the clusters accepted for the sample have fewer than $K + 1$ members of the rare population, there may be a need for weighting, which causes a loss in efficiency (Waksberg 1983).

Identification of members of the rare population can be costly. If so, a two-phase approach may be profitable. For example, at the first phase use a relatively cheap, imperfect, screening method to divide the large sample of units into two or more strata according to their likelihood of being in the rare population. Then subsample the strata with higher likelihoods at a higher intensity using more expensive measurement techniques.

Deming (1977) suggested that the ratio of screening costs in the second to first phase needs to be at least 6 : 1 and should preferably be much larger, 40 : 1 or more. With an initial screening into two strata, namely, members and nonmembers of the rare population, the latter stratum should contain very few members of the rare population and the classification of nonmembers as members becomes less important. One way to take advantage of this is to use less stringent criteria at the first phase for defining the rare population. When the screening criteria are set such that the proportion of positives classified as false negatives is small, there may not be a need to subsample from the negative stratum at the second phase. But if the negative stratum is large, even a small proportion of false negatives can represent a large proportion of the rare population.

Screening can be useful in forest sampling, for example, when assessing lack of seedling vigor or potential mortality in regeneration plantings and exploring the cause of poor vigor or potential mortality. Lennartz and McClure (1979) used characteristics recorded on FIA plots in the Southeast of the USA to identify habitat for the endangered red cockaded woodpecker. The areas identified could then be sampled carefully, recording the number of red cockaded woodpeckers. The two phases taken together would be a good example of the two-phase screening approach discussed.

8.3.2 Disproportionate Sampling

Sometimes strata with higher concentrations of a rare population can be identified. For example, the rare population may be concentrated in certain geographical areas or in strata where lists of sampling units are available. Once identified, the strata with concentrations of the rare population can be sampled at a higher rate.

In the following, we treat separately the prevalence of the rare population and the estimation of its mean using disproportionate sampling:

1. *Prevalence of the Rare Population* ($P = M/N$). If the strata have proportions $P_1, P_2, \ldots, P_l$ that can be estimated roughly beforehand, then an approximation to optimum allocation can be used if simple random sampling (SRS) is applied within strata.
2. *Estimating the Mean for the Rare Population.* Assume there are two strata, with A the proportion of the rare population in stratum 1 and $(1 - A)$ the proportion in stratum 2. If A is assumed known, then $\bar{Y} = A\bar{Y}_1 + (1 - A)\ \bar{Y}_2$ can be estimated by

$$\bar{y} = A\bar{y}_1 + (1 - A)\bar{y}_2$$

where $\bar{Y}_1$, $\bar{Y}_2$ are the sample mean values of the rare members in the two strata.

Assume that both the variances of the y-variable for members of the rare population and the costs of data collection are the same for the two strata. There may be differences in the costs of sampling members of the rare and more common populations because a member of the rare population may need to be interviewed or measured in some other way, while the nonmember only needs to be screened. Thus if c is the ratio of the cost of sampling a member of the rare population to that of sampling a member of the more common population, then the ratio of the variance of $\bar{y}$ under disproportionate and proportionate allocation is approximately

$$R \doteq \frac{[kP - (k-1)W_1P_1][(c-1)\{P + (k-1)W_1P_1\} + (k-1)W_1 + 1]}{kP[(c-1)P + 1]}$$

where P_1 is the proportion of the rare population in stratum 1. The optimum choice of the ratio of the sampling fractions in the two strata is given by

$$k_0^2 = \frac{P_1[(c-1)(P - W_1P_1) + (1 - W_1)]}{(P - W_1P_1)[(c-1)P_1 + 1]} .$$

Even if $\bar{Y}_1$ and $\bar{Y}_2$ are not equal, the values of R and k_0 will often still be good approximations. The gains from disproportionate stratification will be significant only when the strata to be oversampled have high concentrations of the rare population and the strata contain a substantial proportion of the rare population (Kalton and Anderson 1986).

Disproportionate sampling can be particularly appropriate in large areas such as Alaska, for example, where digital Landsat multispectral scanner (LS) data can be used to identify the rare forest areas (commercial and noncommercial). Either the forest areas could be sampled fairly intensely on the ground or an attempt might be made to use Landsat to distinguish commercial from noncommercial forest with the idea of sampling the (perceived) commercial forest stratum more heavily than the noncommercial stratum.

8.3.3 Multiplicity Sampling

Sometimes multiplicity or network sampling can be used to reduce the number of contacts needed to identify members with the rare trait. Multiplicity sampling creates a duplicity of listings and capitalizes on them for collecting data (Kalton and Anderson 1986).

With a multiplicity sample design, in sampling households for example, the informant provides information not only about his or her own household members but also about others who are linked to the household in clearly defined ways, for example, all close relatives of household members.

Linkages have to be specified so that the selection probabilities of sample members can be determined. For example, in an equal probability sample of households, with all members of selected households being included, and with several linkages to siblings, the probability that an individual is included in the sample is proportional to the number of households in which the individual and his or her siblings live. Weights inversely proportional to these selection probabilities are needed in the analysis. An advantage of multiplicity sampling is that an individual living in an institution could be part of the sample, whereas, in a conventional survey, there would be no chance of that individual being selected (Kalton and Anderson 1986).

The critical consideration in choice of linkage is the ability of the members sampled to provide the necessary information, or provide detail on the linkages and accurate information on whether those linked to them have the rare trait. These are the minimal requirements to estimate prevalence of the rare trait. Often other detail is needed (e.g., age and sex of the sampled persons in a prevalence survey of diabetics). Frequently it is not possible for the person to provide all the information on these characteristics for the linked persons. Then direct contact with the linked persons with the rare trait is needed. This must be taken into account when defining the linkages. When interviewing, for example, it may be economically advantageous to restrict linkage to persons in the same geographical area as the initial informant (Kalton and Anderson 1986).

Several factors should be considered in choosing between multiplicity sampling and conventional sampling. Multiplicity sampling requires a smaller sample to yield the required sample size of members of the rare population; but analysis is more complex, the use of informants frequently leads to a significant increase in the level of response error, and the need for weights in the analysis causes an increase in sampling error. In addition, problems may arise in determining linkages, and additional questions might need to be asked to identify them. Nonresponse to these additional questions also creates difficulties. Tracing and contacting the linked persons may involve considerable time and effort and ethical questions may arise; for example, is it appropriate to collect survey data for a person who is not part of the informant's household?

There have not been any applications of multiplicity sampling in forestry, probably due to the lack of obvious linkages such as are available in human populations. However, such linkages might exist and might be useful when defining ownership in surveys of land and forest resources. Possibly, multiplicity sampling could be used in genetics-oriented surveys, for example, in seed orchards (linkage in seed sources).

8.3.4 Multiple Frames

One or more partial lists of the rare population may exist, in which case the sample might comprise two components: the first being a sample from the partial list and the second from the total population to screen for individuals with a particular trait. Since the screening sample is often expensive, it may be prudent to sample the partial list more intensively than the screening sample. When there are multiple frames, some members of the rare population will be included in more than one frame. This gives rise to the problem of duplicates that can be handled by redefining the frames so they do not overlap or by compensating for them in the analysis. Kalton and Anderson (1986) and references therein discussed both approaches in detail.

Multiple lists do not seem to occur frequently in forest populations. They could occur in, say, interspersed government and private land where the government land might be more intensively managed resulting in a partial list about some rare characteristic on the government land and on some adjacent private land, but not on all land in the population.

8.3.5 Snowball Sampling or Reputational Sampling

A necessary condition for snowball sampling is that members of the rare population know each other. This may be true for certain rare ethnic minorities, religious groups, or persons with certain disabilities. One application of snowball sampling is to create a frame of members of the population by identifying a few members, asking them to identify others, contacting those to identify still others, and so on. Once a frame has been compiled, a probability sample can be drawn. The critical issue here is the completeness of the frame. The missing members, if any, are likely to be the socially isolated ones (Kalton and Anderson 1986).

Another application of snowballing is to continue the snowballing process until a sufficient number of members of the rare population has been found for the survey. With this approach, those with many contacts with other members of the rare population are more likely to be included in the sample than those with few contacts. Since the sample is not a probability sample, objective weighting adjustments cannot be used in the analysis to compensate for this problem. Steps may be taken to balance the sample to conform to known or hypothesized distributions for certain background variables, as is done in quota sampling, but distributions of important background variables are seldom known for a rare population. The technique seems more suitable for exploratory and qualitative investigations than for statistical surveys.

The necessary condition for snowball sampling, that members of the population know each other, is certainly not fulfilled for plant or animal populations but may be in ownership surveys in certain instances. There may be some examples where snowball sampling has been used in forest populations to identify infrequently occurring trees with certain characteristics. For example, the identification of "superior" trees and the measurement of their characteristics would fall in this category.

Unfortunately snowball sampling is also defined in a different manner. Goodman (1961) defined an s-stage, k-name snowball sampling scheme as follows: A random sample of n individuals is drawn from a given finite population. Each individual drawn is asked to name k different individuals (best friends, lovers, etc.) in the population with k a specified integer. The individuals named form the first stage. Each of these is then asked to name k individuals with the same criterion. These second stage individuals are then asked to name k individuals, and this is continued for s stages. The data are used to make statistical inferences about various aspects of the relationships present in the population. For example, when $s = k = 1$, the number of mutual relationships present in the population can be estimated.

8.3.6 Sequential Sampling

One sequential sampling approach is to select an initial sample of sufficient size to give the desired sample size (n) of members of the rare population based on the highest estimate of incidence. This will yield n_1 members of the rare population and an estimate of incidence. If $n_1 < n$, a second sample is selected to produce the remaining $n - n_1$ members of the rare population based on the incidence obtained from the first sample.

Another approach is to construct a sample as a set of replicate samples of size large enough to generate the desired number of members of the rare population based on the lowest estimate of incidence. The replicates are then implemented until the desired sample size is achieved. An excellent example of sequential sampling for potentially rare insect populations is given in Waters (1955).

8.3.7 Adaptive Cluster Sampling

An initial set of units is selected probabilistically and additional units in the neighborhood of a unit are added if the variable of interest for that unit satisfies a given criterion. Although these designs in which sample selection can depend on the observed values of the variable of interest introduce bias into conventional estimators, Thompson (1990) considered them to be an extremely effective way of sampling rare clustered populations and he gave design-unbiased estimators for them. More research needs to be done on the utility of this new method and related ones, such as stratified adaptive cluster sampling discussed in Thompson (1991).

Other methods that could be useful include multipurpose surveys where several investigators pool resources to conduct a joint survey to study several rare traits simultaneously. Another method is to cumulate cases with the rare trait over several resurveys. One can also analyze sets of surveys that have collected information needed to identify the members of the rare population and obtain the desired information for these members.

When a sampling list is available for a population, sampling from a rare population is the same as surveying any population. If no adequate list is available, the techniques described above may be useful for selecting a sample from a rare population. Combinations of methods may be useful. When the population is extremely rare or when identification of members of the rare population is expensive, it may not be possible to sample in a satisfactory manner. Note that in some cases the identification of very rare populations is feasible because they are highly distinctive. For example, LS data in the Tanana basin of Alaska showed 1 plot out of 5535 plots covered by "Dryas" (a low shrub with a woody candex and basal rosette of leaves). This shrub covered nearly 100% of the 8-ha plot. Interestingly enough, the same plot was also selected at the three other phases of the AIRIS design described in Chapter 5.

8.4 SPATIAL DISTRIBUTIONS IN FORESTS

An ideal statistical summary of a forest would be a complete description of the spatial distribution of its trees and of the distribution of tree size. The Weibull and the S_B for the univariate case (Hafley and Schreuder 1977) and generalizations of the S_B for the bivariate and trivariate cases are used for the latter (Schreuder *et al.* 1982), that is, the size distribution of tree characteristics such as DBH and height. If equally satisfactory spatial distributions could be established, analytical solutions could be found for comparing the relative efficiencies of various sampling strategies.

As Matérn (1972) noted, when the precision of forest sampling methods is studied empirically using actual populations, the relative merits of different methods vary among forests. Sampling from artificial populations produced by some random mechanism requires an understanding of what may be assumed to be realistic mechanisms. For example, Sukwong *et al.* (1971) made assumptions regarding spatial distribution, diameter distribution, and stand density to generate artificial populations in comparing fixed-radius and

VRP sampling for basal area estimates in a simulation study. They noted that analytical solutions might be available for their situation.

For inventory purposes, spatial distributions may explain the results or facilitate comparison of inventory methods. Matérn (1972) showed that this is true for simple spatial distributions. Holgate (1967) made similar comparisons assuming random and regular spatial patterns. An excellent example where a simple, reliable spatial distribution would have resulted in concrete recommendations is Schreuder (1978), who proposed a combination of sample methods to estimate total basal area, sum of tree diameters, and number of trees in a forest. In all these examples, the key problem is the evaluation of the joint probabilities of selection of individual trees. A related issue is the selection of the optimum angle α to use in VRP sampling. The choice of α also depends on the spatial distribution of the trees, the size distribution of the trees, and stand density. Ord (1978) showed how the precision of the estimate of number of trees in VRP sampling is different for a random or regular tree distribution pattern. Matérn (1947) used models of spatial variation of tree locations to estimate the accuracy of line and plot surveys, and to develop an effective sampling design for the third Swedish National Forest Survey (Matérn 1986, Ch. 5–6). Ranneby (1983) used stochastic models of variation in time and space to help develop a reliable sixth Swedish National Forest Survey. However, realistic spatial stochastic processes are quite different from the classical domain of statistical theory and few useful results are available (Ripley 1988).

Knowledge of spatial distributions in forests is also important from a management point of view. For regeneration purposes, stocking percentage and location of regeneration need to be known to ensure that stocking is adequate over the entire area. Similarly, in older stands, a good understanding of spatial distributions is helpful in designing thinning schedules, optimal harvesting techniques, competition studies, development of individual tree and growth models, and management planning generally.

In section 8.4.1, we focus on a brief description of research directions, explanations of some of the stochastic processes involved, and two attempts in forestry to fit spatial distributions.

8.4.1 Fitting Models

Upton and Fingleton (1985, 1989) defined a spatial pattern as the zero-dimensional characteristic of a set of points that describe the location of these points in terms of the distance between points.

The preferred way of obtaining spatial distributions is to identify a process, determine the form of the distribution generated, and fit its parameters. This is covered first in the discussion which follows. The alternative is to fit flexible distributions and attempt to interpret the parameters afterwards. Examples of this are given subsequently. Different distributions may be needed to describe different situations, for example, in regeneration studies, we may have manual or machine planting versus direct seeding versus natural regeneration, all of which result in different spatial distributions; rock outcrops, swampy ground, streambeds, variation in soil type and depth, unweathered solid wood, and so on, all may affect the spatial distribution of trees in forests; and in any stand, one may have desirable or undesirable trees depending on species, age, quality, and so on, the inclusion or exclusion of which affects the spatial distribution of interest.

Process-Generated Distributions. An understanding of the processes that generate spatial distributions, and an awareness of the types of patterns that are generated are essential if inferential techniques and model building efforts for spatial point processes are to be evaluated critically. Some models studied are the following:

Poisson Model. The two-dimensional homogeneous Poisson point process, often called Poisson forest, is the simplest stochastic model for a spatial point pattern (Diggle 1979). Trees are assumed to be located completely at random in an area. Although this process is implausible, since two trees could be arbitrarily close together, it can serve as a useful approximation or idealized standard. It is a good starting point to explain more complex models. There are two basic postulates:

1. *Uniformity.* The events (tree locations) show no tendency to occupy particular regions in the plane.
2. *Independence.* Location of a given tree is determined without reference to the location of another tree.

Mathematically, if $N(A)$ = number of events in an arbitrary area A, and the probability of $N(A)$ being n is

$$P_n(A) = P\{N(A) = n\} ,$$

then for a constant $\lambda > 0$, called the density (stocking) of the process, for any small area δA,

$$P_n(\delta A) = \left\{ \begin{array}{ll} 1 - \lambda\delta A + o(\delta A) & \text{for } n = 0 , \\ \lambda\delta A + o(\delta A) & \text{for } n = 1 , \\ o(\lambda A) & \text{for } n = 2, 3, \ldots \end{array} \right\} \qquad (8.46)$$

for any two non-intersecting regions A and B; the random variables $N(A)$ and $N(B)$ are statistically independent. Thus

$$P_n(A) = e^{-\lambda A}(\lambda A)^n/n! , \qquad n = 0, 1, \ldots \qquad (8.47)$$

that is, $N(A)$ has a Poisson distribution with mean equal to the density λ times area.

The Poisson forest is simple, which is appealing for theoretical work aimed at developing inferential tools and studying ecological data. It may be a useful approximation initially in a uniform regenerated forest where direct aerial seeding was used. However, it is very limited in explaining most natural phenomena. Hence more plausible models are necessary. These are discussed in the following sections.

Poisson Cluster Processes. Poisson cluster processes are generalizations of the Poisson forest described above. Randomly distributed parents generate random numbers of offspring, spatially distributed around the parent. There are three postulates to the Poisson cluster process:

1. Parent occurrences constitute a Poisson forest with density ρ per unit area.
2. Each parent independently produces a random number N of offspring, with corresponding probability distribution $\{P_n, n = 0, 1, \ldots\}$.
3. Each offspring is, independently, spatially distributed relative to the corresponding parent, with vector displacement relative to the parent governed by a bivariate distribution function $H(\mathbf{x})$ or probability density function (pdf) $h\mathbf{x}$ if the latter exists.

$H(\mathbf{x})$ is usually specified to be radially symmetric, in which case the process is both stationary and isotropic with density $\lambda = \rho E(N)$. The degenerate case where each offspring occupies the same location as its parent gives rise to various generalized Poisson or contagious quadrat count distributions. If N follows a Poisson distribution, the Neyman type A distribution results. If N follows a logarithmic series distribution, the negative binomial distribution results (Johnson and Kotz 1969).

The assumption of randomly distributed point clusters implicit in the use of contagious distributions has not always been recognized, and the fitting of such distributions to quadrat count data is an empirical device that may provide a useful description (Diggle 1979). Quadrat count distributions for nondegenerate cases appear to be very difficult to work with, although in principle the necessary information is available in the form of the characteristic function of a Poisson cluster process (Bartlett 1975, Ch. 1). Analytical results are more likely to be obtained for distributions based on distance sampling methods.

Heterogeneity Processes. Extending the Poisson forest by replacing the previously constant density by a realization of a random function $\lambda(x)$, a class of heterogeneous or doubly stochastic Poisson processes is obtained. Diggle (1979) argued that no method of analysis, based solely on a spatial point pattern, can preserve a formal distinction between true and apparent contagion for distributions such as the negative binomial. He suggested that the meaning of pure heterogeneity is both intuitively clear and distinct from clustering and that it may be useful to incorporate this into the construction of empirical models.

Lattice-Based Processes. Poisson cluster processes, and the even more complicated processes called heterogeneous or doubly stochastic Poisson processes lead to patterns that may be described as aggregated since they comprise a combination of relatively dense and relatively sparse subregions (Diggle 1979). Lattice-based procedures constrain the occurrences to be spaced out more or less regularly over the region under consideration. Extreme forms of regularity are produced by deterministic lattice structures, usually either squares or equilateral triangles. A more flexible model for regular patterns is desirable and may be obtained if a Poisson forest with density ρ per unit area is imposed upon the lattice structure. Patterns range from $\rho = 0$ for extreme regularity to $\rho \to \infty$ for a Poisson forest. Diggle (1975) gave the distribution function $F(x) = 1 - e^{-\pi\rho x^2}\{1 - G(x)\}$ for an arbitrary point-to-event (occasion) nearest neighbor distance where $G(x)$ is the distribution function of the distance from an arbitrary point to the nearest lattice node. Persson (1964) gave $G(x)$ for the square lattice and Holgate (1965) for the equilateral triangular case. Brown and Holgate (1974) discussed a basic square lattice, where random thinning is assumed to occur, to represent a thinned plantation.

The lattice-based process is primarily a device to obtain mathematical tractability and may have limited practical utility except as a yardstick to assess inferential procedures applied to regular patterns. Cormack (1979) discussed lattice-based processes more fully.

Inhibition and Markov Point Processes. Inhibition is assumed so that no two events may occur less than some prescribed distance δ apart. This is a simple inhibition process if, apart from this constraint, the events are located completely at random. Of course, the degree of regularity for a given area A and δ increases with n and fails for very large n. This is a special case of the Markov point process. If there is a finite region A, where B is any subset of A, the environment of B is defined as the set of all points not in B that are at most a distance δ from some point in B for a specific $\delta > 0$. Then a point process is Markov if the distributional properties of the configuration of events in B, given the configuration in the complementary region $A|B$, depend only on the configuration in the environment of B. Simple inhibition processes are the special case where x contains no events. Markov point processes represent a very flexible class of models since they can produce both regular (inhibitory) and aggregated (contagious) patterns. They have been used primarily in the study of regular point patterns. As such, they arise naturally in the study of competitive interactions between individuals (Diggle 1979).

Descriptive Distributions. Two examples of descriptive spatial distributions are discussed below but research in this area does not appear to have yielded practically useful results so far.

Daniels (1978) examined spatial distributions for the squared distance between points in 40 seeded loblolly pine stands 5–12 years old. Specifically, he considered the Poisson and Pearson type XI distributions. For the latter, he considered a special case, the negative binomial distribution with density

$$f(w) = \frac{c}{(1 + cw/k)^{k+1}}, \qquad w > 0, \qquad k > 0, \qquad c > 0 \qquad (8.48)$$

where

w = squared point-to-plant distance,
k = heterogeneity parameter, and
c = density parameter = density in stems/circle of 0.3 m (1-ft) radius.

The stands had been generated by several methods, namely, seed tree/shelterwood system, natural old field regeneration, aerial broadcast seeding, and manual seeding. In each stand 50 random points were located and the distance from each point to the nearest loblolly pine tree was measured. The stocking (c = stems/unit area) was estimated using 3–5 randomly located 0.02–0.04 ha (0.05–0.10-acre) plots in each stand, where c was treated as known without error in each stand. Stand areas were not indicated.

The negative binomial distribution was rejected ($P = .01$) as a satisfactory fit in only 5 out of 40 stands and the Poisson distribution in 22 out of 40 stands. The five stands where the negative binomial did not give a good fit showed peculiarities that may have caused rejection, such as extremes in density, dense hardwood components, and large open spaces.

Cressie (1991) fully discussed analyses of the spatial point pattern of an old-growth longleaf pine (*Pinus palustris*) forest in Georgia. The data consisted of the coordinates and DBH of all longleaf pines at least 2 cm in diameter in a 4-ha forest. Cressie concluded that there was clearly not a completely random spatial distribution but that a Poisson cluster process also did not describe the data well. In his opinion, dynamic modeling of the birth process, the death process, and the growth process need to be incorporated to describe the spatial pattern fully. We think it unlikely that such data will be available for a long time since it would require near-continuous monitoring of forests.

These examples illustrate how difficult it can be to find completely satisfactory spatial distributions.

8.5 REMOTE SENSING IN RESOURCE INVENTORIES

Remote sensing is the science and art of obtaining information about objects, areas such as land use and cover, and phenomena under investigation through analysis of data acquired by some device not in contact with the object, area, or phenomenon under investigation (Lillesand and Kiefer 1987). The utility of the remote sensing data depends on the type of sensor and the part of the electromagnetic spectrum (EMS) used, the quality of the data recorded and processed, and environmental limitations such as platform altitude, topography, variations in solar angle and altitude, and atmospheric interference (Aldrich 1979).

Numerous advantages may be realized by using remotely sensed data instead of or in addition to ground-based data. Remotely sensed data can provide a synoptic view of the study area, be collected quickly over large areas, provide information about land cover in the nonvisible portions of the EMS, and be acquired and processed digitally. Such data is used to determine and identify land cover and to infer physical characteristics of cover types. Remote sensing uses both photographic and nonphotographic data acquisition systems including cameras and multispectral scanners that sense either microwave, radar, thermal infrared (thermal IR), or reflected light. The most commonly used satellite sensors are the thematic mapper (TM), the French system probatoire d'observation de la terre (SPOT), and the advanced very high resolution radiometer (AVHRR). Data may be obtained by manual interpretation, digital data processing, or a combination of the two.

A brief description of the two main types of remote sensing systems, photographic and nonphotographic, is given. Applications of remotely sensed imagery are then discussed, and geographic information systems (GIS) are briefly discussed.

8.5.1 Photographic Sensing Systems

Advantages of aerial photographs over on-the-ground observation include improved vantage point, freeze action, permanent recording, broadened spectral sensitivity, and increased spatial resolution and geometric fidelity (Lillesand and Kiefer 1987). Photographic sensors include both camera and films. Image resolution is usually restricted by the film rather than by the camera optics. Films available for aerial photography include panchromatic black and white (BW), infrared black and white (IR), color, and color infrared (CIR). BW and color films are generally sensitive to the visible portion of the EMS from 0.4 μm (blue) to 0.7 μm (red). IR and CIR films are also sensitive to the visible portion of the EMS, but the sensitivity extends to about 0.9 μm to include some reflected IR. IR and CIR films in combination with a minus-blue (W-12) filter, which eliminates blue and blue-green wavelengths, improve haze penetration and overall image clarity. Using IR film with a Wratten 25A (red) or 89B (dark red) filter gives a lighter tone for broadleaved vegetation and a darker tone with water and moist soil. Heller (1970), Avery and Berlin (1985), and Lillesand and Kiefer (1987) gave detailed descriptions of the theory of imaging with photographic sensors.

8.5.2 Non-Photographic Imaging Systems.

These sensors operate in portions of the EMS from ultraviolet (UV) through the microwave regions. They include optical-mechanical scanners and pushbroom radiometers for multispectral sensing, microwave sensors, and aerial color video systems. Lillesand and Kiefer (1987) provided a good detailed discussion of some of these topics.

Multispectral Scanners. These systems observe the scene under an aircraft or satellite in a number of discrete bands of the EMS. A series of detectors measure reflected energy that may include UV (0.28–0.40 μm), visible (0.4–0.7 μm), near IR (0.7-1.3 μm), medium IR (1.3–3.0 μm), and thermal IR (3.0–13.8 μm) portions of the EMS. Each detector senses the same element of the scene but at different wavelengths. Output signals can be combined to determine the spectral distribution of radiation from each element. Such multispectral data is registered to a coordinate system, easily recorded on photographic film, and directly compatible with computer processing systems (Aldrich 1979).

On July 22, 1972, the National Aeronautical Space Agency (NASA) of the USA launched ERTS 1, later renamed Landsat 1, which carried the first generation of multispectral sensors. The multispectral scanner sensor (MSS) is a scanning radiometer monitoring four spectral bands with approximately 80-m resolution. The MSS was the primary sensor in Landsats 1–3 for agricultural applications (Freden and Gordon 1983). Although it proved useful for resource inventories, it lacked both the spatial and spectral resolution needed for detailed information.

The thematic mapper (TM), the second generation of multispectral sensors, is the primary sensor carried by Landsats 4 and 5. Relative to MSS, TM has improved spatial resolution, additional spectral bands, greater radiometric accuracy and improved geometric fidelity (Hoffer 1988). The spatial resolution is 30 m for six of the seven bands used. Band 6, the thermal IR band, has a spatial resolution of 120 m. The section of the EMS comprising bands 1–6 was chosen primarily for detecting vegetation. Band 7 was added for geological studies. Freden and Gordon (1983) give a very detailed description of the background and capabilities of Landsats 1–4.

Like Landsat MSS data, TM data is available on computer tape for computer processing. It is also easily transferred onto photographic film.

Advanced very high resolution radiometer (AVHRR) sensors have been used regularly since October 13, 1978 by the US National Oceanic and Atmospheric Administration (NOAA) on their polar orbiting meteorological satellites. The multispectral radiometer was designed to produce daily high-spatial-resolution images of regional cloud patterns for weather forecasting. The bands used were designed to discriminate between clouds, water, ice, snow, and land. One band was modified on the operational sensor prior to NOAA-6, which not only allowed discrimination of snow cover (its original intent) but also allowed observations suitable for vegetation studies (Goward *et al.* 1987).

System probatoire d'observation de la terre (SPOT) was launched by the French government in February 1986. A detailed description of the system is given in Courtois and Weill (1985). SPOT carries two identical sensors operating in the visible and IR parts of the spectrum. Three bands of the EMS are recorded with 20-m resolution and a panchromatic band (0.51–0.73 μm) with 10-m resolution. Although this resolution betters that of TM, TM contains more spectral information (Chavez and Bowell 1988). Since SPOT has better spatial resolution and TM has better spectral resolution, the two systems are complementary. The designers of several studies have used data derived from both systems to improve the results (Welch and Ehlers 1987).

Tables 8.1 and 8.2, adapted from Hoffer (1988) and Kidwell (1991), list various satellite sensor systems and many of their characteristics.

Microwave Sensors. Microwave sensors are of two types, active and passive. Active sensors transmit a known signal, such as radar which reflects off the target, and measure the returned signal. The measured response is a function of the power of the transmitter, size of the antenna used, wavelength of the transmitted energy, scattering properties of the target, angle of incidence, width of the signal, and the distance to the target. Measurement is accomplished by converting the returned signal (energy) into a spot of light that is regulated according to its strength to produce a black and white line scan image. Bright and dark tones indicate strong and weak return signals, respectively. Active microwave sensors include radar imagers, scatterometers, and altimeters (Ulaby *et al.* 1986).

Passive microwave sensors, also called microwave radiometers, receive signals emitted by the target itself. Because the signals are often quite weak, the sensors must be extremely sensitive and have little internal noise. Jackson and Schmugge (1991) noted that soil moisture information can be retrieved by passive microwave remote sensing even if vegetation canopy is present. The influence of the vegetation is a function of its amount and type and also of the wavelength of the sensor system. Therefore, corrections need to be made for the effect of the vegetation.

The ground resolution of active microwave sensors is generally poorer than that of visible IR and thermal scanners because it is directly proportional to wavelength and the shortest microwavelength used is 100 times larger than that used for thermal IR scanning. However, microwave sensors can penetrate cloud cover and some types of rain and they can operate day and night. Their response parameters are functions of frequency or wavelength, polarization, and viewing angle. Viewing angles other than straight vertical are usually best (Aldrich 1979). Variables affecting active and passive microwave sensors may be complementary since they represent different, interrelated aspects of the target of interest. This feature may

TABLE 8.1 Comparison of Satellite Sensor Systems[a]

	Satellite		
	Landsats 1, 2, 3	Landsat 4, 5	SPOT 1
Date(s) launched	July 23, 1972 July 21, 1975 March 5, 1978	July 16, 1982 March 1, 1984	Febuary 21, 1986
Altitude (km)	920	705	832
Cycle for complete coverage of the earth (days)	18	16	26[b]
Equatorial crossing time	8:50, 9:08; 9:31 am	9:45 am	10:30 am
Currently functional?	No	Yes	Yes

	Sensor			
	MSS	TM	Multispectral (high resolution visible) (HRV)	Panchromatic HRV
Ground resolution (m)	80	30	20	10
Scene dimension (NS × EW) (km)	170 × 185	170 × 185	60 × 60	60 × 60
Wavelength region and bands (μm)				
Visible	0.5–0.6 0.6–0.7	0.45–0.52 0.52–0.60 0.63–0.69	0.50–0.59 0.61–0.68	0.50–0.73
Near IR (reflective)	0.7–0.8 0.8–1.1	0.76–0.90	0.79–0.89	
Middle IR (reflective)		1.55–1.75 2.08–2.35		
Thermal IR (emissive)		10.4–12.5[c]		

[a] Adapted from Hoffer (1988).
[b] Sensing equipment on SPOT can be "pointed" at adjacent ground tracks, thereby providing a potential for obtaining data from a particular location several times per week.
[c] Ground resolution for the thermal IR wavelength band is 120 m.

TABLE 8.2 AVHRR Sensor Systems and their Characteristics[a]

	TIROS-N	NOAA-6	NOAA-7	NOAA-8	NOAA-9	NOAA-10	NOAA-11	NOAA-12
Launch date	10-13-78	6-27-79	6-23-91	3-28-83	12-12-84	9-24-88	5-14-91	
Altitude	850 km (833–870 km) →							
Cycle	9 days →							
Equatorial Crossing time (hrs)	1500	0730	1430	0730	1420	0730	1340	0730
Currently Functional	No	No	No	No	No	Yes	Yes	Yes
Ground resolution	From 1.1 km at nadir to 3.5 km maximum off nadir							
Scene dimension	Various. Cross track: Up to 2700 km Along track: Up to 4500 km							

[a]Adapted from Kidwell (1991).

allow the collection of data to model the growth and net productivity of vegetation (Daughtry *et al.* 1991).

Considerable research is devoted to microwave sensors as evidenced in the literature. However applications in forestry seem limited at this time. Ulaby *et al.* (1986) gave a brief survey of radar-imaging applications of land cover, classifying crops, and mapping the distribution of soil moisture.

Aerial Color Video Systems. Many types of video cameras, mostly shuttered and in BW or color, can be used for remote sensing. These cameras are easily mounted on a variety of aircraft for either vertical or oblique sensing.

Airborne video imaging is still in an experimental stage. It has weaker spatial resolution than aerial photography but often has better spectral resolution. Images can be digitized easily for computer-aided interpretation. Because the video tape does not need to be developed, it can be used immediately (Ahern and Leckie 1987).

Video systems for remote sensing have several attractive features (Everitt 1988), namely the near-real-time availability of imagery; higher light sensitivity than conventional film cameras so that imaging in narrow spectral bands can be used, for example, in the infrared band; and immediate availability of the electronic signals for digital processing and computer analysis. In addition, video equipment is portable, versatile, easy to use, and can tolerate different light conditions. Such systems are cheaper to operate than photographic systems and the operator can view the imagery on a monitor in the airplane at the time of acquisition and can adjust the exposure settings interactively. Also, comments can be recorded in flight and the high rate of picture acquisition (30 frames/sec) provides extra data.

Disadvantages of video imagery are: its low spatial resolution relative to film; the difficulty of obtaining hard copy images from the data; practical limits on field-of-view because of the small tape format; difficulty in calibrating the cameras because of the automatic exposure control; and vignetting problems with many near-IR video sensors because the camera optics were not designed for that wavelength band (Everitt 1988).

The value of common color video systems for natural resource and agricultural applications is limited because of the difficulty of extracting discrete spectral data from a composite video signal and the lack of spectral bands outside the visible region. Solutions to these difficulties are being sought currently.

The main data sets of nonphotographic images currently comprise those collected by Landsats 1 to 5, SPOT, and AVHRR. This data is available on computer-compatible media and in electronically reconstituted photographs. It is analyzed using both computer-aided and conventional photointerpretation methods. Generally, computer-aided classification using digital data is best suited for classification and mapping. The spectral data, which can indicate bio-geophysical properties of the target, consists of relative values and is affected by atmospheric conditions, but it is system corrected when recorded. Photographic products can be obtained from the recorded data but they may be degraded somewhat in the photographic process.

8.5.3 Spatial Location Systems

A feature critical to the success of remote sensing in forest inventory is having available the technology for spatial location. Great advances are being made in this field. Radio navigation (LORAN), the satellite Global Positioning System (GPS), analytical aerotriangulation, and inertial surveying systems have considerable potential. GPS is particularly promising since the use of markers is not necessary and measurements can be made both day and night regardless of the weather. However, GPS receivers require a clear line of sight to the satellites so discerning location under a dense forest canopy is difficult. GPS can provide positioning accuracies of one part per million or better over distances from a few hundred meters to thousands of kilometers. Using these technologies, the location of aircraft and photo plots can be established rapidly and accurately (Duggin *et al.* 1990). For example, GPS technology has been used to great advantage by the Department of Conservation and Land Management in West Australia, where it forms the basis of a sophisticated navigation and flight recording system controlling the large scale aerial photography associated with a major forest inventory of 1.3-million ha of the state's jarrah forests (Biggs *et al.* 1989, Biggs and Spencer 1990). At the other extreme of sophistication, Wilkie (1989) successfully used a battery-powered, solar-recharged, backpack-mounted GPS receiver to obtain accurate location data in an inaccessible rain forest region in Zaire.

8.5.4 Applications

Aerial photos were first used for forest inventory in Canada in the late 1920's (Heller and Ulliman 1983) and in the USA in the 1930's (Duggin *et al.* 1990). Use increased greatly in the late 1940's partially due to technological developments during World War II and publication of texts on aerial photography such as Spurr (1948). Applications of aerial photography can be divided into three categories: classification and mapping; measurements of resource parameters; and disturbances and change over time.

Classification and Mapping. Homogeneous areas of cover are delineated following observation and interpretation by drawing boundary lines or by point sampling, listing the points by cover type. Land is classified by vegetative cover, land use, and land form.

To be useful for classification and mapping, the best film and scale of photography are dictated by the level of detail wanted in the hierarchy of land cover classification and/or by the degree of detail and accuracy wanted in the mapping.

Successful interpretation of vegetative cover at any scale requires proper timing of the photography to take advantage of phenological developments. In the northern temperate zone, hardwood/conifer discrimination is easiest on CIR film in early spring before leaf flush of the hardwoods or in the fall when the deciduous leaves reflect IR better. The disadvantage of the latter is that this may only be feasible in a rather short period of say 6 weeks and one risks not getting acceptable imagery due to winter conditions such as snow. For high-altitude CIR film, late fall to early spring is the best time to differentiate forest types. Leafless conditions make it possible to differentiate between upland and bottomland deciduous types of trees. Midsummer photographs show little spectral discrimination between hardwood species because reflectance from foliage decreases during the early weeks of the growing season, remains fairly constant until after the middle of summer, and increases rapidly when colors change in the autumn. Autumn is a good time to accentuate the difference between hardwoods and conifers for type mapping because variation in reflectance from deciduous trees is minimized and conifers reflect significantly less energy at the end of the growing season (Aldrich 1979).

Medium-scale photography (1:12 000–1:25 000) can be used to stratify a forested area by forest type and volume and large-scale (1:500 to 1:12 000 with 1:3000 to 1:10 000 being most common) to identify tree species, shrubs, some grasses and forbs, and density and condition classes. Normal color and CIR films are needed to separate individual plant species, normal color generally being best for tree species (Aldrich 1979).

Large-scale aerial photographs can be easily acquired using small format cameras, usually 70-mm focal length. The exact scale of the photographs is not important since detailed information, not precise measurement, is required. Such applications, however, cover very limited areas and a large number of photos is required for complete coverage (Duggin *et al.* 1990). Two recent examples of work involving large-scale aerial photography are discussed below.

Befort (1986) evaluated the utility of 1:1000 color aerial photography in assessing the habitat type to which a forest site belongs by interpreting vegetation in northern Idaho and eastern Washington. Befort defined a habitat type as the set of all sites that produce similar plant communities at climax. In his study, five interpreters assigned 16 field-identified habitat types to 111 stereophoto sample strips using a type-identification key. The average success rate was approximately 75% but occasional extreme misclassification occurred. Interpretations were highly correlated with positions of the habitat types along a bioclimatic gradient ($r = 0.90$). Considering that the test area was very large, no interpreter was familiar with the whole area, interpreters lacked experience in this method of photo-interpretation, and no comparisons, collations, or corrective feedback on the interpretations were permitted, better results are achievable.

Needham and Smith (1987) assessed the accuracy with which species (planted and naturally occurring) in two loblolly pine plantations 17–18 years old could be classified using large-scale (approximately 1:2750) 35 mm aerial photos of sample plots. 95% of the trees were correctly labeled as loblolly pine, Virginia pine, or hardwood, the accuracy increasing to 99.8 percent when only dominant and codominant trees were considered. However, the results should be treated with some caution because Virginia pine and hardwoods only constituted a small portion of the total stand. Needham and Smith concluded that if information on dominant and codominant trees is all that is required, aerial photos of this type may be quite useful, but that a ground survey is needed if information on intermediate or suppressed trees is also desired.

Mapping soils directly from photographs has had only some success, primarily in areas with limited vegetative cover (Aldrich 1979). Maximum variation in reflectance of bare soil occurs in the 0.6–0.7 mm (red) portion of the EMS. IR photographs give increased contrast between wet and dry soils and are superior for mapping drainage patterns. A combination of normal color and IR film is best for gathering soil-related information because the rate and quality of mapping are improved.

In Canada, 1:160 000 CIR photography is used for broad forest classification and mapping. Differences between coniferous and deciduous species on this medium are striking and some species associations and even individual species can be separated. Forest type maps of 1:15 480 scale have been revised based on interpretation of this small-scale photography (Aldrich 1979).

Skylab 190 B color photographs and high altitude CIR photographs can be used to classify range plant communities to the regional level of Daubenmire's (1952) ECOCLASS; for example, classification accuracies of 90% for grassland and conifer and 80% for aspen have been achieved. However, compared to grassland, the success of classifying vegetation depended more on date of photography, scale, and film type (Aldrich 1979). Aldrich also noted that with Skylab 190 B photographs, ten vegetation and land use categories in wetland areas were classified usually with 75–99% accuracy. Forest, wetlands, water, and agricultural lands were classified almost without error, and three saltmarsh grass species, bare soil, sand, cropland, bare mud, and water were separated successfully. In another study using Skylab 190 A photographs, wetland types were classified, freshwater marshes were delineated, and detailed analyses of drainage patterns were made (Aldrich 1979).

Computer-assisted analysis of single-date and multidate MSS data provides fairly accurate and reliable small-scale maps and areas of forest land and can assist in differentiating between deciduous and coniferous cover types. Individual forest cover types can only be identified and delineated sometimes (Hoffer 1986).

Duggin *et al.* (1990), reporting on some studies in the literature (e.g., Anderson *et al.* 1976), noted that MSS data can be used to distinguish forest and other level I land cover classes with accuracy of almost 90%, level II forest classes (basically hardwood vs. softwood) with accuracies from 70 to 85%, and level III forest type classes (species groupings) inconsistently.

TM's improved spatial and spectral resolution over MSS has greatly improved the ability to distinguish vegetation classes. Ekstrand (1986), cited in Ahern *et al.* (1990), was able to map eight vegetation classes with TM in Ethiopia but only four with MSS. He achieved a classification accuracy of 80% with both systems. Moore and Bauer (1990) observed that classifying TM data resulted in a 7–15% absolute increase in the accuracy of forest cover classification relative to MSS data for a state park in Minnesota. Classification accuracies ranged from 26 to 86% depending on the sensor, number of classes, and method of assessment used. The increases in classification accuracy were due more to the higher spectral resolution of the TM imagery relative to MSS than to its greater spatial resolution. TM has also been used to delineate burned areas of forest land, areas of significant tree mortality resulting from insect attack, and areas of regeneration (Ahern *et al.* 1990).

Gervin *et al.* (1985) compared the relative accuracy of classifying level I land cover using MSS and AVHRR data. They found overall accuracies of 77% and 72%, respectively, using unsupervised classifications. AVHRR did as well as MSS in classifying large homogeneous areas but MSS did better for less common and/or spatially heterogeneous conditions. Teuber (1990) compared AVHRR derived estimates of forest area with FIA estimates. The AVHRR estimates were within 1% of the FIA estimates for Louisiana but errors were significantly greater for Arkansas and Mississippi (4.3 and 2.2%, respectively). These discrepancies may be due to the fact that the images of Louisiana were very clear whereas those of the other two states were less so due to small amounts of haze and cloud cover. Evans *et al.* (1992) mapped Mexico's forest lands with AVHRR. Using a supervised training and classification technique, they obtained an accuracy of 78% for the 60 test polygons used.

Measurements of Resource Parameters. Such measurements include distances, heights, numbers, areas, or other expressions of quantitative continuous variables describing basic resources.

Very-large-scale aerial photos (1:500 to 1:1500) may provide accurate estimates of the size and volume of individual trees and stands. This type of photography, called controlled-scale photography by several authors (e.g., Duggin *et al.* 1990), is used extensively in Canada for forest inventory because the forests are often inaccessible from the ground. Although high-quality data can now be acquired because of significant improvements in altimetry, photogrammetry, and location positioning systems, the number of operational surveys using very-large-scale photography is still small because of a general lack of awareness of the capabilities of such photography, doubts about its accuracy, and the high costs for equipment and training (Spencer and Hall 1988).

Aerial photo volume tables were developed in the USA during the 1950's and 1960's for timber inventories. Such tables require photo measurement of tree or stand height, crown size, and percent crown cover of stands on conventional 1:10 000 to 1:20 000 scale aerial photography. With these tables, reliable estimates of net stand volume are achievable. However, developing the tables is costly, no estimate of the distribution of trees by volume, species, or diameter class is obtained, and photo height measurements are time consuming, frequently biased, and require well-trained interpreters (Paine and McCadden 1988). Paine and McCadden suggested that these problems could be overcome by making individual tree measurements on large-scale 70 mm photography (scales 1:1000 to 1:4000) and using tariff volume tables, which do not require measurement of height. In one approach, they discussed the requirements of accurate knowledge of photo scale at each plot location (obtained by using either a laser or radar altimeter) and adjustments for tilt using a tilt indicator. A second approach uses twin cameras mounted a fixed distance apart on a low-flying plane, which are triggered simultaneously to acquire a stereoscopic photo pair for each plot (Paine and McCadden 1988). Using a low-cost modification of the first approach, Paine and McCadden (1988) obtained estimates of mean volume/unit area within 5% of that obtained by an intensive ground inventory in stands consisting predominantly of Douglas fir.

Needham and Smith (1987) assessed the accuracy with which stocking (number of trees per unit area) could be estimated in 17- and 18-year-old loblolly pine plantations. The total number of loblolly pines was overestimated by 1.4% and few interpretative errors occurred in the dominant and codominant crown classes. Trees were more often missed than falsely counted, but the authors do not attribute this to the format of the photography.

It is conventional in inventories of range lands to estimate the percent cover of grasses, forbs, and shrubs. Large-scale aerial color and CIR photographs can be used for this purpose with distinctive species, but for most species the stocking cannot be estimated (Aldrich 1979).

Aerial photos can be used advantageously for sampling both recreation and wildlife because information is captured instantaneously on film. Even at 1:120 000 scales, ski slopes, parking lots, boats, marinas, and the wakes of boats can be detected, but accurate counts of vehicles and people require much larger scales. People counts require at least 1:2000 scale or 0.1-m resolution (Aldrich 1979).

For animal surveys, the best time for photography is a few hours after sunrise and a few hours before sunset when animals are not seeking shade. BW film is probably most acceptable considering its cost and effectiveness, especially when a W-12 filter is used to increase contrast and haze penetration and darken the shadows, but color transparent film is more reliable, especially for identifying animal species. Seber (1992) cited several applications of using photos in animal surveys such as identifying individual whales and dolphins, estimating number of harp seal pups, and so on.

Lee *et al.* (1990) presented an example of mapping forest and estimating timber volume using Landsat 5 TM imagery in the first stage of a two-stage inventory of pine and eucalypt plantations and native forest. The second stage was ground sampling. The area of eucalypt plantations was overestimated by 15% and that of pine plantations underestimated by 18%. Although they met their precision targets for volume, Lee *et al.* recommended that future plantations be stratified by species or groups of species, site, and age.

The use of airborne multispectral scanner imagery in the 1980's has had limited success in surveys, but as computer costs are decreasing and computer power is increasing dramatically, digital information is becoming easier to use. High-resolution multispectral data has been used successfully to classify single trees, and stereo imagery acquired by such systems as the multidetector electro-optical imaging scanner has been used successfully to classify forest stands into height classes (Ahern and Leckie 1987).

SPOT data can be used to derive fairly accurate estimates of the relative height of the ground surface, for example, Gugan and Dowman (1988) used it to map an area at a scale of 1:50 000 with 25-m contours. They concluded that SPOT is a realistic alternative to very-high-altitude aerial photography for mapping at small and medium scales.

Disturbances and Change Over Time. To detect and map disturbances to vegetative cover, reflectance patterns need to be studied carefully. Stresses such as water deficiency caused by insects, disease, and other agents can be detected and mapped on photographs. CIR and color photography must be used to distinguish subtle color differences between healthy and stressed trees.

To detect insect damage, the ground resolution required determines the photographic scale needed. For example, white pine weevils affect only the terminal growth of seedlings and saplings, and this damage can only be reliably assessed on 1:600 scale photography (Aldrich 1979). Color or CIR film with scales 1:8000 or larger can be used for detecting individual coniferous trees killed by bark beetles (Heller 1974). Heller also suggested that if small infestations of one or two trees can be ignored, a scale of 1 : 32 000 using CIR film is efficient. Neither color nor CIR photography is able to detect infestation by bark beetles before symptoms are visible to the human eye.

Aerial photographs are not yet widely used to detect conifer or hardwood defoliators because these pests are not as serious as bark beetles. Video surveys from an aircraft may be more effective in this case mostly because of low cost and ability to operate over a wide range of light conditions. However, color photographs (1:1600) could be used to monitor defoliation by spruce budworm and its effects over time (Aldrich 1979). Photographs taken annually over a 10-year period of spruce-fir plots affected by spruce budworm showed the different levels of damage and were helpful in establishing trends of mortality in the overstory and understory vegetation

(Aldrich and Heller 1969). Large-scale 70 mm color and CIR photography can be used to estimate tree loss and monitor change.

Forest diseases are difficult to detect on aerial photographs because visible symptoms show up slowly. They tend to be patchy over a stand and are usually more subtle than those of insect damage. Dwarf mistletoe, Dutch elm disease, oak wilt, basal canker of white pine, ash dieback, *Fomes annosus*, sulfur dioxide damage, and ozone damage are all detectable to some degree. Scales of 1:1600 and larger of color and CIR film are needed for assessing damage and CIR film of 1:8000–1:16 000 to delineate the boundaries of the disease. 70 mm color and CIR photography can be used in sampling susceptible vegetation types for assessing damage (Aldrich 1979). Inventories of forest decline in both North America and Europe have included the use of CIR photos for acquiring tree-level data (Ciesla 1989). Such photos serve as a permanent record of tree and stand conditions, a useful starting point in monitoring change. These inventories require that information on distribution of vegetation types, damage classes with vegetation types, and tree condition can be obtained from the photos. This in turn requires that the photo interpreters be sufficiently skilled to identify tree species as well as a number of distinct damage types. Ciesla thought that tree-level data can be obtained easily and efficiently from large-and medium-scale CIR aerial photos.

Windstorm, flood, fire, and human-caused disturbances are fairly easy to detect on conventional aerial photography. Detecting and measuring change require photography for both a base year and the current year. Problems faced include: noncoincidence of consecutive flights, differences in shadow direction on consecutive flights, presence of clouds on the photography (can be confused with change), seasonal differences in vegetation, and differences due to film sizes, types, and scales (Aldrich 1979).

Nonvegetative land cover includes surface water and soils. CIR and IR films are preferred for mapping the presence of water which absorbs IR radiation and appears black on IR film and various shades of blue on CIR film, the shade depending on water depth, sedimentation, and pollutants (Aldrich 1979). A useful example of monitoring change using aerial photography was described by Frayer *et al.* (1983). They examined change in wetlands and deepwater habitats of the continental USA from the 1950's to the 1970's based on a stratified random sample using 200 defined geographical areas. A total of 3635 square sample units, each 1036 ha (4 miles^2) in area, was selected at random from the strata proportional to the expected area of wetland and deepwater habitat in each. The photography was BW and primarily 1 : 20 000 (1950's) and 1 : 40 000 (1970's). Sample units in the earlier photographs were interpreted completely but only change was interpreted in the 1970's. All changes in wetland and deepwater habitat were identified as natural or human-induced. Subsequently, Frayer *et al.* (1983) documented major net losses of wetlands in the USA and where they occurred.

There is no evidence that data derived from MSS is useful for insect and disease surveys (Aldrich 1979). Recent work has demonstrated the potential for surveying insect damage with airborne scanners but such problems as the need for increased angular swath width, improved geometric and radiometric corrections, and integration of the data acquired into a reliable and cost effective acquisition and analysis system need to be solved. Spatial resolution is an important criterion in classifying digital data (Ahern and Leckie 1987).

The processes involved in computer-aided classification of vegetation using Landsat MSS digital data were reviewed by Hoffer (1986). He indicated that although land cover classification may be improved when two or more Landsat image dates are combined, not all investigators have found this to be true. Multidate images must be accurately registered pixel to pixel, preferably within 0.5 pixels, or many changes will be missed and false changes will also be observed. Burnt areas, recent clear cuts, and other significant changes in the vegetation can be detected and mapped fairly well.

Zhu and Teuber (1991) used AVHRR for assessing change in the forest area of Alabama from 1982 to 1990. They found a 70% agreement among 5800 FIA survey plots and the corresponding classified AVHRR pixels.

Singh (1989) reviewed digital techniques of detecting change. He noted that an evaluation of results from several procedures produce different patterns of change even under the same conditions. He attributed this to differences in illumination, atmospheric conditions, sensor calibrations, ground moisture conditions, and registration of the images as well as to actual change (the signal). Obviously, models have to be developed to separate out the signal from the temporal variation.

Häme and Rantasuo (1988) studied the use of video imaging equipment to assess insect damage, and for general mapping applications in forest management. The study was not successful because of poor spatial resolution of the video images. They concluded that longer focal length lenses or lower altitude imaging is required than was used in the study, and the video equipment needs to be improved further.

Everitt *et al.* (1990) used video equipment to identify heavy grazing, drought stress, phytomass levels, burned areas, soil salinity, plant species, and gopher infested areas in range land. The technique proved to be inexpensive, quick, and reliable at low altitudes.

Video images appear to present promising alternatives to aerial photography. Aerial photography is still more useful for forest resource inventories than satellite imagery because of the unsatisfactory resolution of the latter. Several studies of species classification have shown that MSS and even TM data provide information on species that is too general for forest inventory requirements (Ahern and Leckie 1987). Given that new satellites are planned by the USA, Japan, and some European countries, several carrying sensors with higher spatial, spectral, and radiometric resolution (Szekielda 1986), the future of satellite data for use in forest inventories looks promising.

8.5.4 Geographic Information Systems (GIS)

Geographic information systems (GIS) derive naturally from remote sensing. They are computerized systems that can play a critical role in inventory, in manipulating and processing data, and in assessing land use and land cover. According to Avery and Berlin (1985), such systems should have the following capabilities:

1. Several formats for data input, such as analog maps and overlay information, tabulations, and digital-image data.
2. Ability to store and maintain information with the necessary spatial relationships.
3. Ability to manipulate data, as in search and retrieval, and to do computations efficiently on the data.
4. Levels of modeling taking into account data interrelationships and cause-and-effect responses of the relevant factors.
5. Ability to present tabulations, video displays, and computer generated maps of data outputs.

Ideally GIS could serve a variety of users and be updated continuously as new data becomes available. Hence, a good GIS depends on high-speed computers, a variety of peripheral input-output devices, and powerful software.

The field of GIS system development is in tremendous flux. A good review of GIS systems and their availability was given by Burrough (1986). There appears to be danger that the use of GIS systems is being oversold as was the utility of satellite imagery 5–10 years ago.

8.6 PROBLEMS

1. What is the difference between a closed and an open population in wildlife sampling?

2. What are the key differences in sampling a population of trees for volume and a population of deer for weight?

3. You are a consultant to a manager of a 100 000-ha forest on the island of Sarawak in Indonesia. This manager wants you to:

(a) Estimate the number of teak trees in the forest (a rare species).

(b) Estimate the number of a rare species of monkey still living in the forest.

Develop a sampling strategy for each parameter individually and for both at the same time.

9 Related Estimation Topics

This chapter deals with topics that should interest forest inventory specialists, but to a lesser extent than those discussed in Chapter 8. Missing data analysis is a topic of considerable relevance in social science inventories and it is likely that its most promising features can readily be adapted to the simpler situations in forest inventories. Measurement errors and their impact on estimation and analysis in forestry have received some attention but not to the degree they deserve. Techniques of Bayesian and likelihood estimation hold considerable promise but practical breakthroughs are only now appearing.

9.1 MISSING DATA ANALYSIS

The phenomenon of missing data is common in surveys. A simple way to deal with missing data is to discard the relevant records, treating the rest of the survey data as a complete data set, and make inference from it. But this can lead to serious bias, and some valid information in those survey units that are missing some data can be lost. Extensive literature is available discussing how to make the best inference when data is missing, for example, Little and Rubin (1987).

9.1.1 Types of Missing Data

There are three categories of possible missing data in any type of survey: noncoverage, total unavailability, and item unavailability. Noncoverage implies a failure to include a section or sections of the targeted population in the sample frame. Total unavailability occurs when no information is collected from a sample unit, for example, when a survey crew fails to locate a particular sample unit. Item unavailability occurs when some but not all the required information is collected from a sample unit, for example, when a record from a sample unit is missing one or more values or when a record is rejected as highly suspect or impossible during the editing process. We often use the term "nonresponse" as it is commonly used in the social science literature, to denote "noncoverage plus total unavailability", even though a more appropriate term in forestry would be "unavailable data." Hence, these terms are used synonymously here.

Almost any type of missing data is conceivable in forestry. For example, aerial photography is often used to increase the accuracy of estimation in forest inventories. Due to cloud cover, some imagery may be of poor quality, making photo interpretation difficult. A common experience in forest surveys is that some landowners refuse access to their property so that part of the population cannot be sampled. Substituting new sample units in this situation does not really solve the problem. Other forestry-related examples are possible: failure to respond to questionnaires (e.g., failure of sawmill operators to report their log usage); failure to locate or relocate sampling units on the ground; failure to identify part of the target population (e.g., small sawmill operators may not be included); and so on.

Typically, noncoverage is compensated by making weighting adjustments based on an external data source. Compensation for total unavailability is also usually compensated by some form of weighting adjustment, while compensation for item unavailability is commonly made by imputation, that is, by assigning sensible values for missing responses.

9.1.2 Methods of Estimating Missing Data

Little and Rubin (1987) discussed statistical analysis with missing data in considerable detail. Chapters 3 and 12 of their book covered quick methods for multivariate data and model-based approaches to deal with survey nonresponse data. We are concerned more in this book with a global overview of how to handle the problem of missing data in inventories, and use Sande (1982) as our main source of reference.

Sande (1982) discussed four ways of dealing with missing data:

1. Ignore all records with missing values. This can result in the loss of much data and leads to biased estimators because the missing values rarely are missing at random.
2. Treat the "unknowns" as a subpopulation. This still ignores the potential information about the missing value available in the other variables but gives a measure of the magnitude of the problem.
3. Adjust each table or estimate, ignoring the missing data in each case. This adjustment, a variation of method 1, can give rise to inconsistent tables or estimates because no complete data set corresponds to the set of estimates, owing to constraints on the data.
4. Make use of imputation. Imputation is defined as the estimation of individual items missing in a survey response (Sande 1982). The imputation process should give values consistent with the nonmissing data to satisfy the editing criteria, reduce the nonresponse bias, and preserve the relationship between items as much as possible. The process should work for most patterns of missing items, be capable of

being set up ahead of time, and be evaluated in terms of impact of the procedure on the bias and precision of the estimates.

Imputation appears to be the most promising of these four procedures and we focus on it here. The most promising methods of imputation use current inventory data as a source of matched individual data records from which one is selected to supply the missing information for a deficient record. Sande (1982) discussed a range of imputation methods used in the social sciences. Generally a combination of procedures is used but the most promising ones are those discussed below.

Hot-Deck Procedures. In general, a "hot-deck" procedure is a duplication process: when a value is missing from a sample, a reported value is duplicated to represent the missing value. The adjective "hot" refers to imputing with values from the current sample. In contrast, a "cold-deck procedure" uses information from data other than the current sample. To reduce the nonresponse bias, a hot-deck procedure usually classifies sampling units into disjoint groups. For each missing value, a reported value is imputed from the same group. Thus it is assumed for each group that the nonrespondents follow the same distribution as the respondents. Therefore, application of a hot-deck procedure in which the classification variables are not well correlated with the missing values is likely to cause misleading results.

Post-stratification is commonly used in the classification process. When the classification variables are quantitative, the value for a missing value can be imputed from the unit that has auxiliary values closest to that of the missing value.

A sequential hot-deck procedure is one in which the sample is put in some type of order within each classification group and, for each missing value, the previous reported value is duplicated. Sequential procedures require an initial value that can be imputed for the first value in the data set if it is missing. Cold-deck values are often used as initial values.

In both sequential and random choice procedures, the choice of matching fields must be made considering likely sources of variation, linkage through edits, and the number of complete or eligible records available as potential donors in each cell. The number of potential donors may be too small if too many fields are used for matching. However, if too few fields are used, there is a risk of a poor match or edit failure in the imputed record.

With hot-deck procedures, the variance of the estimators in simple cases is known to be larger than the variances of the usual expansion estimators of means and totals. But there may be a reduction in bias. Hot-deck procedures should produce imputed data sets that appear more realistic and do a better job of reflecting distributional properties than models. Many hot-deck procedures rely on a common-sense development rather than a rigorous theoretical development. Thus the effectiveness of these procedures is often unknown (Sande 1982).

One of the primary principles governing the construction of hot-deck procedures is to reduce the bias while preserving joint and marginal distributions (Fellegi and Holt 1976). An overview of hot-deck procedures is given in Ford (1983).

Nearest Neighbor Procedures. These procedures use current survey data as a source of individual data records with similar characteristics to estimate missing values. They are appropriate when one is matching strictly numeric data, unlike hot-deck procedures that can also be applied to nonnumeric data. These procedures are called nearest neighbor because the values in the matching fields must be similar and the element of randomness may be absent in the choice of donor.

Numeric matching is determined by the data. This makes it difficult to study the statistical properties of the procedures, especially variance estimation. Nearest neighbor procedures can be changed to hot-deck procedures by randomly choosing the donor record from m nearest neighbors instead of taking the nearest satisfactory record. Both are types of nonparametric regression, that is, regression without an explicit model (Sande 1982).

Post-Stratification Procedure. This procedure uses post-stratification and the post-stratum mean (or another typical value such as the mode for categorical variables), with a variable classified into strata. For numeric data, this is equivalent to item-by-item weighting for estimation of means and totals. But the marginal mean may yield such unrealistic values as 2.3 children in a household. Thus the procedure may need to be modified so as not to conflict with the editing criteria. For example, the criteria may be used to define the strata. This procedure is simple if it works but can create impossible values in the marginal distributions and may be biased if the distribution of the missing data within each cell is not random.

The post-stratification procedure has the advantage that under SRS it is easy to obtain the variance of the post-stratum mean conditional on the post-stratum sizes, and the unconditional variance is not much harder to find.

Sometimes two imputation procedures may be applied in a two-stage process. This is particularly useful when a variable shows a significant number of zeros. The occurrence of zeros is imputed with a model or by a hot-deck procedure and then the nonzero values are imputed separately by another method (Sande 1982).

Predicted Regression Imputation. This method uses respondent data to regress y on the auxiliary variables (Kalton and Kasprzyk 1982). Missing y values are then imputed as the predicted values from the regression equation. Another modified method is to add a residual term to the predicted value of y.

Problems arise in applying imputation procedures (Sande 1982). Great care is required in editing the data and imputing the missing values; and assessing the bias and variance of the estimators is difficult. Though considerable work has been done addressing these problems, its scope is still limited. Most of it deals with cases involving single variables assuming SRS and ignores the constraints imposed by the editing criteria. This reflects the extreme difficulty of the problem, not a deficiency in the research.

Multiple Imputation Procedures. Multiple imputation was first proposed by Rubin (1978) and was treated comprehensively in Rubin (1987). Multiple imputation, in contrast to single imputation, replaces each missing value not with a single value but with a vector of $M \geq 2$ imputed values (Heitjan and Rubin, 1986). The M values are imputed in the sense that M completed data sets are created from the vectors of imputations. Standard complete-data methods are used to analyze each data set. When, for nonresponse, the M sets of imputations are repeated random draws under one model, inference can reflect uncertainty due to nonresponse under that model. However, when the imputations are drawn from two or more models, the inferences can be contrasted across the models to display their sensitivity to the nonresponse. Only procedures employing some random component, such as hot-deck or Bayesian procedures, can be used in this way to obtain imputation errors (Rubin 1978).

Imputation is always based on some model or assumptions. If for a particular survey, no single set of assumptions is valid, a multiple imputation should be used to allow the analyst to check how sensitive the inference is with respect to different models. Multiple imputation uses the data collector's knowledge to incorporate uncertainty and to increase efficiency. Analysis is usually straightforward.

Weighting Adjustments. Weighting adjustments are primarily used to compensate for total nonresponse. The procedures require auxiliary information on either the nonrespondents or the total population. Little and Rubin (1987) discussed several methods of making weighting adjustments. The estimators used were usually Horvitz-Thompson type estimators. Four types of weighting adjustments were reviewed by Kalton and Kasprzyk (1986) and these are covered below.

Population and Sample Weighting Adjustments. To apply these types of adjustment, auxiliary information is needed about the distribution of the population or sample for one or more variables. The difference between the two procedures lies in the auxiliary information used. Population weighting adjustments are based on population distributions obtained externally, whereas sample weighting adjustments employ only data internal to the sample and require information about the nonrespondents. The sample of respondents is divided into a set of weighting classes defined by the available auxiliary information. The weights of all respondents within a class are then adjusted by the same multiplication factor using different factors in different classes. The adjustment for the population weighting procedure is carried out in such a way that the weighted respondent distribution across the classes conforms to the population distribution.

For the sample weighting adjustment procedure, the weight for a particular class is made proportional to the inverse of the response rate in that class. In order to compute the response rate for this procedure, the number of respondents and nonrespondents in each class must be determined. Since very little information about the nonrespondents is usually available, the choice of weighting class is often severely restricted.

Sample weighting adjustments resemble two-phase sampling (Ch. 5). The first phase sample is the total sample of respondents and nonrespondents and the second phase sample is the subsample of respondents, selected with different sampling fractions (response rates) from different strata (weighting classes).

Raking Ratio Adjustments. Raking ratio estimation was first proposed by Deming and Stephan (1940) to assure consistency between complete count and sample data from the 1940 US Census of Population. This is an interactive procedure for scaling sample data to known outside marginal totals (Oh and Scheuren 1983). The term "raking" apparently derives from the analog between an ordinary garden rake and the successive steps in the interactive proportional fitting process done manually in the early days. Oh and Scheuren (1983) described the basic technique and some of the common variants of raking adjustments for nonresponse. Raking develops weights that make the marginal distributions of the auxiliary variables in the sample conform to marginal population distributions (with population weighting) or marginal total sample distributions (with sample weighting) without ensuring that the full joint distribution conforms (Kalton and Kasprzyk 1986). The method of adjustment may be used to obtain weights that satisfy these conditions.

Weighting with Response Probabilities. This method assumes that population elements have positive probabilities of responding to the survey and that these probabilities can be estimated (Kalton and Kasprzyk 1986). The elements are then given nonresponse adjustment weights that are inversely proportional to their estimated response probabilities.

The adjustment procedures proposed for handling missing survey data, namely, imputation, aim chiefly at reducing the biases and MSEs of the estimators while reducing, or at least not unduly increasing, their variances. Unfortunately, evaluating the success of the procedures is rarely possible because it is often difficult to verify the assumed models. Differences between weighting and imputation emerge when one considers the multivariate nature of survey data (Kalton and Kasprzyk 1986). Weighting is a single global adjustment that attempts to compensate for the missing responses to all the items simultaneously. Imputation, on the other hand, is item-specific. This difference has consequences for the way the auxiliary data is used. In forming weighting classes, the focus is on determining classes that differ in their response rates. The choice of auxiliary variables to use in imputation, however, is primarily based on their ability to predict the missing responses.

An assumption underlying all the procedures reviewed is that once the auxiliary variables have been taken into account, the missing values are missing at random (Kalton and Kasprzyk 1986). Thus, for instance, the nonrespondents are assumed to be like the respondents within weighting and imputation classes. Kalton and Kasprzyk noted that all methods of handling missing survey data must depend on untestable assumptions and, if the assumptions are seriously in error, the analyses may give misleading conclusions. It is desirable but difficult and often impossible, to estimate biases both before and after adjustment. However, the effects of adjustments on variances and covariances can usually be estimated, at least approximately.

Chambers (1988) provided a solution to design-adjusted regression estimation that is particularly suited to situations where there is a relationship between the probability of unavailable data and the value of the variable of interest. A censoring model is assumed to generate the unavailability of the data. According to Chambers, the model has been extensively tested in the econometric literature, usually under the topic of "selectivity bias," but the statistical literature has overlooked it in the context of its interaction with sample design effects. Chambers' solution uses a two-stage iterative approach to adjust simultaneously for selection bias and a stratified sample design.

Estimating missing data is facilitated by a computing algorithm called the EM algorithm (Dempster et al. 1977), which can use both maximum likelihood and restricted maximum likelihood estimation. Little and Smith (1987), Laird et al. (1987), and Meng and Rubin (1991) discussed extensive applications of this algorithm and supplements to it.

With forestry data, total nonresponse is more of a problem than item nonresponse and several weighting adjustment procedures may be readily applicable given the availability of the EM algorithm. Certainly, statistically valid missing data techniques should be applied more often when processing forest inventory data. The utility of some of these methods could be demonstrated readily in forestry because data collection locations could be revisited for verification purposes.

9.2 MEASUREMENT ERRORS

Regression models with measurement error were studied even prior to the 20th century (Fuller 1987). In surveys of people, these errors are often called response errors. If we assume model (2.70) where $(x_1, x_2, \ldots, x_n)$ is fixed in repeated sampling, then it is known that the simple least-squares estimator b (eq. 3.157) is the best linear unbiased (BLUE) estimator if $\sigma^2 x_i^k = \sigma^2$ for all i $(i = 1, \ldots, N)$. If the y_i are assumed to be an independent sample from a $N(\mu, \sigma^2)$ distribution, independent from the vector $(e_1, \ldots, e_n)$, then b is the maximum likelihood estimator (mle) of β and is also unbiased under this model.

If instead of observing the x_i $(i = 1, \ldots, n)$ we can only observe

$$X_i = x_i + \delta_i \tag{9.1}$$

where δ_i is called the measurement error for unit i and is assumed to be distributed with mean 0 and variance σ_δ^2, then X_i and x_i are sometimes called the manifest (or indicator) and latent variables, respectively. In addition, models in which x_i is regarded as a fixed constant are called functional models and models in which x_i is regarded as a random variable are called structural models (Fuller 1987). For the remainder of this section we discuss the effects of measurement error in the X_i for the simple linear regression model (2.70) with uniform error variance σ^2.

For the structural error in variables model, exact results can be derived under the assumption that x_i, δ_i, and e_i are all normally distributed and independent. Letting a, b denote the least squares estimators of α, β, then

$$E(a) = \alpha + \beta\mu_x(1 - k) \tag{9.2}$$

$$E[b] = \beta k \tag{9.3}$$

where $k = \sigma_x^2/(\sigma_x^2 + \sigma_\delta^2)$ is the degree of attenuation due to measurement error.

For the functional error in variables model, the least-squares estimator a approaches the right side of eq. (9.2) as a probability limit, and similarly βk is the probability limit of b. Richardson and Wu (1970) derived an approximation to the expected value of b as

$$E[b] \doteq \beta - \beta\sigma_\delta^2(1 - 2(S_x^2/(S_x^2 + \sigma_\delta^2))^2)/(S_x^2 + \sigma_\delta^2)$$

from which an approximation of $E[a]$ may be deduced.

Because $0 \le k < 1$, the least squares estimator of β is attenuated toward zero. The bias of b makes it nearly impossible to construct valid confidence intervals (Gleser and Hwang 1988). Likewise the usual t-test of the null hypothesis H_0: $\beta = \beta_0$, where β_0 is the hypothesized value, will have a significance level different from the nominal level, unless $\beta_0 = 0$ (Carroll 1989).

Introduction of independent measurement errors leads to a reduction in the squared correlation R^2 as it biases b toward zero (Fuller 1987). Despite an extensive literature on the topic spanning many years, only a small percentage of statistical studies use procedures designed for explanatory variables that are assumed to be measured with error (Fuller 1987).

Swindel and Bower (1972) considered rounding errors, a form of measurement error in general regression estimation, and derived a bound for the relative bias in the least-squares estimates of the regression coefficients if the X_i are measured instead of the x_i in model (9.1). If c_1 is the bias divided by the standard error of b, it is bounded by

$$c_1 \leq |\beta|\sqrt{n}\delta/\sigma \tag{9.4}$$

where 2δ is the width of the measurement interval used and the X_i are recorded at the midpoints of these intervals. Although the bound on the relative bias increases with increase in sample size, the absolute bias decreases. The 95% confidence limits are almost unaffected if $c_1 \leq 0.1$, but computing the limits beyond this point becomes misleading (Cochran 1977, Ch. 13). Although δ may be known in rounding error situations, it is usually not known in more general measurement error situations, and Davies and Hutton (1975) gave another bound

$$c_2 \leq |\delta\beta|\sqrt{n}/s \tag{9.5}$$

where s^2 is the estimated mean square error (MSE) of residuals, and δ is the square root of the average squared error. The bound in eq. (9.4) may be easier to estimate than the one in eq. (9.5), since it may often be possible to give an upper bound (δ) for the X-measurements and σ and β can be estimated from the data.

Davies and Hutton gave some practical guidelines on how to deal with measurement errors, namely, calculate

$$\rho_1 = s/\sqrt{\delta^2 v(b)} \tag{9.6}$$

where b is the estimator of the simple linear regression slope given in eq. (3.157). If ρ_1 is not at least "somewhat larger" than $\sqrt{2n}$ (or $\sqrt{n}$ if n is large), then one or both of the estimates a and b have little meaning. If this test is satisfied, then compute

$$\rho_2 = (n\delta|b|)/s\,. \tag{9.7}$$

If ρ_2 is clearly less than 1, the errors in the data matrix can be ignored. If not, and if the measurement errors can be regarded as random and independent, then compute

$$\rho_3 = n\delta^2 b/(\rho_1 s)\,. \tag{9.8}$$

If ρ_3 is clearly less than 1, then the effects of the errors are likely to be negligible especially if n is large. But if ρ_3 is larger than 1, the bias can be a major part of the error in at least one of the coefficients.

Davies and Hutton (1975) also suggested ways of identifying those observations which may have an undue effect on the estimates and determining which coefficient is affected and how serious the bias might be.

Lwin and Spiegelman (1986) used a model similar to that of Swindel and Bower (1972) and Davies and Hutton (1975) but focused on constructing confidence intervals. They provided a procedure that is easy to implement and gives conservative confidence intervals if the errors have a known finite bound.

9.3 BAYESIAN AND LIKELIHOOD ESTIMATION

Much of the previous material in this book has focused on estimating parameters of interest with confidence limits around the estimates to indicate their reliability, but the theory of statistical inference exists to serve those people who collect and use empirical data, either to further natural knowledge or to improve decision making (Barnard 1966). These two alternatives represent two extremes of a broad spectrum of inference. To make decisions that are defensible generally requires fairly well-defined options or hypotheses. However, as empirical data takes many forms and is used in vaguely defined ways, it is often difficult to do much more than summarize the data and give estimates of precision.

The ideal method of inference in most cases is the use of Bayes theorem, discussed below (Edwards 1972).

Theorem 9.1 (Bayes Theorem). Suppose there are two hypotheses H_1 and H_2, which have valid prior probabilities $p(H_1)$ and $p(H_2)$. Then the probability of H_1 occurring given the probabilities of the data (R) given H_1, $p(R|H_1)$, is

$$p(H_1|R) = kp(H_1)p(R|H_1) \tag{9.9}$$

where k is a constant of proportionality given by

$$k = 1/\{p(H_1)p(R|H_1) + p(H_2)p(R|H_2)\} .$$

The theorem can be readily extended to any number of hypotheses as long as a statistical model exists for their occurrence.

The likelihood $L(H|R)$ is closely related to $p(R|H)$ and is defined as follows.

Definition 9.1. The likelihood $L(H|R)$ of the hypothesis H, given data R

and a specific model, is proportional to $p(R|H)$, the constant of proportionality being arbitrary.

With probability, R is the variable and H the constant, but with likelihood, H is the variable and R is the constant. The arbitrary constant of proportionality allows us to use the same definition of likelihood for discrete and continuous variables alike and drops out when likelihoods are compared for different hypotheses with the same data and probability model. Note that

$$\frac{p(H_1|R)}{p(H_2|R)} = \frac{p(H_1)}{p(H_2)} \frac{p(R|H_1)}{p(R|H_2)}$$

$$= \frac{p(H_1)}{p(H_2)} \frac{L(H_1|R)}{L(H_2|R)} \quad \text{(Edwards 1972)}. \tag{9.10}$$

If the odds of two events A_1 and A_2 occurring is defined as the ratio of their probabilities, that is, $Q(A_1, A_2) = p(A_1)/p(A_2)$, then Bayes theorem can be reexpressed in terms of odds as follows.

Theorem 9.2. The posterior odds of two hypotheses on data R occurring is equal to the product of the prior odds and the likelihood ratio, that is,

$$Q(H_1, H_2|R) = Q(H_1, H_2)L(H_1, H_2|R), \tag{9.11}$$

which Edwards (1972) calls the second version of Bayes theorem.

In the case of a continuous parameter θ, two hypotheses might be

$$H_1 : \theta_1 < \theta < (\theta_1 + d\theta_1)$$

$$H_2 : \theta_2 < \theta < (\theta_2 + d\theta_2)$$

where $d\theta_1$ and $d\theta_2$ are not necessarily of equal length. If $p(\theta)$ is the prior distribution function of θ, then

$$p(H_1) = p(\theta_1)d\theta_1$$

$$p(H_2) = p(\theta_2)d\theta_2 ,$$

and the likelihood ratio on data R, which is defined for particular values of a continuous parameter, is written as $L(\theta_1, \theta_2|R)$.

Then, by applying the second version of Bayes Theorem above, we have

$$\frac{p(H_1|R)}{p(H_2|R)} = \frac{p(\theta_1)\, d\theta_1}{p(\theta_2)\, d\theta_2} L(\theta_1, \theta_2|R), \tag{9.12}$$

which can be written as

$$\frac{p_R(\theta_1)\, d\theta_1}{P_R(\theta_2)\, d\theta_2} = \frac{p(\theta_1)\, d\theta_1}{p(\theta_2)\, d\theta_2} L(\theta_1, \theta_2|R), \tag{9.13}$$

where $p_R(\theta)$ is the posterior distribution function of θ and the infinitesimal elements $d\theta_1$ and $d\theta_2$ cancel out. By setting any particular value of θ_2 as a standard and regarding the above as an equation in θ with a series of values $\theta = \theta_1$, we get Bayes Theorem for a continuous variable, as follows.

Theorem 9.3

$$p_R(\theta) = kp(\theta) \cdot L(\theta) \tag{9.14}$$

where k is a constant, which absorbs both the ratio $p_R(\theta_2)/p(\theta_2)$ and the arbitrary constant implicit in $L(\theta)$.

This constant is determined in practice by the fact that both the prior $p(\theta)$ and the posterior distribution $p(\theta)$ of θ integrate to unity. This theorem states that the posterior probability density is proportional to the product of the prior probability density and the likelihood function (Edwards 1972).

Two examples serve to illustrate the utility and danger respectively of Bayes theorem.

Example 1. Let θ, the parameter of a binomial distribution, be chosen at random from the uniform distribution $p(\theta) = d\theta \;\; 0 \leq \theta \leq 1$. Consider a binomial experiment with parameter θ in which a successes and b failures are observed. Based on this outcome, what is the distribution of θ? Note that $p(\theta) = d\theta$, $L(\theta) = \theta^a(1-\theta)^b$, and hence, $p_R(\theta) = k\theta^a(1-\theta)^b$, which is a beta distribution with

$$k = \frac{(a+b+1)!}{a!b!} \; . \qquad \text{(Edwards 1972)} \; .$$

The result is scientifically sound and hence difficult to refute. Note also that no real-life situation was discussed.

Example 2. A suggestion has been made that the probability of a seedling surviving is not a fixed value θ but varies between nurseries and that the between-nursery distribution of θ can be described by the beta-distribution

$$dF = \frac{(m+n+1)!}{m!n!} \, \theta^m(1-\theta)^n d\theta$$

with parameters m and n. Given that a nursery has a seedlings surviving and b dying, what is the probability distribution of θ?

$$p(\theta) = \frac{(m+n+1)!}{m!n!} \, \theta^m(1-\theta)^n \; , \qquad L(\theta) = \theta^a(1-\theta)^b$$

and hence, $p_R(\theta) = k\theta^{m+a}(1-\theta)^{n+b}$, where $k = (m+n+a+b+1)!/ [(m+a)! \; (n+b)!]$.

The prior distribution of θ is an attempt to describe how θ varies between nurseries. But various nurseries could postulate different prior $p(\theta)$ and obtain quite different conclusions. This is a weakness of the application of Bayes theorem. An objective prior distribution of θ is required to avoid differences of opinion. As this distribution can rarely be specified, a number of researchers use likelihood estimation in preference to Bayes theorem. But in situations where decisions have to be made, the decision maker can, and often will, want to incorporate his or her own prior judgment into the decision, in which case Bayes theorem may be appropriate (see, e.g., Raiffa and Schlaifer, 1961). Chaudhuri and Vos (1988) gave a more complete discussion of Bayesian and likelihood estimation in survey sampling than we attempt here.

9.3.1 Bayesian Estimation

Bayesian inference as expressed by Theorem 9.1 appears to offer an attractive method of summarizing information. However, rarely are truly objective prior probabilities $p(H_1)$ and $p(H_2)$ available. Hence, subjective (personal) prior probabilities have to be used, which are often only satisfactory in statistical decision making.

Savage (1972) distinguished three classes of views on probability:

- *Objectivistic*. Repetitive events, such as tosses of a penny, are held to be in close agreement with the mathematical concept of independently repeated random events all having the same probability. Evidence for the degree of agreement between the behavior of the repetitive event and the mathematical concept and the magnitude of the probability that applies (if any) can be obtained by observing some repetitions of the event only.
- *Personalistic*. Probability measures a person's confidence in the truth of a particular proposition such as the possibility that it will rain tomorrow. The person is assumed to be reasonable in this respect but two reasonable individuals may have different degrees of confidence in the truth of the same proposition faced with the same evidence.
- *Necessary*. Probability measures the extent to which one set of propositions confirms the truth of another set out of logical necessity regardless of human opinion. Holders of this view tend to see this as an extension of logic.

Much of our book has been predicated on the objectivistic view, also known as the classical sampling approach. It may be worthwhile to point out that this classical sampling approach does not always yield reasonable results. For example, for a sample of n independent and identically distributed (i.i.d.) observations from $N(\theta, 1)$, the uniformly minimum variance unbiased estimator of θ^2 is

$$\hat{\theta}^2 = (\bar{x}^2 - 1/n)\,,$$

which can be negative for small n even though we know $\theta_2 > 0$ (Press 1989). Other examples of strange results and procedures in the frequentist approach exist. For example, Robinson (1975) gave some examples where the theory of confidence intervals does not work well. People with this objectivistic view would prefer to use Bayes theorem (Theorem 9.1), but this theorem is only easy to apply under the personalistic view.

Empirical Bayes estimation procedures attempt to use Bayes theorem by using objective prior probabilities. The procedures are applicable when an independent sequence of Bayes decision problems having similar structure are available (Kotz and Johnson 1982b, p. 490). Structural similarity includes the assumption that there is an unknown prior distribution on the parameter space, or, less formally, that a prior distribution is used based on other data, for example, past data for the same population. Empirical Bayes procedures violate Bayes theorem, and therefore the laws of probability, because the prior distribution should not depend on the current data set. Hence empirical Bayes estimators are only approximate Bayes estimators and have no natural standard error so that it is difficult to find credible confidence intervals or test hypotheses. Press (1989, p. 43) gave a simple example and noted that the risk of the empirical Bayes estimator is less than that of the maximum likelihood estimator (mle) but always greater than that of the Bayes estimator.

Ericson (1988) noted that the likelihood function

$$L(\mathbf{Y}|(s, y_s)) = \begin{cases} P_x(s)k & \text{for } \mathbf{Y}|\mathbf{S}(\mathbf{Y}) = y_s \\ 0 & \text{otherwise} \end{cases} \tag{9.15}$$

where $k > 0$ is any constant and S is a matrix operator on the vector $\mathbf{Y}$, has disturbed many researchers, since it does not depend on the design by which the sample (s, y_s) was selected and is very similar to the least informative likelihood function, one that is constant for all $\mathbf{Y}$. This has led several statisticians to conclude that from a Bayesian point of view the sampling design is unimportant so that it does not matter how the sample is selected. Other statisticians have adopted the superpopulation approach, viewing $Y = (Y_1, \ldots, Y_N)$ as a random sample from a superpopulation with probability density function (pdf) $f(Y_i, \theta)$ with θ a known or unknown superpopulation parameter.

Ericson (1988) argued that, despite the relatively uninformative nature of the likelihood, one can obtain useful posterior distributions on various functions $g(\mathbf{Y})$ by specifying reasonable prior distributions $p(\mathbf{Y}|\mathbf{X})$. For sufficiently uninformative prior distributions, the resulting Bayes estimators or their limits agree with those obtained by the traditional sampling approach. Various traditional sampling designs, such as stratified and multistage sampling, are optimal from a Bayesian decision-theoretic view for suitable prior distributions. Ericson also saw the superpopulation concept as a partial specification of a prior distribution where the superpopulation is an artificial construct of no real interest *per se*.

Ericson (1988) provided what is, at this time, probably the most useful contribution of Bayesian methodology to survey sampling. He attempted to justify classical survey sampling methodology within the concept of Bayesian methods. The value of this is to sharpen our understanding of why we use designs such as SRS or stratified sampling and specific estimators.

The essence behind SRS from a Bayesian point of view is the idea that the N population units are seen as indistinguishable in that one does not care what subset of n sample units is used for inference. The sample designer is unable or unwilling to specify information about $\mathbf{Y}$ in the unit labels (Ericson 1988).

There is considerable similarity between SRS results concerning inference about the finite population mean μ and those in subjective Bayesian inference with an exchangeable prior distribution on $\mathbf{Y}$. Exchangeability refers to the concept that random variables $Y_1, \ldots, Y_N$ are said to be exchangeable whenever the joint probability of each of the $N!$ permutations of the Y_i is the same. This concept implies that the distribution of any subset of n $(1 \leq n \leq N)$ of the Y_i is the same as that of any other subset of n. Let the subjective Bayesian assume a prior distribution on $\mathbf{Y}$ for which the Y_i are exchangeable with mean $E(Y_i) = m$, variance $V(Y_i) = v$, and covariance $\text{Cov}(Y_i, Y_j) = c$ for all $i \neq j$, and let μ and $\bar{y}$ be the mean of the Y_i in the finite population and in the sample, respectively, and let $\sigma^2 = \Sigma_{i=1}^{N} (Y_1 - \mu)^2/N$ be the finite population variance. Then Ericson, using the condition that the Y_i remain exchangeable when conditioned on any symmetric function of them, gives the following properties for a prior distribution without specifying a sample design.

Theorem 9.4. If the random variables $Y_1, Y_2, \ldots, Y_N$ are exchangeable with means, variances, and covariances as defined above, then

$$E(\mu) = E(\bar{y}) = m\,, \qquad V(\mu) = [v + (N-1)c]/N\,,$$
$$E\sigma^2 = [(N-1)(v-c)]/N\,, \qquad V(\bar{y}) = [v + (n-1)c]/n\,,$$
$$E(\bar{y} \mid \mu, \sigma^2) = \mu \tag{9.16}$$

and

$$V(\bar{y} \mid \mu, \sigma^2) = (N-n)\sigma^2/\{(N-1)n\}\,. \tag{9.17}$$

The conditional mean and variance of $\bar{y}$ are the same values as for SRS. The analogy is even closer because for an exchangeable prior distribution on Y, given the collection of the N population variate values, Y_i, but not the units to which they belong, the probability of any prespecified subset of n of the population units is $1/\binom{N}{n}$, as in SRS.

Ericson concluded that there are not prior grounds for preferring any sample of n units if one's prior knowledge regarding the Y_i can be stated as an exchangeable prior distribution on Y and if costs of selecting and measuring any sample of n units are the same. This does not justify SRS from the Bayesian point of view, but compelling arguments can be advanced anyway for SRS. SRS can be viewed as an easy method of selecting one of the possible samples, as giving an approximately balanced sample and, as such, being a model-robust sample selection scheme. Also, randomization costs little to implement and hence may cost the subjective Bayesian little while potentially increasing the utility of the sample to numerous potential users having different views of inference.

One of the main features of empirical Bayes analysis is the use of information from similar sources to construct estimators and predictors (Ghosh and Meeden 1986). This is feasible in repetitive surveys sampling finite populations, such as household surveys, cropcutting experiments, and FIA surveys. James-Stein (JS) estimation procedures, described under small area estimation in Chapter 8, are one example of empirical Bayes estimation. Another example, by Ghosh and Meeden, is explained in the following.

Let y_i denote the value of a single characteristic for unit i of a population U with units $i = 1, 2, \ldots, N$. The vector $\mathbf{y} = (y_1, \ldots, y_N)'$ is the unknown state of nature and belongs to $\theta = \mathbf{R}^N$. Let S be the set of all possible samples with subsets s(samples) of $\{1, \ldots, N\}$ of size n only.

Consider the model

$$y_i = \theta + e_i \qquad (i = 1, \ldots, N) \tag{9.18}$$

where θ, $e_1, \ldots, e_N$ are independently distributed with $\theta \sim N(\mu, \sigma^2)$ and the e_i i.i.d. $N(0, \tau^2)$. Then the Bayes estimator of $\bar{Y} = \Sigma_{i=1}^{N} y_i/N$ under squared error loss is

$$E[\bar{Y}|s, y(s)] = [n\bar{y} + (N-n)\{B\mu + (1-B)\bar{y}\}]/N \tag{9.19}$$

where $B = M/(M+n)$ and $M = \tau^2/\sigma^2$.

Assume in empirical Bayes estimation that we are at the mth stage of the sampling procedure and that sampling has been repeated (m-1) times. The population size at stage j is denoted by N_j and associated with unit i at that stage is a characteristic $y_i^{(j)}$ ($i = 1, \ldots, N_i$; $j = 1, \ldots, m$). A typical sample at stage j of size n_j is denoted by s_j ($j = 1, \ldots, m$). Assume the model

$$y_i^{(j)} = \theta^{(j)} + e_i^{(j)} \qquad i = 1, \ldots, N_j;\ j = 1, \ldots, m \tag{9.20}$$

where the $\theta^{(j)}$ ~i.i.d. $N(\mu, \sigma^2)$ and the $e_i^{(j)}$ ~i.i.d. $N(0, \tau^2)$. If $B_j = M/$

$(M + \mu_j)$ $(j = 1, \ldots, m)$, then from eq. (9.19) the Bayes estimator of $\bar{Y}^{(m)} = \Sigma_{i=1}^{N_m} y_i^{(j)}/N_j$ is

$$E[\bar{Y}^{(j)}|s_j, y_{sj}] = [n_j\bar{y}_{sj} + (N_j - n_j)(B_j\mu + (1 - B_j))\bar{y}_{sj}]/N_j \quad (9.21)$$

where $\bar{y}_{s_j} = \Sigma_{i\epsilon s_j} y_i^{(j)}/n_j$ $(j = 1, \ldots, m)$. One or both of the parameters M and μ are unknown and need to be estimated from the data. Ghosh and Meeden (1986) proposed one way of estimating M.

As this last example illustrates, empirical Bayes procedures quickly become complex and computationally tedious. Ericson (1988, p. 243) saw the main practical applications of the Bayesian approach in survey sampling, in accounting and auditing, quality control, and in estimating the size of capture-recapture populations. We think that in addition to sharpening the understanding of traditional sampling methods as illustrated earlier, the most promising potential area may be the area of nonresponse, where simply assuming that nonrespondents are similar to respondents is unacceptable. Subjective Bayesian methods may be the best approach for a better solution.

Gelfand et al. (1990) and Gelfand and Smith (1990) discussed the Gibbs sampler, stochastic substitution, and a sampling-importance-resampling algorithm as three Monte Carlo based approaches to calculating numerical estimates of marginal probability distributions. Computation of such marginal densities is usually intractable. All three procedures are said to be straightforward to implement in several frequently occurring practical situations. If computational complexities have indeed been resolved in many situations, empirical Bayes techniques should be considerably more attractive in the future. Such complexities have been a big stumbling block to practical acceptance of subjective Bayesian methods.

9.3.2 Likelihood Estimation

So far, likelihood estimation in survey sampling is primarily of theoretical interest. By treating finite population sampling as the prediction of non-observed values from a hypothetical superpopulation, sampling theory is brought more into line with standard statistical inference. The concepts of likelihood and probability then have the same role in sampling theory as in estimation theory.

The basic problem of inference being studied is to measure the plausibility of future events whose probabilities of occurrence depend on the hypothetical population from which the sample was drawn. For example, by studying the effects of changing the hypothetical distribution of an underlying superpopulation, insight can be gained into the appropriateness of various estimators and the impact on them of distributional properties.

Royall (1976) attempted to clarify the potential role of the likelihood axiom in a non-Bayesian, nondecision-theoretic approach to finite population theory. The axiom is as follows.

The likelihood axiom. Within the framework of a statistical model, all the information that the data provides concerning the relative merits of two hypotheses is contained in the likelihood ratio of those hypotheses on the data, and the likelihood ratio is to be interpreted as the degree to which the data supports one hypothesis over the other.

Royall addressed the basic question: How well are various hypotheses concerning $Y = Y_1 + \cdots + Y_N$ supported by a sample of n units on which the Y-values are observed, given a population consisting of N units, where an unknown value of interest Y_i is associated with each? Clearly, the answer depends on how the question is framed.

After the sample has been observed, there are two groups, the sample group for which the y_i are known, and the group containing the nonsampled units. To draw conclusions about the whole population, the logical relationship between the sample and the unobserved units needs to be considered. This is usually done by representing the relationships with a statistical model, that is, a family of probability distributions, one of which is thought to have generated the observations. With such a model, the relative support provided by the data for various hypotheses can be measured according to the likelihood axiom. If the likelihood function can be determined for Y under a given model, then the ratio of the likelihoods of two hypothesized values of Y measures their relative support. The problem of specifying an appropriate statistical model is fundamental (Royall 1976).

The conventional survey sample approach is straightforward. The unknown values of interest in the population are treated as fixed constants and the vector of values $\mathbf{y} = (y_1, \ldots, y_N)'$ is the unknown parameter. The basic probability distribution is generated by randomly selecting the units to be observed. Superpopulation probability models play only secondary roles if used. Expected values and variances are calculated with respect to the randomization procedure used.

Essentially, the conventional approach includes no probabilistic assumptions except for the sampling design used. Hence it is quite assumption-free unless unequal probability sampling schemes and/or regression or ratio estimation are used, which require certain assumptions for efficiency's sake. For the conventional model, the resulting likelihood function appears to be uninformative regarding the population parameter Y, since it indicates that no single value of the parameter is better supported than any other. Some researchers feel that this casts doubt on the likelihood axiom, but these doubts are not justified (Royall 1976). Royall cited an example of weighing either an axe, an ass, or a box of old horseshoes, where the sample weight obtained gives no information at all about the weight of the other two objects regardless of how the sample is selected. He felt that this is what the likelihood function shows. If stronger inferences are made than the trivial one derived from the likelihood axiom, it is because the conventional model does not adequately describe the relationships among the units in the population.

Another concern arises if both the likelihood and the sufficiency principle are accepted (Basu 1969). These two principles imply that at the analysis stage, the statistician does not need to pay any attention to the nature of the sampling design, that is, the statistician does not need to know how the data were collected. The sufficiency principle is as follows.

Sufficiency Principle. If $T = T(y)$ is a sufficient statistic for a parameter θ, then only the characteristic $T(y)$ of y is relevant for making inference about θ. That is, if $T(y) = T(y')$, then the inference about θ is the same whether the sample is y or y'.

This concern about the relevance of the sampling design has not been satisfactorily resolved. Despite the implication of these principles, Basu (1969) still thinks it is important to select a probabilistic sample. Possibly, a sufficient statistic rarely exists in practice.

In superpopulation predictions in models for finite populations, the numbers $y_1, \ldots, y_N$ are represented as realizations of random variables $Y_1, \ldots, Y_N$. Estimating the population total is equivalent to predicting the sum of the unobserved random variables. The probability model that describes the relation among the random variables $Y_1, \ldots, Y_N$ is central to the inference process in teaching us about the nonsampled units from the observed ones.

Although consideration of likelihoods can lead to prediction models, there are many other approaches to prediction problems, from linear least-squares to Bayesian techniques. Unfortunately, the introduction of prediction models does not lead immediately to nontrivial likelihoods.

Royall (1976) studied the relationship between the likelihood function for the population total and the probability distribution of the minimum variance unbiased estimator under a multivariate normal regression model. The estimator maximized the likelihood function under the most common models used, but not in general. Sedransk (1981) presented some examples of summarizing survey results in terms of likelihoods. The practical difficulties in this are well illustrated by this article.

Considerable theoretical work is being done on likelihood estimation that is applicable when summarizing the results of survey sampling. Existing principles in survey sampling tend to be validated and clarified by this theoretical effort and more powerful summaries of results are likely to be forthcoming but these will require more assumptions. As an example of this, Breckling et al. (1992) developed the theory for maximum likelihood inference about superpopulation parameters based on sample survey data. They included nonignorable sample designs as well as nonignorable nonresponse; but their results are still quite limited and model-dependent and practical implications are not clear yet.

9.3.3 Likelihood and Bayesian Estimation in Forestry

We are only aware of one application of likelihood function estimation for sampling finite populations in forestry. Squillace et al. (1980) used likelihood analysis of monoterpene composition in slash pine to rank possible geographic origins of slash pine plantations. Swindel (1972) discussed principles underlying inference from the Bayesian point of view in a forestry context.

Green and Strawderman (1985) applied James-Stein (JS) estimation (James and Stein 1961; Stein 1955) to estimate the coefficients of a series of volume equations. They concluded that the JS procedures are clearly superior to least-squares in predictive ability and in estimating MSEs of the regression coefficients. Ek and Issos (1978) and Burk and Ek (1982) recommended JS and empirical Bayes procedures for the simultaneous estimation of several means. In the latter paper, analyses for two populations each with four or more groups of means showed consistent reductions in estimated total MSEs relative to maximum likelihood estimates for a standard normal distribution.

Oderwald (1990) noted that an individual's prior knowledge about volume in a stand or area is often not used in estimation except to establish sample size or identify strata. This is even true when survey results for similar areas are available. Oderwald illustrated how such prior knowledge can be used in a Bayesian framework by assuming prior beta distributions for volume, where several estimates of the minimum and maximum possible volume per unit area and most probable volume per unit area are generated prior to sampling and combined with sampling information to generate posterior distributions for plot volume per unit area. He noted that in all cases the posterior distribution had confidence intervals that were narrower than the confidence intervals for the sample data distribution.

10 Future Directions in Multiresource Sampling in Forestry

10.1 UNRESOLVED PROBLEMS IN SAMPLING NATURAL RESOURCES

Considerable work is needed on the development of sound, flexible sampling strategies to accommodate descriptive statistics and analytical uses of survey data, and to identify and establish cause-effect relationships. Schreuder and Czaplewski (1992) attempted to do this, and their work highlights the difficulties involved. The importance of retrospective studies highlights the need for remote sensing, for example, aerial photography over time. Such information allows one to search for past trends that might explain observed changes.

We do not dwell here on identifying where survey sampling research may or should go. Insightful perspectives on the topic have been published by Cochran (1975), Murthy (1981), and Cormack (1988). Instead, we focus on specific opportunities for research that readers might be interested in pursuing, namely, sampling strategies, estimation, data summarization, analytical use of survey data, and instrumentation.

10.1.1 Sampling Strategies

Sampling strategies based on sound statistical principles are well developed and readily available to users. Much current effort is being expended on integrating model-based and design-based sampling strategies. Ghosh and Sinha (1990) showed that, in a hierarchical, Bayes-estimation framework, model-based and design-based estimators are not necessarily incompatible and may often be identical. They illustrated their ideas using the Horvitz-Thompson and the generalized Horvitz-Thompson estimators. An interesting attempt at unifying the theory of finite population and superpopulation sampling is that of Cheng and Li (1987), who showed that the superpopulation and minimax approaches satisfy two extremes in a class of optimality criteria based on a fixed population regression-type model. As they noted, the minimax criteria have sometimes been used to justify the use of SRS in design and sample means in estimation. Further work of this type is needed to give a clearer understanding of the assumptions required to fit each special case into a more general theory. This, in turn, should help in recommending strategies for specific circumstances.

In current work on model-based sampling, much attention is being given to examining the estimators in use, generalizing and comparing them, and developing useful approximations, such as jackknifing or bootstrapping, to estimate the variance of complex sampling strategies. Efforts to unite sampling theory with classical statistical theory have resulted in the assumption of superpopulations that, in turn, have simplified the analytical comparison of estimators used under certain model assumptions. Clear analytical comparisons of the efficiency of sampling strategies for model (2.70) have not been successful so far despite the excellent attempts of Särndal (1980b). Related attempts to understand why particular sampling strategies are efficient have not been particularly successful either. Schreuder (1987) showed why sampling with probability proportional to size (pps) with $\hat{Y}_{HT}$ (eq. 3.7) can be spectacularly inefficient for certain populations, but a similar attempt to explain the difference in efficiency of sampling strategies such as stratified pps (spps), stratified sampling from cumulated x-values (spscx), and sampling based on the range of x (psR) with $\hat{Y}_{wr}$ (eq. 3.139) was not entirely successful (Schreuder *et al.* 1990c).

Model-based sampling procedures can be quite efficient but are susceptible to bias in estimation, unreliability in variance estimation, and are less defensible in potentially controversial situations than probabilistic sampling. Considerable work needs to be done to ensure that the conditions under which such procedures can be used satisfactorily are clearly defined. An example is the acceptance of the biased centroid sampling procedure (Wood *et al.* 1990, Wiant *et al.* 1992b), a special case of importance sampling (Ch. 4 Sect. 4.7), because of the considerable gain in efficiency to be achieved by using this method. There are many situations where the estimation bias of a procedure is negligible and it should be implemented because of its practical advantages, but these situations need to be clearly specified.

Work deriving variance estimators for complex sampling strategies still needs to be integrated better. Considerable theoretical work is being done on bootstrapping and jackknifing, but there is an unfortunate delay in implementing these procedures in practice. Such nonparametric methods are useful too for estimating confidence intervals, and the theory of saddlepoint approximations (Davison and Hinkley 1989; DiCiccio *et al.* 1990) seems to hold particular promise for small sample sizes.

We have mentioned in this book several interesting theoretical developments in survey design which, to our knowledge, have not been applied in practice. Such applications could enrich forest inventory planning and design. Hence, we encourage practitioners to investigate applications of programmed probability sampling (Zöhrer 1978, 1979), Sunter sampling (Sunter 1986), collocated sampling (Brewer and Hanif 1983), adaptive sampling (Thompson (1982), and Satyamurthi's (1979) idea on extending Steiner's rule in pps sampling.

An interesting analogy exists between the parameters of animal and plant populations, that is, removal versus harvesting, recruitment versus ingrowth, survival versus survival growth, and death versus mortality. To our knowledge, nobody has examined this equivalence and its possible implications for improving sampling strategies for growth estimation.

10.1.2 Estimation

Practitioners prefer estimates derived from sampling to be additive. For example in Chapter 4, we describe the considerable effort required to obtain additive estimates for the various components of growth using permanent variable radius plots (VRP). Commitment to this effort is understandable given the need for credibility of the results, but it is costly due to the increased complexity and lower efficiency of the estimation. Research into the magnitude of these costs is desirable.

Srivastava estimation holds considerable promise for increasing the precision of estimation in forest inventory (Ouyang *et al.* 1992b and references therein). However, estimation bias may be a serious problem in approximate Srivastava estimation (Ouyang *et al.* 1992a). Nonetheless, there seems to be ample opportunity for additional development of Srivastava estimation in both theory and practice.

The problem of missing data (nonresponse) is important in natural resource inventories but it is not generally addressed in a statistical framework in forestry. Rubin (1976) showed that ignoring missing data is satisfactory if the data can be assumed to be random, but this assumption is frequently false. In forestry, an investigation is needed of parameter estimation in the presence of nonrandom missing data. The EM algorithm, a maximum likelihood estimation procedure (Dempster *et al.* 1977), is a good example of the practical direction of this type of research. It has potential for application to forestry problems.

A massive amount of information is available from remotely sensed imagery, both photographic and digital. Instrumentation and analytical techniques for extracting the maximum amount of useful data from this imagery need to be developed so as to minimize the amount of costly information to be collected on the ground. Regular remote sensing coverage of entire countries is now feasible. This ensures that up-to-date imagery is always available for many inventory purposes. Efficient processing and analysis of the generated imagery will require reliable, large capacity computers and highly skilled staff. The utility of remote sensing technology in forest surveys has been consistently disappointing. Part of the problem is that some of the variables of interest are highly subjective (such as commercial forest land) even if measured on the ground. Others (such as forest land) may be measured more accurately on low-altitude photos than on the ground, but ground information is generally accepted to be truth. To make information derived from remote sensing more useful, it may be necessary to obtain more objectively defined variables that can be measured readily by remote sensing. For example, forest land can be measured as accurately from some sources of remote sensing as from ground sampling, but this is not true for the more subjective variable, commercial forest land.

10.1.3 Data Summarization

Inventory information should be summarized more completely than it is. Most of the interesting work on likelihood and Bayesian estimation still does not seem to have much practical value. Even for important practical and rather simple designs such as double sampling for stratification or double sampling with ratio or regression estimation, summarizing the information in a likelihood or Bayesian framework seems to be a difficult task (J. Sedransk, personal communication). Perhaps the Gibbs sampler (Gelfand *et al.* 1990) offers a breakthrough!

10.1.4 Analytical Use of Survey Data

Much of this book has focused on enumerative inference, that is, on estimating the total value of one or more characteristics of a population or the mean value per population unit. There is another form of inference, analytical inference, that involves the understanding of relationships between two or more characteristics, and the estimation of parameters of super-population models that is needed to provide that understanding. Whereas enumerative inference is necessarily concerned with features of a specific finite population, analytical inference may be more broadly aimed toward populations similar but not necessarily identical to the one sampled. In a forestry context, for example, there may be interest in modeling past growth trends of a forested ecosystem in order to project trends under an altered set of social or climatic conditions, and to rely on past survey data to fit the model.

Much of the literature on the analytical use of survey sample data is comparatively recent, and much of it deals with the role of sample design in inference. When data is collected according to a design more complex than simple random sampling (SRS), it is crucial to know the conditions under which the selection probabilities need to be included in the estimation of model parameters, or, conversely, when the sample design can be ignored for the purpose of model fitting. The issues of ignorability and informativeness of design for analytical inference from sample survey data have been discussed by Scott (1977), Smith (1989), and Sugden and Smith (1984). As the need for analysis based on forest survey data increases, so too will the need to be aware of these issues and the implications they carry for inference. Even when dealing with ignorable designs, the choice of the model that best accommodates a complex population structure is a complicated decision, especially when the data is collected over time. Pfeffermann and Smith (1985) explored some alternative model forms, and Gregoire (1987) suggested alternative means of accounting for measurement over time when modeling data from permanent plot surveys. An even more critical problem that requires resolution is the use of inventory data for identifying potential cause and effect. Smith and Sugden (1988) suggested an interesting theoretical framework for doing this, but the challenge is to apply it successfully.

10.1.5 Instrumentation

Research is needed into the design and development of light weight, direct reading, moderately priced dendrometers for use in the field under all types of conditions. Also needed is an instrument that remotely measures bark thickness at points on a tree bole out of reach from the ground.

10.2 Opportunities for Technology Transfer

A considerable amount of advanced technology is available in the statistical and forest inventory literature that has not yet been applied to help solve problems in resource inventory. Some specific examples are:

- Using regression estimators instead of ratio-of-means and Horvitz-Thompson estimators for estimation of population totals and means. The regression estimators including Srivastava estimators are more efficient (Schreuder *et al.* 1987b; Ouyang *et al.* 1991) and reliable bootstrap estimators of variance are available (Schreuder and Ouyang 1992).
- Using improved variance estimators, which enable confidence intervals to be specified more reliably, that is, the classical variance should be replaced by more robust ones derived using, for example, jackknife and bootstrap techniques (Royall and Cumberland 1981a,b; Schreuder and Anderson 1984; Schreuder and Ouyang 1992).
- Adjusting for missing data using the EM-algorithm (Dempster *et al.* 1977). This maximum likelihood algorithm, which is commonly used in the social sciences (Laird *et al.* 1987, Little and Smith 1987), should have application in forest inventory

References

Ahern, F. J. and Leckie, D. G. 1987. Digital remote sensing for forestry requirements and capabilities, today and tomorrow. Geocarto Int. 3: 43–52.

Ahern, F. J., Raney, R. K., Dams, R. V., and Werle, D. 1990. Possible contributions of RADARSAT and optical data to tropical forest management. In: Proc. Int. Symp. on Primary Data Acquisition. Manaus, Brazil. June 24–29, 1990. Pp. 141–157.

Alberta Forest Service, 1987. Evaluation and fitting of tree taper functions for Alberta. Dept. Forestry, Lands and Wildlife, Edmonton, Canada. Pub. No. 106, 148 pp.

Alder, D. 1980. Forest Volume Estimation and Yield Prediction, Vol. 2: Yield Prediction. FAO Forestry Paper No. 22/2, Rome, Italy. 194 pp.

Aldrich, R. C. 1979. Remote sensing of wildland resources: A state-of-the-art review. USDA For. Serv., Rocky Mountain For. and Range Exp. Sta., Gen. Tech. Rep. RM-71. 56 pp.

Aldrich, R. C. and Heller, R. C. 1969. Large-scale color photography reflects changes in a forest community during a spruce budworm epidemic. In: Remote Sensing in Ecology (Philip Johnson, Ed.). Univ. of Georgia Press, Athens. Pp. 30–45.

Amidon, E. L. 1984. A general taper functional form to predict bole volume for five mixed-conifer species in California. For. Sci. 30: 166–171.

Amlaner, C. J. and MacDonald, D. W. 1980. A Handbook on Biotelemetry and Radio Tracking. Pergamon Press, Oxford, U.K. 804 pp.

Anderson, J. R., Hardy, E. E., Roach, J. T., and Witmer, R. E. 1976. A land use and land cover classification system for use with remote sensing data. USDI Geolog. Surv. Professional Pap. 964. 28 pp.

Angelone, A., Toledo, J. M., and Burns, J. C. 1980. Herbage measurement in situ by electronics. 1. The multiple-probe-type capacitance meter: A brief review. Grass Forage Sci. 35: 25–33.

Arnab, R. 1978. On strategies of sampling finite populations on successive occasions with varying probabilities. Sankhya C41: 141–155.

Arvanitis, L. G. and Reich, R. M. 1989. Sampling simulations with a microcomputer. Coenoses 4: 73–80.

Assmann, E. and Franz, F. 1963. Preliminary spruce yield tables for Bavaria (in German). Inst. f. Ertragskunde d. Forstl. Forschungsanstalt München. Munich, Germany. 104 pp.

Avery, T. E. and Berlin, G. L. 1985. Interpretation of Aerial Photographs, 4th ed. Burgess, Minneapolis, MN. 554 pp.

Avery, T. E. and Burkhart, H. E. 1983. Forest Measurements, 3rd ed. McGraw-Hill, NY. 331 pp.

Bankier, M. D. 1988. Power allocations: determining sample sizes for subnational areas. Amer. Stat. 42: 174–177.

Barnard, G. A. 1966. The use of the likelihood function in statistical practice. In: Proc. Fifth Berkeley Symp. on Math. Stat. and Probability, Univ. of California Press, Berkeley. Vol. 1, pp. 27–40.

Barnard, G. A. 1971. Discussion of paper by V. P. Godambe and M. E. Thompson. J. Roy. Stat. Soc. B33: 376–378.

Barnard, J. E. 1974. Sampling with partial replacement contrasted with complete remeasurement inventory designs: An empirical evaluation. In: Monitoring Forest Environment through Successive Sampling (T. Cunia, Ed.). State University of NY, College of Environmental Science and Forestry. Syracuse, NY. Pp. 384–390.

Barrett, J. P. 1964. Correction for edge effect bias in point-sampling. For. Sci. 10: 52–55.

Barrett, J. P. and Nevers, H. P. 1967. Slope correction when point sampling. J. For. 65: 206–207.

Bartlett, M. S. 1975. The Statistical Analysis of Spatial Pattern. Chapman and Hall, London, U.K. 90 pp.

Basu, D. 1969. Role of the sufficiency and likelihood principles in sample survey theory. Sankhya A31: 441–454.

Basu, D. 1971. An essay on the logical foundations of survey sampling, Part 1. In: Foundations of Statistical Inference (V. P. Godambe and D. A. Sprott, Eds.) Holt, Rinehart and Winston, NY. Pp. 203–242 (includes discussion).

Battese, G. E., Harter, R. M., and Fuller, W. A. 1988. An error-components model for prediction of county crop areas using survey and satellite data. J. Amer. Stat. Assoc. 83: 28–36.

Bay, B. 1960. Sample plots and the angle count method. N.Z. J. For. 8: 231–237.

Bayes, T. 1763. An essay towards solving a problem in the doctrine of chances. Phil. Trans. Roy. Soc. 53: 370–418. Reprinted in: Studies in the history of probability and statistics IX. Thomas Bayes' essay toward solving a problem in the doctrine of chances (with a biographical note by G. A. Barnard). Biometrika 45: 293–315 (1958).

Beal, S. L. 1989. Sample size determination for confidence intervals on the population mean and on the difference between two population means. Biometrics 45: 969–977.

Bechtold, W. A., Ruark, G. A., and Lloyd, F. T. 1991. Changing stand structure and regional growth reductions in Georgia's natural pine stands. For. Sci. 37: 703–717.

Beck, D. E. 1971. Polymorphic site index curves for white pine in the southern Appalachians. USDA For. Serv. Res. Pap. SE-80. 8 pp.

Becking, J. H. 1954. Einige Gesichtspunkte für die Durchführung von vergleichenden Durchfostungsversuchen in gleichaltrigen Beständen. [Some aspects of the execution of comparative thinning experiments in even-aged stands.] In: Proc. IUFRO Congr. Rome, Italy. 1953. Sect. 23, Pap. No. 3, 2 pp.

Beekhuis, J. 1966. Prediction of yield and increment in *Pinus radiata* stands in New Zealand. N.Z. For. Serv., For. Res. Inst. Tech. Pap. No. 49.

Beers, T. W. 1966. The direct correction for boundary-line slopover in horizontal point sampling. Purdue Univ. Agric. Exp. Sta. Lafayette, IN. Res. Prog. Rept. 224, 8 pp.

Beers, T. W. and Miller C. I. 1964. Point sampling: Research results, theory and applications. Purdue Univ. Agric. Exp. Sta. Lafayette, IN. Res. Bull. 786. 56 pp.

Beers, T. W. and Miller, C. I. 1976. Line sampling for forest inventory. Purdue Univ., Agric. Exp. Sta., Lafayette, IN. Res. Bull. No. 934, 34 pp.

Befort, W. 1986. Large-scale sampling photography for forest habitat-type identification. Photogramm. Eng. Rem. Sens. 52: 101–108.

Bell, J. and Associates. 1990. How ELSE can you set "an appropriate sample size"? Inventory Cruising Newsletter 12: 3–4.

Bellhouse, D. R. 1988. Systematic sampling. In: Handbook of Statistics, Vol. 6: Sampling (P. R. Krishnaiah and C. R. Rao, Eds.). Elsevier Science, Amsterdam, Netherlands. Ch. 6, pp. 125–185.

Belsley, D. A., Kuh, E. and Welsch, R. E. 1980. Regression diagnostics: identifying influential data and sources of collinearity. J. Wiley and Sons, NY. 292 p.

Benson, M. L. and Slinn, R. J. 1964. The penta-prism caliper. Aust. For Res. 1: 34–42.

Berg, S. 1974. A note on the RHC method. Skandinavian Actuarial J. 57: 108–114.

Best, R. G. and Fowler, R. 1981. Infrared emissivity and radiant surface temperatures of Canada and snow geese. J. Wildl. Manage. 45: 1026–1029.

Bickford, C. A. 1956. Proposed design for continuous forest inventory: A system of perpetual forest survey of the Northeast. USDA For. Serv. East. Tech. Meeting, Forest Survey. Cumberland Falls, KY. Oct 8–13, 1956. 38 pp.

Bickford, C. A. 1959. A test of continuous inventory for National Forest management based upon aerial photographs, double sampling, and remeasured plots. Proc. Soc. Amer. For. 1959: 143–148.

Biggs, P. H. and Spencer, R. D. 1990. New approaches to extensive forest inventory in Western Australia using large-scale aerial photography. Aust. For. 53: 182–193.

Biggs, P. H., Pearce, C. J., and Wescott, T. J. 1989. GPS navigation for large-scale photography. Photogramm. Eng. Rem. Sens. 55: 1737–1741.

Biggs, P. H., Wood, G. B., Schreuder, H. T., and Brink, G. E. 1985. Comparison of point-model based and point-Poisson sampling for timber inventory in jarrah forest. Aust. For. Res. 115: 481–493.

Bissell, A. F. 1986. Ordered random selection without replacement. Appl. Stat. 35: 73–75.

Bitterlich, W. 1947. The angle count method (in German). Allg. Forst-u. Holzwirtsch. Ztg. 58: 94–96.

Bitterlich, W. 1949. Das Relaskop. Allg. Forst-u. Holzwirtsch. Ztg. 60: 41–42.

Bitterlich, W. 1952. Das Spiegelrelaskop. Österr. Forst-u. Holzwirtsch. 7: 3–7.

Bitterlich, W. 1962. Relaskop mit Breitskala (in German). Allg. Forstztg. 73: 62–65.

Bitterlich, W. 1980. Gewichtung von Relaskop-Probestämmen am Waldrand (in German). Allg. Forstztg. 91: 343–344.

Bitterlich, W. 1984. The Relascope Idea: Relative Measurements in Forestry. Commonw. Agric. Bur., Farnham Royal, U.K. 242 pp.

Boehmer, D. and Rennie, J. C. 1976. Predicting diameter inside bark for some hardwoods in West Tennessee. Wood Sci. 8: 209–212.

Böhm, W. 1979. Methods of Studying Root Systems. Ecological Studies 33. Springer-Verlag, NY. 188 pp.

Bohrnstedt, G. W. and Goldberger, A. S. 1969. On the exact covariance of products of random variables. J. Amer. Stat. Assoc. 64: 1439–1442.

Born, J. D. and Barnard, J. E. 1983. FINSYS-2: Subsystem TABLE-2 and OUTPUT-2. USDA For. Serv. Gen. Tech. Rep. NE-84. 133 pp.

Bowen, W. D. and Sergeant, D. E. 1983. Mark-recapture estimates of harp seal pup (*Phoco groelandica*) production in the Northwest Atlantic. Can. J. Fish and Aquatic Sci. 40: 728–742.

Bowley, A. L. 1906. Address to the Economic Science and Statistics Section of the British Association for the Advancement of Science. J. Roy. Stat. Soc. 69: 540–558.

Bowley, A. L. 1926. Measurement of the precision attained in sampling. Bull. Int. Stat. Inst. 22: 1–62.

Boyce, S. G. 1975. The use of bole surface in the estimation of woodland productions. Phil. Trans. Roy. Soc. Lond. B 271: 139–148.

Brackstone, G. J. 1987. Small area data: Policy issues and technical challenges. In: Proc. Symp. Small Area Statistics: An Int. Symp. (R. Platek et al., Eds.). John Wiley & Sons, NY. Pp. 3–20.

Breckling, J. V., Chambers, R. L., Dorfman, A. H., Tam, S. M., and Welsh, A. H. 1992. Maximum likelihood inference from sample survey data. Submitted to Intern. Stat. Rev.

Brewer, K. R. W. 1975. A simple procedure for sampling πpswor. Aust. J. Stat. 17: 166–172.

Brewer, K. R. W. and Hanif, M. 1983. Sampling with Unequal Probabilities (Lecture Notes in Statistics). Springer-Verlag, NY. 164 pp.

Brewer, K. R. W. and Särndal, C. E. 1983. Six approaches to enumerative survey sampling. In: Incomplete Data in Sample Surveys, Vol. 3 (W. G. Madow, I. Olkin and D. B. Rubin, Eds.). Academic Press, NY. Pp. 363–405 (includes discussion).

Brewer, K. R. W., Early, L. J., and Joyce, S. F. 1972. Selecting several samples from a population. Aust. J. Stat. 14: 231–239.

Brickell, J. E. 1970. More on diameter tapes and calipers. J. For. 68: 169–170.

Brink, G. B. and Schreuder, H. T. 1992. ONEPHASE: A simulation program to compare one phase sampling strategies. USDA For. Serv. RM Station Pap. RM-302. 8 pp.

Bristol, D. R. 1989. Sample sizes for constructing confidence intervals and testing hypotheses. Stat. Med. 8: 803–811.

Brown, S. and Holgate, P. 1974. The thinned plantation. Biometrika 61: 253–262.

Brownie, C., Anderson, D. R., Burnham, K. P., and Robson, D. S. 1985. Statistical inference from band recovery data—A handbook. US Fish Wildl. Serv., Washington, DC. Resource Publ 156. 305 pp.

Bruce, D. 1925. Forest mensuration today. J. For. 23: 282–286.

Bruce, D. 1975. Evaluating accuracy of tree measurements made with optical instruments. For. Sci. 21: 421–426.

Bruce, D. and Max, T. A. 1990. Use of profile equations in tree volume estimation. Proc. IUFRO Symp. "State-of-the-Art Methodology of Forest Inventory." In: V. J. LaBau and T. Cunia, Tech. Eds. Syracuse, NY. July 30–August 5, 1989. Pp. 213–220. USDA Gen. Tech. Rep. PNW-263.

Bruce, D. and Schumacher, F. X. 1950. Forest Mensuration. McGraw-Hill, NY. 483 pp.

Buckland, W. R. 1951. A review of the literature of systematic sampling. J. Roy. Stat. Soc. B13: 208–215.

Burk, T. E. and Ek, A. R. 1982. Application of empirical Bayes/James Stein procedures to simultaneous estimation problems in forest inventory. For. Sci. 28: 753–771.

Burkhart, H. E. and Tennent, R. B. 1977. Site index equations for radiata pine in New Zealand. N.Z. J. For. Sci. 7: 408–416.

Burnham, K. P. 1980. Is finite population sampling theory always applicable to finite populations? Invited presentation to Amer. Stat. Assoc. Nat. Meeting, Houston, TX. Aug. 1980 (unpublished).

Burnham, K. P., Anderson, D. K., and Lake, J. L. 1980. Estimation of Density from Line Transect Sampling of Biological Populations. Wildl. Monographs 72. The Wildlife Society, Washington DC. 202 pp.

Burrough, P. A. 1986. Principles of Geographic Information Systems for Land Resource Assessments. Oxford Univ. Press, Oxford, U.K. 193 pp.

Byth, K. 1982. On robust distance-based intensity estimators. Biometrics 38: 127–135.

Byth, K. and Ripley, B. D. 1980. On sampling spatial patterns by distance methods. Biometrics 36: 279–284.

Cajander, A. K. 1926. The theory of forest types. Acta For. Fenn. 29: 1–108.

Cajanus, W. 1913. A method for ocular estimation of the growing stock of forests (in Finnish). Tapio 6: 77–79.

Canfield, R. H. 1941. Application of the line interception method in sampling range vegetation. J. For. 39: 388–394.

Cao, V. Q. 1978. Prediction of cubic-foot volume of loblolly pine to any top diameter limit and to any point on the tree bole. M.S. thesis, Virginia Polytech. Inst. and State Univ., Blacksburg, 67 pp.

Cao, V. Q., Burkhart, H. E. and Max, T. A. 1980. Evaluation of two methods for cubic-volume prediction of loblolly pine to any merchantable limit. For. Sci. 26: 71–80.

Carow, J. and Stage, A. 1953. Basal area by the thumb. J. For. 51: 512.

Carroll, R. J. 1989. Covariance analysis in generalized linear measurement error models. Stat. Med. 8: 1075–1093.

Carron, L. C. 1968a. An Outline of Forest Mensuration with Special Reference to Australia. ANU Press, Canberra, Australia. 224 pp.

Carron, L. C. 1968b. A volume tariff system for even-aged forests of *Pinus radiata* (D. Don). Ph.D. thesis, Australian Nat. Univ., Canberra, Australia.

Carron, L. C. 1971. Volume tariff systems. Forestry 44: 145–150.

Cassel, C. M., Särndal, C. E., and Wretman, J. H. 1977. Foundations of Inference in Survey Sampling. Wiley-Interscience, NY. 192 pp.

Castaneda, F. 1979. Converting outside bark to inside bark diameters. Resource Inv. Notes BLM-19: 9–12.

Chakrabarty, R. P. and Rana, D. S. 1977. Joint estimates of change and mean in multi-stage sampling on successive occasions II. Bull. Int. Stat. Inst. 47: 102–107.

Chambers, R. L. 1988. Design-adjusted regression with selectivity bias. Appl. Stat. 37: 323–334.

Chao, M. T. 1982. A general purpose unequal probability sampling plan. Biometrika 69: 653–656.

Chapman, H. H. and Meyer, W. H. 1949. Forest Mensuration. McGraw-Hill, NY. 522 pp.

Chaudhuri, A. and Vos, J. W. E. 1988. Unified Theory and Strategies of Survey Sampling. North-Holland Press, NY. 414 pp.

Chavez, P. S., Jr. and Bowell, J. A. 1988. Comparison of the spectral information content of Landsat thematic mapper and SPOT for three different sites in the Phoenix, Arizona region. Photogramm. Eng. Rem. Sens. 54: 1699–1708.

Chehock, C.R. 1982. Timber volume determination with multi-stage sampling: The application of 3–P sampling and the STXMOD program. USDA For. Serv. Tech. Publ. SA-TP 19. 135 pp.

Cheng, C. S. and Li, K. C. 1987. Optimality criteria in survey sampling. Biometrika 74: 337–345.

Chisman, H. H. and Schumacher, F. X. 1940. On the tree-area ratio and certain of its applications. J. For. 38: 311–317.

Chotai, J. 1974. A note on Rao-Hartley-Cochran method for pps sampling over two occasions. Sankhya C36: 173–180.

Ciesla, W. M. 1989. Aerial photos for assessment of forest decline - A multinational overview. J. For. 87, 2: 37–41.

Clapham, A. R. 1929. The estimation of yield in cereal crops by sampling methods. J. Agric. Sci. 19: 214–235.

Cline, S. 1991. A vertical line intercept method for estimating vegetation structure. In: Forest Health Monitoring: Plot Design and Logistics Study. Methods Manual for Field Measurement and Sample Collection. (K. Druire et al., Eds.). Joint publ. USDA For. Serv. For. Pest Manage. and FIA with Env. Protect. Agency-EMAP, 2nd draft.

Clutter, J. L., Fortson, J. C., Pienaar, L. V., Brister, G. H., and Bailey, R. L. 1983. Timber Management: A Quantitative Approach. John Wiley & Sons, NY. 333 pp.

Cobby, J. M., Ridout, M. S., Bassett, P. J., and Large, R. V. 1985. An investigation into the use of ranked set sampling on grass and grass-clover swards. Grass Forage Sci. 40: 257–263.

Cochran, W. G. 1939. The use of analysis of variance in enumeration by sampling. J. Amer. Stat. Assoc. 34: 492–510.

Cochran, W. G. 1975. Two recent areas of sample survey research. In: A Survey of Statistical Design and Linear Models (J. N. Srivastava, Ed.). North-Holland Press, N.Y. Pp. 101–115.

Cochran, W. G. 1977. Sampling Techniques. 3rd ed. John Wiley & Sons, NY. 428 pp.

Cochran, W. G. and Carroll, S. P. 1953. A sampling investigation of the efficiency of weighting inversely as the estimated variances. Biometrics 9: 447–459.

Cohen, R. 1979. Drug abuse applications: Some regression explorations with national survey data. In: Synthetic Estimates for Small Areas, (J. Steinberg, Ed.). US Nat. Inst. on Drug Abuse Res. Monograph 24. Pp. 194–213.

Connor, W. S. 1966. An exact formula for the probability that two specified sample units will occur in a sample drawn with unequal probability and without replacement. J. Amer. Stat. Assoc. 61: 384–390.

Cooney, T. 1987. Update on portable data recorders. The Compiler 15: 15–17.

Cormack, R. M. 1979. Spatial aspects of competition between individuals. In: Spatial and Temporal Analysis in Ecology, (R. M. Cormack and J. K. Ord, Eds.). Intern. Co-op. Publ. House, Fairland, MD. Pp. 151–211.

Cormack, R. M. 1988. Statistical challenges in the environmental sciences: A personal view. J. Roy. Stat. Soc. A151: 201–210.

Cost, N. D. and McClure, J. P. 1982. Multiresource inventories: Techniques for estimating biomass on a statewide basis. USDA For. Serv., Southeastern For. Exp. Sta. Res. Pap. SE-228. 31 pp.

Courtois, M. and Weill, G. 1985. The SPOT satellite system. In: Monitoring Earth's Ocean, Land, and Atmosphere from Space-Sensors, Systems, and Applications (A. Schnapf, Ed.), Vol. 97, Progress in Astronautics and Aeronautics Series. Amer. Inst. Aeronautics and Astronautics Inc. 1633 Broadway, NY 10019. Pp. 473–523.

Cressie, N. 1991. Statistics for Spatial Data. John Wiley & Sons, NY. 900 pp.

Cumberland, W. G. and Royall, R. M. 1988. Does simple random sampling provide adequate balance? J. Roy. Stat. Soc. B50: 118–124.

Cunia, T. 1965. Continuous forest inventory, partial replacement of samples and multiple regression. For. Sci. 11: 480–502.

Cunia, T. and Chevrou, R. B. 1969. Sampling with partial replacement on three or more occasions. For. Sci. 15: 202–224.

Curtis, R. O. 1967. Height-diameter and height-diameter-age equations for second-growth Douglas-Fir. For. Sci. 13: 365-375.

Czaplewski, R. L. 1989. Graphical analysis of stem taper in model building. Can. J. For. Res. 19: 522–524.

Czaplewski, R. L. and McClure, J. P. 1988. Conditioning a segmented stem profile model for two diameter measurements. For. Sci. 34: 512–522.

Czaplewski, R. L., Brown, A. S., and Guenther, D. G. 1989a. Estimating merchantable tree volume in Oregon and Washington using stem profile models. USDA For. Serv. Res. Pap. RM-286. 15 pp.

Czaplewski, R. L., Brown, A. S. and Walker, R. C. 1989b. Profile models for estimating log end diameters in the Rocky Mountain Region. USDA For. Serv. Res. Pap. RM-284. 7 pp.

Dagum, E. B., Hidiroglou, M. A., and Morry, M. 1987. Sensitivity of small area estimators to misclassification and conceptual differences of variables. In: Proc. Symp. Small Area Statistics: An International Symposium, (R. Platek et al., Eds). John Wiley & Sons, NY. Pp. 175–197.

Dalenius, T. 1957. Sampling in Sweden. Contributions to the Methods and Theories of Sample Survey Practice. Almqvist and Wiksell, Stockholm, Sweden. 247 pp.

Daniels, R. F. 1978. Spatial patterns and distance distributions in young seeded loblolly pine stands. For. Sci. 24: 260–266.

Dargavel, J. G. and Ditchburne, N. 1971. A comparison of the volumes of tree stems obtained by direct measurement and by water displacement. Aust. For. 35: 191–198.

Daubenmire, R. 1952. Forest vegetation of northern Idaho and adjacent Washington, and its bearing on concepts of vegetation classification. Ecol. Monogr. 22: 301–330.

Daughtry, C. S. T., Ranson, K. J., and Biehl, L. L. 1991. C-band backscattering from corn canopies. Int. J. Rem. Sens. 12: 1097–1109.

David, H. A. 1981. Order Statistics. 2nd ed. John Wiley & Sons, NY. 360 pp.

Davies, R. B. and Hutton, B. 1975. The effect of errors in the independent variables in linear regression. Biometrika 62: 383–391.

Davison, A. C. and Hinkley, D. V. 1989. Saddlepoint approximations in resampling methods. Biometrika 75: 417–431.

De Vries, P. G. 1979. A generalization of the Hayne-type estimator as an application of line intersect sampling. Biometrics 35: 743–748.

De Vries, P. G. 1986. Sampling Theory for Forest Inventory. Springer-Verlag, NY. 399 pp.

Dell, T. R. 1969. The theory and some applications of ranked set sampling. Ph.D. dissertation, Univ. of Georgia. Univ. Microfilms Inc., Ann Arbor, MI. 83 pp.

Dell, T. R. and Clutter, J. L. 1972. Ranked set sampling with order statistics background. Biometrics 28: 545–553.

Demaerschalk, J. P. and Kozak, A. 1974. Suggestions and criteria for more effective regression sampling. Can. J. For. Res. 4: 341–348.

Deming, W. E. 1975. On probability as a basis for action. Amer. Stat. 29: 146–152.

Deming, W. E. 1977. An essay on screening, or on two-phase sampling, applied to surveys of a community. Int. Stat. Rev. 45: 29–37.

Deming, W. E. and Stephan, F. F. 1940. On a least square adjustment of a sampled frequency table when the expected marginal totals are known. Ann. Math. Stat. 11: 427–444.

Dempster, A. P., Laird, N., and Rubin, D. B. 1977. Maximum likelihood from incomplete data via the EM algorithm. J. Roy. Stat. Soc. B39: 1–38.

Deng, L. Y. and Wu, C. F. J. 1987. Estimation of variance of the regression estimator. J. Amer. Stat. Assoc. 82: 568–576.

DiCiccio, T. J., Field, C. A., and Fraser, D. A. S. 1990. Approximation of marginal tail probabilities and inference for scalar parameters. Biometrika 77: 77–95.

Diggle, P. J. 1975. Contributions to the discussion of Drs. Cliff and Ord's paper. J. Roy. Stat. Soc. B37: 333–334.

Diggle, P. J. 1979. Statistical methods for spatial point patterns in ecology. In: Spatial and Temporal Analysis in Ecology (R. M. Cormack and J. K. Ord, Eds.). Intern. Co-op. Publ. House, Fairland, MD. Pp. 95–150.

Dixon, B. L. and Howitt, R. E. 1979. Continuous forest inventory using a linear filter. For. Sci. 25: 675–689.

Dress, P. E. 1970. A system for the stochastic simulation of even-aged forest stands of pure species composition. Ph.D. thesis, Purdue Univ. Univ. Microfilms Inc. No. 71-2587, Ann Arbor, MI. 253 pp.

Duggin, M. J., Hopkins, P. F., and Brock, R. H. 1990. A survey of remote sensing methodology for forest inventory. In: Proc. Symp. State-of-the-Art Methodology of Forest Inventory. Syracuse, NY. USDA For. Serv. Pacific Northwest Res. Sta., Gen. Tech. Rep. PNW-GTR-263. Pp. 267–285.

Duncan, G. J. and Kalton, G. 1987. Issues of design and analysis of surveys across time. Int. Stat. Rev. 55: 97–117.

Eberhardt, L. L. 1978. Transect methods for population studies. J. Wildl. Manage. 42: 1–31.

Edwards, A. W. F. 1972. Likelihood. Cambridge Univ. Press, NY. 235 pp.

Efron, E. 1979. Bootstrap methods: Another look at the jackknife. Ann. Stat. 7: 1–26.

Efron, B. and Gong, G. 1983. A leisurely look at the bootstrap, the jackknife, and cross-validation. Amer. Stat. 37: 36–48.

Efron, B. and Stein, C. 1981. The jackknife estimate of variance. Ann. Stat. 9: 586–596.

Ek, A. R. 1971. A comparison of some estimators in forest sampling. For. Sci. 17: 2–13.

Ek, A. R. and Issos, J. N. 1978. Bayesian theory and multi-resource inventory. In Proc. Conf. Integrated Inventories of Renewable Natural Resources. Tucson, AZ. Jan. 8–12, 1978. USDA For. Serv. Gen. Tech. Rep. RM-55, pp. 291–298.

Ekstrand, S. 1986. Thematic mapper in tropical forest inventories: A comparison with Landsat MSS data, panchromatic aerial photography, and colour infrared aerial photography. In: Proc. 20th Int. Symp. Rem. Sens. Environment, Env. Res. Inst. of Mich. Nairobi, Kenya. Dec. 1986. Pp. 1299–1307.

Ellis, J. C. 1982. A three-dimensional formula for coniferous log volumes in New Zealand. N.Z. For. Res. Inst. Bull. 20. 12 pp.

Ellis, J. C. 1984. Forest Research Institute evaluates new computer-based bush skid weighing device. For. Indust. 15(8): 37, 39.

Ellis, J. C. and Duff, P. 1973. Tests of some methods of calculating tree volume from sectional measurements. N. Z. For. Serv., FRI For. Mensuration Rept. 39 (Unpublished).

Empire Forestry Association. 1953. British Commonwealth Forest Terminology, Part 1. Empire Forestry Assoc., London, UK.

Ericksen, E. P. 1973. A method for combining sample survey data and symptomatic indicators to obtain estimates for local areas. Demography 10: 137–160.

Ericksen, E. P. 1974. A regression method for estimating population changes of local areas. J. Amer. Stat. Assoc. 69: 867–875.

Ericson, W. A. 1988. Bayesian inference in finite populations. In: Handbook of Statistics, Vol. 6 (P. R. Krishnaiah and C. R. Rao, Eds.). North Holland Press, NY. Ch. 9, pp. 213–246.

Evans, D. L., Zhu, Z., Eggen-McIntosh, S., Mayoral, P. G. and Ornelas de Anda, J. L. 1992. Mapping Mexico's Forest Lands with Advanced Very High Resolution Radiometer. USDA For. Serv. South. For. Exp. Sta. Res. Note 50–367.4p.

Everitt, J. H. 1988. Introduction to videography: Historical overview, relation to remote sensing, advantages, disadvantages. In: First Workshop on Videography. Terre Haute, IN. May 19–20, 1988. Amer Soc. for Photogrammetry and Rem. Sensing. Pp. 1–4.

Everitt, J. H., Lulla, K., Escobar, D. E., and Richardson, A. J. 1990. Aerospace video imaging systems for rangeland management. Photogramm. Eng. Rem. Sens. 56: 343–349.

FAO. 1954. Jal hypsometer. Forest Equipment Note A.2.54. FAO, United Nations.

Fay, R. E. and R. A. Herriott. 1979. Estimates of income for small places: An application of James/Stein procedures to census data. J. Amer. Stat. Assoc. 74: 269–277.

Fellegi, I. P. and Holt, D. A. 1976. Systematic approach to automatic edit and imputation. J. Amer. Stat. Assoc. 71: 17–35.

Fender, D. E. and Brock, G. A. 1963. Point center extension: A technique for measuring economic growth and yield of merchantable forest stands. J. For. 61: 109–114.

Ferguson, I. S., O'Hara, A. J., Wood, G. B., and Miles, J. A. M. 1984. Calibrating dendrometers for estimating tree volume. Aust. For. Res. 14: 253–263.

Ferree, M. J. 1946. The pole caliper. J. For. 44: 594–595.

Ferree, M. J., Shearer, T. D., and Stone, E. J. 1958. A method of evaluating site quality in young red pine plantations. J. For. 56: 328–332.

Finch, H. D. S. 1957. Plotless enumeration with angle gauges. Forestry 30: 173–192.

Fins, L. and Rust, M. 1988. Advantages of using electronic field data recorders. The Compiler 6: 14–16.

Fisher, R. A. 1922. On the mathematical foundations of theoretical statistics. Phil. Trans. Roy. Soc. Lond. A222: 309–368.

Fisher, R. A. 1925. Statistical Methods for Research Workers. Oliver and Boyd, London, U.K. 239 pp.

Fisher, R. A. and Yates, F. 1957. Statistical Tables for Biological, Agricultural, and Medical Research, 5th ed. Oliver and Boyd, Edinburgh, U.K. 146 pp.

Flewelling, J. W. 1981. Compatible estimates of basal area and basal area growth from remeasured point samples. For. Sci. 27: 191–203.

Fogel, R. 1983. Root turnover and productivity of coniferous forests. Plant and Soil 71: 75–85.

Fogel, R. 1985. Roots as primary producers in below-ground ecosystems. In Ecological Interactions in Soil (A. H. Fitter et al., Eds.). Spec. Publ. No. 4, Brit. Ecol. Soc. Blackwell's Scientific Publications, Oxford, UK. Pp. 23–36.

Ford, D. L. 1983. An overview of hot-deck procedures. In: Incomplete Data in Sample Surveys, Vol. 2, Theory and Bibliographies (W. G. Madow, I. Olkin, and D. B. Rubin, Eds.). Academic Press, NY. Pp. 185–207.

Ford-Robertson, F. C. 1971. Terminology of Forest Science, Technology, Practice and Products. Multilingual Forest Terminology Series No. 1. Soc. Amer. For., Washington, DC. 349 pp.

Forest Resources Economics Research Staff. 1987. Forest Service Resource Inventory: An Overview. USDA For. Serv. Aug. 1987. 29 pp.

Forslund, R. R. 1982. A geometrical tree volume model based on the location of the center of gravity of the bole. Can. J. For. Res. 12: 215–221.

Frayer, W. E. 1966. Weighted regression in successive forest inventories. For. Sci. 12: 464–472.

Frayer, W. E. 1978. Stratification in double sampling—"The easy way out may sometimes be the best way." USDI BLM Resource Inv. Notes BLM-10. Denver Federal Center, Bldg. 50, Denver, CO 80225.

Frayer, W. E. and Furnival, G. M. 1967. Area change estimates from sampling with partial replacement. For. Sci. 13: 72–77.

Frayer, W. E., Monahan, T. J., Bowden, D. C., and Graybill, F. A. 1983. Status and trends of wetlands and deepwater habitats in the conterminous United States, 1950s to 1970s. Rep. Dept. of For. Wood Sci., Colorado State Univ., Ft. Collins. 31 pp.

Freden, S. C. and Gordon, F. Jr. 1983. Landsat satellites. In: Manual of Remote Sensing, Vol. 1, 2nd ed. Amer. Soc. Photogramm., Falls Church, VA. The Sheridan Press. Ch. 12, pp. 517–570.

Fuller, W. A. 1986. Small area estimation as a measurement error problem. In: Proc. Conf. on Survey Res. Meth. in Agric. Amer. Stat. Assoc., Leesburg, VA. June 15–18, 1986. Pp. 7–21.

Fuller, W. A. 1987. Measurement Error Models. John Wiley & Sons, NY. 440 pp.

Furnival, G. M. 1979. Forest sampling-Past performance and future expectations. In: Proc. SAF Workshop on Forest Resource Inventories, Fort Collins, CO. July 23–26, 1979. Vol. 1, pp. 320–326.

Furnival, G. M., Gregoire, T. G., and Grosenbaugh, L. R. 1987. Adjusted inclusion probabilities with 3P sampling. For. Sci. 33: 617–631.

Furnival, G. M., Valentine, H. T., and Gregoire, T. G. 1986. Unbiased estimation of log volume by importance sampling. For. Sci. 32: 1073–1078.

Gambill, C. W., Wiant, H. V., Jr., and Yandle, D. O. 1985. Optimum plot size and BAF. For. Sci. 31: 587–594.

Geary, R. C. 1930. The frequency distribution of the quotient of two normal variates. J. Roy. Stat. Soc. A93: 442–446.

Gelfand, A. E. and Smith, A. F. M. 1990. Sampling-based approaches to calculating marginal densities. J. Amer. Stat. Assoc. 85: 398–409.

Gelfand, A. E., Hills, S. E., Racino-Poon, A., and Smith, A. M. F. 1990. Illustration of Bayesian inference in normal data models using Gibbs sampling. J. Amer. Stat. Assoc. 85: 972–985.

Geron, C. D. and Ruark, G. A. 1988. Comparison of constant and variable allometric ratios for predicting foliar biomass of various tree genera. Can. J. For. Res. 18: 1298–1304.

Gervin, J. C., Kerber, A. G., Witt, R. G., Lu, Y. C., and Sekhon, R. 1985. Comparison of level I land cover classification accuracy for MSS and AVHRR data. Int. J. Rem. Sens. 6: 45–57.

Ghangurde, P. D. and Rao, J. N. K. 1969. Some results on sampling over two occasions. Sankhya A31: 463–472.

Gholz, H. L., Hendry, L. C., and Cropper, W. P., Jr. 1986. Organic matter dynamics of fine roots in plantations of slash pine (*Pinus elliottii*) in north Florida. Can. J. For. Res. 16: 529–538.

Ghosh, M. and Meeden, G. 1986. Empirical Bayes estimation in finite population sampling. J. Amer. Stat. Assoc. 81: 1058–1062.

Ghosh, M. and Sinha, B. K. 1990. On the consistency between model- and design-based estimators in survey sampling. Comm. Stat. Theor. Meth. 19: 689–702.

Ghosh, S. P. 1963. Post cluster sampling. Ann. Math. Stat. 34: 587–597.

Gilbert, J. P., Light, R. J., and Mosteller, F. 1977. In: Statistics and Public Policy, W. B. Farley and F. Mosteller (Eds.). Addison-Wesley, Reading, MA. Pp. 185–241.

Gill, A. M., Brack, C. L., and Hall, T. 1982. Bark probe—An instrument for measuring bark thickness of eucalypts. Aust. For. 45: 206–208.

Gini, C. and Galvani, L. 1929. An application of the representative method in the last Italian census of the population (in Italian). Annali di Statistica 4: 1–107.

Gleser, L. J. and Hwang, J. T. 1988. The nonexistence of 100 $(1-\alpha)\%$ confidence sets of finite expected diameter in errors in variables and related models. Ann. Stat. 15: 1351–1362.

Godambe, V. P. 1955. A unified theory of sampling from finite populations. J. Roy. Stat. Soc. B17: 267–278.

Godambe, V. P. 1966. A new approach to sampling from finite populations. I and II. J. Roy. Stat. Soc. B28: 310–328.

Godambe, V. P. 1978. Discussion of Hansen, Madow, and Tepping. In: Am. Stat. Assoc. Proc. Survey Res. Methods. Pp. 92–93,

Godambe, V. P. and Joshi, V. M. 1965. Admissibility and Bayes estimation in sampling finite populations. Ann. Math. Stat. 36: 1707–1722.

Godman, R. M. 1949. The pole diameter tape. J. For. 47: 585–589.

Goodman, L. A. 1949. On the estimation of the number of classes in a population. Ann. Math. Stat. 20: 572–579.

Goodman, L. A. 1960. On the exact variance of products. J. Amer. Stat. Assoc. 55: 708–713.

Goodman, L. A. 1961. Snowball sampling. Ann. Math. Stat. 32: 148–170.

Goodman, R. and Kish, L. 1950. Controlled selection—a technique in probability sampling. J. Amer. Stat. Assoc. 45: 350–372.

Gordon, A. 1983. Comparison of polynomial taper equations. N.Z. J. For. Sci. 13: 146–155.

Goulding, C. J. 1979. Cubic spline curves and calculations of volume of sectionally measured trees. N.Z. J. For. Sci. 9: 89–99.

Goulding, C. J. 1986. Measurement of tree crops. In: 1986 Forestry Handbook, (H. Levack, Ed.). N.Z. Inst. of Foresters, Wellington, New Zealand. Sect. 6.1.5, pp. 80–81.

Goulding, C. J. and Murray, J. C. 1976. Polynomial taper equations that are compatible with tree volume equations. N.Z. J. For. Sci. 5: 313–322.

Goward, S. N., Dye, D., Kerber, A., and Kalb, V. 1987. Comparison of North and South American biomes from AVHRR observations. Geocarto Int. 1: 27–39.

Gray, H. R. 1956. The form and taper of forest tree stems. Imp. For. Inst., Oxford, UK. Paper 32.

Green, E. J. and Strawderman, W. E. 1985. The use of Bayes/empirical Bayes estimation in individual tree volume equation development. For. Sci. 31: 975–990.

Green, E. J., Thomas, C. E., and Strawderman, W. E. 1987. Stein-rule estimation of timber removals by county. For. Sci. 33: 1054–1061.

Gregoire, T. G. 1982. The unbiasedness of the mirage correction procedure for boundary overlap. For. Sci. 28: 504–508.

Gregoire, T. G. 1984. The jackknife: An introduction with applications in forestry data analysis. Can. J. For. Res. 14: 493–497.

Gregoire, T. G. 1987. Generalized error structure for forestry yield models. For. Sci. 33: 423–444.

Gregoire, T. G. 1992. Estimation of forest growth from successive surveys. For. Ecol. Manage. (in press).

Gregoire, T. G. and Scott, C. T. 1990. Sampling at the stand boundary: A comparison of the statistical performance among eight methods. In: Proc. XIX World Forestry Congress IUFRO. Montreal, Canada. Aug. 5–11, 1990. Publ. No. FWS-3–90, Virginia Polytech. Inst. and Univ., Blacksburg, pp. 78–85.

Gregoire, T. G., Valentine, H. T., and Furnival, G. M. 1986. Estimation of bole volume by importance sampling. Can. J. For. Res. 16: 554–557.

Gregoire, T. G., Valentine, H. T., and Furnival, G. M. 1987. Sampling methods for estimating stem volume and volume increment. For. Ecol. Manage. 21: 311–323.

Gregoire, T. G., Valentine, H. T., and Furnival, G. M. 1992. Estimation of bole surface area and bark volume with Monte Carlo methods. Biometrics (in press).

Gregoire, T. G., Zedaker, S. M., and Nicholas, N. S. 1990. Modeling relative error in stem basal area estimates. Can. J. For. Res. 20: 496–502.

Greig-Smith, P. 1964. Quantitative Plant Ecology. 2nd ed. Butterworths, London, U.K. 256 pp.

Grosenbaugh, L. R. 1948. Improved cubic volume computation. J. For. 46: 299–301.

Grosenbaugh, L. R. 1952. Plotless timber estimates—New, fast, easy. J. For. 50: 32–37.

Grosenbaugh, L. R. 1954. New tree-measurement concepts: Height accumulation, giant tree, taper and shape. USDA For. Serv., Southern For. Exp. Sta., Asheville, NC. Occ. Pap. No. 134. 32 pp.

Grosenbaugh, L. R. 1958. Point-sampling and line sampling: Probability theory, geometric implications, synthesis. USDA For. Serv. Southeastern For. Exp. Sta., Asheville, NC. Occ. Pap. No. 160, 34 pp.

Grosenbaugh, L. R. 1963. Optical dendrometers for out-of-reach diameters: A conspectus and some new theory. For. Sci. Monograph 4. 47 pp.

Grosenbaugh, L. R. 1964. Some suggestions for better sample-tree measurement. In: Proc. Soc. Amer. Foresters, Boston, MA. 1963. Pp. 36–42.

Grosenbaugh, L. R. 1967a. The gains from sample-tree selection with unequal probabilities. J. For. 65: 203–206

Grosenbaugh, L. R. 1967b. STX-FORTRAN-4 PROGRAM for estimates of tree populations from 3P sample-tree-measurements. USDA For. Serv., Pac. S.W. For. and Range Expt. Sta. Res. Pap. PSW-13. 76 pp.

Grosenbaugh, L. R. 1967c. Choosing units of measure. In: Proc. Conf. "Young-growth forest management in California." Univ. of California Agr. Extension Serv., Berkeley. Mar. 1–3, 1967. Pp. 143–146.

Grosenbaugh, L. R. 1971. STX 1–11–71 for dendrometry of multistage 3P samples. USDA For. Serv. Publ. FS-277. 63 pp.

Grosenbaugh, L. R. 1974. STX 3–3–73: Tree content and value estimation using various sample designs, dendrometry methods, and V-S-L conversion coefficients. USDA For. Serv. Res. Pap. SE-117.

Grosenbaugh, L. R. 1976. Approximate sampling variance of adjusted 3–P estimates. For. Sci. 22: 173–176.

Grosenbaugh, L. R. 1979. 3P sampling theory, examples, and rationale. USDA BLM Tech. Note 331. 18 pp.

Guenther, W. C. 1965. Concepts of Statistical Inference. McGraw-Hill, NY. 353 pp.

Gugan, D. J. and Dowman, I. J. 1988. Topographic mapping from SPOT imagery. Photogramm. Eng. Rem. Sens. 54: 1409–1414.

Hafley, W. L. and Schreuder, H. T. 1976. Some non-normal bivariate distributions and their potential for forest applications. In: Proc. IUFRO World Forestry Congr. Div. VI. Pp. 104–114. Oslo, Norway.

Hafley, W. L. and Schreuder, H. T. 1977. Statistical distributions for fitting diameter and height in even-aged stands. Can. J. For. Res. 7: 481–187.

Hájek, J. 1949. Representative sampling by the method of two phases (in Czechoslovakian). Statisticky Obzor. 29: 384–395.

Hájek, J. 1957. Some contributions to the theory of probability sampling. Bull. Int. Stat. Inst. 36: 127–133.

Hájek, J. 1964. Asymptotic theory of rejective sampling with varying probabilities from a finite population. Ann. Math. Stat. 35: 1491–1523.

Hájek, J. 1981. Sampling from a Finite Population. Statistics: Textbooks and Monographs, Vol. 3. Marcel Dekker, Inc., NY. 247 pp.

Halls, L. K. and Dell, T. R. 1966. Trial of ranked-set sampling for forage yields. For. Sci. 12: 22–26.

Häme, T. 1987. Satellite image aided change detection. Remote sensing forest inventory. Univ. of Helsinki, Dept. of For. Mens. Manage., Helsinki, Finland. Res. Note. 19: 47–60.

Häme, T. and Rantasuo, M. 1988. Shuttered camera—Aerial color video imaging in the visible and near infrared. Photogramm. Eng. Rem. Sens. 54: 1735–1738.

Hamilton, D. A., Jr. 1970. Precision requirements for some information in timber management decisions. Ph.D. dissertation, Iowa State Univ., Ames. Iowa. 114 pp.

Hamilton, D. A., Jr. 1979. Setting precision for natural resource inventories: The manager and the mensurationist. J. For. 77: 667–670.

Hamilton, G. D. 1975. Forest Mensuration Handbook. Forestry Commission Booklet No. 39. Her Majesty's Stationery Office, London, U.K. 275 pp.

Hammersley, J. M. and Morton, K. W. 1956. A new Monte Carlo technique: Antithetic variates. Proc. Cambridge Philos. Soc. 52: 449–475.

Hammersley, J. M. and Handscomb, D. C. 1964. Monte Carlo Methods. Methuen, London, UK. 178 pp.

Hanif, M. and Brewer, K. R. W. 1980. Sampling with unequal probabilities without replacement: a review. Intern. Stat. Rev. 48: 317–335.

Hansen, M. H. 1987. Some history and reminiscences on survey sampling. Stat. Sci. 2: 162–179.

Hansen, M. H. 1990. The eastwide forest inventory database. In: Proc. Symp. State-of-the-Art Methodology of Forest Inventory. Syracuse, NY. USDA For. Serv. Pacific Northwest Res. Sta., Gen. Tech. Rep. PNW-GTR-263. Pp. 545–552.

Hansen, M. H. and Hurwitz, W. N. 1943. On the theory of sampling from finite populations. Ann. Math. Stat. 14: 333–362.

Hansen, M. H. and Madow, W. G. 1976. Some important events in the historical development of sample surveys. In: On the History of Statistics and Probability, D. B. Owen (Ed.). Marcel Dekker, NY. Pp. 75–102.

Hansen, M. H., Hurwitz, W. N., and Madow, W. G. 1953. Sample Survey Methods and Theory, Vol. I and II. John Wiley & Sons, NY. 638 pp. 332 pp.

Hansen, M. H., Madow, W. G., and Tepping, B. J. 1983. An evaluation of model-dependent and probability-sampling inferences in sample surveys. J. Amer. Stat. Assoc. 78: 776–807 (includes discussion).

Harris, M., Horvitz, D. G., and Mood, A. M. 1948. On the determination of sample size in designing experiments. J. Amer. Stat. Assoc. 43: 391–402.

Hart, H. M. J. 1928. Stem number and thinning (in Dutch). Proefstation Boschwesen, Batavia, Indonesia Mededelingen 21.

Hart, M. E. 1990. US Forest Service experiences with field computers. Paper presented to SAF Southeastern Field Comput. Conf., Atlanta, GA. March 6–7, 1990 (unpublished).

Hartley, H. O. and Rao, J. N. K. 1962. Sampling with unequal probabilities and without replacement. Ann. Math. Stat. 33: 350–374.

Hartley, H. O. and Rao, J. N. K. 1975. Some comments on labels: A rejoinder to a section on Godambe's paper, "A reply to my critics." Sankhya 37: 163–170.

Hartley, H. O. and Ross, A. 1954. Unbiased ratio estimators. Nature 174: 270–271.

Hasel, A. A. 1942a. Estimation of volume in timber stands by strip samplings. Ann. Math. Stat. 13: 179–206.

Hasel, A. A. 1942b. Sampling error of cruises in the California pine region. J. For. 40: 211–217.

Hazard, J. 1977. Estimating area in sampling forest populations on two successive occasions. For. Sci. 23: 253–267.

Hazard, J. and Promnitz, L. C. 1974. Design of successive forest inventories: Optimization by convex mathematical programming. For. Sci. 20: 117–127.

Heeringa, S. G. 1981. Small area estimation: Prospects for the survey of income and program participation. A review of the literature and interim report on the methodology. Inst. For. Sci. Res., Univ. of Michigan. Ann Arbor. 100 pp.

Heitjan, D. F. and Rubin, D. B. 1986. Inference from coarse data using multiple imputation. Paper presented to the Interface Conf. of Computer Science and Statistics, Fort Collins, CO (unpublished).

Heller, R. C. 1970. Imaging with photographic sensors. In: Remote Sensing with

Special Reference to Agriculture and Forestry. Nat. Acad. of Sci., Washington, DC. Pp. 35–72.

Heller, R. C. 1974. Remote sensing as an inventory tool for forest insect survey. In: Proc. Monitoring Forest Environment Through Successive Sampling. State Univ. of New York, Syracuse. June 24–26, 1974. Pp. 321–343.

Heller, R. C. and Ulliman, J. J. 1983. Forest resource assessment. In: Manual of Remote Sensing, Vol. 2, 2nd ed. Amer. Soc. Photogramm., Falls Church, VA. The Sheridan Press. Ch. 34, pp. 2229–2324.

Hermann, R. K. 1977. Growth and production of tree roots: A review. In: The Below Ground Ecosystem: A Synthesis of Plant-Associated Processes (J. K. Marshall, Ed.). Range Sci. Dept., Colorado State Univ., Fort Collins. Science Series No. 26, pp. 7–28.

Hidiroglou, M. A. and C. E. Särndal. 1985. An empirical study of some regression estimators for small domains. Survey Meth. 11: 65–77.

Hinkley, D. V. 1988. Bootstrap method. J. Roy. Stat. Soc. B50: 321–337.

Hirata, T. 1955. Height estimation through Bitterlich's method vertical angle-count sampling (in Japanese). Jap. J. For. 37: 479–480.

Hoffer, R. M. 1986. Digital analysis techniques for forestry applications. In: Remote Sensing Reviews. Harwood Academic Press, NY. Pp. 61–110.

Hoffer, R. M. 1988. Remote sensing from space—Two decades of change. In: Proc. Resource Technology 88, Int. Symp. Advanced Techniques in Natural Resource Management, Ft. Collins, CO. June 20–23, 1988. Published by Amer. Soc. Photogramm. and Remote Sens., Falls Church, VA. Pp. 4–17.

Hoffer, R. M., Fleming, M. D., Bartolucci, L. A., Davis, S. M., and Nelson, R. F. 1979. Digital processing of Landsat MSS and topographic data to improve capabilities for computerized mapping of forest cover types. Purdue University, W. Lafayette, IN. LARS Tech. Rep. 011579. 159 pp.

Holgate, P. 1965. The distance from a random point to the nearest point of a closely packed lattice. Biometrika 52: 261–263.

Holgate, P. 1967. The angle count method. Biometrika 54: 615–623.

Holt, D. and Skinner, C. J. 1989. Components of change in repeated surveys. Int. Stat. Rev. 57: 1–18.

Honer, T. G. and Hegyi, F. 1990. Forest inventory—Growth and yield in Canada: Past, present and future. For. Chron. 66, 2: 112–118.

Horvitz, D. G. and Thompson, D. J. 1952. A generalization of sampling without replacement from a finite universe. J. Amer. Stat. Assoc. 47: 663–685.

Hubback, J. A. 1927. Sampling for rice yield in Bihar and Orissa. Imp. Agric. Res. Inst., Pusa, India. Bull. 166. (Reprinted in: Sankhya A7: 281–294 (1946)).

Hummel, F. C. 1951. Instruments for the measurement of height, diameter and taper on standing trees. For. Abstr. 12: 261–269.

Hummel, F. C. 1954. Definition of thinning treatments. In: Proc. IUFRO Congr. Rome, Italy. 1953. Sect. 23, Pap. No. 4. 6 pp.

Hummel, F. C. 1955. The volume-basal area line: A study in forest mensuration. U.K. For. Comm. Bull. No. 24. 84 pp.

Husch, B. 1963. Forest Mensuration and Statistics. Ronald Press, NY. 474 pp.

Husch, B., Miller, C. I., and Beers, T. W. 1982. Forest Mensuration. 3rd ed. John Wiley & Sons, NY. 402 pp.

Iachan, R. 1982. Systematic sampling: A critical review. Int. Stat. Rev. 50: 293–303.

Iachan, R. 1985. Robust designs for ratio and regression estimation. J. Stat. Plan. Inf. 11: 149–161.

Iles, K. 1979. Systems for the selection of truly random samples from tree populations and the extension of variable plot sampling to the third dimension. Ph.D. Dissertation, Univ. British Columbia, Vancouver, Canada. 144 pp. (Diss. Abstr. 40/04B).

Iles, K. 1981. Permanent variable plots for forest growth. Paper presented to meeting of Western Forest Mensurationists, Sun Valley, ID. Unpublished manuscript.

Iles, K. 1988. Growth on permanent angle-count plots, historical estimation procedures through critical height sampling. In: Proc. IUFRO Conf. Forest Growth Modeling and Prediction. Minneapolis, MN. Aug. 23–27, 1987. USDA For. Serv. NC For. Exp. Sta. Gen. Tech. Rep. NC 120. Vol. 2; pp. 1105–1113.

Iles, K. 1990. Critical height sampling: A workshop on the current state of the technique. In: Proc. IUFRO Symp. "State-of-the-Art Methodology of Forest Inventory," Syracuse, NY. August 1989. USDA For. Serv. Gen. Tech. Rep. PNW-GTR-263. Pp. 74–83.

Iles, K. and Beers, T. W. 1983. Growth information from variable plot sampling. In: Proc. Int. Conf. Renewable Resource Inventories for Monitoring Changes and Trends (J. F. Bell and T. Atterbury, Eds.). Corvalis, OR.

Iles, K. and Wilson, W. H. 1988. Changing angle gauges in variable plot sampling: Is there a bias under ordinary conditions? Can. J. For. Res. 18: 768–773.

IMSL. 1987. International Mathematical Statistical Library Version 1.0. 2500 Park West Tower 1, 2500 City West Boulevard, Houston, Texas, 77042-3020, USA.

Ilvessalo, Y. 1923. Research into the condition of the private forests in the middle part of government district Häme (Tawastland) (in German). Acta For. Fenn. 26.

Innes, J. L. 1988a. Forest health surveys: Problems in assessing observer objectivity. Can. J. For. Res. 18: 560–565.

Innes, J. L. 1988b. Forest health surveys—A critique. Env. Poll. 54: 1–15.

IUFRO. 1959. The standardization of symbols in forest mensuration. Int. Union of For. Res. Organizations. FAO, Rome, Italy. 32 pp.

Jackson, T. J. and Schmugge, T. J. 1991. Vegetation effects on the microwave emission of soils. Rem. Sens. Env. 36: 203–212.

James, W. and Stein, C. 1961. Estimation with quadratic loss. In: Proc. Fourth Berkeley Symp. Math Stat. Prob. Vol. 1. Univ. of California Press, Berkeley. Pp. 361–379.

Jerram, M. R. K. 1939. Elementary Forest Mensuration. Thomas Murby and Co., London, U.K. 124 pp.

Jessen, R. J. 1942. Statistical investigation of a sample survey for obtaining farm facts. Iowa Agr. Exp. Sta. Res. Bull. 304. 104 pp.

Johnson, N. L. and Kotz, S. 1969. Discrete Distributions. Houghton Mifflin Co., Boston. 328 pp.

Johnson N. L. and Kotz, S. 1970a. Continuous Univariate Distributions—1. Houghton Mifflin, Boston, MA. 300 pp.

Johnson, N. L. and Kotz, S. 1970b. Continuous Univariate Distributions—2. Houghton Mifflin, Boston, MA. 306 pp.

Johnson, N. L. and Kotz, S. 1972. Distributions in Statistics: Continuous Multivariate Distributions. John Wiley & Sons, NY. 333 pp.

Johnson, T. E. and Wood, G. B. 1987. Simple linear model reliably predicts bark thickness of radiata pine in the Australian Capital Territory. For. Ecol. Manage. 22: 173–183.

Jolly, G. M. 1965. Explicit estimates from capture–recapture data with both death and immigration–stochastic model. Biometrika 52: 225–247.

Jolly, G. M. 1982. Mark-recapture models with parameters constant in time. Biometrics 38: 301–321.

Joshi, V. M. 1988. Randomization versus the prediction approach in survey sampling. J. Stat. Plan. Inf. 20: 215–219.

Jukola-Sulonen, E. L. and Salemaa, M. 1985. A comparison of different sampling methods of quantitative vegetation analysis. Silva Fennica 19: 325–337.

Kaiser, L. 1983. Unbiased estimation in line-intercept sampling. Biometrics 39: 965–976.

Kalbfleisch, J. D. and Sprott, D. A. 1969. Applications of likelihood and fiducial probability to sampling finite populations. In: New Developments in Survey Sampling. Eds. N.L. Johnson and H. Smith. Wiley–Interscience, NY. pp. 358–389.

Kaltenberg, M. C. 1978. Evaluation of regeneration sampling methods: A Monte Carlo analysis using simulated stands. State of Washington Dept. of Natural Resources, Olympia. DNR Rep. No. 39. 50 pp.

Kalton, G. and Anderson, D. W. 1986. Sampling rare populations. J. Roy. Stat. Soc. A149: 65–82.

Kalton, G. and Kasprzyk, D. 1982. Imputing for missing survey responses. In: Proc. Conf. Amer. Stat. Assoc., Sect. on Survey Research Methods. Aug. 1982. Pp. 22–31.

Kalton, G. and Kasprzyk, D. 1986. The treatment of missing survey data. Survey Meth. 12: 1–16.

Kidwell, K. B. (Ed.) 1991. NOAA Polar Orbiter Data Users Guide (TIROS-N, NOAA-6, NOAA-7, NOAA-8, NOAA-9, NOAA-10, NOAA-11, and NOAA-12). Ch. 1, 3. National Oceanic and Atmospheric Admin. Washington DC.

Kieser, R. and Mulligen, T. J. 1984. Analysis of echo counting data. Can. J. Fish Aquatic Sci. 41: 451–458.

Kilpatrick, D. J. 1981. Optimum allocation in stratified sampling of forest populations on successive occasions. For. Sci. 27: 730–738.

Kimura, D. K. and Lemberg, N. A. 1981. Variability of line intercept density estimates (A simulation study of the variance of hydroacoustic biomass estimates). Can. J. Fish Aquatic Sci. 38: 1141–1152.

Kitamura, M. 1962. On an estimate of the volume of trees in a stand by the sum of critical heights (in Japanese). 73 Kai Nichi Rin Ko: 64–67.

Klapp, E. 1934. Methods of studying grassland stands (in German). Rep. 3rd Grassland Conf. Zurich, Switzerland. Pp. 193–202.

Klitscher, J. A. 1983. Permanent Sample Plot System: ICL 2980 Data Entry User Manual. For. Res. Inst., N.Z. For. Serv., Rotorua, New Zealand.

Knoebel, B. R. and Burkhart, H. E. 1991. A bivariate distribution approach to modeling forest diameter distributions at two points in time. Biometrics 47: 241–253.

Kott, P. S. 1986. Some asymptotic results for the systematic and stratified sampling of a finite population. Biometrika 73: 485–491.

Kotz, S. and Johnson, N. L. (Eds.). 1982a. Encyclopedia of Statistical Sciences, Vol 1. John Wiley & Sons, NY. 480 pp.

Kotz, S. and Johnson, N. L. (Eds.) 1982b. Encyclopedia of Statistical Sciences, Vol. 2. John Wiley & Sons, NY. 613 pp.

Kotz, S. and Johnson, N. L. (Eds.). 1983. Encyclopedia of Statistical Sciences Vol. 3. John Wiley & Sons, NY. 722 pp.

Kotz, S. and Johnson, N. L. (Eds.). 1985. Encyclopedia of Statistical Sciences Vol. 5. John Wiley & Sons, NY. 741 pp.

Kotz, S. and Johnson, N. L. (Eds.) 1988. Encyclopedia of Statistical Sciences. Vol. 8. John Wiley & Sons, NY. 870 pp.

Kotz, S. and Johnson, N. L. (Eds.). 1991. Encyclopedia of Statistical Sciences, Vol. 9. John Wiley & Sons, NY. 762 pp.

Kovar, J. G., Rao, J. N. K., and Wu, C. F. J. 1988. Bootstrap and other methods to measure errors in survey estimates. Can. J. Stat. 16, Supplement 1988: 25–45.

Kozak, A., Munro, D. D., and Smith, J. H. G. 1969. Taper functions and their application in forest inventory. For. Chron. 45: 278–283.

Krajicek, J. E., Brinkman, K. A., and Gingrich, S. F. 1961. Crown competition—A measure of density. For. Sci. 7: 35–42.

Kramer, P. J. and Kozlowski, T. T. 1979. Physiology of Woody Plants. Academic Press, NY. 811 pp.

Kruskal, W. and Mosteller, F. 1980. Representative sampling, IV: The history of the concept in statistics, 1895–1939. Int. Stat. Rev. 48: 169–195.

Kumar, S. and Lee, J. 1983. Evaluation of composite estimation for the Canadian Labour Force Survey. Survey Meth. 9: 178–201.

Kupper, L. L. and Hafner, K. B. 1989. How appropriate are popular sample size formulas? Amer. Stat. 43: 101–105.

LaBau, V. J. and Winterberger, K. C. 1989. Use of a four-phase sampling design in Alaska multiresource vegetation inventories. Univ. of Helsinki, Dept. of For. Mens. and Manag. Res. Notes No. 21: 89–102.

Lahiri, D. B. 1951. A method of sample selection providing unbiased ratio estimates. Bull. Int. Stat. Inst. 33: 133–140.

Laird, N., Lange, N., and Stram, D. 1987. Maximum likelihood computation with repeated measures: Application of the EM algorithm. J. Amer. Stat. Assoc. 82: 97–105.

Lebreton, J.-D., Burnham, K. P., Clobert, J., and Anderson, D. R. 1992. Modeling survival and testing biological hypotheses using marked animals: A unified approach with case studies. Ecol. Monographs 62: 67–118.

Lee, D. C. L., Novaes, R. A., Hernandez Filko, P., and Ponzoni, F. J. 1990. A practical example of forest mapping and timber volume estimation using Landsat-5 TM imagery. In: Proc. Symp. State-of-the-Art Methodology of Forest Inventory. Syracuse, NY. USDA For. Serv. Pacific Northwest Res. Sta., Gen. Tech. Rep. PNW-GTR-263. Pp. 303–309.

Leech, J. W. 1984. Estimating crown width from diameter at breast height for open-grown radiata pine trees in South Australia. Aust. For. Res. 14: 333–337.

Lehmann, E. L. 1983. Theory of Point Estimation. John Wiley & Sons, NY. 506 pp.

Lennartz, M. R. and McClure, J. P. 1979. Estimating the extent of red cockaded woodpecker habitat in the Southeast. In: Proc. SAF Workshop, Forest Resource Inventories. Ft. Collins, CO. July 23–26, 1979. Vol. 1, pp. 48–62.

Lewis, N. B., Keeves, A., and Leech, J. W. 1976. Yield regulation in South Australian *Pinus radiata* plantations. South Aust. Woods and Forests Dept. Bull. No. 23. 174 pp.

Lewis, N. B., McIntyre, G. A., and Leech, J. W. 1973. Regional volume table for *Pinus radiata* in South Australia (Metric edition). Woods and Forests Dept., South Australia. Bull. 20. Pp. 64.

Lexen, B. 1943. Bole area as an expression of growing stock. J. For. 41: 883–885.

Li, H. G., Schreuder, H. T., and Bowden, D. C. 1984. Four-phase sampling estimation for the Alaska survey. In: Proc. Soc. Amer. For. Conf. "Inventorying Forest and Other Vegetation of the High Latitude and High Altitude Regions." Fairbanks, AL. July 23–26, 1984. Pp. 61–67. Bethesda, MD.

Li, H. G., Schreuder H. T., and Scott, C. T. 1990. Combining estimates that are both in error subject to marginal constraints. Can. J. For. 20: 1675–1679.

Li, H. G., Schreuder, H. T., Van Hooser, D., and Brink, G. E. 1992. Estimating strata means in double sampling with corrections based on second-phase sampling. Biometrics. 48: 189–199.

Lievesley, D. and Waterton, J. 1985. Measuring individual attitude change. In: British Social Attitudes. The 1985 Report (R. Jowell and S. Witherspoon, Eds.). Gower, Aldershot, Hampshire, UK. Pp. 177–194.

Lillesand, T. M. and Kiefer, R. W. 1987. Remote Sensing and Image Interpretation, 2nd ed. John Wiley & Sons, NY. 721 pp.

Little, R. J. A. 1983a. Comment on paper by M. H. Hansen, W. G. Madow and B. J. Tepping. J. Amer. Stat. Assoc. 78: 797–799.

Little, R. J. A. 1983b. Estimating a finite population mean from unequal probability sampling. J. Amer. Stat. Assoc. 78: 596–604.

Little, R. J. A. and Rubin, D. B. 1987. Statistical Analysis with Missing Data. John Wiley & Sons, NY. 278 pp.

Little, R. J. A. and Smith, P. J. 1987. Editing and imputation for quantitative survey data. J. Amer. Stat. Assoc. 82: 58–68.

Loetsch, F. and Haller, K. E. 1964. Forest Inventory Vol. 1. Statistics of Forest Inventory and Information from Aerial Photographs. BLV Verlagsgesellschaft München, Munich, Germany. 436 pp.

Loetsch, F., Zöhrer, F., and Haller, K. E. 1973. Forest Inventory. Vol. II. BLV Verlagsgesellschaft, München, Germany. 469 pp.

Lundström, S. 1987. An evaluation of small area estimation methods: The case of estimating the number of nonmarried cohabiting persons in Swedish municipalities. In: Proc. Symp. Small Area Statistics: An Int. Symp. (R. Platek et al., Eds.). John Wiley & Sons, NY. Pp. 239–253.

Lwin, T. and Spiegelman, C. H. 1986. Calibration with working standards. Appl. Stat. 35: 256–261.

Lynch, T. B. 1986. Critical height sampling and volumes by cylindrical shells. For. Sci. 32: 829–834.

Mackisack, M. S. and Wood, G. B. 1990. Simulating the forest and the point-sampling process as an aid in designing forest inventories. For. Ecol. Manage. 36: 79–103.

Madow, W. G. 1948. On the limiting distributions of estimates based on samples from finite universes. Ann. Math. Stat. 19: 535–545.

Madow, W. G. 1949. On the theory of systematic sampling-II. Ann. Math. Stat. 20: 333–354.

Mandallaz, D. 1990. Superpopulation models in forest sampling. In: Forest Statistics, Proc. Conf. on Forest Statistics (D. R. Pelz, Ed.). Sept. 1989. Freiburg, Germany. IUFRO Sect. 6.02. Pp 10–19.

Mandallaz, D. 1991. A unified approach to sampling theory for forest inventory based on infinite population and superpopulation models. Ph.D. thesis. Published by Chair of Forest Inventory and Planning, Dept. of Forest and Wood Science, Swiss Federal Institute of Technology. Zurich, Switzerland. 256 pp.

Marshall, J. D. and Waring, R. H. 1985. Predicting fine root production and turnover by monitoring root starch and soil temperature. Can. J. For. Res. 15: 791–800.

Martin, A. J. 1981. Taper and volume equations for selected Appalachian hardwood species. USDA For. Serv., Northeastern For. Expt. Sta., Res. Pap. NE-490. 22 pp.

Martin, G. L. 1982. A method for estimating ingrowth on permanent horizontal sample points. For. Sci. 28: 110–114.

Martin, G. L. 1983. The relative efficiency of some forest growth estimators. Biometrics 39: 639–650.

Martin, G. L., Sharik, T. L., Oderwald, R. G., and Smith, D. W. 1980. Evaluation of ranked set sampling for estimating shrub phytomass in Appalachian oak forests. Virginia Polytechnical Inst., Blacksburg. Publ. No. FWS-4–80.

Masuyama, M. 1953. A rapid method of estimating basal area in timber survey—An application of integral geometry to aerial sampling problems. Sankhya 12: 291–302.

Matérn, B. 1947. Methods of estimating the accuracy of line and sample plot surveys (in Swedish, English summary). Medd. Stat. Skogsforsk. Inst. 36. 138 pp.

Matérn, B. 1956. On the geometry of the cross-section of a stem. Medd. Stat. Skogsforskn. Inst. 46 (11). 28 pp.

Matérn, B. 1972. The precision of basal area estimates. For. Sci. 18: 123–125.

Matérn, B. 1982. Statistics in theory and practice. Swedish Univ. of Agric. Sci., Section of Forest Biometry. Uméa, Sweden.

Matérn, B. 1986. Spatial Variation, 2nd ed. Springer-Verlag, NY. 151 pp.

Matérn, B. 1990. On the shape of the cross-section of a tree stem: An empirical study of the geometry of mensurational methods. Swedish Univ. Agric. Sci., Sec. For. Biometry, Rep. 28. 46 pp.

Matis, K. G., Hetherington, J. C., and Kassab, J. Y. 1984. Sampling with partial replacement—A literature review. Commonw. For. Rev. 63: 193–206.

Max, T. A. and Burkhart, H. E. 1976. Segmented polynomial regression applied to taper equations. For. Sci. 22: 283–289.

McArdle, R. E. 1928. Relative accuracy of calipers and diameter tapes in measuring Douglas fir trees. J. For. 26: 338–342.

McCarthy, P. J. and Snowden, C. B. 1985. The bootstrap and finite population sampling. Vital and Health Statistics, Public Health Service, US Dept. of Health and Human Services. NCHS Series 2, No. 95. 23 pp.

McClure, J. P., Schreuder, H. T., and Wilson, R. L. 1983. A comparison of several volume table equations for loblolly pine and white oak. USDA For. Serv. Res. Pap. SE-240. 8 pp.

McConnell, M. J. and David, H. A. 1976. Order statistics and their concomitants in some double sampling situations. In: Essays in Probability and Statistics. S. Takeda et al. (Eds.). Shinko Tshusho, Tokyo, Japan. Pp. 451–466.

McIntyre, G. A. 1952. A method of unbiased selective sampling, using ranked sets. Aust. J. Agr. Res. 3: 385–390.

McLintock, T. F. and Bickford, C. A. 1957. A proposed site index for red spruce in the northeast. USDA For. Serv., Northeast. For. Exp. Sta., Pap. No. 93. 30 pp.

McTague, J. P. and Bailey, R. L. 1985. Critical height sampling for stand volume estimation. For. Sci. 31: 899–911.

Mead, B. R. and Setzer, T. S. 1984. Aerial photo stand volume tables for interior Alaska. In: Proc. Soc. Amer. For. Conf. "Inventorying Forest and Other Vegetation of the High Latitude and High Altitude Regions." Fairbanks, AL. July 23–26, 1984. Pp. 172–177. Bethesda, MD.

Mendenhall, W. R., Scheaffer, L., and Wackerly, D. D. 1986. Mathematical Statistics with Applications. Duxbury Press, NY. 750 pp.

Meng, X.-L. and Rubin, D. B. 1991. Using EM to obtain asymptotic variance-covariance matrices: The SEM algorithm. J. Amer. Stat. Assoc. 86: 899–909.

Mesavage, C. 1969. Converting dendrometer estimates of outside-bark diameters to wood diameters on major southern pines. USDA For. Serv., Res. Note SO-93. 4 pp.

Midzuno, H. 1950. An outline of the theory of sampling systems. Ann. Inst. Stat. Math. 1: 149–156.

Midzuno, H. 1951. On the sampling system with probabilities proportionate to sum of signs. Ann. Inst. Stat. Math. 2: 99–108.

Miller, R. G. 1974. The jackknife—A review. Biometrika 61: 1–15.

Monserud, R. A. 1984. Problems with site index: An opinionated review. In: Proc. Symp. "Forest land classification: Experiences, problems, perspectives" (J. Bockheim, Ed.). Univ. Wisconsin, Dept. Soil Sci., Madison. Pp. 167–180.

Mood, A. M., Graybill, F. A., and Boes, D. C. 1974. Introduction to the Theory of Statistics, 3rd ed. McGraw-Hill, NY. 564 pp.

Moore, M. H. and Bauer, M. E. 1990. Classification of forest vegetation in north-central Minnesota using Landsat multispectral scanner and thematic mapper data. For. Sci. 36: 330–342.

Müller, G. 1936. The average ocular estimation error in wood volume measurement in Sachsen, Tharandt (in German). Forstl. Jahrb. 87: 327–344.

Müller, G. 1953. Das Baumzählrohr (in German). Allg. Forstztg. 64: 249–251.

Murthy, M. N. 1967. Sampling Theory and Methods. Statistical Publishing, Calcutta, India. 684 pp.

Murthy, M. N. 1981. Future directions in survey sampling. Bull. Int. Stat. Inst. 49: 792–812.

Muttlak, H. A. and McDonald, L. L. 1990. Ranked set sampling with size-biased probability of selection. Biometrics 46: 435–445.

Nadelhoffer, K. J., Aber, J. D., and Melillo, J. M. 1985. Fine roots, net primary production and soil nitrogen availability: A new hypothesis. Ecology 66: 1377–1390.

Näslund, M. 1930. On standard error calculations of line surveys (in Swedish). Svenska Skogsv. För. Tidskr. 28: 309.

Näslund, M. 1941. A new method to decide the strip width of a survey. (In Swedish). Medd. Stat. Skogsforsöksanst. No. 32.

Needham, T. D. and Smith, J. L. 1987. Stem count accuracy and species determination in loblolly pine plantations using 35-mm aerial photography. Photogramm. Eng. Rem. Sens. 53: 1675–1678.

Newnham, R. M. 1968. The generation of artificial populations of points (spatial patterns) on a plane. Forest Management Institute, Ottawa, Canada. Inf. Rep. FMR-X-10.

Newton, C. M., Cunia, T., and Bickford, C. A. 1974. Multivariate estimators for sampling with partial replacement on two occasions. For. Sci. 20: 106–116.

Neyman, J. 1934. On the two different aspects of the representative method: The method of stratified sampling and the method of purposive selection. J. Roy. Stat. Soc. A109: 558–606 (includes discussion).

Neyman, J. 1938. Contribution to the theory of sampling human populations. J. Amer. Stat. Assoc. 33: 101–116.

Nichols, J. D. and Pollock, K. H. 1983. Estimating taxonomic diversity, extinction rates, and speciation rates from fossil data using capture–recapture models. Paleobiology 9: 150–163.

Nieman, T. C. 1990a. Development and use of portable data recorders in North America. Bull. Recherches Agronomiques Gembloux 25: 139–151. Proc. IUFRO Workshop, "Forest Growth Data: Capture, Retrieval and Dissemination." Gembloux, Belgium. April 3–5, 1989.

Nieman, T. C. 1990b. Choosing the field computer. Paper presented to SAF Southeastern Field Computer Conf. Atlanta, GA. Mar. 6–7, 1990 (unpublished).

Nilsson, S. and Duinker, P. 1987. The extent of forest decline in Europe. Environment 29: 4–31.

Nylinder, P. 1967. Wood measurement in Scandinavia. In: Proc. Wood Measurement Conf., Fac. For., Univ. Toronto, Toronto, Canada. Tech. Rep. No. 7, pp. 29–46.

O'Brien, R. and Van Hooser, D. D. 1983. Understory vegetation inventory: An efficient procedure. USDA For. Serv. Interm. For. and Range Exp. Sta., Ogden, UT. Res. Pap. INT-323.

Oderwald, R. G. 1981. Comparison of point and plot sampling basal area estimators. For. Sci. 27: 42–48.

Oderwald, R. G. 1990. Personal knowledge as prior information in volume estimates. In: FWS-2–90. School of For. Wildl. Res. Virginia Polytech. Inst. and State Univ., Blacksburg.

Ogus, J. L. and Clark, D. F. 1971. The annual survey of manufactures: A report on methodology. US Dept. of Comm. Tech. Pap. 24. 107 pp.

Oh, H. L. and Scheuren, F. J. 1983. Weighting adjustment for unit nonresponse. In: Incomplete Data in Sample Surveys, Vol. 2, Theory and Bibliographies (W. G. Madow, I. Olkin, and D. B. Rubin, Eds.). Academic Press, NY. Pp. 143–184.

Olkin, I. 1958. Multivariate ratio estimation for finite populations. Biometrika 45: 154–165.

O'Muircheartaigh, C. and Wong, S. T. 1981. The impact of sampling theory on survey sampling practice: A review. Proc. Int. Stat. Inst. 43rd Session. Vol. XLIX, No. 1, pp. 465–493.

Ord, K. 1978. How many trees in a forest? Math. Scientist 3: 23–33.

Osborne, J. G. 1942. Sampling errors of systematic and random surveys of cover-type areas. J. Amer. Stat. Assoc. 37: 256–264.

Ouyang, Z. 1990. Investigation on some estimators and strategies in sampling proposed by Srivastava. Ph.D. dissertation, Colorado State Univ., Fort Collins. 83 pp.

Ouyang, Z. and Schreuder, H. T. 1992. Srivastava estimation in forestry. For. Sci. (in press).

Ouyang, Z., Schreuder, H. T., and Li, J. 1991. Regression estimation under sampling with one unit per stratum. Comm. Stat. Theory Meth. 20: 2431–2449.

Ouyang, Z., Schreuder, H. T., Max, T., and Williams, M. 1992a. Poisson-Poisson and binomial-Poisson sampling. Survey Meth. (in press).

Ouyang, Z., Srivastava, J. N., and Schreuder, H. T. 1992b. A general ratio estimator and its application to regression model sampling. Ann. Inst. Stat. Math. (in press).

Packard, J. M., Summers, R. C., and Barnes, L. B. 1985. Variation of visibility bias during aerial surveys of manatees. J. Wildl. Manage. 49: 347–351.

Paine, D. P. and McCadden, R. J. 1988. Simplified forest inventory using large-scale 70-mm photography and tariff tables. Photogramm. Eng. Rem. Sens. 54: 1423–1427.

Palley, M. N. and Horwitz, L. G. 1961. Properties of some random and systematic point sampling estimators. For. Sci. 7: 52–65.

Pearson, K. 1903. On the influence of natural selection on the variability and correlation of organs. Phil. Trans. Roy. Soc. A200: 1–66.

Pearson, E. S. 1929. Some notes on sampling tests with two variables. Biometrika 21: 337–360.

Pennington, M. 1983. Efficient estimators of abundance for fish and plankton surveys. Biometrics 39: 281–286.

Persson, O. 1964. Distance Methods. Studia Forestalio Suecica 15. 68 pp.

Persson, H. A. 1983. The distribution and productivity of fine roots in boreal forests. Plant and Soil 71: 87–101.

Petterson, H. 1955. Yield of coniferous forests. Medd. Stat. Skogsförsöksanst. 45,1B. 189 pp.

Pfeffermann, D. and Smith, T. M. F. 1985. Regression models for grouped populations in cross-section surveys. Int. Stat. Rev. 53: 37–59.

Pfeiffer, K. 1967. Analysis of methods of studying operational efficiency in forestry. M.F. thesis, Fac. For., Univ. of British Columbia, Vancouver, Canada. 94 pp.

Philip, M. S. 1983. Measuring Trees and Forests. Div. Forestry, Univ. of Dar es Salaam, Tanzania. 338 pp.

Phillips, D. R. and Cost, N. D. 1979. Estimating the volume of hardwood crowns, stems, and the total tree. USDA For. Serv., Southeast. For. Exp. Sta., Res. Note SE-276. 6 pp.

Pickands, J. and Raghavachari, M. 1987. Exact and asymptotic inference for the size of a population. Biometrika 74: 355–363.

Pierce, L. L. and Running, S. W. 1988. Rapid estimation of coniferous forest leaf area index using a portable integrating radiometer. Ecology 69: 1762–1767.

Pinkham, R. S. 1987. An efficient algorithm for drawing a simple random sample. Appl. Stat. 36: 370–372.

Pollanschütz, J. 1965. A new method to determine form factor and volume of standing trees (in German). Bundesversuchsanst. Wien. 68: 1–186.

Pollock, K. H., Nichols, J. D., Brownie, C., and Hines, J. E. 1990. Statistical inference for capture–recapture experiments. Wildl. Monographs. The Wildlife Society, Washington, DC. 97 pp.

Prasad, N. G. N. and Rao, J. N. K. 1990. The estimation of the mean square error of small-area estimators. J. Amer. Stat. Assoc. 85: 163–171.

Press, S. J. 1989. Bayesian Statistics: Principles, Models and Applications. John Wiley & Sons, NY. 237 pp.

Purcell, N. J. and Kish. L. 1979. Estimation for small domains. Biometrics 35: 365–384.

Purcell, N. J. and Linacre, A. 1976. Techniques for the estimation of small area characteristics. Paper presented to 3rd Australian Stat. Conf., Melbourne, Australia. Aug. 18–20, 1976 (unpublished).

Quenouille, M. H. 1949. Problems in plane sampling. Ann. Math. Stat. 20: 355–375.

Quenouille, M. H. 1956. Notes on bias in estimation. Biometrika 43: 353–360.

Raiffa, H. and Schlaifer, R. 1961. Applied Statistical Decision Theory. Harvard Univ. Press, Cambridge, MA. 356 pp.

Raj, D. 1965. On sampling over two occasions with probability proportionate to size. Ann. Math Stat. 36: 327–330.

Ranneby, B. 1980. The precision of the estimates of the growth in basal area. Swed. Univ. Agr. Sci., For. Biometry Sect., Rep. No. 19. 25 pp.

Ranneby, B. 1983. Stochastic models of variation in time and space. In: Statistics in

Theory and Practice: Essays in Honour of Bertil Matérn. (B. Ranneby Ed.). Swedish Univ. of Agri. Sci. Sect., For. Biometry S-901 83, Umea, Sweden. Pp. 227–245.

Ranneby, B., Cruse, T., Häggland, B., Jonasson, H., and Swärd, J. 1987. Designing a new national forest survey for Sweden. Studia Forestalia Suecica No. 177, Swedish Univ. of Agr. Sci. Faculty of Forestry. Uppsala, Sweden. 29 pp.

Rao, J. N. K. 1969. Ratio and regression estimators. New developments in survey sampling. (N. L. Johnson and H. Smith, Jr., Eds.). John Wiley & Sons, NY. Pp. 213–214.

Rao, J. N. K. 1980. Estimating the common mean of possibly different normal populations—A simulation study. J. Amer. Stat. Assoc. 75: 447–453.

Rao, J. N. K. 1986. Synthetic estimators, SPREE and best model-based predictors of small area means. In: Proc. Conf. on Survey Res. Meth. in Agric. Amer. Stat. Assoc. Leesburg, VA. June 15–18, 1986. Pp. 1–6.

Rao, J. N. K. and Webster, J. T. 1966. On two methods of bias reduction in the estimation of ratios. Biometrika 53: 571–577.

Rao, J. N. K. and Wu, C. J. W. 1988. Resampling inference with complex survey data. J. Amer. Stat. Assoc. 83: 231–241.

Rao, J. N. K., Hartley, H. O., and Cochran, W. G. 1962. On a simple procedure of unequal probability sampling without replacement. J. Roy. Stat. Soc. B24: 482–491.

Rao, P. S. R. S. 1981. Estimation of the mean square error of the ratio estimator. In: Current Topics in Survey Sampling (D. Krewski, R. Platek, and J. N. K. Rao, Eds.). Academic Press, NY. Pp. 305–315.

Rao, P. S. R. S. and Rao, J. N. K. 1971. Small sample results for ratio estimators. Biometrika 58: 625–630.

Reddy, V. N. 1974. On a transformed ratio method of estimation. Sankhya 36: 59–70.

Reid, N. 1988. Saddlepoint methods and statistical inference. Stat. Sci. 3: 213–238.

Reineke, L. H. 1926. The determination of tree volume by planimeter. J. For. 24: 183–189.

Reineke, L. H. 1933. Perfecting a stand-density index for even-aged forests. J. Agric. Res. 46: 627–638.

Rennolls, K. 1981. The use of superpopulation-prediction methods in survey analysis, with applications to the British National Census of woodlands and trees. In: Proc. In-Place Resource Inventories: Principles and Practices. Univ. of Maine, Orono. Aug. 9–14, 1981. Pp. 395–401.

Reynolds, R. T., Scott, J. M., and Nussbaum, R. A. 1980. A variable circular-plot method for estimating bird numbers. Condor 82: 309–313.

Richardson, D. H. and Wu, D-M. 1970. Least squares and grouping methods estimators in the errors in variables models. J. Amer. Stat. Assoc. 65: 724–748.

Ridout, M. S. and Cobby, J. M. 1987. Ranked set sampling with non-random selection of sets and errors in ranking. Appl. Stat. 36: 145–152.

Ridpath, M. G., Begg, R. J., Dudzinski, M. L., Forbes, M. A., and Graham, A. 1983. Counting the same populations of large tropical mammals from the ground and from the air. Aust. Wildl. Res. 10: 487–498.

Ripley, B. D. 1988. Statistical Inference for Spatial Processes. Cambridge Univ. Press, NY. 148 pp.

Robbins, W. C. and Young, H. E. 1968. An evaluation of the McClure and Wheeler optical calipers. For. Chron. 44: 16–18.

Robinson, G. K. 1975. Some counterexamples to the theory of confidence intervals. Biometrika 62: 155–161.

Robinson, P. M. and Särndal, C. E. 1983. Asymptotic properties of the generalized regression estimator in probability sampling. Sankhya 45: 240–248.

Roesch, F. A., Green, E. J., and Scott, C. T. 1989. New compatible estimators for survivor growth and ingrowth from remeasured horizontal point samples. For. Sci. 37: 136–145.

Royall, R. M. 1970a. Finite population sampling-on labels in estimation. Ann. Math. Stat. 41: 1774–1779.

Royall, R. M. 1970b. On finite population sampling theory under certain linear regression models. Biometrika 57: 377–387.

Royall, R. M. 1971. Discussion of paper by D. Basu. In: Foundations of Statistical Inference (V. P. Godambe and D. A. Sprott, Eds.). Holt, Rinehart and Winston, NY. pp. 238–239.

Royall, R. M. 1976. Likelihood functions in finite population sampling theory. Biometrika 63: 605–614.

Royall, R. M. and Cumberland, W. G. 1978. Variance estimation in finite population sampling. J. Amer. Stat. Assoc. 73: 351–358.

Royall, R. M. and Cumberland, W. G. 1981a. An empirical study of the ratio estimator and estimators of its variance. J. Amer. Stat. Assoc. 76: 66–87 (includes discussion).

Royall, R. M. and Cumberland, W. G. 1981b. The finite population linear regression estimator and estimators of its variance—An empirical study. J. Amer. Stat. Assoc. 76: 924–930.

Royall, R.M. and Eberhardt K.R. 1975. Variance estimates for the ratio estimation. Sankhya C 37: 43–52.

Royall, R. M. and Herson, J. 1973a. Robust estimation in finite populations I. J. Amer. Stat. Assoc. 68: 880–889.

Royall, R. M. and Herson, J. 1973b. Robust estimation in finite populations II: Stratification on a size variable. J. Amer. Stat. Assoc. 68: 890–893.

Rubin, D. B. 1976. Inference and missing data. Biometrika 63: 581–592.

Rubin, D. B. 1978. Multiple imputations in sample surveys: A phenomenological Bayesian approach to nonresponse. In: Proc. Amer. Stat. Assoc., Sect. on Survey Research Methods. Pp. 20–34.

Rubin, D. B. 1987. Multiple Imputation for Nonresponse in Surveys. John Wiley & Sons, NY. 258 pp.

Rubinstein, R. Y. 1981. Simulation and the Monte Carlo Method. John Wiley & Sons, NY. 278 pp.

Rubinstein, R. Y. and Samorodnitsky, G. 1985. Variance reduction by the use of common and antithetic varieties. J. Stat. Comput. Simul. 22: 161–180.

Sahai, A. 1979. An efficient variant of the product and ratio estimators. Stat. Neerlandica 33: 27–35.

Sande, I. G. 1982. Imputation in surveys: Coping with reality. Amer. Stat. 36: 145–152.

Santantonio, D. and Grace, J. C. 1987. Estimating fine-root production and turnover from biomass and decomposition data: A compartment-flow model. Can. J. For. Res. 18: 900–908.

Särndal, C. E. 1978. Design-based and model-based inference in survey sampling. Scand. J. Stat. 5: 27–52.

Särndal, C. E. 1980a. A two-way classification of regression estimation strategies in probability sampling. Can. J. Stat. 8: 165–177.

Särndal, C. E. 1980b. On π-inverse weighting versus best linear unbiased weighting in probability sampling. Biometrika 67: 639–650.

Särndal, C. E. 1982. Implications of survey design for generalized regression estimation of linear functions. J. Stat. Plan. Inf. 7: 155–170.

Särndal, C. E. 1984. Design-consistent versus model-dependent estimators for small domains. J. Amer. Stat. Assoc. 79: 624–631.

Särndal, C. E. and Hidiroglou, M. A. 1989. Small domain estimation: A conditional analysis. J. Amer. Stat. Assoc. 84: 266–275.

Särndal, C. E., Swensson, B., and Wretman, J. H. 1989. The weighted residual techniques for estimating the variance of the general regression estimator of the finite population total. Biometrika 76: 527–537.

Satyamurthi, K. R. 1979. Sampling with a rope—A novel and useful method. Indian Forester 105: 50–56.

Savage, L. J. 1972. The Foundations of Statistics, 2nd ed. Dover Publications, Inc. NY. 310 pp.

Schmid-Haas, P. 1969. Sampling at the forest edge (in German). Mitt Schweiz Anst Forstl. Versichswes. 45: 234–303.

Schmid-Haas, P. 1982. Sampling at the forest edge. In: Statistics in Theory and Practice, Essays in Honour of Bertil Matérn (B. Ranneby, Ed.). Swedish Univ. Agric. Sci. Umea, Sweden. Pp. 263–276.

Schmid-Haas, P. and Winzeler, K. 1981. Efficient determination of volume and volume growth. In Proc. 17th IUFRO World Congr. Kyoto, Japan, Sept. 6–12, 1981. Pp. 231–257.

Schmidt, A. 1967. Der rechnerische Ausgleich von Bestandeshönkurven (in German). Forstwiss. Centralbl. 86: 370–382.

Schmidt, K. 1891. The circular plot sampling method of forest supervisor Zetscho (in German). Allg. Forst-u. Jagdztg. 67: 73–76.

Schreuder, H. T. 1970. Point sampling theory in the framework of equal-probability cluster sampling. For. Sci. 16: 240–246.

Schreuder, H. T. 1975. Two stratified sampling alternatives to 3–P sampling. In: Monitoring Forest Environment through Successive Sampling, (T. Cunia, Ed.). Proc. Symp. State Univ. New York. Coll. Environ. Sci. and For., Syracuse. 1974. Pp. 55–68.

Schreuder, H. T. 1978. Count sampling in forestry. For. Sci. 24: 267–272.

Schreuder, H. T. 1981. Multiresource inventory sampling strategies. Bull. Int. Stat. Inst. 49: 250–270 (includes discussion).

Schreuder, H. T. 1984. Poisson sampling and some alternatives in timber sales sampling. For. Ecol. Manage. 8: 149–160.

Schreuder, H. T. 1987. Model-dependent versus design-dependent procedures in timber sale and updating surveys. In: Proc. Soc. Amer. For. Meeting, Birmingham, AL. Oct. 1986. Publ. SAF 87.02, pp. 74–76.

Schreuder, H. T. 1990. The gain in efficiency in πps sampling over Poisson sampling. For. Sci. 36: 1146–1152.

Schreuder, H. T. and Anderson, J. 1984. Variance estimation for volume when D^2H is the covariate in regression. Can. J. For. Res. 14: 818–821.

Schreuder, H. T. and Brink, G. E. 1983. The jackknife—a useful statistical tool. In: Proc. Amer. Soc. Amer. For. Workshops "Renewable Resources Inventories for Monitoring Changes and Trends." College of For., Oregon State Univ. Corvallis, OR. Pp. 531–535.

Schreuder, H. T. and Czaplewski, R. 1992. Long-term strategy for the statistical design of a forest health monitoring system. Env. Monitoring and Assessment (in press).

Schreuder, H. T. and McClure, J. P. 1991. Modifying forest survey procedures to establish cause–effect. Should it be done? Invited position paper. IUFRO World Forestry Congress. Sept. 1991. Paris, France. Revue Forestière Francaise. Hors Série No. 4. Sect. D. pp 67–78.

Schreuder, H. T. and Ouyang, Z. 1992. Optimal sampling strategies for weighted linear regression estimation. Can. J. For. Res. 22: 239–247.

Schreuder, H. T. and Thomas, C. E. 1985. Efficient sampling techniques for timber sale surveys and inventory updates. For. Sci. 31: 857–866.

Schreuder, H. T. and Thomas, C. E. 1991. Establishing cause–effect relationships using forest survey data. For. Sci. 37: 1497–1525 (includes discussion).

Schreuder, H. T. and Wood, G. B. 1986. The choice between design-dependent and model-dependent sampling. Can. J. For. Res. 16: 260–265.

Schreuder, H. T., Banyard, S. G., and Brink, G. E. 1987a. A comparison of three sampling methods in estimating stand parameters for a tropical forest. For. Ecol. Manage. 21: 119–128.

Schreuder, H. T., Li, H. G., and Hazard, J. 1987b. Pps and random sampling estimation using some regression and ratio estimators for underlying linear and curvilinear models. For. Sci. 33: 997–1009.

Schreuder, H. T., Li, H. G., and Scott, C. T. 1987c. Jackknife and bootstrap estimation for sampling with partial replacement. For. Sci. 33: 676–689.

Schreuder, H. T., Bhattacharyya, H. T., and McClure, J. P. 1982. Towards a unified distribution theory for stand variables using the S_{BBB} distribution. Biometrics 38: 137–142.

Schreuder, H. T., Brink, G. E., and Wilson, R. L. 1984a. Alternative estimators for point-Poisson sampling. For. Sci. 30: 803–812.

Schreuder, H. T., Brink, G. E., Schroeder, D. I., and Dieckman, R. 1984b. Model-based sampling versus point-Poisson sampling on a timber sale in the Roosevelt National Forest in Colorado. For. Sci. 30: 652–656.

Schreuder, H. T., Li, H. G., and Sadooghi-Alvandi, S. M. 1990a. Sunter's pps without replacement sampling as an alternative to Poisson sampling. USDA For. Serv., RM For. and Range Exp. Sta. Res. Pap. RM-290. 6 pp.

Schreuder, H. T., Li, H. G., and Wood, G. B. 1990b. Model-dependent and design-dependent procedures—A simulation study. USDA For. Serv., RM For. and Range Exp. Sta. Res. Pap. RM-291. 19 pp.

Schreuder, H. T, Ouyang, Z., and Li, H. G. 1990c. Understanding the efficiency of model-dependent procedures with the weighted regression estimator. Paper presented to IUFRO XIX World Forestry Congr. Montreal, Canada. Aug. 5–11, 1990. School of For. Wildl. Resources, Virginia Polytech. Inst. and State Univ., Blacksburg, VA. Publ. No. FWS-3–90, pp. 218–229.

Schreuder, H. T., Rennie, J. C., and Williams, M. 1992a. Comparison of three sampling schemes for estimating frequency and D^2H by diameter class—a simulation study. For. Ecol. Manage. 50: 117–131.

Schreuder, H. T., LaBau, V. J., and Hazard, J. W. 1992b. The Alaska four-phase sampling design. Submitted to Photogramm. Eng. Rem. Sens.

Schreuder, H. T., Ouyang, Z., and Williams, M. 1992c. Point-Poisson, point-pps, and modified point-pps sampling: Efficiency and variance estimation. Can. J. For. Res. 22: 725–728.

Schreuder, H. T., Schreiner, D. S., and Max, T. A. 1981. Ensuring an adequate sample at each location in point sampling. For. Sci. 27: 567–573.

Schreuder, H. T., Sedransk, J., and Ware, K. D. 1968. 3–P sampling and some alternatives, I. For. Sci. 14: 429–454.

Schreuder, H. T., Sedransk, J., and Ware, K. D. 1971. Forest sampling in the absence of a well-defined frame. In Statistical Ecology, Vol. 2: Sampling and Modeling Biological Populations and Population Dynamics. Pennsylvania State Univ. Press. Pp. 119–150.

Schreuder, H. T., Tyre, G. L., and James, G. A. 1975. Instant- and interval-count sampling: Two new techniques for estimating recreation use. For. Sci. 21: 40–44.

Schreuder, H. T., Clerke, W. H., Barry, P., and Holland, D. 1980. Two-stage stratified sampling with regression to assess southern pine beetle damage. USDA For. Serv. SE For. Exp. Sta. Res. Pap. SE-212. 6 pp.

Schreuder, H. T., Li, H. G., Ryan, M. G., and Scott, C. T. 1988. Adjusting estimates in two-way tables by incorporating outside information. Can. J. For. Res. 18: 1280–1285.

Schumacher, F. X. and Bull, H. 1932. Determination of the errors of estimate of a forest survey, with special reference to the bottom-land hardwood forest region. J. Agric. Res. 45: 741–756.

Schumacher, F. X. and Chapman, R. A. 1942. Sampling methods in forestry and range management. Duke Univ. School of Forestry, Durham, NC. Bull. 7. 213 pp.

Scott, A. J. 1977. On the problem of randomization in survey sampling. Sankhya, Series C, 39: 1–9.

Scott, C. T. 1981. Simplified estimators for sampling with partial replacement on multiple occasions. Dept. of Forest Resources, Univ. of Minnesota, St. Paul. Staff Paper Series No. 23. 48 pp.

Scott, C. T. 1986a. An evaluation of sampling with partial replacement. In: Proc. Use of Auxiliary Information in Natural Resource Inventories. Blacksburg, VA. Oct. 1985. Soc. Amer. For. SAF 86–01. Pp. 74–79.

Scott, C. T. 1986b. Proposed northeastern forest survey emphasizing disturbance detection. In: Proc. Use of Auxiliary Information in Natural Resource Inventories. Blacksburg, VA. Oct. 1985. Soc. Amer. For. SAF 86–01. Pp. 146–151.

Scott, C. T. 1990a. An overview of fixed versus variable-radius plots for successive inventories. In: Proc. IUFRO Symp. "State-of-the-Art Methodology of Forest Inventory," Syracuse, NY. Aug. 1989. USDA For. Serv. Gen. Tech. Rep. PNW-GTR-263. Pp. 97–104.

Scott, C. T. 1990b. Computers in the field. For. Farmer 49: 8–10.

Scott, C. T. 1992. Optimal design of a plot cluster for monitoring. In: Proc. IUFRO Inventory Sect. London, U.K. Sept. 1991 (in press).

Scott, C. T. and Alegria, J. 1990. Fixed versus variable-radius plots for change estimation. In: Proc. IUFRO Symp. "State-of-the-Art Methodology of Forest Inventory," Syracuse, NY. Aug. 1989. USDA For. Serv. Gen. Tech. Rep. PNW-GTR-263. Pp. 126–132.

Seber, G. A. F. 1965. A note on the multiple-recapture census. Biometrika 52: 249–259.

Seber, G. A. F. 1982. The Estimation of Animal Abundance. 2nd ed. Griffin, London, U.K. 654 pp.

Seber, G. A. F. 1986. A review of estimating animal abundance. Biometrics 42: 267–292.

Seber, G. A. F. 1992. A review of estimating animal abundance II. Int. Stat. Rev. 60: 129–166.

Sedransk, J. 1981. New developments in statistics to more completely summarize survey results for analytical purposes. In: Proc. National Workshop on In-Place Resource Inventories. Univ. of Maine, Orono. Aug. 9–14, 1981. Pp. 421–427.

Seng, Y. P. 1951. Historical survey of the development of sampling theories and practice. J. Roy. Stat. Soc. A114: 214–231.

Seppälä, R. 1971. Variable probabilities in sample-tree selection. Comm. Inst. For. Fenniae. The Finnish For. Res. Institute. 74.4 Helsinki, Finland. 29 pp.

Setzer, T. S. and Mead, B. R. 1986. Verification of aerial photo stand volume tables for southeast Alaska. USDA For. Serv. Res. Pap. PNW-RP 396. 13 pp.

Shirahata, S. 1982. An extension of the ranked set sampling theory. J. Stat. Plan. Inf. 6: 65–72.

Singh, A. 1989. Review article: Digital change detection techniques using remotely-sensed data. Int. J. Rem. Sens. 10: 989–1003.

Singh, H. P., Iachan, R., and Upadhyaya, L. N. 1985. Almost unbiased ratio and product estimators based on interpenetrating subsamples. Comm. Stat. Theory Meth. A14: 963–978.

Skinner, C. J., Holt, D., and Smith, T. M. F. (Eds.). 1989. Analysis of complex surveys. John Wiley and Sons, N.Y. 306 pp.

Smith, T. M. F. 1976. The foundations of survey sampling: A review. J. Roy. Stat. Soc. A139: 183–204.

Smith, T. M. F. 1981. Regression analysis from complex surveys. In: Current topics in survey sampling. Eds.: D. Krewski, R. Platek, J. N. K. Rao. Academic Press: New York. pp. 268–302.

Smith, T. M. F. 1983a. A comment: An evaluation of model-dependent and probability-sampling inferences in sample surveys. J. Amer. Stat. Assoc. 78: 801–802.

Smith, T. M. F. 1983b. On the validity of inferences from non-random samples. J. Roy. Stat. Soc. A. 146: 394–403.

Smith, T. M. F. 1989. Introduction to Part B. In: Analysis of Complex Surveys (C. J. Skinner, D. Holt, and T. M. F. Smith, Eds.). Wiley, Chichester. Pp. 135–147.

Smith, T. M. F. and Sugden, R. A. 1988. Sampling and assignment mechanisms in experiments, surveys and observational studies. Int. Stat. Rev. 56: 165–180.

Smith, W. H. 1990. The health of North American forests: stress and risk assessment. J. For. 88: 32–35.

Society of American Foresters. 1984. Regeneration surveys. In: New Forests for a Changing World. Proc. 1983. SAF National Conv. Portland, OR. 48 pp.

Space, J. C. 1974. Predicting product recovery in southern pine using dendrometry and STX. For. Prod. J. 24: 85–87.

Spencer, R. D. and Hall, R. J. 1988. Canadian large-scale aerial photographic systems (LSP). Photogramm. Eng. Rem. Sens. 54: 475–482.

Spurr, S. H. 1948. Aerial Photographs in Forestry. Ronald Press, NY. 340 pp.

Spurr, S. H. 1952. Forest Inventory. The Ronald Press Co., NY. 476 p.

Squillace, A. E., Schreuder, H. T., and Bhattacharyya, H. T. 1980. Seed provenance identification using likelihoods. Silvae Genetica 29, 3/4: 152–154.

Srivastava, J. N. 1985. On a general theory of sampling, using experimental design. Concepts I: Estimation. Bull. Int. Stat. Inst. 51: 1–16.

Srivastava, J. N. and Ouyang, Z. 1992a. Studies on a general estimator in sampling, utilizing extraneous information through a sample weight function. J. Stat. Plan. Inf. 31: 177–196.

Srivastava, J. N. and Ouyang, Z. 1992b. Some properties of a general estimator in finite population sampling. Sankhya (in press).

Srivastava, S. K. 1967. An estimator using auxiliary information in sample surveys. Calcutta Stat. Assoc. Bull. 16: 121–132.

Srivenkataramana, T. 1980. A dual to ratio estimator in sample surveys. Biometrika 67: 199–204.

Srivenkataramana, T. and Tracy, T. S. 1979. On ratio and product methods of estimation in sampling. Stat. Neerlandica 33: 37–49.

Stein, C. 1955. Inadmissibility of the usual estimator for the mean of a multivariate normal distribution. Proc. Third Berkeley Symp. Math. Stat. Prob. 1: 197–206. Univ. of Calif. Press, Berkeley, Calif.

Stephan, F. F. 1948. History of the uses of modern sampling procedures. J. Amer. Stat. Assoc. 43: 12–39.

Stephan, F. F. and McCarthy, P. J. 1958. Sampling Opinions. John Wiley & Sons, NY. 451 p.

Sterba, H. 1980. Stem curves—A review of the literature. For. Abstr. 41: 141–145.

Stokes, S. L. 1980. Estimation of variance using judgment ordered ranked set samples. Biometrics 36: 35–42.

Stokes, S. L. and Sager, T. W. 1988. Characterization of a ranked-set sample with application to estimating distribution functions. J. Amer. Stat. Assoc. 83: 374–381.

Strand, L. 1957. Relascopic height and cubic volume determination (in Norwegian). Norsk. Skogbruk. 3: 535–538.

Stuart, A. 1964. Some remarks on sampling with unequal probabilities. Bull. Int. Stat. Inst. 40: 773–780.

Student. 1908. On the probable error of the mean. Biometrika 6: 1–25.

Sture, S. 1987. Precision of volume and volume increment estimates. Scand. J. For. Res. 2: 379–387.

Sudman, S. 1985. Efficient screening methods for the sampling of geographically clustered special populations. J. Mark. Res. 22: 20–29.

Sudman, S., Sirken, M. G., and Corvan, C. D. 1988. Sampling rare and elusive populations. Science 240: 991–995.

Sugden, R. A. and Smith, T. M. F. 1984. Ignorable and informative designs in survey sampling inference. Biometrika 71: 495–506.

Sukhatme, P. V. 1944. Moments and product moments of moment-statistics for samples of the finite and infinite populations. Sankhya 6: 363–382.

Sukhatme, P. V., Sukhatme, B. V., Sukhatme, S., and Asok, C. 1984. Sampling Theory of Surveys with Applications. Iowa State Univ. Press. Iowa. 526 pp.

Sukwong, S., Frayer, W. E., and Mogren, E. W. 1971. Generalized comparisons of the precision of fixed-radius and variable-radius plots for basal-area estimates. For. Sci. 17: 263–271.

Sunter, A. 1986. Solutions to the problem of unequal probability sampling without replacement. Int. Stat. Rev. 54: 33–50.

Swank, W. T. and Schreuder, H. T. 1974. Comparison of three methods of estimating surface area and biomass for a forest of young eastern white pine. For. Sci. 20: 91–100.

Swindel, B. F. 1972. The Bayesian controversy. USDA For. Serv. Res. Pap. SE-95. 12 pp.

Swindel, B. F. and Bower, D. R. 1972. Rounding errors in the independent variables in a general linear model. Technometrics 14: 215–218.

Swindel, B. F. and Yandle, D. O. 1972. Allocation in stratified sampling as a game. J. Amer. Stat. Assoc. 67: 684–686.

Szekielda, K. H. (Ed.). 1986. Satellite Remote Sensing for Resource Development. Graham and Trotman. Gaithersburg, MD. 326 pp.

Takahasi, K. and Wakimoto, K. 1968. On unbiased estimates of the population mean based on the sample stratified by means of ordering. Ann. Inst. Stat. Math. 20: 1–31.

Tam, S. M. 1987. Analysis of repeated surveys using a dynamic linear model. Int. Stat. Rev. 55: 63–73.

Teuber, K. B. 1990. Use of AVHRR imagery for large-scale forest inventories. For. Ecol. Manage. 33/34: 621–631.

Thomas, C. E. 1986. Stein Rule estimation of county level inventory statistics. In:

Proc. Symp. Use of Auxiliary Information in Natural Resource Inventories. Blacksburg, VA. Oct. 1–2, 1985. Soc. of Amer. For. Publ. No. SAF 86–01, pp. 45–54.

Thompson, S. K. 1982. Adaptive sampling of spatial point processes. Ph.D. thesis. Oregon State Univ., Corvallis.

Thompson, S. K. 1990. Adaptive cluster sampling. J. Amer. Stat. Assoc. 85: 1050–1059.

Thompson, S. K. 1991. Stratified adaptive cluster sampling. Biometrika 78: 389–397.

Tikkiwal, B. D. 1967. Theory of multiphase sampling from a finite or infinite population on successive occasions, 1, 2. Rev. Int. Stat. Inst. 35: 247–263.

Tschuprow, A. A. 1923. On the mathematical expectation of the moments of frequency distributions in the case of correlated observations. Metron 2: 461–493; 646–683.

Tucker, C. J. 1980. A critical comparison of remote sensing and other methods for non-destructive estimation of standing crop biomass. Grass Forage Sci. 35: 177–182.

Turnbull, K. J. and Hoyer, G. E. 1965. Construction and analysis of comprehensive tree volume tariff tables. State Dept. Nat. Res., Olympia, Washington. Resource Manage. Rep. No. 8.

Ueno, Y. 1978. On the nature of form factors and its application for stand volume estimation (I) (in Japanese). J. Jap. For. Soc. 60: 87–93.

Ueno, Y. 1979. A new method of estimating stand volume I (in Japanese). J. Jap. For. Soc. 61: 346–348.

Ueno, Y. 1980. A new method of estimating stand volume (II) (in Japanese). J. Jap. For. Soc. 62: 315–320.

Ulaby, F. T., Moore, R. K., and Fung, A. K. 1986. Microwave Remote Sensing. Active and Passive, Vol. III, From Theory to Applications. Book-Mart Press, North Bergan, NY. 2162 pp.

Upton, G. and Fingleton, B. 1985. Spatial Data Analysis by Example, Vol. 1, Series in Probability and Mathematical Statistics. John Wiley & Sons, NY. 410 pp.

Upton, G. and Fingleton, B. 1989. Spatial Data Analysis by Example, Vol. 2, Series in Probability and Mathematical Statistics. John Wiley & Sons, NY. 416 pp.

Ure, J. 1950. The natural vegetation of the Kiangoroa plains as an indicator of site quality for exotic conifers. N.Z. J. For. 6: 112–123.

USDA Forest Service. 1992. Forest Service Resource Inventories: An Overview. USDA For. Serv. Forest Inventory Economics, and Recreation Research Staff, Wash. DC. 39 pp.

Valentine, H. T., Bealle, C., and Gregoire, T. F. 1992. Comparing vertical and horizontal modes of importance and control-variate sampling for bole value. For. Sci. (in press).

Valentine, H. T., Tritton, L. M., and Furnival, G. M. 1984. Subsampling trees for biomass, volume, or mineral content. For. Sci. 30: 673–681.

Van Deusen, P. C. 1987. 3–P sampling and design versus model-based estimates. Can. J. For. Res. 17: 115–117.

Van Deusen, P. C. 1989. Multiple-occasion partial replacement sampling for growth components. For. Sci. 35: 388–400.

Van Deusen, P. C. 1990. Critical height versus importance sampling: Does critical height prevail? For. Sci. 36: 930–938.

Van Deusen, P. C. and Grender, J. M. 1989. An alternative estimator for horizontal point clusters. For. Sci. 35: 257–262.

Van Deusen, P. C. and Lynch, T. B. 1987. Efficient unbiased tree-volume estimation. For. Sci. 33: 583–590.

Van Deusen, P. C. and Meerschaert, W. J. 1986. On critical-height sampling. Can. J. For. Res. 16: 1310–1313.

Van Deusen, P. C., Dell, T. R., and Thomas, C. E. 1986. Volume growth estimation from permanent horizontal points. For. Sci. 32: 415–422.

Vanclay, J. K. and Henry, N. B. 1988. Assessing site productivity of indigenous cypress pine forest in southern Queensland. Commonw. For. Rev. 67: 53–64.

Vos, J. W. E. 1980. Mixing of direct, ratio, and product method estimators. Stat. Neerlandica 34: 209–218.

Wakeley, P. C. 1954. The growth intercept method of site classification. In: Proc. 3rd Ann. For. Symp. Louisiana State Univ., Baton Rouge. Pp. 32–33.

Wakeley, P. C. and Marrero, J. 1958. Five-year height intercept as site index in southern pine plantations. J. For. 56: 332–336.

Waksberg, J. 1983. A note on "Locating a special population using random digit dialing." Publ. Opinion Quarterly 47: 576–579.

Wald, A. 1950. Statistical Decision Functions. John Wiley & Sons, NY. 179 pp.

Walker, D. A., Acevedo, W., Everett, K. R., Gaydos, L., Brown, J., and Webber, P. J. 1982. Landsat-assisted environmental mapping in the Arctic National Wildlife Refuge, Alaska. US Army Corps of Engineers, Cold Regions Res. and Eng. Lab. Hanover, NH. CRREL Rep. 82–37. 59 pp.

Walters, P. E. 1985. An examination of the conservative nature of 'classical' confidence limits for a proportion. Biometrical J. 7: 851–856.

Ward, D. 1991. Triangular tessellation: A new approach to forest inventory. For. Ecol. Manage. 44: 285–290.

Ware, K. D. and Cunia, T. 1962. Continuous forest inventory with partial replacement of samples. For. Sci. Monog. 3. 30 pp.

Waring, R. H. 1983. Estimating growth and efficiency in relation to canopy leaf index. Adv. Ecol. Res. 13: 327–354.

Waring, R. H. 1985. Imbalanced forest ecosystems: Assessments and consequences. For. Ecol. Manage. 11: 93–112.

Warren, W. G. 1990. Line intersect sampling: An historical perspective. In: Proc. IUFRO Symp., "State-of-the-Art Methodology of Forest Inventory," Syracuse, NY. Aug. 1989. USDA For. Serv. Sta. Gen. Tech. Rep. PNW-GTR-263. Pp. 33–38.

Waters, W. E. 1955. Sequential sampling in forest insect surveys. For. Sci. 1: 68–79.

Watson, D. J. 1937. The estimation of leaf areas. J. Agric. Sci. 27: 474–483.

Welsh, R. and Ehlers, M. 1987. Merging multiresolution SPOT, HRV and Landsat TM data. Photogramm. Eng. Rem. Sens. 53: 301–303.

Wensel, L. C. 1975. Computer generation of points on a plane. Hilgardia 43: 131–142.

Wensel, L., Leritan, J., and Barber, K. 1980. Selection of basal area factor in point sampling. J. For. 78: 83–84.

West, P. W. 1983. Comparison of stand density measures in even-aged regrowth eucalypt forest of southern Tasmania. Can. J. For. Res. 13: 22–31.

White, G. C. 1983. Numerical estimation of survival rates from land-recovery and biotelemetry data. J. Wildl. Manage. 47: 716–728.

Whittaker, R. H. 1966. Forest dimensions and productivity in the Great Smoky Mountains. Ecology. 47: 103–121.

Whittaker, R. H. and Woodwell, G. M. 1967. Surface area relations of woody plants and forest communities. Amer. J. Bot. 54: 931–939.

Whyte, A. G. D. 1971. Sectional measurement of trees: A rationalized method. N.Z. J. For. Sci. 1: 74–79.

Whyte, A. G. D. and Tennent, R. B. 1975. Improving estimates of stand basal area in working plan inventories. N. Z. J. For. 20: 134–147.

Wiant, H. V., and Koch, C. B. 1974. Predicting diameters inside bark from outside bark measurements on some Appalachian hardwoods. J. For. 72: 775.

Wiant, H. V., Wood, G. B., and Forslund, R. R. 1991. Comparison of centroid and paracone estimates of tree volume. Can. J. For. Res. 21: 714–717.

Wiant, H. V., Wood, G. B., and Furnival, G. M. 1992a. Estimating log volume using the centroid position. For. Sci. 38: 187–191.

Wiant, H. V., Wood, G. B., and Gregoire, T. G. 1992b. Practical guide for estimating the volume of a standing sample tree using either importance or centroid sampling. For. Ecol. Manage. 49: 333–339.

Wiant, H. V., Wood, G. B., and Miles, J. A. 1989. Estimating the volume of a radiata pine stand using importance sampling. Aust. For. 52: 286–292.

Wilkie, D. S. 1989. Performance of a backpack GPS in a tropical rain forest. Photogramm. Eng. and Rem. Sens. 55: 1747–1749.

Wingerd, D. E. and Wiant, H. V., Jr. 1982. Variables for predicting inside-bark diameters of upper stems of Appalachian hardwoods. J. For. 88: 791–792.

Winkworth, R. E. and Goodall, D. W. 1962. A crosswire sighting tube for point quadrat analysis. Ecology 43: 342–343.

Wishart, J. and Clapham, A. R. 1929. A study in sampling technique. The effect of artificial fertilizers on the yield of potatoes. J. Agric. Sci. 19: 600–618.

Wolter, K. M. 1985. Introduction to Variance Estimation. Springer-Verlag, NY. 427 pp.

Wood, G. B. 1990a. Ground sampling methods used to inventory tropical mixed/ moist forests. For. Ecol. Manage. 35: 199–206.

Wood, G. B. 1990b. Choosing a portable data recorder. Aust. For. 53: 173–181.

Wood, G. B. and Schreuder, H. T. 1986. Implementing point-Poisson and point-model based sampling in forest inventory. For. Ecol. Manage. 14: 141–156.

Wood, G. B. and Wiant H. V. 1990. Determining the volume of logs in Australian hardwood trees using centroid sampling. In: Proc. XIX IUFRO World Congr. (H. E. Burkhart, G. M. Bonnor, and J. J. Lowe, Eds.). Montreal, Canada. Aug. 5–11, 1990. Virginia Polytech. Inst. and State Univ., Blacksburg. Publ. No. FWS-3-90. Pp. 240–246.

Wood, G. B. and Wiant, H. V. 1992a. Comparison of point-3P and modified point-list sampling for inventory of mature native hardwood forest in southeastern New South Wales. Can. J. For. Res. 22: 725–728.

Wood, G. B. and Wiant, H. V. 1992b. Test of application of centroid and importance sampling in a point-3P forest inventory. For. Ecol. Manage. 53: 107–115.

Wood, G. B., Schreuder, H. T., and Brink, G. E. 1985. Comparison of a model-based sampling strategy with point-Poisson sampling on a timber management area in the Arapahoe-Roosevelt National Forest in Colorado. Canad. J. For. Res. 15: 83–86.

Wood, G. B., Wiant, H. V., Loy, R. J., and Miles, J. A. M. 1990. Centroid sampling: A variant of importance sampling for estimating the volume of sample trees of radiata pine. For. Ecol. Manage. 36: 233–243.

Wright, W. G. 1925. Statistical methods in forest-investigation work. For. Branch, Dept. Interior. Ottowa, Canada. Bull. 77. 36 pp.

Wu, C. F. J. 1985. Variance estimation for the combined ratio and combined regression estimators. J. Roy. Stat. Soc. B47: 147–154.

Wu, C. F. J. 1986. Jackknife, bootstrap and other resampling methods in regression analysis. Ann. Stat. 14: 1261–1295 (includes discussion).

Wu, C. F. J. and Deng, L. Y. 1983. Estimation of variance of the ratio estimator: An empirical study. In: Scientific Inference, Data Analysis, and Robustness (G. E. P. Box, T. Leonard, and C. F. J. Wu, Eds.), Academic Press, NY. Pp. 245–277.

Yanagawa, T. and Chen, C.-H. 1980. The MG-procedure in ranked set sampling. J. Stat. Plan. Inf. 4: 33–44.

Yanagawa, T. and Shirahata, S. 1976. Ranked set sampling theory with selective probability matrix. Aust. J. Stat. 18: 45–52.

Yandle, D. O. 1982. Binomial sampling of forest populations. In: Proc. Soc. Amer. For., Nat. Workshop, "In-Place Resource Inventories: Principles and Practices." Orono, ME. Aug. 9–14, 1981. Pp. 362–370.

Yandle, D. O. and White, F. M. 1977. An application of two-stage forest sampling. South. J. Appl. For. 1: 27–32.

Yang, Y.-C. and Wang, C.-H. 1987. Expectations and variances of basal area estimates using remeasured horizontal line sampling. For. Sci. 33: 174–184.

Yates, F. 1946. A review of recent statistical developments in sampling and sampling surveys. J. Roy. Stat. Soc. A109: 12–43 (includes discussion).

Yates, F. and Grundy, P. M. 1953. Selection without replacement from within strata with probability proportional to size. J. Roy. Stat. Soc. B15: 253–261.

Yates, F. and Zacopanay, I. 1935. The estimation of the efficiency of sampling with special reference to sampling for yield in cereal experiments. Agric. Sci. 25: 545–577.

Young, H. E., Robbins, W. C., and Wilson, S. 1969. Errors in volume determination of primary forest products. In: Proc. IUFRO 14th Congr. Munich, Germany. Sect. 25, VI, pp. 546–562.

Zahner, R. 1962. Loblolly pine site curves by soil groups. For. Sci. 8: 104–110.

Zhu, Z. and Teuber, K. B. 1991. AVHRR data for forest area change in Alabama.

In: Proc. 1991 ACSM-ASPRS Annu. Conv., Baltimore, MD and Bethesda, MD. Vol. 1, pp. 516–522.

Zöhrer, F. 1978. Fundamental sampling concept of forest inventory. I (in German). Allg. Forstztg. 89: 180–185.

Zöhrer, F. 1979. Fundamental sampling concept of forest inventory. II (in German). Allg. Forstztg. 90: 112–115.

Answers to the Problems

CHAPTER 3

1. Unit 1, unit 6, unit 10
2. Unit 5, unit 9, unit 10
3. Unit 1, unit 4, unit 10
4. $\hat{Y}_{HT} = 369320.6$
5. $\hat{Y}_{HT} = 573237.5$
6. ($\bar{y}_{HT} = 1532.7$, $\bar{y}_{rm} = 1524.4$, $\bar{y}_{mr} = 1523.1$, $\bar{y}_{p1} = 1541.0$, $\bar{y}_{p2} = 2586.66$).
7. $a = 191.4$, $b = 0.925$, $a_{pi} = 389.2$, $b_{pi} = 0.845$, $a_{piwr} = 584.9$, $b_{piwr} = 0.745$, $a_{wr} = 330.8$, $b_{wr} = 0.874$; $y_{lr} = 1518.5$, $y_{pi} = 1606.3$, $y_{piwr} = 1653.2$, $y_{wr} = 1584.9$

10. Yandle (1982) gives the solution.

18. ($\hat{Y} = 40$), ($\hat{Y}_A = 40$), ($v(\hat{Y}_{(a)}) = 14373.3$), ($v(\hat{Y}_{(a)}) = 106.7$), ($\hat{Y} = 24$ for SRS; $\hat{Y} = 32$ based on stratified sampling).

28. Both sampling techniques should produce unbiased estimators for both populations, but $\hat{Y}_{HT}$ will be more efficient for the positively correlated data found in Population 1. For the negatively correlated data (Population 2), $v(\hat{Y}_{HT})$ with pps sampling will be greater than $v(\hat{Y}_{HT})$ with SRS.

CHAPTER 4

1. and 2. Linear Regression—Data set 1, Stratified Sampling—Data set 2, Ranked Set Sampling—Data set 3. Not obvious.

CHAPTER 6

1. The critical issues are whether $\alpha = 0$ as model (5.1) assumes and whether $V(Y_i) = \sigma^2 x_i$. For variable v_1, $\alpha \neq 0$ but is likely to be close to zero. For v_2, α may be very different from zero depending on what the merchantability limits are. Generally one can assume $V(Y_i) = \sigma^2 x_i^k$ where $1 \leq k \leq 2$. Simulation results on efficiency have been shown to be fairly insensitive to the actual values of k.

CHAPTER 7

1. Six; three or four; infinite; one or two; infinite; three or four; three or four.
2. Two. The actual area has been rounded to the nearest hundred ha.

   (a) four; **(b)** infinite; **(c)** infinite; **(d)** one; **(e)** 1. × 10.

4. To avoid disseminating data likely to mislead, i.e., making the data appear to be more precise than they actually are.
5. 0.999 853 68; 0.999 853 7; 0.999 854; 0.999 85; 0.9999 if rounding off from 9 decimal places and 0.9998 if rounding off from 5; 1.000; 1.00; 1.0
6. Measure: on uphill side; from mineral soil level; on each fork at 1.3 m (1.4 m) above ground; at breast height.
7. **(a)** (i) Tree: Non-circular cross-section.

   Instrument: Wear in the shoe of the sliding arm causing play
Arms slightly bent, i.e., not flush when closed.

   Operator: Not placing the long axis of the caliper perpendicular to the long axis of the tree.
Misreading the scale
Measuring at the incorrect height
Measuring on one axis only or on the wrong axis (axes)
Failing to lay caliper arms flush against the tree surface/incorrect pressure.

   (ii) Tree: Non-circular cross-section.

   Instrument: Nil, if calibrated correctly.

   Operator: Not laying the tape perpendicular to the long axis of the tree.
Misreading the scale
Measuring at the incorrect height
Incorrect pressure.

   (b) Datum stick (if used) is too long or other consistent flaw in procedure for locating breast height. Stretched tape, e.g., much used plastic or fiberglass tape.

   (c) Stem with elliptical cross-section. Measurements taken on major and minor axes. Caliper properly used and in perfect working condition.

   (d) Stem with circular cross-section. Instruments in perfect working condition and properly used.

8. (a) 25.8 cm

   (b) By trial and error, locate a point on the stem of diameter 25.8 cm and mark it permanently with a scribe or paint mark.

9. Purpose of the project and how the data will be used; availability of instruments, and cost (if they have to be purchased); size range of trees to be measured; and time constraints.
10. (i) and (ii)

Medium	Advantages	Disadvantages
Graduated stick	No cost for equipment Aesthetically more pleasing than paint	Less precise points of measurement lead to less precise estimates of growth
Paint or scribe mark	Precise point of measurement Faster measurements after year 1 Easier for measure crew to find plot (if paint used). Unrepresentative points of measurement much less of a problem after year 1	More costly initially (year 1) Periodic maintenance of marks is required Alerts other people to presence of plot (if paint used), e.g., routine forest crews, vandals, etc. Clothes spattered/soiled (if paint used)

11. Thoroughly train the measure crews in correct procedures and error recognition. Standardize the point of measurement on each tree (paint or scribe mark). At each measure, provide the crew with a copy of previous measure data. Randomly check about 5% of all measurements taken.
12. (a) Purpose of the measurement
    (b) Where to measure
    (c) How to measure—methods (direct and indirect), instruments
    (d) Costs and limitations on measurement
13. **(a)** 34′23″; **(b)** 62 cm; **(c)** 38.5 m.
14. (i) tree 4; (ii) tree 1.
    Justification: The instrument can be read to the nearest 1/5 of a narrow band.
    Precision(%) = (10 × Horiz. dist. (m))/dbh (cm)
15. (a) On level ground, set up a target (preferably flat) of known diameter (width) at a known horizontal distance from the eye of the observer.
    (b) With the Relaskop mounted on a tripod and the brake freed, estimate the number of bands and fractions of a band covering the target, and deduce the diameter. Compare this diameter with true.
    (c) Repeat (a) and (b) on ground of various slopes up to about 30°.

16. The recording implies that measurements can be made to the nearest decimeter. This is unrealistic. Precision of the Haga is ±2.5%, thus, the reading should be expressed to the nearest half-meter.
17. Precise, unbiased, accurate.
18. Tree 1: 19 1/2 m (19.4 m); Tree 2: 8 3/4 m (8.8 m); Tree 3: 22 m; Tree 4: 32 m (32.1 m).
19. Tallest tree: #3; shortest tree: #4.
20. **(a)** Unacceptable. Acceptable range for the measurement is 25.0 ± 2.5%; **(b)** No; **(c)** No; **(d)** Forgot to sight to tree base, instrument bias, tree leaning away from observer (undetected), horizontal distance incorrectly set out, incorrect reading to tip (operator D); **(e)** Precise – if result of operator D is ignored; **(f)** Nil; **(g)** Nil; **(h)** As for **(d)** but excluding instrument bias; **(i)** Operator D is incompetent.
21. The students have overlooked two things, viz. checking for tree lean (away from the observer) and that readings were taken to the tree bases (as well as to the tips).
22. **(a)** 46 1/2°, 4 1/2°; **(b)** 18 m, 36 m; **(c)** 48 m, 24 m.
23. Check the readings. If both are correct, check the calibration of the Abney. Reason: A reading of 47 on a Suunto clinometer is consistent with an Abney reading of 25°10′.
24. Measurement procedures are faulty. Check particularly for parallax error but also for errors in counting the sticks, stick orientation (ferrules point down), measuring the odd length at the tree base, and calculations.
25. Tree leaning to the south.
26. (i) The nature of the damage suffered would be leaning and windblown (not broken) trees. Thus, the measurement equipment required should include height sticks, plumb bobs, a hypsometer (e.g., Haga, Blume Leiss, etc.), and linear tapes:
    (ii) As for (i) but delete the height sticks.
27. Purpose of survey including precision of estimate required; time allowed; map of the area (location, topography, access); variability of tree height and height range; survey design/sample frame; crew size and its composition; equipment available/needed; measurement methods to be applied; training required; transport.
28. Locate an upstanding object, e.g., cornice of a building, plumb it to ground, and measure accurately its height above ground (=true height). Set off from the plumb point a series of horizontal distances, e.g., 20, 25, 30 m and, for each distance, record the height of the cornice by sighting to both it and the plumb point. Compare the resulting estimates with true height. The difference should not exceed 2.5%.

29. (i) Stratify the population by age class (and site class if boundaries are known).
 (ii) Determine the size of each stratum.
 (iii) Determine or specify sample size allowing for time and cost constraints.
 (iv) Allocate samples to strata proportional to stratum size
 (v) Consider appropriate bark thickness or tree volume models and specify the sample tree variables requiring measurement.
 (vi) Locate sample trees within strata, measure the required tree variables, record and sort.
30. Because the form of tree boles from ground to tip and inflection points along the bole are so variable.
31. Enumerate the stand for d (cm) and determine the quadratic mean diameter, d_q. Sample a proportion of the trees for total height (m) and establish the stand height curve (h vs. d). Estimate stand mean height (h_m) by reading off the height equivalent to d_q. Then,

$$\text{stand bole area (m}^2) = d_q \times h_m \times 63 \times 0.0172\,.$$

32. Modified taper step method. Regulating the drop in diameter (taper steps) minimises errors in the basic assumptions regarding the solid shape of sections.
33. The same method of measurement should be used as that applied to the sample trees used to compile the table. Normally, the choice is between the standard sectional and modified taper step method both of which give acceptable precision and accuracy at reasonable cost.
34. $0.023\ \text{m}^3$ (conoid); $0.034\ \text{m}^3$ (2nd degree paraboloid).
35. Several procedures are feasible, viz.:
 (a) Measure the DBHOB and total height of each tree in the plot, derive individual tree volume using the function, and sum the results.
 (b) Enumerate the plot for diameter. Derive the v_o/g line for the plot from sample trees (each measured for DBHOB and total height with volume being derived from the function). Then derive plot volume using the parameters of the v_o/g line, plot stocking and G (derived from the enumeration).
 (c) Enumerate the plot for diameter and compile a stand table. Select sample trees across the range of diameter, measure them for diameter and total height, and then derive the stand height curve. Read from the curve the height of a tree at the mid-point of each diameter class. Using the function, derive the volume of each mid-class tree, multiply each volume by class frequency, and sum the results.
36. Measure the appropriate tree and crown variables (let d = DBHOB, w = crown width, r = derived crown radius, l = length of the living

crown, s = sloping length of the living crown (derived from l and r), and h = total tree height. Then,

Crown ratio: = w/d

Crown length ratio: = $100l/h$

Crown surface area: Formula to apply depends on the assumed solid shape of the crown, e.g.,

Conical:	Surface area = $\pi rs + 2\pi r$
Paraboloid:	Surface area $= 2\pi r / 12h^2[(r^2 + 4h^2)^{3/2} - r^3] + \pi r^2$
Green crown percent:	Synonymous with 'crown length ratio'.

37. Difficulties include: false rings induced either by unseasonable drought, early and late frosts or other causes; complete absence of a ring in very severe drought years; poor colour differentiation between early and late wood; ring distortion arising from branch whorls, defects, compression and tension wood, etc.
38. Advise the member that the source of most error in VRP sampling is not the method but the operator. Unless operators are thoroughly trained in all aspects of VRP procedure including calibration of instruments and direct checking of doubtful trees, serious errors will occur.
39. *Level ground*: Set up a flat target of known width (w) and, while sighting to it with the brake freed, move away from it until it is perfectly covered by Band 4 of the Relaskop. Then measure the distance (s) from the observer's eye to the target and determine the F by substituting for s and w in the formula. The result should be 4.0.
40. Sweep the stand with one of the standard bands, viz. Band 1, 2, 4 (m^2/ha) and record the count. Derive G and divide it by 9 (mid-point of 6–12). Select the band or band combination whose F most closely approximates the result.
41. Wedge prism: measure slope distance from the centre of the tree at breast height to the wedge which must be held vertically above the angle count point.

    Relaskop: measure slope angle, slope distance from the centre of the tree at breast height to the eye of the observer (who must be standing on the angle count point), and correct the slope distance to horizontal before determining the limiting distance.
42. Describe the external and internal features of the Relaskop giving particular attention to the bands and band combinations and their respective Fs, and use of the sun visor and brake button. Explain the tapering of the bands, action of the brake, and the associated slope correction feature. Stress the changes in procedure required when using the Relaskop, viz. position of the operator when sweeping, slope correction, checking doubtful trees.

43. $F = 2500\ [0.66/(16.75\ \mathrm{Cos}\ 10°]^2 = 4$, i.e., Band 4 is being used.
44. 42.9 m^2/ha. The first doubtful tree is 'out' and the second is a true borderline (counts 0.5).
45. $n = (\mathrm{Ct}/\mathrm{E})^2 = [(6/36)(100)(\underline{2})/5]^2 = 44.4$. Thus, locate 44 or 45 sample points and estimate $\bar{G}$ at each.
46. Density in the area is variable, viz. low (~16 m^2/ha) in the immediate vicinity of the sample point, high (~40 m^2/ha) in the outer fringe of the area swept by Band 2, and low again (~19 m^2/ha) in the outer fringe of the area swept by Band 1.
47. Frequency $= 4.35/(\pi 0.623^2) = 14.3$ trees/ha.
48. The intensity of sampling with F 1 is five times greater than for F 4. Thus, provided operator is eliminated from the count, a better precision of estimate is to be expected with F1.
49. Density of the plantation is uniform and the counts are made without error. If density is variable, more sample points would be required with the 4° prism.
50. D, C, A, B. The intensity of sampling is: D >> C = A >> B. C can be expected to be more precise than A because of a better distribution of sampling units.
51. Radius of plot $= \sqrt{[0.050 \times 10\,0000/\pi]} = 12.62$ m.
    As $F = 2500\ (\mathrm{w}/\mathrm{d})^2$, $w = \sqrt{[1 \times 12.62^2/2500]} = 0.252$ m.

    Thus, stand the 25.2 cm tubing vertically at the sample point. Then move in a circle about the point and periodically sight to the tubing using Band 1 of the Relaskop. Establish the borderline position in each case. Define the plot boundary by paint marks on appropriate trees.
52. **(a)** Divide the plot into four 12.5 m × 20 m sub-plots, locate by eye the two tallest trees in each, and measure the total height of these trees. Average the heights of the tallest tree in each sub-plot.
    (b) (i) Determine or decide the number of sample points (n) required.
    (ii) Assume that a systematic sample with random start is used, dimensions (sides) of the orthogonal grid will be: $\sqrt{[16 \times 10\,000/n]}$ metres.
    (iii) Establish a random point on a map of the compartment. Draw the grid to the same scale as the map and superimpose the two ensuring that one of the grid intersections lies on the random point.
    (iv) By measurement on the map and ground survey, locate on the ground each grid intersection point. Establish a small circular plot, e.g., 0.05 ha (12.62 m radius) at each point and determine its predominant height. Average the predominant heights of all the plots.

53. Derive either from the quadratic mean dbhob (d_g) of the block or from its stand height curve.

d_g: enumerate the block for diameter and determine d_g. Locate the 30 trees whose diameters (dbhob) most closely approximates d_g, measure the total height of each and average the results.

Stand height curve: Enumerate the block for diameter and determine d_g. Sample the stand to establish the stand height curve and read from the curve the height corresponding to d_g.

54. Based on a systematic sample with random start, enumerate 20–30% of the stand for dbhob identifying each measured tree in the process. Compile a stand table for the sample and, using the diameter classes as strata, select trees for measurement of total height by weighting the sample by the frequency of occurrence of trees within strata. Locate these trees in the field and measure and record their total heights. Process the sample tree data (dbhob, total height) by computer fitting selected regression equations, to produce formulas stating the explicit relationship between the predicted variable, total height, and the predictor variable, dbhob.

55. Enumerate all or a sample of the stand for dbhob and compile the stand table (based on dbhob). Sample the stand to establish the stand height curve. Read off this curve (or calculate from the derived regression equation) the height equivalent to the mid-diameter of each diameter class. Access the volume function to derive the volume of trees of this height and diameter. Multiply the derived volume by class frequency to derive class volume. Repeat the process for each diameter class, and then sum the class volumes to derive stand volume.

Alternatively, if the relationship of v_0 vs. g for the stand is known to be linear, sample the stand to establish both the volume-basal area line (measure sample trees for dbhob and total height and derive volume from the 2-variable volume function) and stand density (G-by VRP sampling), count or estimate stocking (N-trees/ha), and derive stand volume from the volume-basal area line, viz.:

$$V = 1.5[\mathrm{Na} + \mathrm{bG}] .$$

56. 94 m^3.

57. **(a)** $a = -0.129$ and $b = 9.003$.
Thus, stand volume $= 0.7\ [275 \times (-0.129) + 9.003 \times 30] = 164\ m^3$.
(b) The slope 'b' would increase, because height for a given diameter class increases with time, and the y-intercept 'a' would become more negative because volume lines (merchantable volume vs. g) tend to converge on or near the x-axis at a slight positive basal area (0.008 m^2 for *P. radiata*).

58. **(a)** Independently regressing b_0 and b_1 on h_{dom},
$b_0 = 0.0126 - 0.0034h_{dom}$ and $b_1 = 1.2900 + 0.3033h_{dom}$
Thus, stand volume $= N\ (0.0126 - 0.0034h_{dom}) + (1.2900 + 0.3033h_{dom})G$
$= 0.0126N - 0.0034Nh_{dom} + 1.2900G + 0.3033h_{dom}G$

(b) Volume $= 40[0.0126 \times 800 - 0.0034 \times 800 \times 22 + 1.2900 \times 25 + 0.3033 \times 22 \times 25]$
$= 6000\ m^3\ (5972\ m^3)$

59. **(a)** The relationships of b_0 and b_1 plotted against h_{dom} are curvilinear and linear respectively.
(b) Both relationships referred to in (*a*) are linear.

60. Such a relationship for a forest permits stand volume to be estimated quickly and cheaply at any point within it from an angle count sweep (to determine G) and an estimate of h_{dom}.

61. **(a)** Both lines would be linear but dispersion of points about the regression line would be much less for d^2h as it accounts for more of the variation in v_o.
(b) The coefficients b_0 and b_1 of the v_o versus d^2h relationship would remain relatively stable (tree form would remain unchanged) whereas those for the v_o versus d^2 relationship would decrease and increase respectively (for the reasons given in Question 57(b)).

INDEX